Breast Cancer 2

Advances in Research and Treatment

Experimental Biology

Breast Cancer

Advances in Research and Treatment

Edited by **WILLIAM L. McGUIRE, M.D.**

Volume 1 • **CURRENT APPROACHES TO THERAPY**

Volume 2 • **EXPERIMENTAL BIOLOGY**

A Continuation Order Plan is available for this series. A continuation order will bring delivery of each new volume immediately upon publication. Volumes are billed only upon actual shipment. For further information please contact the publisher.

Breast Cancer 2

Advances in Research and Treatment

Experimental Biology

Edited by

William L. McGuire, M. D.
University of Texas Health Science Center
San Antonio, Texas

Springer Science+Business Media, LLC

Library of Congress Cataloging in Publication Data

Main entry under title:
Breast cancer.
Includes bibliographical references and index.
CONTENTS: v. 1. Current approaches to therapy, edited by W. L. McGuire. v. 2. Experimental biology.
1. Breast–Cancer. [DNLM: 1. Breast neoplasms. 2. Breast neoplasms–Therapy. WP870 B8245]
RC280.B8B673 T 616.9'94'49 77-23505

ISBN 978-1-4757-4675-4 ISBN 978-1-4757-4673-0 (eBook)
DOI 10.1007/978-1-4757-4673-0

Originally published by Plenum Publishing Corporation in 1798.
Softcover reprint of the hardcover 1st edition 1798

Contributors

Arthur E. Bodgen, Mason Research Institute, Worcester, Massachusetts

David Colcher, National Cancer Institute, National Institutes of Health, Bethesda, Maryland

William Drohan, National Cancer Institute, National Institutes of Health, Bethesda, Maryland

Ronald B. Herberman, Laboratory of Immunodiagnosis, National Cancer Institute, Bethesda, Maryland

Kathryn B. Horwitz, Department of Medicine, University of Texas Health Science Center at San Antonio, San Antonio, Texas

Donald Kufe, National Cancer Institute, National Institutes of Health, Bethesda, Maryland

Marc E. Lippman, Medicine Branch, Division of Cancer Treatment, National Cancer Institute, National Institutes of Health, Bethesda, Maryland

Daniel Medina, Department of Cell Biology, Baylor College of Medicine, Houston, Texas

William L. McGuire, Department of Medicine, University of Texas Health Science Center at San Antonio, San Antonio, Texas

C. Kent Osborne, Medicine Branch, Division of Cancer Treatment, National Cancer Institute, National Institutes of Health, Bethesda, Maryland; present address: Department of Medicine, University of Texas Health Science Center at San Antonio, San Antonio, Texas

Jeffrey Rosen, Department of Cell Biology, Baylor College of Medicine, Houston, Texas

Lewis M. Schiffer, Cancer Biology Section, Cancer Research Unit, Clinical Radiation Therapy Research Center, Allegheny General Hospital, Pittsburgh, Pennsylvania

Jeffrey Schlom, National Cancer Institute, National Institutes of Health, Bethesda, Maryland

Albert Segaloff, Section on Endocrine Research of the Richard W. Freeman Research Institute, Alton Ochsner Medical Foundation, New Orleans, Louisiana

Yoshio A. Teramoto, National Cancer Institute, National Institutes of Health, Bethesda, Maryland

Preface

Breast cancer continues to be the focus of intense basic and clinical research. In Volume 1 of this series we dealt exclusively with topics concerned with therapy. In the present Volume 2, we turn our attention to the experimental biology which is the foundation for our understanding of problems concerned with breast cancer etiology, mechanisms of hormone action, cell kinetics, experimental chemotherapy, and markers of tumor burden. The contributors to the volume are all noted scholars who are personally investigating these problems.

The first chapter addresses the question, do hormones cause breast cancer? Segaloff provides us with a rational up-to-date overview of the existing data. He concludes that hormones by themselves are not tumor initiators but rather alter the host environment so that other carcinogens are effective. It is pointed out that the selection of the model test system is critical; one can almost assure any desired result by choosing an appropriately biased test system.

The question of the role of viruses in the etiology of human breast cancer remains unanswered despite elegant studies in mouse systems. Schlom and associates using powerful molecular hybridization techniques have shown that 95% of viral nucleic acid sequences are very similar in several strains of mouse mammary tumors that appear very early, while viral sequences from late-appearing tumors in the same strain may be substantially different. They also show that some viral sequences are not transmitted via the germ line but may be passed by the placenta, milk, or seminal fluids. Type-specific and group-specific immunochemical reactivities can be shown to differ between tumor viruses from different tumor strains. Although candidate viruses have been identified from human breast tumors, the authors point out that even if viruses are eventually found to be necessary in the genesis of human breast cancer, they are probably not sufficient in themselves to cause the disease.

Turning to preneoplastic lesions, the mammary gland offers one of the few systems where defined preneoplastic lesions can be studied during progression to neoplasia. Medina has used the mouse mammary gland for such studies. This model depends on the important point that mouse mammary tumorigenesis is characterized by the presence of a precursor stage, the hyperplastic alveolar nodule, which has a greater probability of developing into mammary carcinoma than does normal epithelial tissue. A variety of etiologic agents, viral, chemical, and physical, can interact at both the normal and preneoplastic stages to induce and promote tumorigenesis. This fact may be particularly important since Medina points out that many of the chemotherapeutic agents commonly used in human breast cancer therapy are among those compounds which can enhance the rate of tumor formation from precursor lesions. With regard to effects of known carcinogens on precursor lesions, Medina proposes that precursor lesions are a mixed population of normal cells, unaltered precursor cells, and carcinogen-altered precursor cells which are regulated by cell–cell interactions. Carcinogens may either act directly on precursor cells to alter them and increase their tumorigenicity and/or indirectly by disrupting the microenvironment and consequently disrupting normal cell–precursor cell interaction. The result in either case would be an increased probability of expression of potentially tumorigenic precursor cells.

Model systems to study the effects of hormones on breast tumor growth and regression have been available for many years, but have been fraught with difficulties because of complex interactions between the various endocrine glands, normal tissues, and the tumor cells. Now with the availability of a human breast cancer cell line in long-term tissue culture many of the above problems can be entirely avoided. Osborne and Lippman use this to show unequivocally the direct effects of estrogens and antiestrogens on breast tumor cells. They show that androgen, glucocorticoid, progesterone, insulin, etc. can be studied in a similar manner. Their approach probably represents the best currently available model for studying hormone action in human breast cancer cells. Clearly it has been the stimulus for a number of other laboratories to use these tissue culture cell lines in unravelling the mechanisms of hormone action.

Antiestrogens represent a relatively new approach to the treatment of breast cancer. Horwitz and McGuire review the current state of knowledge regarding what is known about these remarkable compounds. They are clearly as effective in producing objective breast tumor regressions as any other form of endocrine therapy. Within the tumor cell they bind to and translocate the estrogen receptor to nuclear sites and somehow prevent endogenous estrogens from stimulating tumor growth; they have several other possible sites of action such as the pituitary and ovaries as well. There are certain features of these compounds which demand further investigation. For example, it has been shown that certain doses of antiestrogens can directly stimulate progesterone receptor synthesis, a property thought previously to be exclusively restricted to active estrogens. It is likely that these compounds will replace most current forms of endocrine therapy because of their low associated morbidity, and they will no doubt also play an important role in adjuvant therapy of primary breast cancer.

The search for biological markers of the presence of breast tumor cells continues. Heberman suggests that there are several promising leads for the application of immunologic or biochemical tests to the diagnosis and management of breast cancer. They include CEA, ferritin, calcitonin, breast-cyst fluid protein, and urinary hydroxyproline. To these might be added some measures of immune function and reactivity, particularly lymphocyte counts and subpopulations of rosette-forming cells, lymphocyte-proliferative responses, and leukocyte-migration inhibition. Most of the techniques which are currently being used need to be refined considerably. Most studies reported to date have concentrated on demonstrating differences between populations of women with breast cancer and control populations; more of the markers now need to be thoroughly evaluated for their practical use in staging and monitoring of patients. Procedures still need to be developed for application to the problems of individual patients.

A great deal of new information is available regarding cell kinetics in experimental breast cancer. Schiffer reviews this subject and provides us with the following generalizations. Transplantable tumors are more highly proliferative than their spontaneous counterparts and this difference is exaggerated with increasing transplant generations. Metastases also appear to be more highly proliferative than their primary tumors. In general, small tumors are more highly proliferative than large ones, and tumors near blood vessels appear to be more proliferative than those further away. However, we cannot extrapolate cell kinetic data from one experimental breast-tumor system to another because the variability is very wide even within a single system. Schiffer points out that clinicians may have thought that a standard chemotherapy regimen based on average cell kinetic data could be designed for groups of patients, while in fact, in view of the variability of the data from one tumor to the next, the possibility of one overall treatment scheme seems to have little chance of success. It is apparent that to properly design a chemotherapeutic strategy for a single patient, cell kinetic data must be obtained from that patient's tumor. Now that rapid *in vitro* techniques, designed so that the results are known soon enough for clinical utilization, are becoming available, combination chemotherapy may soon become less empirical and much more rational.

It has been stated on occasion that clinical trials of chemotherapy are usually first performed on human patients and then repeated on mice and rats for illustrative purposes. This is often close to the truth in the case of breast cancer and can be explained by the lack of an overall model breast cancer system in mice or rats. This is discussed in detail by Bogden in the chapter on therapy of experimental breast cancer. He demonstrates that if the experimental question is carefully defined and the animal model is carefully chosen, valuable information can be obtained which can be of considerable benefit to the clinician. This is illustrated by actual experiments designed to compare various doses of X-ray therapy and to determine the optimum time to change from one drug regimen to another. The latter point prompts a discussion of late intensification therapy which is based on the fact that very small tumors may be more resistant to therapy and thus require more intensive therapy than larger tumors at the inflection point of the Gompertzian curve. It now remains for the practicing chemotherapist to pay some attention to these experimental models and use the information to design innovative trials in human breast cancer patients.

The molecular biologists have finally begun to investigate hormone action in breast cancer. Rosen provides convincing evidence that modern biochemical technology can be applied to biologically relevant questions. For example, it has been well demonstrated that both normal and neoplastic breast tissues contain receptors for prolactin. We also know that prolactin stimulates the synthesis of specific milk proteins in these tissues. The ability to rapidly induce specific mRNAs and to perform pulse-chase experiments in mammalian cells will permit an elucidation of the primary site of steroid and peptide hormone action, i.e., whether they act at transcriptional, posttranscriptional, or translational levels. With the advent of recombinant DNA technology it should be possible to isolate probes for each of the mild protein mRNAs which can then be used to characterize the genes and their primary transcription products. An understanding of the primary gene structure and the availability of larger pieces of native DNA containing both the specific genes and their regulatory regions should provide the basis for future studies of the precise mechanism of hormone action in the mammary gland, which in turn may help explain how these regulatory mechanisms have deviated in hormone-dependent breast cancer.

In summary, Volume 2 of this series covers a wide range of topics, all of which should provide stimulating reading for basic biologists as well as for clinicians caring for breast cancer patients.

William L. McGuire

Contents

1. Hormones and Mammary Carcinogenesis

Albert Segaloff

5. Antiestrogens: Mechanism of Action and Effects in Breast Cancer

Kathryn B. Horwitz and William L. McGuire

6. Biological Markers in Breast Cancer

Ronald B. Herberman

8. Therapy in Experimental Breast Cancer Models

Arthur E. Bogden

9. Gene Expression in Normal and Neoplastic Breast Tissue

Jeffrey Rosen

1

Hormones and Mammary Carcinogenesis

ALBERT SEGALOFF

1. Introduction

In the classic sense of carcinogenesis, such as that produced by ionizing radiation or "carcinogenic" hydrocarbons, hormones are probably not carcinogenic. They are either procarcinogens or cocarcinogens. In this sense, they produce more substrate either for initiation or for promotion by other substances, but of themselves they are probably not carcinogens. In the intact organism, the effects of administration of various hormones at crucial periods of the host's life cycle may result in carcinogenic effects because of profound lifelong changes in endocrine balance that can be induced by a single hormone administration. This latter phenomenon might be referred to as the inducement of unfavorable hormonal imbalance, and this imbalance has a procarcinogenic effect.

2. History

The real beginning of great interest in the hormone induction of mammary cancer starts with the work of Professor Anton Lacassagne[1] in 1932, when he administered an extract, folliculine benzoate, to three male and two female RIII mice and found that all but one female de-

ALBERT SEGALOFF • Section on Endocrine Research of the Richard W. Freeman Research Institute, Alton Ochsner Medical Foundation, 1516 Jefferson Highway, New Orleans, Louisiana 70121.

veloped mammary cancers. In the RIII strain of mice, the females, particularly on forced breeding, developed a high incidence of mammary cancer, but neither intact nor castrated males developed this lesion, and the female that developed cancer did so in 4 months, while the earliest spontaneous cancer seen in untreated females was at 5 months. This result was therefore a demonstration of a new kind of cancer in an animal that does not develop mammary cancer. I remember the impressive Professor Lacassagne telling us that he kept the mice in a battery jar on his desk so that he could observe them closely. This would be impossible to do in studies of thousands of animals in the pursuit of highly significant results, but it does not make his observation any less important nor ours any more.

2.1. Studies in Mice

The scientific soil on which Professor Lacassagne's substantial contribution was planted should be clarified. It was obvious at that time that mice were easy to breed in captivity. They lent themselves well to genetic studies, and most stocks, under laboratory conditions in which they lived longer than wild mice, survived into cancer age, and had substantial numbers of breast cancers.[(2)] In addition, mice have essentially flat mammary glands, and even a tiny nodule of beginning cancer produces a bulge on the ventral surface, so that even the most unsophisticated, uneducated animal handler can look for mammary tumors as he transfers the mice by picking them up by their tails. For these reasons, highly inbred, homozygous strains of mice were already available with incidences of spontaneous mammary cancer in the females, either virginal or breeding, from zero to essentially 100%.

The seeds of Dr. Lacassagne's observations grew luxuriantly in this soil, and there rapidly grew a literature documenting that estrogens were capable of inducing increases in mammary cancer in many mice providing they had the proper genetics. The mice without spontaneous mammary cancers did not appear to develop them when treated with estrogens, and the continuous administration of estrogen was considerably more effective in increasing the incidence of breast cancers than intermittent administration.

It was about this time that Bittner[(3,4)] observed the "milk factor" or mammary tumor virus that is responsible for the high incidence of mammary tumor in some stocks of mice. Transmission of this virus is not genetic or DNA transmission, but is vertical transmission by the mother's milk. It can be stopped by foster-nursing of the offspring, or the virus can be transmitted to non-virus-carrying strains by foster-nursing their offspring on virus-bearing mice.

Huseby and Bittner[5] suggested that the presence of the "milk factor" in mice has a significant effect on their endocrine balance, as evidenced by vaginal smears. This hypothesis was strengthened by the preliminary report that the milk agent altered fecal excretion of steroids.[6] Bittner and co-workers suggest that it may be through the endocrine system that the mammary tumor virus increases the incidence of breast cancer in mice that harbor it. Further work did not strengthen evidence of the relationship.[7]

2.2. Studies in Rats

These studies were then extended to another common laboratory rodent, the rat. Not as many strains were available, and many studies employed random-bred stocks of various designations. Even to this day, the bulk of observations on hormonal effects on carcinogenesis in rats are made in heterozygous stocks. The problem in many of the rat studies exaggerates problems seen in the mouse studies, namely, a difference of opinion regarding the microscopic appearance of the cancers induced and also the rarity of metastasis from induced tumors. Although there does not appear to be a mammary tumor virus in the rat, all the other observations mentioned above are similar. In addition, in the rat, it was first observed[8] that the continuous administration of estrogen leads to chromophobe adenomas of the pituitary gland that sometimes reach large size and cause the death of the animal.[8] We now know that these chromophobe adenomas produce pituitary hormones, at least prolactin, and may indeed be part of the genesis of the observed breast cancers.[9]

Huggins *et al.*,[10] in looking for a model of carcinogenesis, found that the females of a particular random-bred rat stock, the Sprague–Dawley, subsequently developed a high incidence of breast cancer when they administered 7,12-dimethylbenz(a)anthracene (DMBA) to these animals at a sharply restricted time in their development. In addition, various hormonal manipulations altered this incidence of cancer.

2.3. Studies in Dogs

Even though dogs may have American Kennel Club registration, they are not inbred stocks and are heterozygous; there is, however, a great variation in the incidence of mammary tumors in various breeds of dogs. The characteristic mammary cancer of dogs is a mixed adenocarcinoma.[11] It generally grows slowly, and frequently metastasizes, but rarely kills the host. It is most common in beagles, but estrogens do not increase the incidence of this type or other types of breast cancers. Indeed, estrogens are quite toxic to beagles, and it appears that this toxicity is mediated through the bone marrow.[12] In this species, however, progestational hormones did lead to a high incidence of mammary cancers, generally of different histology than the spontaneous cancers.[13] Despite this knowledge, the insistence of the United States Food and Drug Administration on the use of beagles for long-term studies has deprived us of the ability in this country to use progesterone derivatives as therapeutic agents because of the beagles' high rate of tumor formation on this type of therapy.

2.4. Studies in Primates

Extensive studies[14] have not demonstrated that steroid hormones in primates increase the incidence of mammary cancer, and the single case observed by Kirschstein *et al.*[15] should probably be considered a spontaneous rather than an induced tumor.

2.5. Incidence in Women

Ever since the introduction of estrogen into our clinical armamentarium, occasional reports of breast cancers in women receiving estrogen for short or long periods of time have appeared. When one considers that the incidence of breast cancer in women in our Western urban society is 8–10% of all women, one can expect a high correlation between the finding of breast cancer and women who have received estrogen. Indeed, there is such a correlation. Large numbers of women who are postmenopausal and are or were taking estrogens as therapy develop breast cancer every year. It was recently suggested[16] that this may be a cause–effect relationship; however, a great many studies with huge numbers of subjects will be required for statistically significant evidence of this relationship because of the high background incidence of breast cancer. I personally doubt that the postmenopausal use of estrogens, even continuously, is going to increase the incidence of breast cancer enough that we can show a significant increase in the incidence.

3. The Rat as a Model of Hormone-Induced Mammary Cancer

Since most of my work on mammary-gland carcinogenesis has been done in the rat, I will deal extensively with rat mammary tumors.

3.1. Morphology of Mammary Tumors

Many classifications of the morphology of mammary tumors and cancers in the rat have been proposed, but one by Shellabarger[9] appears to be most useful and simple and to agree most nearly with my own classification.

Fundamentally, mammary tumors in rats and other rodents are divided into benign tumors (most of which are classified as adenofibromas) and cancers (most of which are classified as adenocarcinomas). Efforts at classifying the cancers as other than adenocarcinomas seem to confuse the issue. The benign tumors may transplant readily and grow rapidly in syngeneic animals, and such tumors have been employed for extensive study.[17] The cancers also grow progressively, can frequently be transplanted into estrogen-conditioned, syngeneic animals, and have a high incidence of metastases, particularly when they are allowed to continue until they cause the death of the animal. As they do in man, these metastases first occur in the regional lymph nodes. Successive generations of transplants into conditioned animals may lead to the development of autonomous tumor lines.[18]

Such cancers and benign tumors can readily be induced in appropriate strains of rats by the continuous administration of estrogenically active hormones. McEuen[19] and Geschicter[20] reported having induced mammary cancer in the rat by prolonged treatment with estrogen. The latter author believes that carcinogenesis requires a superphysiologic dose of estrogen and that an orderly course of progression then occurs, consisting of hyperplasia, cyst formation, adenosis, benign tumor formation, and finally cancer. He reported rare distant metastases. In our experience, and that reported by Nelson,[18] all these lesions tend to appear simultaneously in different areas of even the same mammary gland in animals chronically treated with estrogens, and proof of the orderly progression from hyperplasia to cancer is lacking. These early studies antedate our present ability to measure hormone levels in the host animals by the use of radioimmunoassay (RIA).

No uniform agreement has been reached as to whether or not these similar-appearing tumors were indeed cancerous or on their ability to metastasize and kill the host. For example, Noble[21] at one time believed that the tumors were not unequivocally cancers, but he has now changed his view.[22]

3.2 Continuous Estrogen Administration to A × C Rat

Having the good fortune of working with Dunning and Curtis, we were able to study the induction of mammary cancer in their superb stocks of inbred rats. The A × C rats in our laboratory live to an older age than our other rats and have occasional benign mammary tumors, but they ordinarily fail to get mammary cancers. When estrogen is given continuously to the intact A × C female rat, however, this rat is probably the most consistently susceptible rat to estrogen induction of mammary gland cancers.

Other strains, such as the Copenhagen, appear resistant to the induction of mammary cancers. The Fischer rat develops considerably fewer mammary cancers on the continuous administration of estrogen than the A × C or most other rats. The Fischer rat, however, is a more fragile animal; it tends to have a shorter survival time, and much more important, it is prone to the estrogen induction of huge chromophobe adenomas of the pituitary that produce and secrete prolactin. The continuous administration of estrogen in ample amount to Fischer rats leads to their demise from the huge induced symptomatic pituitary chromophobe adenomas. This observation leads to a truism that must never be forgotten: In biological studies, particularly in studies of carcinogenesis in which prolonged periods of time intervene between initiation of the experiment and production of cancer, the carcinogenic effect will be missed if the compound or physical agent to which the host is exposed is sufficiently toxic that the animal fails to survive to the minimum tumor age. In other words, *in vivo* carcinogenesis represents a race between mortality from toxic compounds that are carcinogenic and mortality from the carcinomas induced by the compounds.

3.3. Estrogen Administration and Radiation to the A × C Rat

An illustration of this important concept is our own work in which animals were exposed to graded doses of radiation to the mammary glands and continuous estrogen administration.[23] In the early part of the experiment, when the minority of the mammary cancers had occurred, there was a direct dose–response relationship between the amount of radiation given and the numbers of tumors induced. There was, however, a flexion point in the curve when the intermediate dose of radiation continued to induce progressively increasing numbers of mammary cancers, but the highest dose of radiation failed to produce a greater number of cancers because the animals died at too great a rate. Therefore, for experiments of long duration, we must use the intermediate amount of radiation to induce the largest number of cancers. I presume we must follow the biblical commandment "Thou shalt not kill" if we want optimal carcinogenesis.

In our own experiments with the cancer-inducing effect of continuous estrogen administration and the synergism with radiation, it became apparent that we produced the greatest number of tumors in the greatest number of animals if we employed intact female A × C rats. Since we had assumed that it was the continuous estrogenization that was most important in this carcinogenesis, we believed we would be using our estrogens in a more standardized animal if we removed their gonads. We attempted to replace the other ovarian steroid, progesterone, either continuously in the form of pellets or cyclically by injections of progesterone. However, neither the continuous nor the cyclic administration of progesterone restored the sensitivity of the A × C rat to the levels seen in intact animals.[24]

3.4. Effects of Relaxin

Of the various substances from the ovary that might explain the effects of oophorectomy, relaxin, a polypeptide that originates in the corpus luteum, seemed a possibility. Cutts had studied estrone-induced mammary cancers in the hooded rats previously reported on by Noble and his colleagues. Cutts[25] demonstrated that relaxin was capable of accelerating the growth rate of his estrone-induced rat mammary cancers.

In our hands, administration of the NIH crude relaxin to intact or spayed A × C female rats temporarily accelerated the growth rate of mammary cancers induced by estrogen or by estrogen and radiation. These tumors then decreased rapidly in their growth and virtually disappeared despite continued administration of the relaxin. The serum of the animals at this time had a high titer of antibodies against pure porcine relaxin, which is what the NIH preparation contains as the active material. Thus, the possibility remains that the missing nonsteroidal factor from the rat ovary is indeed relaxin.

Porcine relaxin originates in the corpus luteum; it is a polypeptide composed of two chains that, like insulin, are joined by two disulfide bonds.[26] Relaxin contains no tyrosine, so that RIA requires the addition of tyrosine, which can be iodinated and used for RIA. With our best technique, we have only rarely been able to demonstrate relaxin in the blood of nonpregnant rats. During pregnancy, however, as has been reported by others,[27] there is an initial low level of radioimmunoassayable relaxin that reaches its peak at the time of delivery. The physiological role of relaxin in lower mammals, particularly in the pocket gopher, is to produce sufficient relaxation of the pelvic ligaments to permit the animal to deliver a fetus that is larger than its birth canal. The physiological role of relaxin in primate pregnancy is still unknown. Our studies in human pregnancy indicate that the curve for relaxin levels (measured by RIA with porcine relaxin) in peripheral blood is highest in early pregnancy and lowest at term.[28] The usual human infant does not require separation of the pelvic bones for delivery. Although we do not have sufficient evidence to prove it, we currently believe that relaxin is the ovarian nonsteroidal substance required for the greatest carcinogenesis and the greatest growth rate of the tumors.

3.5. Estrogen: Substrate Enhancer of Carcinogenesis

Little has been learned about the effect of hormone administration on mammary carcinogenesis since the early studies. All known estrogenic compounds appear capable of producing mammary cancers, whether because they produce the substrate for the other carcinogenic activities or because they are themselves carcinogenic. Estriol has been said to be not only an antiestrogen but also possibly a protector against carcinogenesis. Nothing could be further from the truth. Even before our present knowledge of the extremely short activity of estriol on the estrogen receptor,[(29)] we knew that for the greatest estrogenic expression of estriol, it had to be administered frequently,[(30,31)] and carcinogenesis requires continuous estrogenization. Therefore, to show that a highly water-soluble estrogen such as estriol is *not* carcinogenic to the mammary gland, one would give small amounts intermittently and show its lack of carcinogenicity. Noble[(32)] and Rudali *et al.*,[(33)] who are aware of these properties as a necessity for the display of estrogenic activity of estriol, did do appropriate studies and reported that it is as carcinogenic as other estrogens in their hands.

The difficult decision now is left: how can one classify the obvious ability of estrogens to increase mammary carcinogenesis? To my knowledge, a single acute dose of any hormone has never been reported to act as an initiator of mammary carcinogenesis. Therefore, when hormonal administration leads to carcinogenesis, it is not playing the role of an initiator. The possible exception is the administration to the neonatal mouse of a single dose of estrogen (it generally takes more), which initiates a chain of events that may lead to mammary carcinogenesis.[(34)] In this instance, however, the initiating event is an imprintation by the estrogen on the mouse's immature hypothalamus that changes the characteristic development of the hypothalamus so that the normal female cyclic pattern does not develop. This imprintation can also be done with hormonal agents other than estrogen, such as progesterone or testosterone. This is a dose-related phenomenon, and vaginal tumors can be prevented by oophorectomy in the animals that have gotten lower doses neonatally; if larger doses are employed, oophorectomy does not prevent the development of the characteristic vaginal changes. If this experiment is done in animals that are susceptible to the estrogen induction of mammary carcinogenesis, these animals also develop mammary cancers late in life. This is the only context I know of in which the hormones might be considered as initiators. It is my interpretation of these data, however, that this is not initiation.

Estrogens seem to play the role of providing the substrate for carcinogenesis in animals that already have a genetic or other means of initiation. Does the provision of a substrate for carcinogenesis constitute promotion of carcinogenesis? The sense in which *promotion* is used for chemical carcinogenesis may indeed apply, but I prefer to use a different term—"substrate enhancer" or "substrate enhancement." This term more nearly describes the role of hormones in carcinogenesis as I see it, yet does not prohibit us from considering the role as a special type of promotion.

3.6. Strain and Species Susceptibility

The active hormone acts as part of the endocrine system, and it is impinged on by a whole host of external influences, such as stress, pregnancy, diet, and drugs. This leads to an examination of strain and species differences in susceptibility to the estrogen induction of mammary cancers. Animals may lack a genetic initiator, and in this case, the initiation could be supplied by some nonhormonal, nongenetic means. We can best illustrate this dilemma again from our studies of the A × C rat. These animals, as indicated above, have the genetic initiation for mammary carcinogenesis, but apparently do not have any endogenous promotors. Exposure of these animals to ionizing radiation increases the number of mammary tumors, particularly the benign tumors, after a prolonged latent period. Thus, these animals have some endogenous promotion after initiation by the ionizing radiation.

3.7. Dependence

These radiation/estrogen-induced mammary cancers illustrate the problems of dealing with the endocrine system. If the cancers are induced either with or without radiation and with the implantation of a pellet from which estrogen is absorbed continuously, removal of the pellet induces substantial regression or disappearance of the mammary cancers. If, however, the pellet is left in place so that the source of estrogen is still available, the tumors can be made to regress by removal of the ovaries. This function of the ovary cannot be replaced by the further administration of progesterone. The required presence of the hormonal stimulus for the continued growth, indeed for the continued presence, of the induced malignancy is another dimension that is peculiar almost exclusively to hormonally sensitive tumors. This phenomenon is known as dependence and means, at least in the case of the A × C rat, that if the hormonal stimulus or the induced changes in the hormonal milieu are removed, the tumor with its metastases will regress and often disappear, and may remain dormant unless the stimulus is reapplied.

Dependence is more clearly shown in rat tumors than in human tumors, in which instance the term is more loosely applied to tumors that show temporary or partial regression on removal of a normal hormone. If dependence requires that the tumor with its metastases disappear on removal of the stimulus, then most human tumors do not fulfill the requirements. A handmaiden of dependence is independence, and the independent tumor is termed "autonomous." If successive generations of induced tumors in rodents are transplanted into hormonally conditioned or treated hosts, independent, usually rapidly growing, tumors that no longer require the presence of the hormonal stimulus may develop.

3.8. Dual Hormone Requirement in the Sprague–Dawley Rat Model

Although many of the characteristics of hormonal carcinogenesis have been illustrated in a single model, we must employ an additional model system to flesh out our background data. Huggins[10] demonstrated many years ago that the administration of DMBA to Sprague-Dawley rats at a crucial time in their life (50- to 60-day-old females) led to a high incidence of mammary tumors generally having the microscopic appearance of adenocarcinomata. These tumors do not appear to require the exogenous administration of hormones. Instead, there may be some peculiarity about the mammary gland of the Sprague–Dawley rat at this crucial time that makes it more susceptible to the carcinogenic effect of DMBA. These tumors are dependent on the continued presence of both estrogen and prolactin for their continued growth. In common with many hormonally responsive human tumors, the DMBA-induced rat tumors contain an estrogen receptor in the cytoplasm and also a specific prolactin receptor in the cell membranes. These tumors have a dual hormonal requirement for continued growth. They must have adequate amounts of both estrogen and prolactin, or growth ceases and tumors regress. The estrogen may be removed by oophorectomy, adrenalectomy, hypophysectomy, or a combination of all three. The prolactin may be removed through hypophysectomy or by the administration of drugs. Some of the drugs that inhibit the synthesis and release of prolactin are L-dopa and many of the ergot derivatives, such as α-bromo-ergocryptine or ergoline mesylate. The tumors also regress when estrogen activity is inhibited by the administration of an antiestrogen, such as tamoxifen.

3.9. Genetic Substrate on Which Hormonal Agents Work

The hormonal balance differs for all animals, and in addition, animals differ in their responses, both qualitatively and quantitatively, to drug administration. The different handling of hormones or drugs by different animals is because of the difference in tissues with respect to receptor proteins, or to their ability to metabolize the hormone or drug to either an inactive or superactive form, or to both factors. For example, if high levels of prolactin are obligatory for hormonally induced mammary carcinogenesis, a test for estrogen carcinogenicity might be hampered if the animal selected is one of those that fail to respond to estrogen administration with an increase in prolactin, as in the dog. On the other hand, the selection of an animal that is exquisitely sensitive to the estrogen induction of high prolactin levels may enhance the carcinogenic effect of the administered estrogen.

Although I know of no species of mammal in which the mammary gland does not contain estrogen receptors, the analogy of the human disease of feminizing testes, in which the receptors are either defective or absent in the end organs, makes one believe that it is entirely possible that we will find an animal with mammary glands that are resistant to estrogen because of the absence of receptors. Animals in which administered hormones exhibit an extremely short half-life are also not suitable candidates for demonstrating the carcinogenicity of hormones.

3.10. Choice of Animal Models

The choice of animal species to test these hypotheses therefore becomes crucial. Taking the factors mentioned above into consideration, an experimenter can select his test animal and agent in such a way that the administered hormone will or will not exhibit carcinogenic effects as desired. If one wishes to show that progestational agents are carcinogenic, one can administer them continuously for long periods of time to beagle dogs, and lo and behold, progesterone derivatives are carcinogenic. On the other hand, if one wishes to show that estrogens are not carcinogenic to the mammary gland, one may select the same animal (the beagle), administer estrogen continuously, and lo and behold, estrogens are *not* carcinogenic. Many of the experiments that influence our thinking profoundly were either deliberately or inadvertently designed without regard for these principles. Whatever the reasons for the decisions, the experiments are misleading. Thus, one must consider any experiment as incomplete unless it is applied to several model systems.

We are barely on the threshold of adequate testing of hormonal agents as procarcinogens, and a great deal of difficult and thoughtful work is still needed. In some animals, the guinea pig, for example, the mammary gland does not seem to serve as a main target for carcinogenesis. The continuous administration of estrogen to the guinea pig was studied thoroughly by Lipschutz[35] and his colleagues in Santiago. Their work followed the original demonstration by Nelson[36] that in this species, the major estrogenic effect is the production of benign fibrous tumors generally within the peritoneum or pleural cavities. Although the production of such tumors is not mammary carcinogenesis, the knowledge of the interactions between estrogens and other steroidal hormones gained from the experiments of this group is incalculable. I recommend Lipschutz's book, *Steroid Hormones and Tumors*.[35]

Is it possible to temper the carcinogenic effect of a hormonal agent by the administration of another hormonal agent? Indeed it is. We have illustrated the profound effects on the endocrine system of the administration of single agents. Some of the "deleterious" effects of estrogen administration can be counteracted by the administration of progesterone at the same time. Indeed, in the example of the estrogen-treated A × C female rat, we can see that the administration of progesterone has a distinct protective effect for the breast.[24]

4. Specific Hormones and Mammary Carcinogenesis

4.1. Thyroid Substance

The interest in thyroid substance and its effect on breast cancer goes back to Beatson,[37] who introduced the use of castration for the treatment of human breast cancer. He was working at the same time that thyroid substance was first introduced as treatment for myxedema, and it appeared that most of his patients who obtained good objective regressions were also given thyroid substance. The effect of thyroid substance on breast cancer remained a question, however, until the recent study by the Cooperative Breast Cancer Group[38] showed that thyroid supplementation had no significant effect on the number of objective regressions following castration in advanced human breast cancer.

The possibility that the administration of thyroid substance to women could be etiological in breast cancer was suddenly raised in 1976 by Kapdi and Wolfe.[39] These authors from the Hutzel Hospital in Detroit studied women who had mammograms at that hospital, and found that breast cancer occurred more often in women receiving thyroid hormone therapy than in those who were not. They said that the incidence of breast cancer increased with greater duration of treatment with thyroid hormone, particularly in nulliparous women. This anecdotal article, which ignored such other questions as those of underlying thyroid disease and other medications, produced a number of letters to the *Journal of the American Medical Association* and received wide publicity in the press. The unfortunate consequence of the wide publicity was that many patients in unquestionable medical need of thyroid therapy have discontinued its use because of their greater fear of breast cancer.

Much more careful attempts to link the incidence of breast cancer to thyroid disease have led to conflicting results. In a careful study by Mittra and Hayward,[40] it appears that there is a lower level of thyroid function in breast cancer patients as a group than in their matched controls. There has also been a suggestion that hyperthyroidism may be more common among women with breast cancer than among control subjects, but I am aware of no adequate study in this area. In experimental carcinogenesis, and particularly in dealing with DMBA carcinogenesis in rats, either iodine restriction or the induction of hypothyroidism increases the incidence of tumorigenesis, whereas thyroid feeding decreases the numbers of induced tumors. These findings would seem to contradict the findings of Kapdi and Wolfe,[39] but are consistent with the Mittra and Hayward[40] study.

Mustacchi and Greenspan,[41] with the provisional inference that both breast cancer and the duration of thyroid therapy are age-dependent, question the correlation suggested by Kapdi and Wolfe,[39] and point out that the true relationship between these two variables can be determined only by careful prospective study. The American Thyroid Association, which is the scientific body of physicians and basic scientists in the United States who are most concerned with the thyroid and its diseases, warned patients taking thyroid hormones for well-established indications to continue to do so, and urged that controlled studies be made to determine the possible relationship between thyroid substance and human breast carcinogenesis.[42]

The questionable role of thyroid hormone in carcinogenesis brings up the question of interactions among hormones and the real difference between the administration of a given hormone and the delivery of that hormone and/or its metabolites to the mammary gland by the general metabolic processes of the host. Not only may thyroid hormone have a direct effect on the mammary gland, but also the level of thyroid hormone in the host has a profound effect on the biotransformation of estrogens. Most of the studies in this regard have been done in man, and the pioneering studies of Fishman *et al.*[43] are noteworthy. Persons with hyperthyroidism, either idiopathic or iatrogenic, show a sharp increase in the fraction of carbon-labeled estradiol-17-β that is converted to 2-methoxyestrone. This increase is accompanied by an equally sharp decrease in the fraction that is converted to estriol. In further studies, Fishman *et al.*[44] observed that in myxedema, the level of conversion to 2-hydroxyestrone is greatly diminished and the conversion to estriol is increased. With hyperthyroidism, the formation of 2-hydroxyestrone makes it the major estrogen metabolite.[44]

Thus, it is possible that the increase in mammary cancers following DMBA administration in Sprague–Dawley rats that are either iodine-deficient or hypothyroid is due to the increase of estriol as a metabolite. On the other hand, the catechol estrogens (2-hydroxyestrone and 2-methoxyestrone) are increased in hyperthyroidism at the expense of estriol and estradiol-17-β, and these catechol estrogens may well be responsible for the sparing effect on mammary cancers due to the thyroid feeding. This seems to be an even more logical explanation when we consider that 2-hydroxyestrone is the characteristic urinary estrogen found in anorexia nervosa, in which genital and mammary atrophy are the rule, whereas an increase in estriol in the urine is characteristic of obese individuals, who are also more susceptible to breast cancer.[45] Along this same line of thought, progesterone, when given simultaneously with estradiol-17-β, profoundly decreases the peripheral estrogenic activity of the administered estradiol-17-β while altering the hepatic biological inactivation of the same estrogen in the opposite direction.[46]

4.2. Estrogens

Regarding estrogens and mammary carcinogenesis, continuous estrogen action on the mammary gland seems to be required for a prolonged period of time and in an appropriate amount for each animal species. There is undoubtedly an optimal level at which the estrogenic activity produces the greatest number of mammary cancers without undue toxicity to the host. There appears to be no particular predilection for any certain animal and any particular estrogen, natural or synthetic, to produce breast cancer as long as the compound possesses the necessary estrogenic activity in the test animal.

4.3. Insulin

The studies of Cohen and Hilf[47] indicate that insulin is required for lactation, which probably indicates that insulin is also required for mammary carcinogenesis, at least for promotion to identifiable clinical size.

4.4. Progesterone

Studies regarding a possible carcinogenic role for progesterone show even greater disagreement than those regarding estrogens, but in the case of progesterone it is largely related to the species in which the studies are done. In our experience, estrogen-induced carcinogenesis of the mammary gland is reduced in rats through the simultaneous administration of progesterone[24]; on the other hand, in the dog, in which estrogens do not appear to induce mammary cancer, progesterone and its derivatives produce mammary cancers without the concomitant administration of estrogen.[13]

Sherman and Korenman[48] suggested that breast cancer occurs in women who have a short luteal phase of the menstrual cycle and are therefore exposed to unopposed estrogen longer during any given cycle. Kodama *et al.*[49] have been studying urinary steroids in Japanese patients with breast cancer and in normal control subjects. The breast-cancer patients have shown a significant decrease in metabolites, particularly progesterone, which Kodama and associates term "menstruation-dependent steroids." They interpret their data to indicate a disturbed corpus luteum function in women who develop breast cancer. This disturbance is reflected not only by the excretion studies, but also by the fact that they have a lower number of live births. Hoover *et al.*[16] observed, in private practice, a group of patients who had a significant increase in endometrial cancer after being given estrogens for long periods of time and an increase in the incidence of breast cancer in these same patients. The latter increase was not statistically significant, but since the basal incidence of breast cancer is so great, it is doubtful that there will ever be a statistically identifiable increase of breast cancer in women treated with estrogens. In none of these studies was there prolonged treatment with estrogens in combination with progesterone.

Various studies have failed to connect contraceptive pills with a significant increase or decrease in malignant breast tumors, but these compounds, which at present are a combination of estrogen and a 19-nortestosterone derivative and formerly were a combination of an estrogen and a progesterone, are said to decrease the incidence of benign breast tumors requiring biopsy.

4.5. Pituitary Hormones

Moon *et al.*[50] showed that the prolonged administration of both bovine and ovine growth hormone led to neoplasms of most of the organ systems in the Long–Evans rat, and that one of the common systems involved was the mammary gland. As far as I know, this is the sole well-executed study showing a tumorigenic effect from a single pituitary hormone. Most other studies are not reliable—they employ either (1) impure mixtures of drugs such as hormones active on the central nervous system, which have a major effect on the release of pituitary hormones, and which lead to increases in prolactin secretion; or (2) the transplantation of pituitary tumors that produce predominantly prolactin; or (3) the transplantation of pituitary anterior lobes to ectopic sites, which leads to the increased secretion of prolactin because it separates the pituitary from the pituitary stalk. The stalk is the source of prolactin-inhibiting factor, which holds prolactin secretion in check in normal animals.

When we analyze the effect of pituitary hormones, we should recall that the endocrine system is a system of checks and balances influencing through a series of hypothalamic hormones the secretion of pituitary hormones, and that both the pituitary gland and the hypothalamus are affected by the hormones, both steroidal and nonsteroidal, produced by the end organs. None of the hormones involved in carcinogenesis acts by itself. We are not speaking of carcinogenesis in *in vitro* studies, but in an intact, living organism the endocrine system of which responds to the administration of hormones or to the removal of the organs by which they are synthesized.

For example, the chronic administration of estrogen, such as is required for carcinogenesis, carries with it an appropriate host reaction to decrease gonad-stimulating hormones (follicle-stimulating hormone and luteinizing hormone) and to increase prolactin levels. In some animals, the administration of estrogen appropriately produces the status of pseudo-pregnancy and the consequent additional production of progesterone and relaxin. The belief has been espoused not only that prolactin is a fundamental requirement for the carcinogenic effect of estrogen, but also the opposite, that the complete absence of prolactin, either through interference with its release by drugs such as L-dopa or by hypophysectomy, stops the carcinogenic effect of estrogen. The possibility also exists that the progesterone inhibition of estrogen carcinogenesis is mediated through its effect on pituitary gonadotropin and its inhibition of the prolactin-releasing effect of estrogens. Thus, all endocrine manipulations have reflections on the rest of the systems.

Meites[(51)] summarized a great deal of his and his colleagues' work on the relationship of prolactin to mammary tumorigenesis. Spontaneous, benign mammary tumors in old female Sprague–Dawley rats occur at a time when their prolactin levels are higher than in young animals. Meites and his co-workers therefore made surgical lesions in the median eminence of young, adult Sprague–Dawley rats. These lesions produced a rapid and sustained rise in serum prolactin concentrations. The rats bearing these lesions, as opposed to the control rats with operations, were larger and heavier, and had serum prolactin levels almost four times greater. There was greater development of the mammary glands in three times as many lesion-bearing rats and five times as many tumors. These findings indicate that prolactin in the Sprague–Dawley rat is etiological at least in the production of benign mammary tumors. In further studies, these workers showed that if the serum prolactin is raised by a wide variety of methods before the Sprague–Dawley rat is exposed to DMBA, there is a significant decrease in the number of mammary cancers. I believe that these observations indicate that prolactin in the Sprague–Dawley rat may be increasing the number of benign tumors, but decreasing the carcinogenic response to a chemical carcinogen. This view is in rather sharp contrast to studies by the Furth group (see Section 5.2).

5. Interaction between Hormonal Change and Carcinogens in Mammary Carcinogenesis

We have learned a great deal about the effects of hormones and carcinogenesis through the study of the interaction between hormonal change and carcinogens, particularly radiation and carcinogenic hydrocarbons.

5.1. Estrogen and Irradiation

In our laboratory, we have studied the synergism between estrogen and radiation in mammary carcinogenesis. The synergism is optimal in the A × C female rat given diethylstilbestrol continuously in diethylstilbestrol–cholesterol pellets and exposed to γ radiation. Our observations have been extended by others to show that neutron radiation is equally effective. For radiation–estrogen synergism, we found the dosage–response curve for optimal production of mammary cancers to be 150 rads in a single dose; larger doses produce excessive deaths, and smaller doses produce fewer mammary cancers. The presence of the ovary is essential for optimal carcinogenesis, and the other major ovarian steroid, progesterone, does not return the ability for this synergistic effect to occur. It is our belief that the missing factor from the ovary that is essential for carcinogenesis may well be relaxin.

5.2. Mammotrophic Hormone and Irradiation or Chemical Carcinogen

Furth[52] and his school have contributed much to our knowledge of the hormonal genesis of mammary tumors. They have developed in the pituitary of rats transplantable tumors that secrete large amounts of what he refers to as "mammotrophic hormone" (MtH). In general, it appears that these tumors produce large amounts of prolactin. In the Furth experiments, if rats are exposed to 50 rads of irradiation, they get no tumors. If they bear the MtH-producing pituitary tumor alone, they get no mammary tumors. If, however, they bear the tumor and are exposed to 50 rads of irradiation, 58% of the animals develop mammary tumors. A similar result is obtained by the single administration of 10 mg 3-methylcholanthrene. By itself, this substance produces no mammary tumors in the rats, but when it is given to rats bearing the MtH-producing, transplantable pituitary tumors, 85% of the animals have mammary tumors. The mixed results seem to indicate that the pituitary hormones coming from the transplantable pituitary tumor are able to synergize with a dose of radiation or chemical carcinogen, insufficient of itself to produce tumors in these animals, such that a substantial number of the animals with a combined exposure have mammary tumors.

Yokoro and Furth[(53)] attempted to extend the observations by transplanting the induced benign and malignant tumors. When these tumors were transplanted into female rats not bearing the MtH-secreting tumor, only an occasional tumor grew. However, when they were transplanted into female rats bearing the MtH-secreting tumor, all the mammary tumors, both benign and malignant, grew.

The hormone required for mammary carcinogenesis in the Sprague–Dawley rat given DMBA has also been the subject of many studies. This carcinogenesis is inhibited by hypophysectomy, by the drugs that inhibit prolactin release, by castration, by pregnancy, and by hyperthyroidism. Thus, we have some evidence that rodent carcinogenesis, either by hydrocarbons such as DMBA or by various forms of radiation, may serve as a good model for the endocrine background of breast cancer.

5.3. Prolactin and Various Carcinogens

Clifton *et al.*[(54)] have evidence that the prolactin production from transplantable, functional pituitary tumors enhances development of mammary carcinomas, but further alteration in the animals is necessary. When a functional tumor was used in Fischer rats, few carcinomas developed in the animals bearing the functional pituitary tumor or after exposure to γ rays or fission neutrons. When these animals were adrenalectomized and maintained with deoxycorticosterone, however, 86% of them developed mammary carcinomas. If the Fischer female rats had a graft of an anterior pituitary gland as their presumed source of increased prolactin, the same type of irradiation led only to a significant increase in mammary fibroadenomas.[(55)] Unfortunately, data are not available in these papers to compare the levels of prolactin attained by the transplanted gland as opposed to the functional tumor. The authors do suggest that the functional tumor produces much more prolactin.

Bulbrook *et al.*[56] conclude from their own studies, and those in the literature, that prolactin is important, not in the etiology of human breast cancer, but only in very specialized animal models. They believe that the similar protective effects of early pregnancy and early castration in humans indicate that large amounts of estrogen given early are protective, and that women who undergo early oophorectomy probably had hyperestrogenism previously. This is a very ingenious suggestion and should be considered.

Welsch and Nagasawa[57] reviewed the data on prolactin and rodent mammary carcinogenesis. In a concluding note, which they confirmed in their laboratory, they reported that although ovine prolactin does not stimulate human breast tissue in organ culture, human prolactin (or human placental lactogen) shows evidence of doing so. They suggest that there may be some human carcinomas that have been influenced by prolactin, but proof of this suggestion requires further work.

5.4. Hormones and Nutrition

There is increasing evidence that nutrition has an effect on mammary carcinogenesis largely because of its effect on the endocrine system. A common observation is that mammary cancer is more common among obese women and that when it does occur in obese women, it has a worse prognosis than in lean women. Obesity is frequently associated with a deranged menstrual cycle and sterility, both of which are associated with a higher incidence of breast cancer. Obese women have excessive conversion of androstenedione to estrone, which may be the basis for the irregular menses and the sterility and which exposes the mammary gland to continuous estrogen stimulation. Chan *et al.*[58] report that breast cancer is more readily induced in animals fed a diet high in saturated fat and that obese rats have higher prolactin levels. Obese women, like hypothyroid women, excrete more estriol in their urine. Very thin women with anorexia nervosa excrete excessive 2-hydroxyestrone in their urine. This may be the real link between obesity and breast cancer and the hypothyroid women with breast cancer reported by Hayward.[37]

6. References

1. A. Lacassagne, Apparition de cancer de la mamelle chez la souris male, soumise des injections de folliculine, *C. R. Acad. Sci. Paris* **195,** 630–632 (1932).
2. J. Staats, The laboratory mouse, in: *Biology of the Laboratory Mouse* (E. L. Green, ed.), 2nd Ed., p. 1, McGraw-Hill, New York (1966).

3. J. J. Bittner, Tumor incidence in reciprocal F_1 hybrid mice: A × D high tumor stocks, *Proc. Soc. Exp. Biol. Med* **34,** 42–44 (1936).
4. J. J. Bittner, Some possible effects of nursing on the mammary gland tumor incidence in mice, *Science* **84,** 162–166 (1936).
5. R. A. Huseby and J. J. Bittner, Studies on the inherited hormonal influence, *Acta Unio Int. Contra Cancrum* **6,** 197–205 (1948).
6. L. T. Samuels and J. J. Bittner, Excretion of steroids in the feces of mice of various strains with and without the mammary tumor milk agent, 38th Annual Meeting of the American Association of Cancer Research (1947) (abstract).
7. R. A. Huseby, Personal communication, August 27 (1977).
8. B. Zondek, Tumor of pituitary induced with follicular hormone, *Lancet* **1** (April 4), 776–778 (1936).
9. C. J. Shellabarger, in: *Biology of Radiation Carcinogenesis* (J. M. Yuhas, R. W. Pennant, and J. D. Reagan, eds.), pp. 31–43, Raven Press, New York (1976).
10. C. Huggins, L. C. Grand, and F. P. Brillantes, Mammary cancer induced by a single feeding of polynuclear hydrocarbons and its suppression, *Nature (London)* **189,** 204–207 (1961).
11. A. C. Andersen, Parameters of mammary gland tumors in aging beagles, *J. Am. Vet. Med. Assoc.* **147,** 1653–1654 (1965).
12. R. Tyslowitz and E. Dingemanse, Effect of large doses of estrogens on the blood picture of dogs, *Endocrinology* **29,** 817–827 (1941).
13. R. Hill and K. Dumas, The use of dogs for studies of toxicity of contraceptive hormones, in: *Pharmacological Models in Contraceptive Developments,* p. 74, World Health Organization, Stockholm (1974).
14. V. A. Drill, D. P. Martin, and E. R. Hart, Effect of oral contraceptives on the mammary glands of rhesus monkeys: A preliminary report, *J. Natl. Cancer Inst.* **52,** 1655–1657 (1974).
15. R. L. Kirschstein, A. S. Rabson, and G. W. Rusten, Infiltrating duct carcinoma of the mammary gland of a rhesus monkey after administration of an oral contraceptive: A preliminary report, *J. Natl. Cancer Inst.* **48,** 551–556 (1972).
16. R. Hoover, L. A. Gray, Sr., P. Cole, *et al.,* Menopausal estrogens and breast cancer, *New Engl. J. Med.* **295,** 401–405 (1976).
17. E. M. Glenn, S. L. Richardson, and B. J. Bowman, A method of assay of antitumor activity using a rat mammary fibroadenoma, *Endocrinology* **64,** 379–389 (1959).
18. W. O. Nelson, The induction of mammary carcinoma in the rat, *Yale J. Biol. Med.* **17,** 217–228 (1944).
19. C. S. McEuen, Occurrence of cancer in rats treated with oestrone, *Am J. Cancer* **34,** 184–195 (1938).
20. C. F. Geschicter, *Diseases of the Breast,* J. B. Lippincott, Philadelphia (1943).
21. R. L. Noble, C. S. McEuen, and J. B. Collip, Mammary tumours produced in rats by the action of oestrone tablets, *Can. Med. Assoc. J.* **42,** 413–417 (1940).
22. R. L. Noble, A new approach to the hormonal cause and control of experimental carcinomas including those of the breast, *Ann. R. Coll. Physicians Surg. Can.* **9,** 169–180 (1976).
23. A. Segaloff and H. Pettigrew, Effect of radiation dosage on the synergism between radiation and estrogen in the production of mammary cancer in the rat *Cancer Res.* (in press).
24. A. Segaloff, Inhibition by progesterone of radiation-estrogen-induced mammary cancer in the rat, *Cancer Res.* **33,** 1136–1137 (1973).
25. J. H. Cutts, Estrogen-induced breast cancer in the rat, Sixth Canadian Cancer Conference, London (1966), p. 50.
26. F. Schiotzhe and J. K. McDonald, Relaxin: A disulfide homolog of insulin, *Science* **197,** 914–915 (1977).

27. E. M. O'Byrne and B. G. Steinetz, Radioimmunoassay (RIA) of relaxin in sera of various species using an antiserum to porcine relaxin, *Proc. Soc. Exp. Biol. Med.* **152,** 272–276 (1976).
28. E. M. O'Byrne, B. T. Carriere, L. Sorensen, A. Segaloff, C. Schwabe, and B. G. Steinetz, Plasma immunoreactive relaxin levels in pregnant and non-pregnant women, *J. Clin. Endocrinol. Metab.* (in press) (1978).
29. J. H. Clark, Z. Paszko, and E. J. Peck, Jr., Nuclear binding and retention of the receptor estrogen complex: Relation to the agonistic and antagonistic properties of estriol, *Endocrinology* **100,** 91–96 (1977).
30. B. G. Miller, The relative potencies of oestriol, oestradiol and oestrone on the uterus and vagina of the mouse, *J. Endocrinol.* **43,** 563–570 (1969).
31. C. W. Emmens and L. Martin, Estrogens, in: *Methods in Hormone Research* (R. I. Dorfman, ed.), Vol. 3, pp. 1–80, Academic Press, New York (1964).
32. R. L. Noble, Hormonal control of growth and progression in tumors of Nb rats and a theory of action, *Cancer Res.* **37,** 82–94 (1977).
33. G. V. Rudali, F. Apiou, and B. Muel, Mammary cancer produced in mice with estriol, *Eur. J. Cancer* **11,** 39–41 (1975).
34. H. A. Bern, The neonatal mouse—Tumorigenesis after short-term exposure to hormones and its possible relevance to human syndrome, in: *Proceedings: Symposium on Endocrine-Induced Neoplasia,* Eppley Institute for Research in Cancer, Omaha, Nebraska (1977).
35. A. Lipschutz, *Steroid Hormones and Tumors,* Williams and Wilkins, Baltimore (1950).
36. W. O. Nelson, Endometrial and myometrial changes, including fibromyomatous nodules, induced in the uterus of the guinea pig by the prolonged administration of oestrogenic hormone, *Anat. Rec.* **68,** 99–102 (1937).
37. G. W. Beatson, On the treatment of inoperable cases of carcinoma of the mammary gland: Suggestions for a new method of treatment with illustrative cases, *Lancet* **2,** 104, 162 (1896).
38. R. M. O'Bryan, G. S. Gordan, R. M. Kelley, R. G. Ravdin, A. Segaloff, and S. Taylor, Does thyroid substance improve response of breast cancer to surgical castration?, *Cancer* **33,** 1082–1085 (1974).
39. C. C. Kapdi and J. N. Wolfe, Breast cancer: Relationship to thyroid supplements for hypothyroidism, *J. Am. Med. Assoc.* **236,** 1124–1127 (1976).
40. I. Mittra and J. L. Hayward, Hypothalamic–pituitary–thyroid axis in breast cancer, *Lancet* **1** (May 11), 885–889 (1974).
41. P. Mustacchi and F. Greenspan, Thyroid supplementation for hypothyroidism: An iatrogenic cause of breast cancer?, *J. Am. Med. Assoc.* **237,** 1446–1447 (1977).
42. American Thyroid Association, Breast cancer and thyroid therapy, *J. Am. Med. Assoc.* **237,** 1459–1460 (1977).
43. J. Fishman, L. Hellman, B. Zumoff, and T. F. Gallagher, Influence of thyroid hormone on estrogen metabolism in man, *J. Clin. Endocrinol. Metab.* **22,** 389–392 (1962).
44. J. Fishman, L. Hellman, B. Zumoff, and T. F. Gallagher, Effect of thyroid on hydroxylation of estrogen in man, *J. Clin. Endocrinol. Metab.* **25,** 365–368 (1965).
45. J. Fishman, R. M. Boyar, and L. Hellman, Influence of body weight on estradiol metabolism in young women, *J. Clin. Endocrinol. Metab.* **41,** 989–991 (1975).
46. A. Segaloff, The "sparing" effect of progesterone on the hepatic inactivation of alpha-estradiol, *Endocrinology* **40,** 44–46 (1947).
47. N. D. Cohen and R. Hilf, Influence of insulin on growth and metabolism of 7,12-dimethylbenz(a)anthracene induced mammary tumors, *Cancer Res.* **34,** 3245–3252 (1974).
48. B. M. Sherman and S. S. Korenman, Hormonal characteristics of the human menstrual cycle throughout reproductive life, *J. Clin. Invest.* **55,** 699–702 (1975).
49. M. Kodama, T. Kodama, S. Miura, and M. Yoshida, Hormonal status of breast cancer.

III. Further analysis of ovarian, adrenal dysfunction, *J. Natl. Cancer Inst.* **59,** 49–54 (1977).
50. H. D. Moon, M. E. Simpson, C. H. Li, and H. M. Evans, Neoplasms in rats treated with pituitary growth hormone. III. Reproductive organs, *Cancer Res.* **10,** 549–556 (1950).
51. J. Meites, Relation of prolactin to mammary tumorigenesis and growth in rats, in: *Tenovus Workshop, Fourth, Cardiff, Wales, Prolactin and Carcinogenesis* (A. R. Boyns and K. Griffiths, eds.), pp. 54–63, Alpha Omega Alpha, Cardiff, Wales (1972).
52. J. Furth, The influence of hormones on initiation and the maintenance of tumorous growths by radiation and other carcinogens, in: *Symposium on Tumorigenic and Genetic Effects of Radiation* (G. Woalinder, ed.), pp. 57–106, National Swedish Environment Protection Board, Solna (1976).
53. K. Yokoro and J. Furth, Relation of mammotropes to mammary tumors. V. Role of mammotropes in radiation carcinogenesis, *Proc. Soc. Exp. Biol. Med.* **107,** 921–924 (1961).
54. K. H. Clifton, B. N. Sridharan, and E. B. Dupole, Mammary carcinogenesis-enhancing effect of adrenalectomy in irradiated rats with pituitary tumor MtT-F 4, *J. Natl. Cancer Inst.* **55,** 485–487 (1975).
55. K. H. Clifton, E. B. Dupole, and B. N. Sridharan, Effects of graft of single anterior pituitary gland on the incidence and type of mammary neoplasm in neutron or gamma irradiated Fischer female rats, *Cancer Res.* **36,** 3732–3735 (1976).
56. R. D. Bulbrook, D. Y. Wang, and M. C. Swain, Prolactin and breast cancer, in: *Tenovus Workshop, Fourth, Cardiff, Wales, Prolactin and Carcinogenesis* (A. R. Boyns and K. Griffiths, eds.), pp. 143–148, Alpha Omega Alpha, Cardiff, Wales (1972).
57. C. W. Welsch and H. Nagasawa, Prolactin and murine mammary tumorigenesis: A review, *Cancer Res.* **37,** 951–963 (1977).
58. Po-Chueu Chan, J. F. Head, L. A. Cohen, and E. L. Wynder, Effect of high fat diet on serum prolactin levels and mammary cancer development in ovariectomized rats, 68th Annual Meeting of the American Association of Cancer Research, May (1977) (abstract #753).

2

Viruses and Mammary Carcinoma

JEFFREY SCHLOM, DAVID COLCHER, WILLIAM DROHAN, DONALD KUFE, AND YOSHIO A. TERAMOTO

1. Introduction

Viruses have long been known to be involved in the etiology of a variety of neoplasms of animals ranging from fowl to nonhuman primates. This involvement has been demonstrated by the isolation of these agents from spontaneous tumors, their inoculation into animals of the same species, and the reisolation of the particular virus from the subsequent tumor that developed.

The viruses belong to a group called "RNA tumor viruses," because they contain RNA as their genetic information. They are also referred to as "retroviruses"[1] because they contain an RNA-directed DNA polymerase enzyme that is capable of reversing the process of transcription, converting the viral RNA into DNA; this DNA eventually becomes integrated into the host genome as provirus. In general, members of this group of viruses cause tumors only in their species of isolation. This specificity, of course, raises immediate problems in the isolation and identification of putative human viruses. Moral and ethical considerations rule out the kind of inoculation experiments described above that are performed on laboratory animals. How then does one identify viruses suspected of being involved in the etiology of human breast cancer? An additional point that should be considered in any discussion of viruses and breast cancer is that whereas viruses may be necessary in the genesis of a mammary tumor, they may not, and indeed probably

JEFFREY SCHLOM, DAVID COLCHER, WILLIAM DROHAN, DONALD KUFE, AND YOSHIO A. TERAMOTO • National Cancer Institute, National Institutes of Health, Bethesda, Maryland 20014.

Table I
Characterization of MMTVs from Various Mouse Strains

Mouse strain	Mammary tumor incidence (%)	Latent period (months)	Type of mammary tumor	Virus strain	Virions in tumor[a]
C3H	100	7	Fast-growing, hormone-independent	MMTV-S	++
RIII	96	9	Fast-growing, hormone-independent	MMTV-S	++
GR	100	3	Hormone-dependent plaques, progress to hormone-independent	MMTV-P	++
C3HfC57BL	35	19	Slow-growing, hormone-independent	MMTV-L	+
BALB/c	10	14	Slow-growing, also acanthomas	MMTV-O	–
C57BL	< 1	24	Hormone-independent	MMTV-Y	–

[a] As determined by electron microscopy.

are not, sufficient in themselves to cause the disease. To be more specific, it has been demonstrated that genetics, hormones, chemicals, radiation, diet, as well as many other factors may play a role, along with viruses, in the etiology of the extensively studied murine mammary tumor model.[2,3] The studies outlined in this chapter make use of recent techniques of molecular biology and immunochemistry to help answer questions relating viruses and mammary neoplasia. The central question to be answered is: what is the variation (if any), biochemically, immunologically, and in mode of transmission, of viruses involved in the etiology of mammary carcinoma of a given species? The answer to this question is fundamental if one is to draw conclusions concerning data relating viruses and human breast cancer.

The murine model is widely used to study factors involved in the etiology of mammary carcinoma. It became evident from the early studies of Bittner[4] that a filterable agent, i.e., a virus, is involved in at least some mouse strains in the causation of mammary cancer. Over the past four decades, experimental systems have been developed in numerous mouse strains, and in almost every strain studied, a murine mammary tumor virus (MMTV) has been revealed. These studies have been reviewed elsewhere.[2,3] Table I surveys some of the mouse strains that have been extensively used, and reveals the variations in incidence of spontaneously occurring mammary tumors, latent period to tumor, types of tumors produced, and whether or not murine mammary tumor virions are observed in mammary tumors by electron microscopy. Also included are the various designations given to the MMTVs that have been isolated from the various mouse strains. It should be noted that mammary tumors appear "early" (before 1 year) in the high-incidence mouse strains C3H, RIII, and GR, and "late" (after 1 year) in moderate- and low-incidence strains C3HfC57BL and BALB/c.

A question that has remained unanswered concerning the origin of mammary oncogenesis in the mouse is: how many MMTVs are there? There are at least two possible answers to this question: (1) each mouse strain contains its own MMTV or MMTVs, which are distinct from other viruses in other mouse strains; or (2) there is only one MMTV, and each mouse strain exerts its own control over various properties of this virus, such as its expression or its virulence. It is essential that this question be answered if we are to understand the etiology of this disease.

2. Growth of Murine Mammary Tumor Viruses in Cell Culture

A major problem in the study of mouse mammary tumors has been the inability to obtain pure MMTV preparations. Until recently, the only sources of MMTV available were mouse-milk and tumor-cell homogenates. These are both highly complex media, containing numerous normal-cell or tumor-cell components, and it is thus extremely difficult to distinguish viral from host-cell components.

When mouse mammary tumors are placed in culture, MMTV is produced in many instances, but these cultures are often contaminated by the spontaneous release of murine type-C leukemia viruses. The studies of Cardiff *et al.*,[5] McGrath,[6] and Young *et al.*,[7] demonstrated that it is possible to grow mammary tumors of the BALB/cfC3H mouse strain in densely seeded primary mammary tumor cultures containing insulin and hydrocortisone. Moreover, we recently demonstrated[8,9] that this method allows the production of MMTV from all mouse strains examined. These include mammary tumors of strains C3H, RIII, GR, DD, BALB/cfC3H, and BALB/c. These cultures have been extensively examined by a variety of criteria to demonstrate the purity of MMTV produced, and have been shown to contain little or no contaminating type-C murine leukemia virus or other adventitious viruses. Furthermore, these cultures are an excellent source of both MMTV 60–70 S RNAs and radioactive MMTV cDNAs (DNAs complementary to the RNA of the MMTVs). These viral RNAs and cDNAs can now be employed in molecular hybridization studies to answer questions concerning differences among MMTVs and their natural distribution in the murine population.

3. Nucleic Acid Studies of Murine Mammary Tumor Viruses

3.1. Differences in Genomes of Murine Mammary Tumor Viruses

Early studies on the distribution of the MMTV, using molecular hybridization, reported that MMTV sequences were present in the DNA of all mice examined.[10] These studies, unfortunately, may have given the impression that on a molecular level, there is only one MMTV, and that it is equally present in all mouse strains regardless of their incidence of mammary tumors. These studies were conducted, however, with double-stranded [^{3}H]-cDNA probes consisting of an unequal representation of the MMTV genome[11]; therefore, only a portion of the MMTV genome may have actually been followed.

The use of primary cultures of mouse mammary tumors from various mouse strains, as described above, made possible the radioactive labeling of the 60–70 S RNAs of the MMTVs from various mouse strains. These ^{3}H- or ^{32}P-labeled 60–70 S RNAs were first used in competitive molecular hybridization experiments to determine whether differences exist among various MMTVs.[12] In these experiments, increasing amounts (0.3–1200 ng) of unlabeled RNA from GR, RIII, and C3H

MMTVs were added to a hybridization reaction between radioactively labeled 60–70 S RNA of MMTV(RIII) and DNA from RIII early mammary tumors. The addition of the unlabeled RNAs resulted in a complete competition of the hybrid formation between the labeled 60–70 S RNA of MMTV(RIII) and its homologous RIII tumor-cell DNA. These results indicated that the 60–70 S RNAs of the GR and C3H mouse strains are very similar to the 60–70 S RNA of the RIII mouse strain. It is important to point out that in all these studies, the source of MMTVs used was cultures of early-occurring mammary tumors.

To extend these studies, increasing amounts of 60–70 S RNAs from GR, RIII, and C3H MMTVs were added to hybridization mixtures between radioactively labeled 60–70 S RNA of MMTV(GR) and DNA from early GR mammary tumors. Again, complete competition was observed. Finally, increasing amounts of unlabeled RNAs of GR, RIII, and C3H MMTVs were added to the hybridization between radioactively labeled 60–70 S RNA of MMTV(C3H) and DNA from early C3H mammary tumors; this again resulted in a similar complete displacement of all hybrid formations. These data indicate that the viral genomes of the 60–70 S RNAs of MMTVs released from early mammary tumors of the RIII, C3H, and GR mouse strains are very similar in nucleic acid sequences within the limits of the assay employed; since these limits constitute approximately 5%, one can state that these viruses are at least 95% related in their nucleic acid sequences.

Mammary tumors were also cultivated from the C3HfC57BL mouse. This strain arose[(3,13)] by foster-nursing of C3H newborns on C57BL mothers (C57BL mice do not express detectable MMTV in their milk). As can be seen in Table I, these tumors arise at a moderate incidence late in the life of the animal. MMTV was obtained from these tumors, and the RNA of the virus was extracted. When this RNA was used[(12)] as competitor in the hybrid formation between radioactive MMTV(C3H) 60–70 S RNA and the DNA of early mammary tumors of the C3H mouse, approximately 75–80% displacement of the hybridization was observed. These results indicated that there is a substantial difference in nucleic acid sequences between the MMTV released from early C3H mammary tumors and the MMTV released from the late mammary tumors of that same mouse strain. This also constituted the first evidence that mice harbor MMTVs that may differ significantly in their nucleic acid sequences.

3.2. Distribution of Murine Mammary Tumor Virus Sequences in Murine DNAs

MMTVs have been shown to be transmitted in different mouse strains either by the milk or via the germ line.[(2,3)] Occasionally, MMTV may also be transmitted by male seminal fluids to females, which in turn can transfer the virus to their progeny via the milk. Other modes of transmission are, of course, possible. The question that we set out to answer is: can one distinguish whether an MMTV has been introduced into a given mouse via the germ line (i.e., as a germinal provirus or virogene) or via some non-germ-line mechanism, such as via the placenta, milk, or seminal fluid, or as a plasmid? The term "horizontal transmission" is not used here due to the confusion that would arise from such modes of viral transmission as via the placenta or as a plasmid in a germ cell. If an MMTV were introduced into a mouse via the germ line, one would expect to find MMTV proviral sequences equally distributed in the DNA of all tissues of that mouse. On the other hand, if an MMTV were introduced into a mouse by some other mechanism, one would expect to see an uneven distribution of MMTV sequences in the DNA of different tissues of that mouse. To address these points, we used the technique of molecular hybridization.

3.3. Kinetics of Hybridization of Murine Mammary Tumor Virus [^{125}I]-RNA to Cellular DNAs

MMTV(C3H) was isolated from supernatant fluids of the Mm5mt/c_1 C3H mammary tumor cell line. The 60–70 S RNA from these virions was purified as described by Drohan *et al.*,[14] and was iodinated to a specific activity of approximately 2×10^7 cpm/μg; this RNA was 99% TCA precipitable and 98% RNase sensitive. This RNA was then hybridized at various C_0t values to DNA from C3H mammary tumor cells and DNA from an apparently normal C3H liver; hybridization to sheep DNA was used as a control. (C_0t is defined as the product of the DNA concentration in moles of deoxyribonucleotide per liter and time in seconds.) As assayed by resistance to ribonuclease (RNase A and T_1) digestion, hybridization to sheep DNA remained consistently at 5–6% up to a C_0t of 35,000 (Fig. 1), and was thus scored as nonspecific background. Hybridization between the iodinated MMTV(C3H) 60–70 S RNA and DNA extracted from the C3H mammary tumor cell line from which the virus was produced reached a maximum of 60% (Fig. 1). This value was about 10% more than the maximum extent of hybridization between this MMTV(C3H) RNA and DNA from livers of C3H mice (Fig. 1).

The $C_0t_{\frac{1}{2}}$ value of the hybridization between MMTV(C3H) [^{125}I]-RNA and the C3H mammary tumor cell line DNA was approximately 380, and the $C_0t_{\frac{1}{2}}$ value of the hybridization to DNA from C3H liver was approximately 440. For comparison, poly-A-enriched C3H cellular RNA (selected by poly-U sepharose chromatography) representing mes-

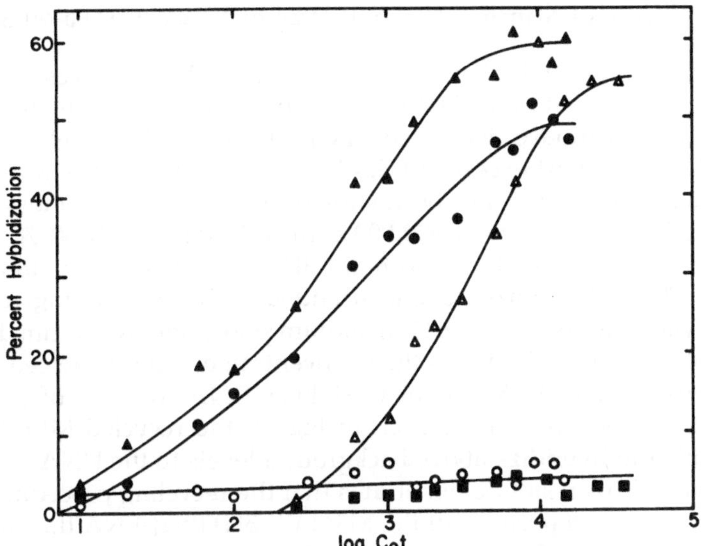

Fig. 1. Hybridization of MMTV(C3H) [^{125}I] 60–70 S RNA to various cellular DNAs. Hybridization conditions were as described by Drohan *et al.*(14) The hybridization mixtures contained MMTV(C3H) [^{125}I] 60–70 S RNA and DNA from the C3H mammary tumor cell line Mm5mt (▲), from normal C3H liver (●), and from sheep lung (○). For comparison, hybridizations were performed between ^{125}I-labeled poly-A-enriched mouse RNA and C3H liver DNA (△) and calf thymus DNA (■).

senger RNA was also iodinated and hybridized to C3H liver DNA. As can be seen in Fig. 1, the $C_0t_{\frac{1}{2}}$ value obtained using this RNA was approximately 3100, thus representing the value obtained with "unique" DNA. As an additional control, the poly-A-enriched [^{125}I]-RNA was also hybridized to calf thymus DNA; no significant hybridization was observed up to a C_0t of 35,000. The results depicted in Fig. 1 demonstrate that both the C3H mammary tumor cell line and C3H liver contain MMTV proviral sequences in the low repetitive range.(14,15) The lower $C_0t_{\frac{1}{2}}$ value obtained with the C3H mammary tumor cell line DNA suggests that there are more MMTV proviral sequences in this DNA than in the DNA of the C3H liver. The differences in final percentage of hybridization—i.e., approximately 60% for the C3H tumor cell line DNA and approximately 50% for the C3H liver DNA—may, however, be indicative of two phenomena: (1) there are quantitatively more MMTV proviral sequences in the mammary tumor DNA than in the liver DNA; or (2) the DNA of the C3H mammary tumor cells contains a portion of the MMTV genome that is not found in the DNA of C3H liver cells. To answer this question, recycling experiments were performed.

3.4. Recycling of Murine Mammary Tumor Virus [^{125}I] 60–70 S RNA

To determine whether there are any MMTV sequences in tumors that are not present in the DNA of an apparently normal organ, i.e., the liver of a C3H mouse, iodinated MMTV (C3H) 60–70 S RNA was first hybridized to a vast excess of C3H liver DNA. Liver DNA was chosen because murine livers have been shown to be negative for most MMTV markers.(2,3,16) A sample of 300,000 cpm of MMTV(C3H) 60–70 S RNA was first annealed to 30 mg normal C3H liver DNA at 68°C to a C_0t of 20,000. The unhybridized single-stranded [^{125}I]-RNA eluting from the hydroxylapatite column at 0.14 M sodium phosphate was termed "recycled RNA."(14) This RNA was then concentrated and reannealed to C3H mammary tumor DNA and to C3H liver DNA to a C_0t of 20,000 as described above. As can be seen in Fig. 2, the recycled MMTV(C3H) RNA failed to hybridize above background levels to the DNA of normal C3H livers. This result demonstrates that the recycling procedure effectively removed all portions of the MMTV(C3H) [^{125}I]-RNA that are complementary to the DNA of normal C3H liver. The same low level of hybridization can be seen between the recycled MMTV RNA and sheep lung DNA. The recycled [^{125}I]-RNA hybridizes more than 50%, however, with DNA from C3H mammary tumor cells (Fig. 2).

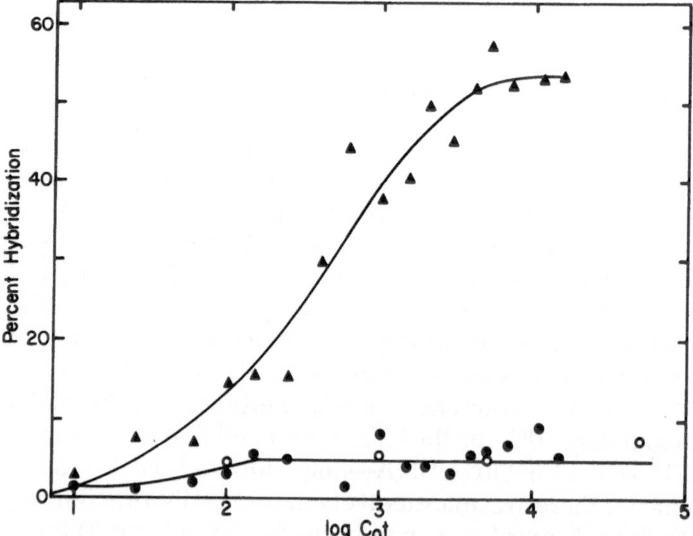

Fig. 2. Hybridization of recycled MMTV(C3H) [^{125}I]-RNA to murine cellular DNAs. MMTV(C3H) iodinated 60–70 S RNA was extensively hybridized to normal C3H liver, and the unhybridized fraction was recovered by hydroxylapatite column chromatography as described by Drohan *et al.*(14) The recycled RNA was then hybridized to the following cellular DNAs: (▲) C3H mammary tumor cell line Mm5mt (●) C3H liver; (○) sheep lung.

The results described above were obtained with DNA from a C3H mammary tumor cell line and DNA from the liver of an apparently normal animal. To determine whether similar results could be obtained with naturally occurring "early" mammary tumors, DNA was extracted from spontaneous mammary tumors of C3H mice and from livers of those same tumor-bearing animals. Livers were examined grossly and demonstrated no evidence of metastatic lesions. No significant difference was observed in percentage hybridization (to a C_0t value of 15,000) when MMTV(C3H) [^{125}I]-RNA was hybridized to the DNA of spontaneous C3H mammary tumors before or after recycling against C3H liver DNA (Table II). This same [^{125}I]-RNA, however, hybridized 49% to the DNA livers from the tumor-bearing animals before recycling, but hybridized to only background levels to this same DNA after recycling (Table II). These data again demonstrate that there are proviral sequences in "early" C3H mammary tumors that are non-germ-line transmitted.[14,17]

To determine whether the single-stranded [^{125}I]-RNA recovered by recycling is indeed part of the MMTV(C3H) genome, purified unlabeled 60–70 S RNAs of MMTV(C3H) and murine leukemia virus from C3H mice [MuLV(C3H)] were added to the hybridization between recycled MMTV(C3H) [^{125}I]-RNA and mammary tumor DNA. Addition of 0.2 or 1.5 μg MuLV(C3H) RNA did not inhibit the hybridization. However, 0.2 μg MMTV(C3H) 60–70 S RNA inhibited this hybridization by 87%, and addition of 1.5 μg of this RNA resulted in a greater than 98% inhibition of the hybridization. These results further demonstrated that the radioactive sequences obtained by recycling against C3H liver DNA are non-germ-line-transmitted MMTV(C3H) sequences.

Table II
Hybridization of MMTV(C3H) ^{125}I-Labeled 60–70 S RNA, before and after Recycling against C3H Liver DNA, to Murine DNAs

Source of DNA	Hybridization (%)[a]	
	Before recycling	After recycling
C3H spontaneous mammary tumor	55	53
C3H mammary tumor cell line (Mm5mt)	61	58
C3H liver from mammary-tumor-bearing animal	49	6
C3H liver from a non-tumor-bearing animal	49	5
Bovine ovary	6	6

[a] All values refer to an average percentage of hybridization of duplicate points assayed at a C_0t of 15,000.

3.5. Distribution of Non-Germ-Line-Transmitted Murine Mammary Tumor Virus Sequences

MMTV(C3H) [^{125}I]-RNA was recycled against DNA from C3H livers because liver cells do not appear to express detectable amounts of known MMTV antigens. To determine whether other organs of C3H would be suitable for recycling experiments, a DNA preparation was made from a pool of five tissues of apparently normal C3H mice: brain, spleen, lung, kidney, and heart. A sample of 300,000 cpm of MMTV(C3H) [^{125}I]-RNA was hybridized to 30 mg DNA from these tissues, and the recycled single-stranded RNA was purified and concentrated as described by Drohan *et al.*[14] This recycled RNA hybridized substantially to DNA from the mammary tumor cell line both before and after recycling. After recycling, however, the [^{125}I]-RNA failed to hybridize above background levels to DNA from livers or kidneys from normal C3H mice. Apparently, normal C3H pooled organs can therefore also be used for recycling to obtain the non-germ-line-transmitted MMTV sequences found in the DNA of C3H mammary tumors.

The recycled MMTV(C3H) [^{125}I]-RNA was also hybridized to DNA

Table III
Hybridization of MMTV(C3H) Recycled [^{125}I]-RNA to DNAs of Mammary Tumors and Apparently Normal Tissues of Different Strains of Mice[a]

Source of DNA	Hybridization (%)
C3H mammary tumor	44
C3H liver	7
RIII mammary tumor	47
RIII liver	7
GR mammary tumor	40
GR liver	38
BALB/c mammary tumor	7
BALB/c liver	7
C57BL-6N liver	4
C57BL/10SCN liver	6
C3HfC57BL mammary tumor	7
Ovine lung	6

[a] A sample of 600 cpm of MMTV(C3H) recycled [^{125}I]-RNA was hybridized to 500 μg cellular DNA to a C_0t of 15,000 (in 0.4 M NaPB, pH 6.8, and 0.05% SDS). The hybridizations were assayed using RNase as described by Drohan *et al.*[14]

from RIII mammary tumors and RIII livers. Complementary sequences were found in the DNA of mammary tumors, but not in the DNA of livers (Table III).

The GR strain of mice is of interest due to the transmission of the MMTV as a one-gene dominant characteristic.[2] The recycled MMTV(C3H) [^{125}I]-RNA was hybridized to GR mammary tumor and liver DNAs. Complementary sequences are present in the DNAs of both GR mammary tumors and GR livers (Table III). DNA of mammary tumors of the low- and moderate-incidence strains BALB/c and C3HfC57BL (see Table I) were also analyzed for the presence of the recycled sequences in their DNA and were negative. Since the C3HfC57BL strain was originated by C3H mice foster-nursed on C57BL mothers, a strain of mice devoid of overt MMTV in its milk, this finding is further evidence that the recycled sequences are part of the milk-transmitted MMTV(C3H) and are not germ-line transmitted.

Livers of BALB/c, C57BL/6N, and C57BL/10SCN mice were shown to contain some MMTV proviral information[17]; they do not, however, contain sequences homologous to the MMTV(C3H) recycled RNA (Table III).

It thus appears that the virus responsible for causing early mammary tumors in C3H is substantially different from the virus causing late mammary tumors in C3HfC57BL and BALB/c mice. Furthermore, a virus similar to the highly oncogenic non-germ-line-transmitted C3H virus appears to be integrated as a germinal provirus in the DNA of the GR strain.

3.6. Thermal Analyses of Hybrids

It is important in the interpretation of hybridization experiments to determine whether or not the hybrids formed between viral RNA and cellular DNAs are composed of well-matched complementary nucleic acid sequences. The DNA–RNA hybrids formed were therefore analyzed for thermal stability by hydroxylapatite column chromatography.

Thermal analysis of hybrids formed between [^{125}I] MMTV(C3H) 60–70 S RNA and DNAs from C3H, RIII, and GR mice is shown in Table IV. ΔT_m values of 3.0–3.3°C were demonstrated with RIII DNAs, and a value of 3.8°C was seen for GR DNA. Since a 1°C ΔT_m corresponds to approximately 1.5% mismatching,[18] these ΔT_m values indicate that the RNA genomes of the highly oncogenic MMTVs of RIII, C3H, and GR mice are approximately 4–6% divergent in nucleic acid sequence homology.

Table IV
Thermal Stability of Hybrids of MMTV(C3H) [^{125}I] 60–70 S RNA and DNAs of Various Murine Tissues

Mouse strain	Source of DNA	ΔT_m[a]
C3H	Mammary tumor (Mm5mt/c_1) cell line	0
C3H	Mammary tumor	0.3
C3H	Liver	0.5
RIII	Mammary tumor	3.3
RIII	Liver	3.0
GR	Liver	3.8
BALB/c	Liver	2.2
C57BL	Liver	1.9

[a] The ΔT_m for a given hybrid is defined as the difference in degrees centigrade between (1) the degrees centigrade of 50% dissociation of cellular DNA–DNA duplexes minus the degrees centigrade of 50% dissociation of cellular DNA–viral RNA duplexes, employing DNA from the cell in which the virus was grown, and (2) the same as (1) but employing DNA from a different tissue or strain of mouse, i.e., (2) − (1).

3.7. Comments on Nucleic Acid Studies

The results reported here demonstrate that substantial differences exist in nucleic acid sequence homology between the RNA genomes of the endogenous MMTV of C3H mice and the non-germ-line-transmitted MMTV of that mouse strain. Whereas mammary tumors of C3HfC57BL mice contain B-particles and intracytoplasmic A particles,[2,13,19] they do not contain the portion of the MMTV genome that is found in the DNA of early mammary tumors of C3H mice. The mammary tumors of the C3H, RIII, and GR mice that arise at high frequency early in the life of the animal all contain these particular sequences, whereas the mammary tumors of BALB/c and C3HfC57BL mice, low- and moderate-mammary-tumor-incidence strains, do not. It is of interest to note that the only mouse strain thus far examined in which these "recycled" MMTV(C3H) sequences are found in apparently normal livers is the GR mouse, a strain in which genetic evidence has been presented for a one-gene dominant characteristic for MMTV.[2,20] One possible explanation, therefore, is that a virus similar to the non-germ-line-transmitted viruses of C3H and RIII has become integrated as an endogenous virus of GR. This virus is substantially different, however, from the endogenous MMTVs of C3HfC57BL or BALB/c.

Qualitative differences between MMTVs of different mouse strains were also determined by the thermal analysis of hybrids. The differences in ΔT_m values observed indicate that there is a difference of approximately 5% in nucleic acid sequence homology between MMTV(C3H) and the MMTVs of RIII and GR mice. This finding is in accord with previous competitive molecular hybridization studies in which a nucleic acid sequence homology of 95% (the limit of detection of the assay) was reported for the MMTVs from early occurring mammary tumors of C3H, RIII, and GR mice,[12] and with minor differences seen in partial sequence analysis[21] of the RNAs of MMTVs of GR and BALB/cfC3H. These studies thus demonstrate that both major differences [as detected with recycled MMTV(C3H) RNA] and minor differences (as detected by ΔT_m values) exist in MMTVs of various mouse strains. Searches for a viral involvement in mammary cancer of other experimental animals or in human breast cancer should consider the complexities in both the mode of transmission and the distribution of nucleic acid sequences of the various MMTVs.

4. Immunological Studies of Murine Mammary Tumor Viruses

4.1. Introduction

We have also addressed the question of diversity of MMTVs from the immunological view. Immunological assays including virus neutralization,[22] immunization,[23] and immunodiffusion[24] previously indicated a possible antigenic difference among MMTVs isolated from different strains of mice. These serological data are difficult to interpret, however, principally due to the lack of pure-cell-culture sources of MMTV virions and MMTV proteins, free of contaminating cellular debris, as well as to the lack of MMTVs grown in nonmurine cells.

Highly purified preparations of MMTVs have been obtained from primary cultures of mouse mammary tumor cells and from mouse mammary tumor cell lines. The polypeptide and antigenic structures of MMTVs purified from culture fluids were described previously.[25,26] The major surface component of the MMTV virion envelope is a 52,000-dalton glycoprotein (gp52). The other major protein components of the MMTV virion are a 36,000-dalton glycoprotein (gp36), and the 28,000-, 14,000-, and 10,000-dalton polypeptides (p28, p14, and p10, respectively).

With the development of sensitive radioimmunoassays (RIAs) for the whole MMTV virion[25] or for purified MMTV polypeptides,[27–29] it has become possible to analyze precisely, both quantitatively and qualitatively, similarities or differences among MMTVs from different mouse strains. The propagation in nonmurine cells of MMTV from RIII mouse milk[30] and other mouse strains[31] has made possible the tasks of discriminating (1) viral-coded from host-coded reactivities and (2) different antigenic reactivities among MMTVs derived from different mouse strains.

Reactivities among immunologically related proteins were categorized by Hunter[32] as (1) reactions of identity, (2) complete cross-reactions, and (3) incomplete cross-reactions. Reactions of identity are shown by proteins that compete in RIA in a manner that is indistinguishable from that of the antigen being assayed. The slope of competition curves of proteins showing reactions of "identity" are exactly the same as that of the antigen being assayed. Proteins showing "complete cross-reaction" are capable of competing for all the antibodies binding to the antigen; the affinity, however, is reduced. More of the cross-reacting protein than antigen is required for an equal displacement of labeled antigen, resulting in a shallower slope of the competition curve in comparison with that of the antigen. "Incompletely cross-reacting" proteins will compete for some but not all of the antibodies binding to the antigen. Competition curves of incompletely cross-reacting proteins will plateau at some level above complete displacement of labeled antigen. The first category of reactivity is characteristic of a group-specific reaction, while the latter two can be designated as type-specific reactions. We set out to determine whether both type and group specificities are associated with the 52,000-dalton major envelope glycoprotein (gp52) of MMTV virions.

4.2. Development of Radioimmunoassays

Highly purified MMTV preparations were used in virion RIAs.[33] Purified virions were routinely monitored for the presence of intact MMTV type-B virions and the absence of murine leukemia virus type-C virus by double diffusion in agar, by divalent cation preference of virion-associated reverse transcriptase, and by electron microscopy.[8,9,25] By all these criteria, no evidence of type-C retroviruses was detected in any of the MMTV preparations used. Sodium dodecyl sulfate–polyacrylamide gel electrophoresis (SDS-PAGE) profiles of unlabeled virions revealed polypeptide patterns consistent with those reported earlier[34] for MMTV. In lactoperoxidase-labeled intact MMTV-(RIII) or MMTV(C3H), greater than 90% of ^{125}I label migrated as a 52,000-dalton protein (Fig. 3A). Purified MMTV(RIII) or MMTV(C3H) gp52, labeled with ^{125}I by the lactoperoxidase method, was also found to migrate as a homogenous species in SDS-PAGE (Fig. 3B). The specificity of the goat anti-MMTV(RIII) serum and its ability to react with native gp52 on the surface of the virion was determined by direct precipitation

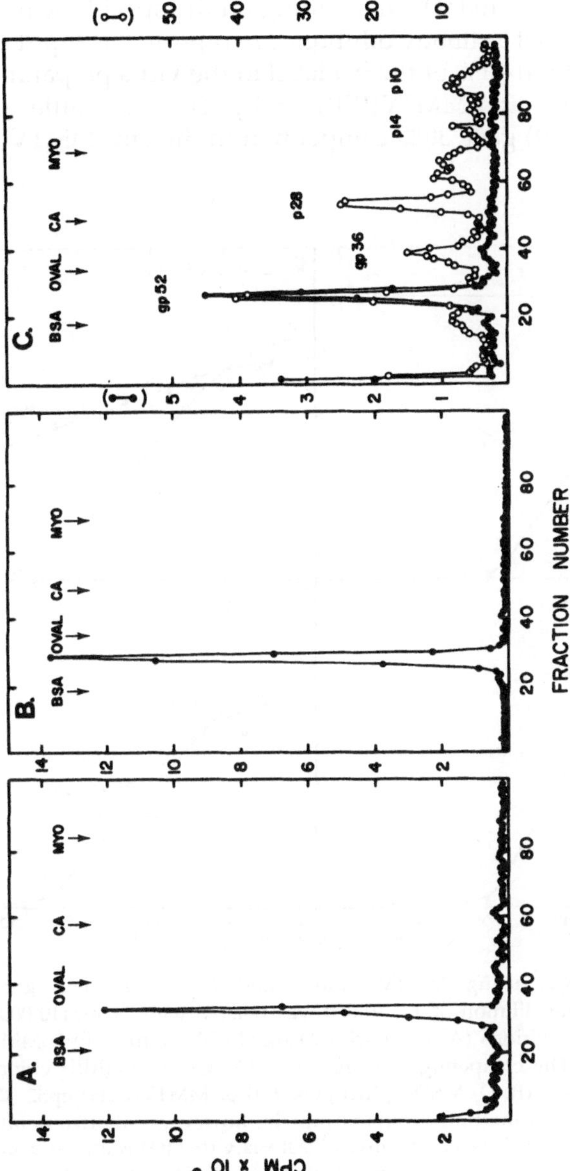

Fig. 3. Electrophoresis in 10% polyacrylamide gels in the presence of 0.1% SDS.[37] (A) Externally iodinated intact MMTV (RII) virions; (B) lactoperoxidase-labeled purified MMTV (RIII) gp52; (C) Triton-X-100s-disrupted and iodinated MMTV (RIII) (O); immunoprecipitate by anti-MMTV (RIII (●). The designation of the MMTV polypeptides and markers are as described by Teramoto *et al.*[33]

of Triton-X-100-disrupted ^{125}I-labeled MMTV virions.[33] Only the gp52 protein was precipitated by this antiserum (Fig. 3C).

The 50% end-point titer of the anti-MMTV(RIII) usually ranged from 1 : 50,000 to 1 : 100,000, depending on the specific activity of the virus preparation. All antibody dilutions are reported as input dilutions in 10 μl. Greater than 90% of the ^{125}I label in the virus preparations was precipitable by the anti-MMTV(RIII) or by TCA. As little as 10 ng purified MMTV(RIII) gave 30% competition in the anti-MMTV(RIII) vs.

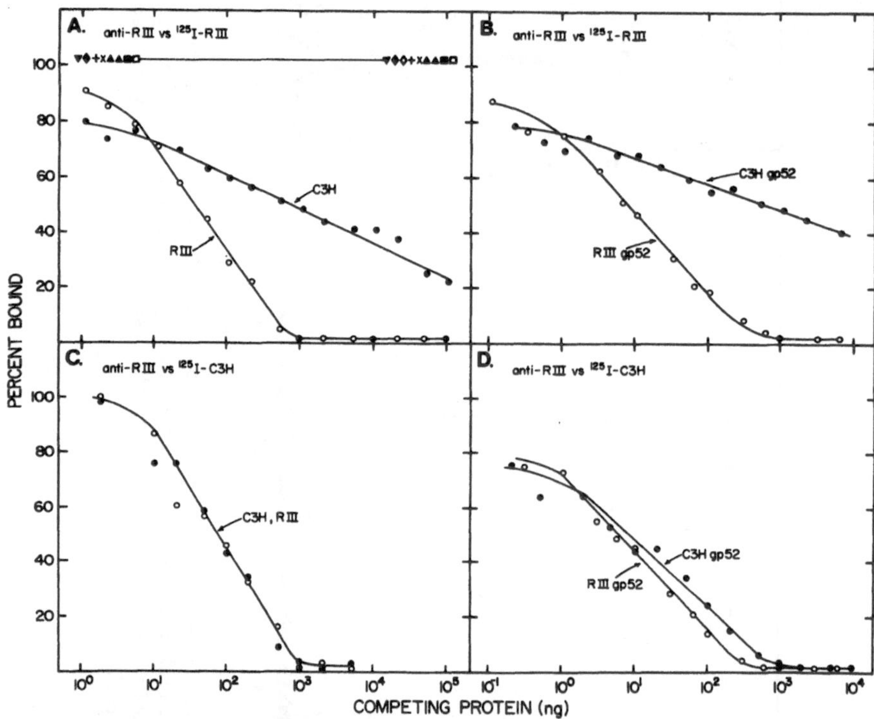

Fig. 4. Competition RIA using MMTV virions and MMTV purified gp52. Anti-MMTV(RIII) at an input dilution of 1 : 10,000 was reacted with 24 ng (10,000 cpm) of [^{125}I]-MMTV(RIII) intact virions (A, B) or with 24 ng (10,000 cpm) of [^{125}I]-MMTV(C3H) intact virions (C, D). The competing proteins were: (A, C) MMTV(RIII) virions (○) or MMTV(C3H) virions (●); (B, D) MMTV(RIII) gp52 (○) or MMTV(C3H) gp52 (●). Other competitors were tested at multiple inputs of protein; however, only two points are depicted for clarity: (▼) RD-114; (◆) avian myeloblastosis virus; (◇) Mason–Pfizer monkey virus; (+) C3H type-C virus; (×) squirrel monkey retrovirus; (▲) feline kidney cell extract; (△) C57BL lactating mammary gland extract; (■) C57BL clarified milk; (□) fetal calf serum proteins.[33]

[^{125}I]-MMTV(RIII) assay (Fig. 4A). No significant qualitative differences were observed whether virion preparations from mouse milk or from tissue-culture fluid were used in the competition RIA.

The specificity of the RIA was further examined by using other retroviruses, mouse and feline tissues, and fetal calf serum proteins as competitors in the binding of [^{125}I]-MMTV(RIII) virions to the goat anti-MMTV(RIII) antiserum (Fig. 4A). Rauscher murine leukemia virus, a type-C virus from C3H cells (C3H/T10), avian myeloblastosis virus, RD-114, Mason–Pfizer virus, and squirrel monkey retrovirus did not inhibit the RIA. Similarly, mouse milk and lactating mammary gland extracts from C57BL mice, cell-free extracts of CrFK cells, and fetal calf serum proteins did not inhibit.

4.3. Type and Group Reactivities of Murine Mammary Tumor Viruses

The antigenicity of gp52s associated with MMTV virions from RIII and C3H mice were compared by RIA.[33] The anti-MMTV(RIII) serum precipitated greater than 90% of ^{125}I-labeled MMTV(RIII) or MMTV(C3H) virions, indicating the presence of cross-reactive antigens on both virions. In addition, the 50% end-point titer of the anti-MMTV(RIII) was similar with either MMTV(RIII) or MMTV(C3H), suggesting comparable specific activities and concentrations of cross-reacting antigens for both virion preparations. The precipitation of MMTV(C3H) by the anti-MMTV(RIII) serum was consistent with previous data describing group-specific antigens of MMTV.[8,28,35,36]

In addition to the reactivities observed by direct precipitation of MMTV(RIII) and MMTV(C3H) virions, differences between the RIII and C3H MMTVs were observed when MMTV(C3H) was used to compete for the binding of anti-MMTV(RIII) serum to [^{125}I]-MMTV(RIII).[33] In this homologous MMTV(RIII) virion assay, a reduction in the slope of the competition curve generated by increasing amounts of MMTV(C3H) was observed as compared with the curve generated with MMTV(RIII) as competitor (Fig. 4A). While complete competition was obtained with 1 μg MMTV(RIII), only 80% inhibition was achieved by adding 100 μg of competitor MMTV(C3H). To eliminate the possibility that the reduced competition by MMTV(C3H) was due to incomplete purification of the virus or to an alteration in antigenic structure, both MMTV(C3H) and MMTV(RIII) were analyzed in a RIA using the same antiserum and MMTV(C3H) as the labeled antigen. In this system, identical competition curves were obtained with MMTV(C3H) or MMTV(RIII) used as competitor (Fig. 4C). Thus, Figs. 4A and 4C demonstrate type-specific determinants between MMTV(RIII) and MMTV(C3H) as well as show that both virion preparations contain similar amounts of group-specific determinants.

4.4. Reactivities of Purified Murine Mammary Tumor Virus Glycoproteins

The gp52s from MMTV(RIII) and from MMTV(C3H) were purified and analyzed to determine whether the observed type- and group-specific reactivities were associated with this molecule. Results similar to those observed with whole virions as competitors were obtained when the purified gp52s were used as competitors. Increasing amounts of MMTV(C3H) gp52 competed in the anti-MMTV(RIII) vs. [^{125}I]-MMTV(RIII) assay with a reduced slope as compared with purified MMTV(RIII) gp52 as competitor (Fig. 4B). Only 60% competition was obtained with 5 μg MMTV(C3H) gp52 as competitor, while less than 0.02 μg MMTV(RIII) gp52 was required for the same extent of competition. In the assay employing anti-MMTV(RIII) serum and [^{125}I]-MMTV(C3H), however, both MMTV(C3H) gp52 and MMTV(RIII) gp52 competed completely with very similar slopes and with comparable quantities (Fig. 4D).

4.5. Reactivities of Murine Mammary Tumor Viruses Grown in Heterologous Cells

To exclude the possibility that the observed type- or group-specific differences were due to murine cellular antigens, MMTV(RIII), MMTV(C3H), and MMTV(GR) were grown in the identical feline cell line (CrFK) and were analyzed in the same RIAs described above. These MMTV preparations were first tested by a variety of techniques and shown to be free of detectable murine and feline type-C viruses including RD-114.[31] The results obtained with the MMTVs grown in feline cells were similar to those obtained with the murine-grown MMTVs in regard to slope differences. MMTV(C3H)Fe [i.e., MMTV(C3H) virus grown in feline cells] and MMTV(GR)Fe competed in the anti-MMTV(RIII) vs. [^{125}I]-MMTV(RIII) RIA with reduced slopes (Fig. 5A), and much more competing protein was required as compared with the complete competition observed with MMTV(RIII)Fe preparations. In the assay using anti-MMTV(RIII) vs. [^{125}I]-MMTV(C3H), however, all three MMTVs grown in feline cells competed with identical quantities and with identical slopes (Fig. 5B). The preservation of both the type- and group-specific reactivities with the MMTVs grown in feline cells demonstrates that these reactivities are viral-coded and are not due to antigenic differences of different strains of mice.

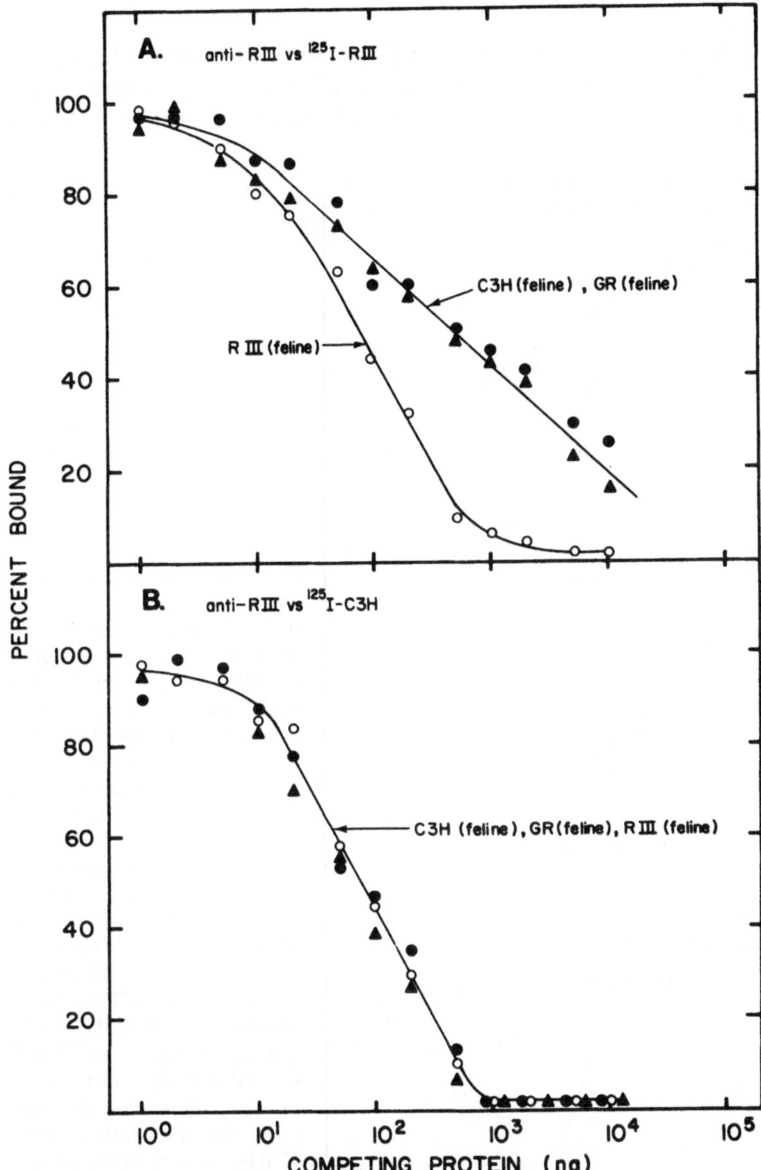

Fig. 5. Competition RIA with MMTV grown in feline cells. Anti-MMTV(RIII) at an input dilution of 1 : 10,000 was used to precipitate (A) [^{125}I]-MMTV(RIII) whole virions (24 ng, 10,000 cpm) or (B) [^{125}I]-MMTV(C3H) whole virions (24 ng, 10,000 cpm). Feline-grown MMTV(RIII) (○), feline-grown MMTV(C3H) (●), or feline-grown MMTV(GR) (▲) was used as competitor at the designated inputs of protein.[33]

4.6. Reciprocal Cross-Reactivities

To determine whether the type-specific differences observed were uniquely detected by the anti-MMTV(RIII) serum, reciprocal experiments were carried out with anti-MMTV(C3H) serum.[33] MMTV(RIII)

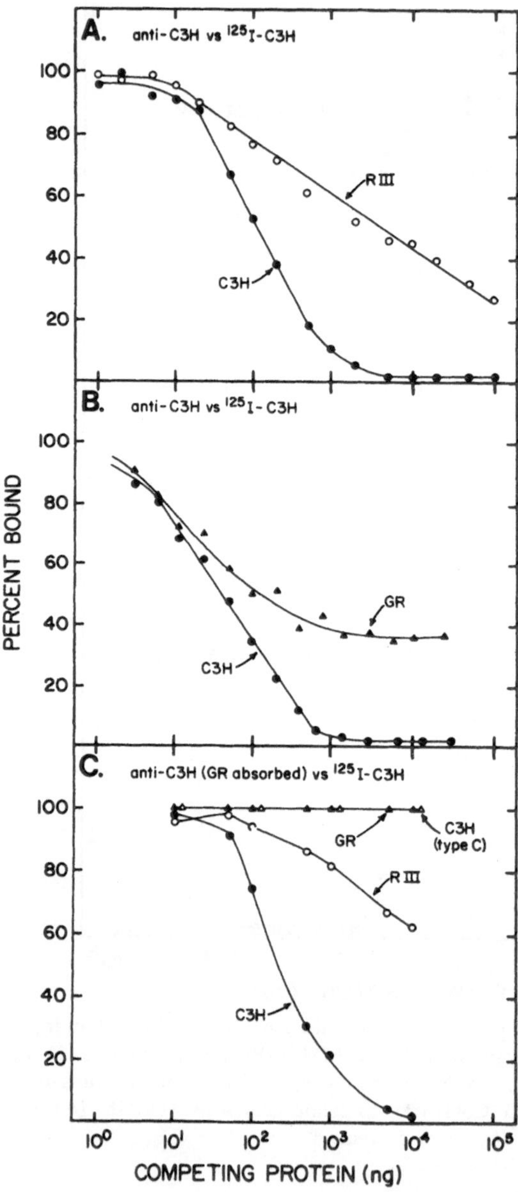

Fig. 6. Competition radioimmunoassay for MMTV. (A) Anti-MMTV (C3H) serum at a 1:10,000 input dilution was used to precipitate 24 ng (10,000 cpm) of [^{125}I]-MMTV (C3H) intact virions. MMTV(RIII) (○) or MMTV(C3H) (●) was used as competitor. (B) Anti-MMTV(C3H) serum at a 1:10,000 input dilution was used to precipitate [^{125}I]-MMTV(C3H) whole virions. MMTV(C3H) (●) or MMTV-(GR) (▲) was used as competitor. (C) Anti-MMTV(C3H) serum, absorbed with MMTV(GR), was used at a 1:10,000 dilution to precipitate [^{125}I]-MMTV(C3H) whole virions. MMTV(C3H) (●), C3H type-C virus (△), MMTV(RIII) (○) or MMTV(GR) (▲) was used as competitor. For absorption of antiserum, 1 ml of the anti-MMTV-(C3H) serum, at a 1:500 dilution, was mixed with an equal volume of MMTV(GR) containing 1 mg of virus and incubated at 37°C for 1 hour. Immune complexes and residual MMTV(GR) virions were removed by centrifugation at 20,000*g* for 35 min.[33]

and MMTV(C3H) were used to compete for the binding of anti-MMTV(C3H) to [^{125}I]-MMTV(C3H). The results were analogous to those of the previous experiments described. In this system, increasing amounts of MMTV(RIII) competitor gave a shallower slope than that given by the homologous MMTV(C3H) (Fig. 6A). At the highest input of competing MMTV(RIII) protein employed, i.e., 100 μg, only 75% inhibition was obtained, while less than 1 μg MMTV(C3H) resulted in the same competition. In assays using anti-MMTV(C3H) vs. [^{125}I]-MMTV(RIII), both viruses competed identically, i.e., with comparable inputs and with the same slope.

The addition of increasing amounts of MMTV(GR) into the anti-MMTV(C3H) vs. [^{125}I]-MMTV(C3H) system did not cause complete inhibition of the precipitation of [^{125}I]-MMTV(C3H) (Fig. 6B). Even at high inputs of protein, MMTV(GR) was incapable of competing for all the antibodies binding to MMTV(C3H). The anti-MMTV(C3H) serum therefore appears to contain an antibody population that is directed toward antigenic determinants that are present in MMTV(C3H) but not in MMTV(GR). To further amplify the type-specific reactions observed, anti-MMTV(C3H) serum was absorbed with MMTV(GR) and the immune precipitate was removed by centrifugation (see the Fig. 6 caption). The resulting MMTV(GR)-absorbed anti-MMTV(C3H) serum retained its ability to bind [^{125}I]-MMTV(C3H). This binding could be completely inhibited by the addition of MMTV(C3H) or MMTV(C3H) gp52, but was not inhibited by the addition of up to 10 μg MMTV(GR) competitor (Fig. 6C). The altered slope with MMTV(RIII) as competitor was retained using this antiserum. Additional type-specific reactivities among the various MMTVs also exist. These include differences between MMTV(C3H) and the endogenous C3H virus obtained from C3HfC57BL mice.

4.7. Comments on Immunological Studies

We have demonstrated here that MMTVs derived from different mouse strains can differ and contain both type-specific and group-specific reactivities. Both the reactivities observed are associated with the 52,000-dalton major external glycoprotein of the MMTV virion. The type and group specificities of MMTVs grown in feline cells were indistinguishable from the reactivities observed with murine-grown MMTVs. The maintenance of the antigenic differences and similarities of the feline-grown MMTV provides strong evidence that the MMTV gp52 antigens are coded for by the virus. The analysis of feline-grown MMTV further excludes the possibilities that the observed antigenic differences were due to either differences in murine antigenic determinants of the different mouse strains producing the virus or to host-coded differences in glycosylation of virions.

The three categories of antigenic reactivities described by Hunter[32] (see Section 4.1) have been demonstrated here to be present on the gp52 of MMTVs: (1) "reactions to identity," i.e., group-specific reactions between MMTVs from C3H, RIII, and GR mice, and two kinds of type-specific reactions: (2) "complete cross-reactions," resulting in shallower slopes of the competition curve, and (3) "incomplete cross-reactions," resulting in plateaus at some level above complete displacement. The type differences among MMTV variants were further magnified by absorption of antisera that removed antibodies against group-specific determinants. It was possible under the appropriate conditions to completely eliminate the antigenic reactivity to MMTV(GR) from an antiserum made against MMTV(C3H).

The identification of the type-specific differences for different MMTVs should now be of great help in monitoring the host's immune response to mammary tumorigenesis, as well as in studies seeking transspecies reactivities with MMTVs. These studies further delineate the molecular diversity of viruses that can be involved in the etiology of mammary carcinoma within a given species.

ACKNOWLEDGMENTS: We thank E. Clark, P. Hand, L. Kiefer, F. Nicholson, and J. Norman for assistance in these studies. These studies were funded, in part, by Contract NO1CP43223 from the Virus Cancer Program, National Cancer Institute.

5. References

1. A. J. Dalton, J. L. Melnick, H. Bauer, G. Beaudreau, P. Bentvelzen, D. Bolognesi, R. Gallo, A. Graffi, F. Haguenau, W. Heston, R. Huebner, G. Todaro, and U. I. Heine, The case for a family of reverse transcriptase viruses: Retraviridae, *Intervirology* **4,** 201–206 (1974).
2. P. Bentvelzen, The biology of the mouse mammary tumor virus, *Int. Rev. Exp. Pathol.* **11,** 259–297 (1972).
3. S. Nandi and C. M. McGrath, Mammary neoplasia in mice, *Adv. Cancer Res.* **17,** 353–414 (1973).
4. J. J. Bittner, Some possible effects of nursing on the mammary gland tumor incidence in mice, *Science* **84,** 162 (1936).
5. R. D. Cardiff, P. B. Blair, and P. Nakayama, *In vitro* cultivation of mouse mammary tumor virus: Detection of MTV production by radioisotope labeling and identification by immune precipitation, *Proc. Natl. Acad. Sci. U.S.A.* **59,** 895–902 (1968).
6. C. M. McGrath, Replication of mammary tumor virus in tumor cell cultures: Dependence on hormone-induced cellular organization, *J. Natl. Cancer Inst.* **47,** 455–467 (1971).

7. L. J. T. Young, R. D. Cardiff, and R. Ashley, Long-term primary culture of mouse mammary tumor cells: Production of virus, *J. Natl. Cancer Inst.* **54,** 1215–1221 (1975).
8. P. C. Kimball, M. B. Truitt, G. Schochetman, and J. Schlom, Characterization of mouse mammary tumor viruses from primary cell cultures. I. Immunological and structural studies, *J. Natl. Cancer Inst.* **56,** 111–117 (1976).
9. P. C. Kimball, R. Michalides, D. Colcher, and J. Schlom, Characterization of mouse mammary tumor viruses from primary tumor cell cultures. II. Biochemical and biophysical studies, *J. Natl. Cancer Inst.* **56,** 119–124 (1976).
10. H. E. Varmus, J. M. Bishop, R. C. Nowinski, and N. Sarkar, Mammary tumor virus specific nucleotide sequences in mouse DNA, *Nature (London) New Biol.* **238,** 189–191 (1972).
11. H. Varmus, N. Quintrell, E. Medeiros, J. M. Bishop, R. C. Nowinski, and N. H. Sarkar, Transcription of mouse mammary tumor virus genes in tissues from high and low tumor incidence mouse strains, *J. Mol. Biol.* **79,** 663–679 (1973).
12. R. Michalides and J. Schlom, Relationship in nucleic acid sequences between mouse mammary tumor virus variants, *Proc. Natl. Acad. Sci. U.S.A.* **72,** 4635–4639 (1975).
13. W. E. Heston and G. Vlahakis, C3H-A^{vy}—A high hepatoma and high mammary tumor strain of mice, *J. Natl. Cancer Inst.* **40,** 1161–1166 (1968).
14. W. Drohan, R. Kettmann, D. Colcher, and J. Schlom, Isolation of the mouse mammary tumor virus sequences not transmitted as germinal provirus in the C3H and RIII mouse strains, *J. Virol.* **21,** 986–995 (1977).
15. J. O. Bishop, in: *Vth Karolinska Symposium on Research Methods in Reproductive Endocrinology* (A. Diczfalusy, ed.), pp. 247–276, Karolinska Institutet, Stockholm (1972).
16. J. H. M. Hilgers, G. J. Theuns, and R. Van Nie, Mammary tumor virus (MTV) antigens in normal and mammary tumor-bearing mice, *Int. J. Cancer* **12,** 568–576 (1973).
17. J. Schlom, D. Colcher, W. Drohan, R. Kettman, R. Michalides, G. Vlahakis, and J. Young, Differences in mouse mammary tumor viruses—Relationship to early and late occurring mammary tumors, *Cancer* **39,** 2727–2733 (1977).
18. N. A. Straus and T. I. Bonner, Temperature dependence of RNA : DNA hybridization kinetics, *Biochim. Biophys. Acta* **277,** 87–95 (1972).
19. D. R. Pitelka, H. A. Bern, S. Nandi, and K. Deome, On the significance of virus-like particles in mammary tissues of C3Hf mice, *J. Natl. Cancer Inst.* **33,** 867–885 (1964).
20. P. Bentvelzen and J. H. Daams, Hereditary infections with mammary tumor viruses in mice, *J. Natl. Cancer Inst.* **43,** 1025–1035 (1969).
21. R. Friedrich, V. L. Morris, H. M. Goodman, J. M. Bishop, and H. E. Varmus, Differences between genomes of two strains of mouse mammary tumor virus as shown by partial RNA sequence analysis, *Virology* **72,** 330–340 (1976).
22. P. B. Blair and D. W. Weiss, Immunology of the mouse mammary tumor virus: Comparison of mammary tumor virus with the agent found in C3Hf/Crgl mice, *J. Natl. Cancer Inst.* **36,** 423–429 (1966).
23. D. H. Lavrin, P. B. Blair, and D. W. Weiss, Immunology of spontaneous mammary carcinomas in mice. III. Immunogenicity of C3H preneoplastic hyperplastic alveolar nodules in C3Hf hosts, *Cancer Res.* **26,** 293–304 (1966).
24. P. B. Blair, Strain specificity in mouse mammary tumor virus virion antigens, *Cancer Res.* **31,** 1473–1477 (1971).
25. R. D. Cardiff, M. J. Puentes, Y. A. Teramoto, and J. K. Lund, Structure of the mouse mammary tumor virus: Characterization of bald particles, *J. Virol.* **14,** 1293–1303 (1974).
26. C. Dickson and J. J. Skehel, The polypeptide composition of mouse mammary tumor virus, *Virology* **58,** 387–395 (1974).
27. W. P. Parks, R. S. Howk, E. M. Scolnick, S. Oroszlan, and R. V. Gilden, Immunochemical characterization of two major polypeptides from murine mammary tumor virus, *J. Virol.* **13,** 1200–1210 (1974).

28. A. A. Verstraeten, R. van Nie, H. G. Kwa, and Ph. C. Hagemen, Quantitative estimation of mouse mammary tumor virus (MTV) antigens by radioimmunoassay, *Int. J. Cancer* **15**, 270–281 (1975).
29. E. Ritzi, A. Baldi, and S. Spiegelman, The purification of a gs antigen of the murine mammary tumor virus and its quantitation by radioimmunoassay, *Virology* **75**, 188–197 (1976).
30. E. Y. Lasfargues, J. C. Lasfargues, A. S. Dion, A. E. Greene, and D. H. Moore, Experimental infection of a cat kidney cell line with the mouse mammary tumor virus, *Cancer Res.* **36**, 67–72 (1976).
31. D. K. Howard, D. Colcher, Y. A. Teramoto, J. M. Young, and J. Schlom, Characterization of mouse mammary tumor viruses propagated in heterologous cells, *Cancer Res.* **37**, 2969–2704 (1977).
32. W. M. Hunter, The radioimmunoassay, in: *Handbook of Experimental Immunology* (D. M. Weir, ed.), pp. 17.1–17.36, Blackwell Scientific Publications, Oxford (1973).
33. Y. A. Teramoto, D. Kufe, and J. Schlom, Multiple antigenic determinants on the major surface glycoprotein of murine mammary tumor viruses, *Proc. Natl. Acad. Sci. U.S.A.* **74**, 3564–3568 (1977).
34. Y. A. Teramoto, M. J. Puentes, L. J. T. Young, and R. D. Cardiff, Structure of the mouse mammary tumor virus: Polypeptides and glycoproteins, *J. Virol.* **13**, 411–418 (1974).
35. M. C. Noon, R. G. Wolford, and W. P. Parks, Expression of mouse mammary tumor viral polypeptides in milks and tissues, *J. Immunol.* **115**, 653–658 (1975).
36. J. B. Sheffield, T. Daly, A. S. Dion, and N. Taraschi, Procedures for radioimmunoassay of the mouse mammary tumor virus, *Cancer Res.* **37**, 1480–1485 (1977).
37. J. V. Maizel, *Fundamental Techniques in Virology*, pp. 334–362, Academic Press, New York (1969).

3

Preneoplasia in Breast Cancer

DANIEL MEDINA

1. Introduction: Significance of Preneoplasias*

The pathogenesis of murine mammary cancer has been extensively studied over the past 25 years. One concept that illustrates the course of mammary tumorigenesis was proposed and developed by DeOme and co-workers.[1–4] The concept states that mammary tumors arise from morphologically discrete epithelial lesions that are altered from normal mammary epithelial cells. This concept of progressive stages in the development of mammary neoplasia was recognized at the turn of the century by Haaland,[5] who stressed the biological significance of hyperplastic changes that preceded neoplasia. The concept of multistage development of neoplasia gained general acceptance after the elegant experiments on experimental skin and liver tumorigenesis[6–19] and for neoplasias arising from most epithelial tissues.[20] A variety of human cancers are thought to progress through several stages, as evidenced by the terms "carcinoma *in situ*" and "precancerous cystic hyperplasia,"[21]

*Abbreviations used in this chapter: (BSA) bovine serum albumin; (Con A) concanavalin A; (CP) *Corynebacterium parvum;* (CTX) cytoxan; (DH) ductal hyperplasia; (DMBA) 7,12-Dimethylbenz(a)anthracene; (5-FU)5-fluorouracil; (GPDH) glucose-6-phosphate dehydrogenase; (HAN) hyperplastic alveolar nodule; (KN) keratinized nodule; (LDH) lactate dehydrogenase; (LNC) lymph node cell(s); (L-PAM) phenylalanine mustard; (MCA) 3-methylcholanthrene; (MTV) mammary tumor virus; (MTX) methotrexate; (NIV) nodule-inducing virus; (NMR) nuclear magnetic resonance; (PRL) prolactin; (VC) vincristine; (WGA) wheat germ agglutinin.

DANIEL MEDINA • Department of Cell Biology, Baylor College of Medicine, Houston, Texas 77030.

although the concept is muddied by the lack of clearly defined criteria to characterize the various stages biologically.

The concept of preneoplasia as a biologically significant entity has been deeply engrained in scientific literature and thought over the past half-century. The description of preneoplastic mammary lesions by morphological, biochemical, and biological criteria has developed extensively over the past 25 years. The relationship between preneoplastic lesions in experimental animals and suspected analogous lesions in humans has remained debatable, however, and only in the past few years has evidence accumulated that supports the hypothesis that similar lesions do exist in humans. Previous discussion of the concept of the significance of preneoplastic lesions in humans has been clouded by the attempt to develop an exact analogy between experimental mammary tumorigenesis and human tumorigenesis. Like all analogies, these attempts are fraught with inherent difficulties. One cannot and should not expect an exact analogy between human and experimental tumorigenesis with respect to etiology, hormonal dependence, metastatic frequency, invasive potentials, morphology, and other criteria. The significant features of experimental tumor systems are to provide concepts and models by which one can understand the pathogenesis, on both the cellular and the molecular level, of human mammary tumors. The murine mammary tumorigenesis model provides a concept by which human tumorigenesis can be understood and analyzed. Basically, the model stresses the important point that mouse mammary tumorigenesis is characterized by the presence of a precursor stage, which has a greater probability of developing into mammary carcinomas than do normal mammary epithelial cells. Second, a variety of etiological agents, viral, chemical, and physical, can interact at both the normal and preneoplastic stages to induce and promote mammary tumorigenesis.

It is the purpose of this review to define, illustrate, and discuss the preneoplastic state. The majority of the chapter will concentrate on the cellular and cell-population characteristics of precursor mammary populations. The methodology involved in the induction, observation, and transplantation of mammary preneoplastic nodules, the biological properties of mammary preneoplastic nodules, and the effects of carcinogens on different nodule cell populations, particularly the information available before 1973, was thoroughly reviewed by Medina.[(4)] In addition, recent reviews have covered the general topics of comparative aspects of mammary preneoplasia[(22)] and the roles of mammary tumor viruses[(23)] and endocrine hormones in mammary tumorigenesis.[(24)] These topics will therefore be mentioned only to complete discussion of particular aspects in this review.

2. Induction of Preneoplasias

2.1. Etiological Agents

2.1.1. *Viral*

The principal preneoplastic lesion in mouse mammary glands is the hyperplastic alveolar nodule (HAN), which is a focus of hyperplastic lobuloalveolar development in an area of nonstimulated mammary gland (i.e., ductal). HANs are found in high-incidence murine mammary tumor virus strains (i.e., C3H, A, DBA, RIII), which carry the murine mammary tumor virus (MTV*) and the nodule-inducing virus (NIV[25,26]); in high-incidence mammary tumor strains, which carry GR virus (i.e., GR/A[27]); in low-mammary-tumor strains, which carry just NIV (i.e., C3Hf[25,28]); and in MTV-free, NIV-free mice (i.e., BALB/c[3]). HANs can be induced in susceptible strains by MTV if MTV is introduced by foster-nursing,[29,30] by intraperitoneal injection of cell-free extracts of virus-containing normal mammary tissues and tumors,[29,30] or by intraperitoneal injection of blood from virus-positive mice.[31,32] HANs can be induced by NIV in susceptible strains only by mating the susceptible strain with an NIV-positive strain.

The frequency and tumor-producing capabilities of HANs are correlated with the oncogenic properties of the inducing virus. Thus, MTV infection leads to the early appearance of HANs and subsequently the early appearance of tumors and a high tumor incidence. For instance, hormonally stimulated, MTV-free, NIV-free BALB/c and C3H/StWi mice produce few HANs and tumors by 15 months of age, but hormonally stimulated MTV-positive BALB/c and C3H/StWi mice have a 90% incidence of HANs by 5 months and 100% incidence of tumors by 9 months. HANs found in MTV-positive BALB/c and C3H mice have very high tumor potentials (Table I). In contrast, NIV-positive C3Hf mice contain a low incidence of HANs and subsequently a low incidence of tumors, and the tumor potential of the nodules is low (Table I). In pathogen-

*The terminology of the various mammary tumor viruses is complex because of different terminology used by different investigators and confused due to the lack of terminology that distinguishes between exogenous viruses, which are always expressed as complete viral particles, and endogenous viruses, which may or may not be expressed as complete viral particles (i.e., endogenous virus carried in C3Hf/He compared with viral DNA sequences carried in BALB/c, respectively). Therefore, for this discussion, MTV refers to MTV-S or Bittner virus, NIV is the same as MTV-L, and MTV-free, NIV-free refers to those strains presumably carrying viral DNA information that is not expressed as intact viral particles except under very unusual and generally poorly documented conditions (i.e., BALB/c).

Table I
Tumor-Producing Capabilities of Hyperplastic Alveolar Nodules

Strain	Transplant[a]	Type of MTV	Tumors/ transplants	Percentage	Mean time of tumor appearance (weeks)[a]
C3H	HAN	MTV-S, MTV-L	31/44	70	20.6
C3H	NMG	MTV-S, MTV-L	5/47	11	27.5
C3Hf	HAN	MTV-L	3/29	10	45.0
BALB/cfC3H	HAN	MTV-S	31/38	82	29.5
BALB/c	HAN[b]	None	8/8	100	N.D.

[a] (HAN) Hyperplastic alveolar nodules; (NMG) normal mammary gland; (N.D.) not determined.
[b] Experiments performed using nodule outgrowth lines rather than primary HAN.

defined C3Hf mice that live until 32 months of age, the incidence of HANs (70% of mice have HANs by 15 months) and tumors (76% by 32 months) is high, and the survival curve is parallel to that of C3H but extended in time by 16 months.[33] These data suggest that NIV-induced C3Hf HANs can be tumorigenic under specific conditions. When C3Hf HANs have been tested directly for tumorigenicity by transplantation experiments, however, they have exhibited low tumor potentials.[34,35] Thus, although individual HANs induced by NIV have relatively low tumorigenic potential, a high frequency of HANs induced under specific conditions (i.e., pathogen-defined mice) resulted in a high tumor incidence. The oncogenic potential of NIV is low if expressed in terms of time, but high if expressed in terms of incidence (cases).

2.1.2. *Chemical*

HANs can be induced by a variety of chemical carcinogens, including 3-methylcholanthrene,[36] 7,12-dimethylbenz(a)anthracene (DMBA), urethane, and 1,2-benz(α)pyrene (ref. 37 and unpublished observations). However, the frequency of HANs induced by a potent carcinogen, such as DMBA, is very low.[37] Table II shows the results of several experiments examining the effects of DMBA on noduligenesis in mice of BALB/c and other low-incidence strains. Surprisingly, the frequency of HANs was very low, although the mammary tumor incidence was moderate. The HANs, which were isolated and transplanted from carcinogen-fed mice, had morphological and growth properties similar to those of HANs isolated from MTV-infected BALB/c mice.[38] The primary lesions found in DMBA-fed mice were ductal hyperplasias, which produced mammary tumors on transplantation.

Table II
Incidence of Mammary Dysplasia and Neoplasia in Chemical-Carcinogen-Treated Mice

		Incidence of dysplasias[a]				Neoplasias	
Strain	Carcinogen (mg)	HANs	%	DH	%	Neoplasms/ total mice	%
BALB/c	DMBA (4)	0/12	0	7/12	58	11/30	37
BALB/c	DMBA (6)	2/30	7	23/30	71	22/50	44
C3H/StWi	DMBA (6)	0/10	0	6/10	60	17/37	46
C57BL[b]	DMBA (6)	1/5	20	3/5	60	11/19	58
(C57BL × DBAf)F_1[b]	Urethane (200)	0/13	0	11/13	85	9/13	69

[a] Incidence is the number of mice with dyplasias over the total number of mice examined by observation of stained whole mount preparations of the mammary glands.
[b] All mice were virgins except the C57BL and (C57BL × DBAf)F_1, which carried a pituitary isograft for 3–6 months.

2.1.3. *Hormone Stimulation*

Prolonged hormone stimulation by pituitary isografts or forced breeding will induce HANs in MTV-free BALB/c mice[3,39] and increase the effectiveness of MTV and NIV in inducing HANs in susceptible strains.[35,40] In BALB/c mice, nodules are very rare and appear late in the second year of life. The morphological and biological characteristics of HANs induced under these conditions are very similar to those of HANs induced by MTV and DMBA.[35] Hormone stimulation enhances markedly the noduligenic and tumorigenic effect of both MTV and DMBA. Thus, virgin, MTV-positive BALB/c mice produce 10% tumors by 12 months of age, whereas MTV-positive BALB/c mice stimulated by pituitary isografts produce 90% or greater HAN and tumor incidence.[35] Analogous results are found with pituitary isograft stimulation of DMBA-treated BALB/c mice, although the morphological variant of HANs differs as compared with those produced by MTV.[35]

2.2. Morphological Types

The predominant preneoplastic lesions in mice are HANs, which are morphologically similar whether induced by MTV, NIV, or DMBA, or whether they appear spontaneously. However, two other types of lesions that occur predominantly in chemical-carcinogen-fed mice have been described. The frequency and morphological appearance of these ductal hyperplasias have been described in detail[37,38,41] and will only be reviewed here. The ductal hyperplasias were found in DMBA-treated BALB/c, C57BL, (C57BL × DBAf)F_1, and C3H/StWi (MTV-free) female mice. The incidence varied from 50 to 85%. The ductal hyperplasias were divided into four categories on the basis of their general morphological type: (1) simple duct hyperplasia, (2) lobular hyperplasia, (3) papillary hyperplasia, and (4) end-bud hyperplasia. The first three types of ductal hyperplasias were characterized by intraluminal epithelial hyperplasia and have been shown to give rise to mammary carcinomas either *in situ* or by transplantation into the mammary gland free fat pads of syngeneic mice.[38,41]

The other type of dysplastic lesion found in DMBA-fed mice are foci of keratinizing alveoli, termed "keratinized nodules."[37,38,42] These lesions are numerous in DMBA-treated, pituitary-isograft-bearing BALB/c mice and have limited growth and tumorigenic potential.[37] When transplanted into control unstimulated mice, these lesions produced ductal outgrowths and only 7% tumors by 56 weeks after transplantation. Such dysplasias transplanted into mice bearing pituitary isografts exhibited persistent lobuloalveolar development and produced a higher incidence of tumors (32%) on transplantation. The low tumor potential of these lesions supports similar results reported in an abstract by Liebelt and Liebelt.[43] These lesions are absent in DMBA-treated BALB/c mice without pituitary isografts[37,38] and in DMBA-treated C57BL mice bearing pituitary isografts.

A majority of keratinized nodules (KNs) appear to be reversible lesions. By 6 weeks after removal of the pituitary isograft, 72% of DMBA-treated BALB/c mice had KNs (X = 29 KNs/mouse), but by 12 weeks, only 26% of the mice had KNs (X = 5 KNs/mouse). The behavior of KNs on transplantation supports the observations based on mammary gland whole mounts that KNs are reversible dysplasias. On transplantation, 75% survived and all produced ductal outgrowth, but an average of only 48% of the mammary fat pad was filled by 9 weeks after transplantation. The presence of a pituitary isograft led to full lobuloalveolar development, which persisted in a minority of cases after removal of the pituitary isograft. This result suggested that the growth of existing nodule cells was selectively enhanced by hormone stimulation, and it was these cells that eventually produced tumors. KNs can be conceived as a mixed population of HAN cells and desquamating cells, which are nontumorigenic.

2.3. Comparative Aspects of Noduligenesis

Since this subject was reviewed recently by Cardiff *et al.*,[22] only a few points will be discussed here. Hyperplastic lesions occur in mice,[4] rats,[44,45] dogs,[46,47] and humans.[48–50] The predominant lesions in DMBA-fed rats are alveolar hyperplasias similar to HANs and ductal hyperplasias (DHs).[44,45] The former have a very low tumor potential,[51–54] and mammary tumors can be induced in the absence of any observable HANs.[52] Recently, Haslam[53] demonstrated that the alveolar hyperplasias found in DMBA-treated rats and transplanted into the mammary gland free fat pads of untreated female rats have little tumor potential. However, similar lesions transplanted into the mammary fat pads of rats that were subsequently treated with DMBA produced a high incidence of mammary tumors.[54] These experiments suggested that DMBA-induced HANs may be high-risk lesions in the rat, although their tumor potential may be relatively low compared with that of other types of hyperplasias in the DMBA-treated rat. Several questions remain unanswered in these experiments—in particular, the tumor potential of normal-appearing duct taken from DMBA-treated rats and transplanted into rats subsequently treated with DMBA. Second, the morphological type of tumor (adenoma vs. adenocarcinoma) was not clearly documented in these experiments. Third, the relatively long latent period for the appearance of HANs as compared with that for DHs would suggest that HANs represent only a minor pathway in tumor progression in the rat. Finally, the observation that HANs and HAN outgrowths do not regress after ovariectomy,[55] although the vast majority of DMBA-induced mammary tumors are ovarian-dependent, has not been explained satisfactorily. Either HANs give rise to ovarian-independent mammary tumors relatively late in the lifetime of the rat or the HANs are composed of mixed cell populations of hormone-sensitive and hormone-independent cells, with the former cell type being more tumorigenic than the latter.

Russo *et al.*[56,57] looked at the early changes in the DMBA-treated rat mammary glands and described DHs similar to the simple DHs described in the mice.[56,57] These hyperplasias progressed to mammary tumors based on *in situ* observations rather than transplantation experiments. These lesions represent a possible intermediate stage in DMBA-induced rat mammary cancer, although Sinha and Dao[52] maintain that the majority of rat mammary tumors arise directly, rather than indirectly from normal mammary gland. The recent description by Russo *et al.*[57] of three cytologically different cell types in the normal mammary epithelium offers the possibility that HANs and DHs may be composed of different cell types derived from different normal cell progenitors.

The recent experiments by Wellings and co-workers[48–50] described early lesions in human breast cancer. On the basis of their experiments, they postulate that the earliest changes occurred in the intralobular terminal duct of the mammary gland. They presented a scheme to suggest that these changes occurred in the terminal ducts, and eventually the altered ducts coalesced to present a structure that could be interpreted as a single larger duct showing intraluminal epithelial hyperplasia. They considered the alterations in the terminal ducts to progress to atypical DH, carcinoma *in situ*, and subsequently infiltrating ductal carcinoma. The alterations in the terminal ducts were found more frequently in cancerous breasts and in contralateral breasts of breast cancer patients, and increased with the age of the patient. So far, transplantation of various ductal atypias into the mammary fat pads of nude mice has not shown that these lesions progress to more atypical forms, although the results were preliminary.[49,50]

2.4. Model for Experimental Tumorigenesis

The data discussed in the preceding sections have been interpreted to provide the model in Fig. 1. The model states that MTV induces primarily HANs (reaction 1) in both virgin and hormone-stimulated mammary glands. The tumor potential of HANs depends on the oncogenic potential of the virus and, more important, on the innate characteristic of the individual HAN. Furthermore, MTV can enhance the tumor potentials of existing HANs (reaction 4). In contrast, DMBA leads to primarily DHs in virgin mammary glands (reaction 3a) and to HANs (reaction 2) and KNs (reaction 3b) in hormone-stimulated mammary glands. DMBA can increase further the tumor potential of existing HANs (reaction 5); however, the effect of DMBA on existing DHs or KNs has not been fully investigated. This model is based on experiments in C3H, BALB/c, BALBcfC3H, and C57BL mice. It omits consideration of plaques, which are probably pregnancy-dependent tumors and not preneoplastic lesions in the context of this discussion. The model considers that one-step transformations from normal to neoplastic cells are rare.

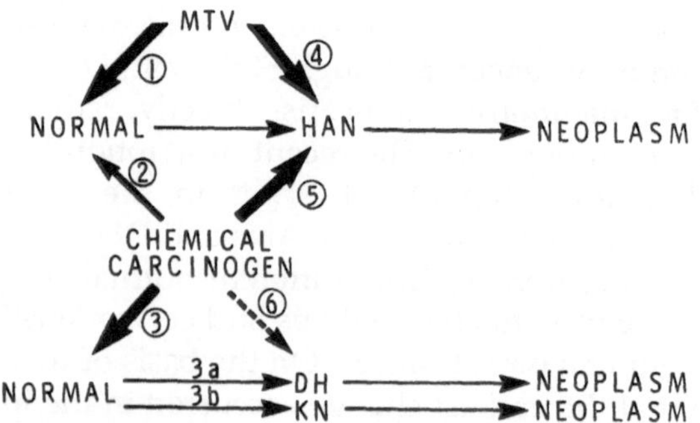

Fig. 1. Proposed model for murine mammary tumorigenesis.

The concept of precursor populations is probably valid for DMBA-induced rat mammary tumors and human breast carcinoma, although any particular pathway may be emphasized over another in any given species. The model differs from earlier models[2-4] in stressing multiple pathways to mammary tumorigenesis, but is similar in stressing the existence of discrete preneoplastic precursor populations.

3. Behavioral Properties *in Vivo*

3.1. Establishment of Nodule Outgrowth Lines

In investigating the morphological, hormonal, immunological, and biochemical properties of mammary preneoplasias under minimal variability, it has been advantageous to examine established *in vivo* cell lines of preneoplastic lesions rather than multiple primary lesions. These established lines are referred to in general terms as preneoplastic outgrowth lines and specifically as either nodule or ductal outgrowth lines. The majority of the analyses have been done on nodule outgrowth lines established from primary HANs. One of the elegant simplicities of the mammary tumor system is the ability to transplant normal and dysplastic mammary tissues into their normal anatomical site, the mammary fat pad. Since the normal mammary parenchyma grows and extends into the mammary fat pad up to 12 weeks of age, the areas of the mammary fat pads that contain the growing ductal elements in 3-week-old weanlings can be dissected from the host, leaving the remaining 80% of the mammary fat pad free of host mammary parenchyma. This gland-free mammary fat pad serves as a natural and ideal transplantation site for normal and dysplastic tissues.[1] Normal ductal and lobuloalveolar tissues give rise to ductal outgrowth, whereas preneoplastic tissues give rise to hyperplastic nodule or hyperplastic ductal outgrowths. Samples of the preneoplastic outgrowth tissues can be serially transplanted from one fat pad to another indefinitely, always giving rise to preneoplastic tissue that fills the mammary fat pad. By this technique, preneoplastic outgrowth lines can be established. These lines can be considered *in vivo* equivalents of *in vitro* cell lines.[1,4] The preneoplastic lines can be characterized with respect to numerous criteria that might distinguish them from normal or neoplastic tissues.

3.1.1. *Transplantability*

Normal cells *in vitro* and *in vivo* have a finite life span that can be estimated by serially passaging the cells.[58-61] Under these conditions, normal cells have a limited division potential. Normal mammary cells can be serially transplanted *in vivo* for 5 or 6 transplant generations.[58-61] The transplantability of normal mammary cells is not influenced by hormones, but is decreased by increasing age of the host.[60] Neoplastic cells can be transplanted indefinitely. Preneoplastic populations, whether alveolar or ductal, behave like neoplastic populations, since they exhibit an indefinite division potential.[61] The unlimited division potential is not influenced by the etiological agent inducing the original preneoplastic lesion. Certain dysplastic variants, which arise as primary lesions or as variants within a preneoplastic population, exhibit a limited division potential. These variants include KNs, cystic alveolar nodules, and end-bud DHs.

In contrast to neoplastic cells, which will grow subcutaneously, intraperitoneally, and in other white fat depots (i.e., gonadal fat organ), preneoplastic nodule cells, like normal cells, will grow only in the mammary fat pad. Cells from nodule outgrowth line D2 transplanted into the gonadal fat organ and into subcutaneous fat failed to grow, although they remained viable. The possibility existed that the environment inside the peritoneal cavity might be harmful to growth; however, the gonadal fat organ exteriorized on the abdominal wall failed to support active growth of nodule cells. Therefore, some intrinsic factors exist in the mammary fat pad that are conducive to mammary nodule growth. Since both the mammary fat pad and the gonadal fat pad are considered to be part of a systemic fat organ that behaves physiologically and genetically as a unit, some local factors peculiar to mammary fat pads must play an important regulatory role.[62] Whatever the factors, preneoplastic tissues, unlike neoplastic cells, are subject to them.

Similarly, the spatial pattern of normal mammary ducts and preneoplastic nodule outgrowths is regulated by local growth regulatory factors produced by the ducts.[63] Neoplastic cells are unresponsive to these local growth regulatory factors.

3.1.2. *Stability*

The morphological, hormonal, growth, and tumorigenic properties of preneoplastic lesions are generally stable over long periods of serial transplantation.[4,64] It should be emphasized, however, that this stability is partly artifactual, since a selective pressure is exerted at each transplantation generation for those characteristics that the investigator wishes to maintain in each line. Thus, the characteristics of each line are consciously selected for by the investigator. Several examples can be used to illustrate this point. First, over the first 10 transplant generations of nodule line D2, the overall morphology of the line was lobuloalveolar and the tumorigenic potential was moderate (45% tumors by 12 months). During transplant generation 6, however, two distinct sublines emerged. One line was characterized by large cystic alveoli, and the tumor potential decreased to less than 5% (2/50). This line was selected against and not transplanted. This variant reappeared twice over the next 30 transplant generations and was always selected against. In two other nodule lines, this variant was selected for with the result that the tumor potential dropped; subsequently, the nodule lines could no longer be serially transplanted after 3 transplant generations. Second, the tumor potentials of nodule lines vary over each transplant generation. In at least two cases, a low oncogenic line D1 produced sublines with high tumor potentials. Similarly, high oncogenic line D2 produced sublines with low tumor potentials. The stability of the biological characteristics of nodule outgrowth lines is often misunderstood because the unconscious selective pressures exerted by the investigator that determine this stability often go unrecognized or unstated.

3.1.3. *Tumorigenicity*

The tumor potential of primary nodules and established nodule outgrowth lines is determined at the time of formation of the lesion by the etiological agent. If one examines the tumor potential of a variety of primary HANs from different strains with differing tumor incidences, early results clearly demonstrate that HANs from MTV-positive strains produce tumors at a high frequency and with short latent periods, whereas HANs from NIV-positive strains produce tumors at a low frequency.[34,35]

The establishment of nodule outgrowth lines from primary HANs found in untreated, in chemical-carcinogen-treated, and in virus-

Table III
Tumor-Producing Capabilities of Preneoplastic Outgrowth Lines

Preneoplastic line and strain	Type	Origin	TG[a]	Tumors/ transplants	Percentage	TE_{50} (weeks)[a]
D1 BALB/c	Alveolar	Hormonal	36–39	14/122	11.5	—
D2 BALB/c	Alveolar	Hormonal	18–26	45/89	51	40
C3 BALB/c	Alveolar	DMBA	8–16	88/109	81	26
C4 BALB/c	Alveolar	DMBA	8–10	39/48	81	29
C6 BALB/c	Alveolar	DMBA	3–11	46/83	55	48
CD-3 BALB/c	Alveolar	DMBA	4–8	50/62	80	20
CD-7 BALB/c	Ductal	DMBA	4	14/20	70	21
HD-4 (C57BL × DBAf)F_1	Ductal	Urethane	2–6	24/68	35	—
HD-7 (C57BL × DBAf)F_1	Ductal	Urethane	2–7	50/110	45	—

[a] (TG) Transplant generation; (TE_{50}) 50% tumor endpoint.

Table IV
Enhancement of Mammary Tumor Formation by Viral and Chemical Carcinogens

Preneoplastic line	Agent	Tumors/ transplants	Percentage	TE_{50} (weeks)
D1	—	10/250	4	—
D1	MTV-S	91/122	75	44
D1	DMBA[a]	63/89	71	28
D2	—	41/92	45	—
D2	MTV-S	43/53	81	26
D2	DMBA	32/40	80	17
C3	—	26/30	87	23
C3	DMBA	21/22	95	11
C4	—	21/25	84	32
C4	DMBA	29/30	97	13
CD-3	—	21/28	75	18
CD-3	DMBA	21/26	81	11

[a] Total dose = 1.5 mg.

positive mice clearly shows that the tumor potentials of HANs vary from very low to very high. Table III shows the tumor potential of several established lines from untreated and carcinogen-treated mice. The tumor potential ranges from a low of 12% at 12 months after transplantation to a high of 81% at 6 months. A similar range of tumor potential was seen in ductal lines established from DHs, with HD-4 and HD-7 showing a low tumor potential and CD-7 a very high tumor potential.

The tumor potential of both nodule and ductal lines can be enhanced by subsequent infection with MTV and treatment with chemical carcinogens. Table IV gives some selected examples of these experiments that demonstrate convincingly that the tumor potential is not a fixed property of a cell population, but can be significantly altered by a variety of agents. Subsequent discussion will also demonstrate that agents can significantly inhibit as well as enhance tumorigenesis.

3.2. Hormonal Responsiveness

Since the hormonal responsiveness of nodule populations was extensively reviewed by Banerjee,[24] this discussion will summarize only the principal findings and will present some new results. The hormones necessary for noduligenesis (induction of nodules) are the same for mammogenesis, i.e., estrogen plus luteoid or corticoid plus growth hormone or prolactin (PRL).[65] The hormones necessary for HAN maintenance are similar to those for HAN induction (luteoid or corticoid plus growth hormone or PRL); however, estrogen is no longer obligatory, but only facilitative. The hormones necessary for tumorigenesis (formation of tumors) are the same for nodule maintenance; however, once tumors are formed, mammary carcinomas become hormone-independent as a rule. Large doses of PRL can maintain nodules in the absence of steroid hormones from the ovary and adrenal,[66] thus overriding, under these specific conditions, the requirement for a corticoid.

Recent experiments by Welsch and Gribler[67] and Welsch[68] demonstrated that inhibition of PRL synthesis and secretion by 2-bromo-α-ergocryptine inhibited the nodule to tumor transformation. Thus, the mammary tumor incidence in 14-month-old C3H virgin mice decreased from 27 to 1%, and the nodule incidence decreased from 90 to 30%.[67] The inhibition by 2-bromo-α-ergocryptine was as effective, if not more so, than the inhibition of mammary tumorigenesis by ovariectomy.[68]

While C3H nodules and nodule outgrowths are very sensitive to growth inhibition by the absence of estrogen or PRL or both, BALB/c nodule outgrowth lines show a more diverse response. Of eight BALB/c nodule outgrowth lines examined, the growth of five lines was unaffected by ovariectomy, whereas none of them was affected by 2-bromo-α-ergocryptine administration. Interestingly, the three nodule lines whose growth was inhibited by ovariectomy were all induced by chemical carcinogens, whereas four of five unresponsive lines originated in hormone-stimulated mice.[38,69] The sensitivity to hormonal treatments was more varied when one considers other hormones; for instance, administration of testosterone led to inhibition of growth and tumorigenesis in three of five nodule lines tested, but there was no correlation between sensitivity to ovariectomy and to testosterone among the lines tested.[69] These results clearly demonstrate the remarkable variety of responses to hormonal treatments seen in preneoplastic populations.

If preneoplastic mammary populations exhibit diverse hormonal responsiveness, carcinomas derived from preneoplastic nodules exhibit a very uniform response. Of 40 tumors tested for their responsiveness to ovariectomy, nafoxidine, testosterone, and CB-154, none showed any responsiveness.[69] The tumors were derived from nodule lines that were both responsive and unresponsive to the agents listed above. Not all murine mammary tumors are hormone-unresponsive. Mammary carcinomas arising from plaques in GR mice are pregnancy-dependent. Recently, mammary carcinomas arising from DMBA- and urethane-induced DHs in BALB/c and (C57BL × DBAf)F_1 mice were shown to be

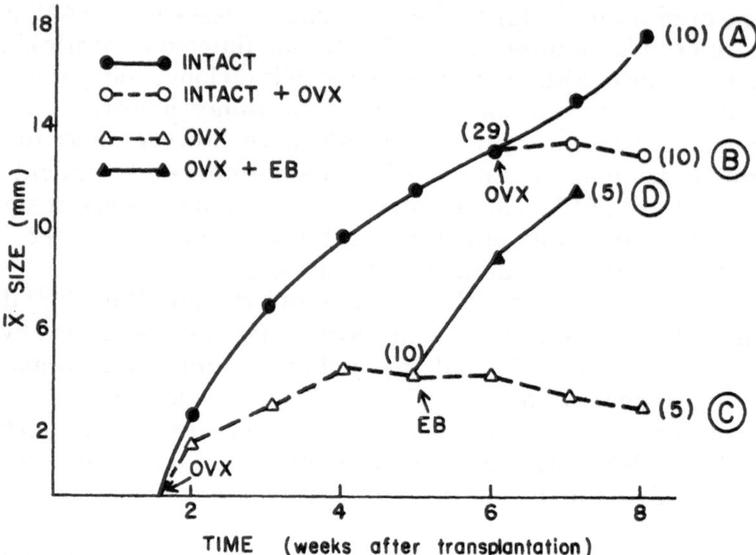

Fig. 2. Hormone responsiveness of chemical-carcinogen-induced mammary tumors in (C57BL × DBA)F_1 mice.

ovarian-responsive. Of 11 tumors transplanted subcutaneously into ovariectomized mice, 10 failed to grow progressively.[70] Two of these tumors were investigated further, and the data are illustrated in Fig. 2. Ovariectomy markedly inhibited growth of both small and large tumors, and estrogen-replacement therapy stimulated tumor growth in ovariectomized mice.[70] Progesterone also stimulated tumor growth in ovariectomized mice, but to a greater extent than estrogen. Recent experiments showed that these tumors have large quantities of cytoplasmic estrogen receptors that are translocatable to the nucleus and that exhibit a high level of specific estrogen binding.[70] These tumors are unlike tumors arising in C3H mice or BALB/c mice, which have estrogen receptors that are either untranslocatable[71] or undetectable.[72] These tumors were classified as papillary ductal carcinomas or well-differentiated ductal carcinomas, in contrast to the adenocarcinomas arising in C3H and BALB/c mice.

3.3. Immunogenicity

Several investigators have demonstrated that mammary adenocarcinomas in mice are immunogenic.[73–75] The primary antigens responsible for the immunogenicity of mammary adenocarcinomas and HANs are related to the MTV.[76,77] Some MTV-positive mammary tumors confer resistance in MTV-positive syngeneic hosts that is not cross-reactive with other MTV-positive tumors.[78–80] Cross-reactive and non-cross-reactive antigenic types were also demonstrated by *in vitro* immunological techniques.[81]

MTV-negative BALB/c mammary tumors have been reported to be weakly immunogenic in BALB/c mice by both *in vivo*[82,83] and *in vitro*[84] techniques. Recent experiments demonstrated that the strength of the immunogenicity of BALB/c tumors is correlated with the nature of the inductive stimulus. Tumors arising from HANs induced in a hormone-stimulated environment were weakly immunogenic (ref. 84 and unpublished observation), whereas tumors arising from HANs or DHs induced by chemical carcinogens were strongly immunogenic[85] Table V). The latter tumors did not show cross-reactive antigens in two experiments *in vivo*. Prehn[86] recently demonstrated that the immunogenicity of induced tumors was related to the concentration of the carcinogen. The immunogenicity of tumors decreased with low doses of carcinogen even when the latent period was constant. At very low concentrations, the immunogenicity of tumors was like that of spontaneous tumors.

Some limited experiments have examined the immunogenicity of HANs. Treatment of MTV-positive C3H mice with MER, a nonspecific immunological enhancement substance, produced protection against growth of tumor isografts, spontaneous tumor development, and spontaneous nodule development.[74] Sensitization of MTV-negative C3Hf hosts with MTV-positive C3H nodules resulted in protection against growth of subsequent challenge implants of C3H nodules and

Table V
Immunogenicity of Mammary Tumors Arising from Preneoplastic Lines

Preneoplastic line	Ethiological agent		Tumors immunogenic/ tumors tested (%)	Degree of immunogenicity[a]
	Preneoplasia	Neoplasia		
D1	Hormonal	Spontaneous	2/9 (22)	Weak
D1	Hormonal	DMBA/MCA[b]	6/15 (40)	Weak
D2	Hormonal	Spontaneous	7/14 (50)	Weak
C4	DMBA	Spontaneous	8/8 (100)	Strong
C5	DMBA	Spontaneous	3/3 (100)	Strong
CD-3	DMBA	Spontaneous	4/4 (100)	Strong

[a] Weak: decrease in *either* the percentage takes *or* the mean tumor size; strong: decrease in both the percentage takes and the mean tumor size.
[b] Tumors arise in mice bearing D1 transplants in the mammary fat pad and treated with either DMBA or MCA.

tumors.[87,88] Rejected nodule outgrowths grew poorly and/or were devoid of alveolar cells characteristic of nodule tissues.

Nodules occurring in mice treated with 3-methylcholanthrene (MCA) are highly immunogenic.[85] Nodule outgrowths derived from these carcinogen-induced nodules grew out as ductal outgrowths in appropriately sensitized syngeneic mice. The normality of the ductal cells remains unclear. Limited experiments demonstrated that the nodules exhibited non-cross-reactive immunogenicity and could sensitize the host against subsequent challenge of homologous tumors.

In vitro studies on the immunogenicity of MTV-negative HANs and tumors have supported the *in vivo* studies showing that these lesions are weakly immunogenic. Furthermore, they suggested that the preneoplastic HANs and tumors derived from them contained some common antigens, and different HAN lines also shared common antigens.[84] These antigens were not shared by tumors derived from fibroblasts.[84] One of the surprising results from these experiments was the enhanced survival of target cells in the presence of sensitized lymphocytes. Lymph node cells (LNC) from mice bearing HAN implants, or from mice whose implants had been removed less than 7 days prior to testing, failed to inhibit significantly the survival of HAN cells in culture, but instead enhanced survival. Specific and significant inhibition of HAN cells was found with LNC from mice whose HAN implants had been removed for more than 10 days. Enhanced but not decreased survival cross-reacted between D1 and D2 HANs, between D1 and D2 tumors, and between D1–D2 tumors and MTV-induced BALB/c tumors, but not between D1–D2 and chemically induced mammary tumors or fibrosarcomas.[84,89] Recent results in our laboratory showed that this phenomenon of LNC-mediated enhancement followed by LNC-mediated resistance applies to DMBA-induced mammary tumors when measured by the microcytotoxicity test *in vitro*. Interestingly, these tumors are highly immunogenic *in vivo* and confer strong resistance to tumor challenge (Table V).

In summary, the preneoplastic HAN shared some of the same antigens found in tumors originating from the nodules, particularly for MTV-positive HANs and the hormonally induced D series of HANs. Non-cross-reacting antigens at the HAN and tumor level have been shown to be easily demonstrable in chemical-carcinogen-induced lesions and are a minority component in MTV-positive lesions. The nature of the cross-reactive antigens seen in the D series of lesions is not understood, and organ-specific antigens have not been ruled out at present. The possibility that HAN may contain unique stage-specific antigens has not been examined thoroughly and has not been ruled out.

3.4. Response to Cytostatic Drugs

The effects of cytostatic drugs on experimental mammary cancer have been examined recently in several systems. Studies by Martin and co-workers[90,91] using transplantable tumors derived from MTV-positive hybrid mice demonstrated the chemotherapeutic effectiveness of a variety of drugs such as phenylalanine mustard (L-PAM), cytoxan (CTX), methotrexate (MTX), 5-fluorouracil (5-FU), and adriamycin. Fisher *et al.*[92] examined the effects of similar drugs given singly, in combinations, and with the nonspecific immune enhancement agent, *Corynebacterium parvum* (CP). Generally, the combination treatment was more effective than any drug given singly.[92] The addition of CP to any combination of drugs showed the same effects as the combination alone, although CP enhanced the effects of drugs given singly. Bogden and Taylor[93] reported extensive and similar results using an immunogenic, transplantable rat mammary tumor sensitive to various chemotherapeutic drugs.

The response of preneoplastic mammary lesions to cytostatic drugs has not been addressed until recently. Mice bearing the weakly immunogenic, moderate-tumor-potential nodule line D2 and the highly immunogenic, moderate-tumor-potential nodule line C4 were transplanted into syngeneic BALB/c female mice that were subsequently treated with a variety of cytostatic drugs including L-PAM, 5-FU, CTX, MTX, vincristine (VC), and prednisone. The results are shown in Figs. 3 and 4. The results were unexpected, since L-PAM, 5-FU, CTX, and VC enhanced the rate of tumor formation, whereas MTX and prednisone inhibited the rate of tumor formation in nodule line D2. For nodule line C4, L-PAM enhanced and 5-FU, CTX, MTX, and prednisone inhibited the rate of tumor formation. VC had no effect on line C4. There was no correlation between the effects on nodule growth rate in filling the mammary fat pads after transplantation (measurement of the cytotoxic effect of the drugs) and the rate of tumor formation for any of these drugs. Furthermore, L-PAM, 5-FU, CTX, and MTX inhibited the growth rate of subcutaneously transplanted tumors derived from both nodule lines.[94]

These results suggest that preneoplastic mammary populations can respond differently from neoplastic mammary populations to the same cytostatic drugs, and that some cytostatic drugs can act to promote mammary tumorigenesis. The possible implications of these results for human breast cancer are worthwhile to consider. Human breast cancer is a disease affecting both breasts, and persons with prior breast cancer are high-risk patients for subsequent breast cancer in the contralateral

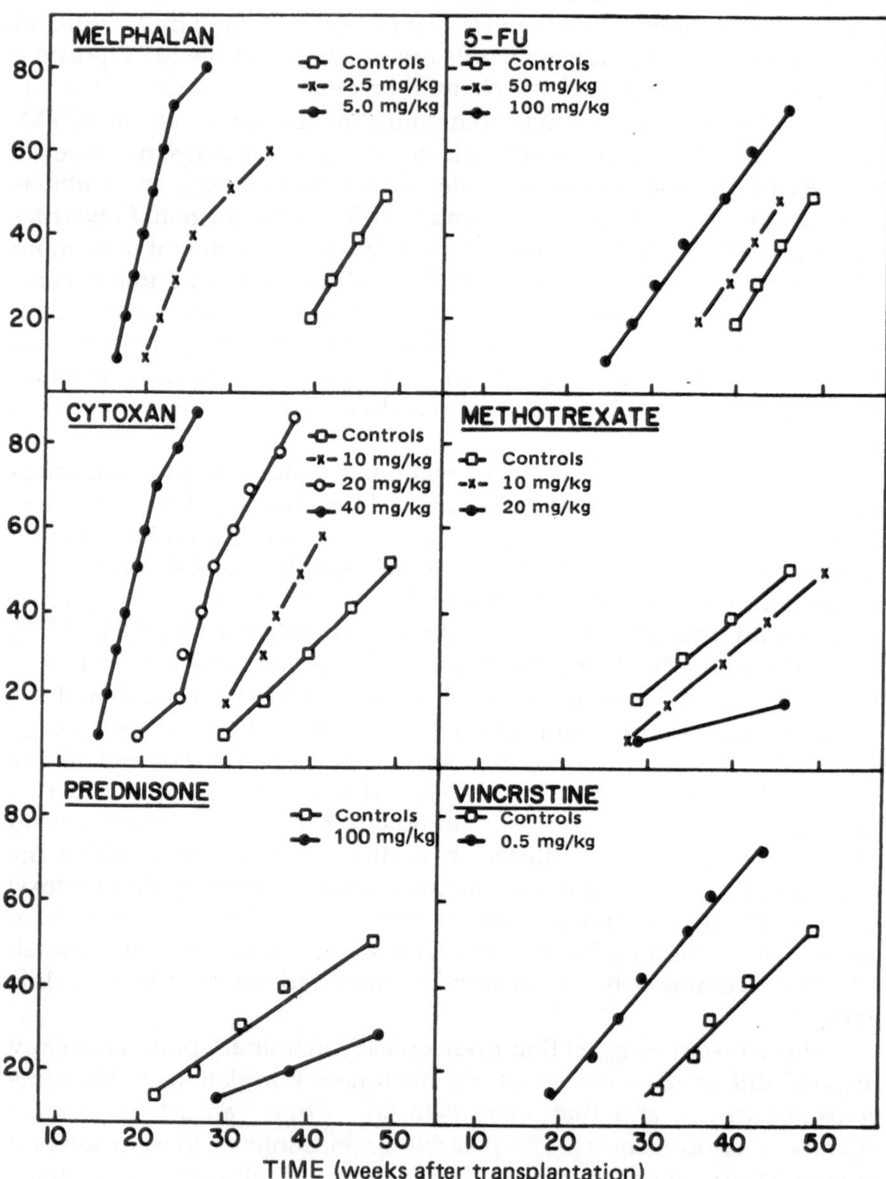

Fig. 3. Effects of cytostatic drugs on tumor potential of nodule line D2.

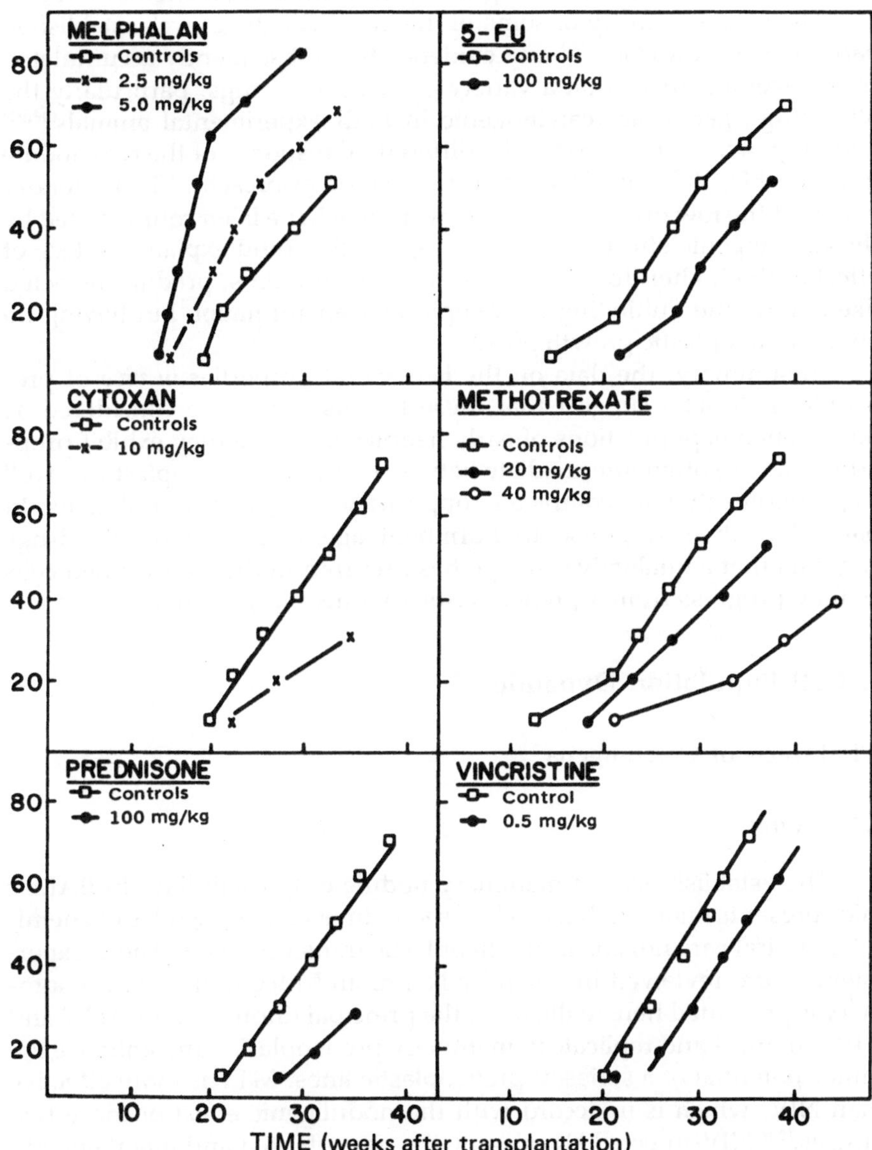

Fig. 4. Effects of cytostatic drugs on tumor potential of nodule line C4.

breast. In addition, the disease is suspected of passing through preneoplastic stages.[48–50] The results stated above suggest that MTX rather than L-PAM would be the safer drug to use in combination chemotherapy.

The carcinogenicity of some of the cytostatic drugs was established early by Haddow *et al.*,[95] and evidence has subsequently accumulated from several sources that a variety of cytostatic drugs, particularly the alkylating agents, are carcinogenic in both experimental animals[96,97] and humans.[98] The basis for the differential response of the two nodule lines to5-FU, VC, and CTX is not understood at present. The inhibitory effects of the low dose of CTX, however, may have been counteracted by the carcinogenic effects of high doses, which would explain the lack of effect at the higher dose levels. In these experiments, prednisone acted like a hormone, inhibiting the neoplastic transformation but having no effect on neoplastic growth *per se.*

In summary, the data on the behavioral properties *in vivo* of preneoplastic lesions suggest that these lesions cannot be considered as homogeneous populations of early neoplasms, since they exhibit properties that are often intermediate between normal and neoplastic as well as properties that are completely opposite to neoplastic populations. In particular, their response to hormonal agents and cytostatic drugs suggests that a qualitative change has occurred in the transformed cells as they progress from a preneoplastic to a neoplastic state.

4. Cell-Population Dynamics

4.1. Effects of Carcinogens

4.1.1. Viral

The establishment of mammary nodule outgrowth lines in BALB/c mice presents material that can be used to investigate the roles of chemical and viral carcinogens in the neoplastic transformation. These experiments were reviewed in detail by Nandi and McGrath,[23] but a summary is presented here to illustrate the principal findings. Both MTV and NIV can infect and replicate in mammary preneoplasias and enhance the tumor potential of a series of preneoplastic lines. MTV is more effective than NIV, which is in accord with the noduligenic effect of these two viruses.[23] NIV-infected D1 outgrowths transplanted and maintained in BALB/cCrgl mice continued to produce a low incidence of tumors in comparison with MTV-infected outgrowths.[39] Prior infection with NIV, however, inhibited subsequent infection by MTV.[99] The mechanism behind this inhibition is unclear, although it is possible that the inhibition may occur at the blood level by blocking the blood stage of the life cycle of MTV or at the cellular level by blocking absorption of MTV.

One experiment, which investigated possible differences between the effects of the two viruses, demonstrated by electron microscopy that MTV infects or replicates, or both, in preneoplastic nodule cells very quickly (by 8–10 weeks after transplantation), whereas NIV infects or replicates, or both, in the preneoplastic cells by 40 weeks after transplantation. The delayed appearance of NIV expression correlated with its low oncogenic potential. PRL stimulation of the mammary gland, which enhances NIV-induced nodules and tumors, also leads to early expression of NIV in the preneoplastic tissue.[(100)] These data demonstrated that both types of MTV can infect nodule tissues and enhance their tumor potential as well as induce primary HANs. The altered tumor potentials of such populations are stable over serial transplantation, which may be due to the continued presence of the virus. This system has not been studied with temperature-sensitive mutants to determine whether viral presence is obligatory for the permanent increase of tumor potential. *In vitro* infectivity has not been successful at this level, although MTV was shown to infect and replicate in normal mammary tissues.[(101)] The frequency of cells infected in a given preneoplastic population has not yet been determined. Preneoplastic mammary lines in BALB/c mice are free of B-type virus particles,[(2)] have undetectable levels of gp52 viral antigen,[(102)] have insignificant levels of MTV reverse transcriptase, which is not increased by dexamethasone,[(103)] and have levels of endogenous viral DNA (4 copies/genome) similar to normal intact BALB/c mammary gland.[(104)] The significance of endogenous MTV DNA is not clear; however, it will be important to determine whether chemical carcinogens increase the number of these copies per cell. The relationship between endogenous MTV, NIV, and exogenous MTV in this type of system would be of considerable interest to determine from the standpoint of the similarities and dissimilarities of the degree of MTV information contained in the genome of each of the virus DNA. In this respect, Drohan *et al.*[(105)] recently demonstrated that the RNA of exogenous MTV contains information in addition to that of the endogenous virus, although McGrath *et al.*[(104)] could not confirm this observation. Finally, such information might help to explain why BALB/c preneoplasias that contain putative endogenous MTV are infectable by exogenous MTV and how NIV can inhibit infectivity by MTV.

4.1.2. *Chemical Carcinogens*

The oncogenic effect of several chemical carcinogens, MCA, DMBA, and urethane, as well as of γ-irradiation, has been tested on several series of BALB/c nodule outgrowth lines. All four agents increased the tumor potential of nodule lines derived from nodules induced originally by hormones or DMBA. The preneoplastic stage is much more sensitive to the effects of chemical carcinogens like DMBA than the normal stage. For instance, 0.5 mg DMBA enhances the tumor potential of line D1 from 10 to 62%, whereas it requires 4 mg DMBA to induce a low incidence of preneoplastic lesions.[4,37] The effectiveness of chemical carcinogens on the preneoplastic stage may be due to several conditions: (1) The nodule cell population, containing many alveolar cells, may be more susceptible than ductal cells. In this respect, hormone-stimulated normal mammary gland produces a high incidence of dysplastic lesions with 1.0 mg DMBA, although these are mainly self-limiting dead-end squamating alveolar lesions (KNs). (2) Tumorigenesis by chemical carcinogens is a function of cell division rather than cell number. Recent experiments showed that the number of cells undergoing DNA synthesis in preneoplastic nodule cells is greater than normal 2-day lactating mammary gland (a peak period of cell division in normal mammary gland). These experiments were done by examining the frequency of cells incorporating a pulse dose of radiolabeled thymidine after 60 min. (3) The carcinogens may act indirectly on the local population to disrupt the local environment and cell–cell interactions, thereby allowing high-risk preneoplastic cells to express themselves. On the basis of recent experiments examining the response of nodule outgrowth lines to enzymatic agents, this latter possibility is not unreasonable.

4.1.3. *Direct vs. Indirect Effects of Chemical Carcinogens*

The effects of chemical carcinogens at the target-organ and systemic levels have been examined intensively over the past decade. Experiments by Sachs and Huberman,[106–108] Heidelberger and Chen,[109] and DiPaolo *et al.*[110] leave little doubt that the initial interactions between carcinogen and the target cell are the critical events in the initiation of neoplasia. Such experiments have not been done critically in mammary tumorigenesis, although Brennan *et al.*[111] demonstrated that exposure of an intact mammary gland *in vitro* to DMBA initiated alterations in the mammary cells that were expressed as mammary tumors on *in vivo* transplantation. Preneoplastic alveolar or ductal mammary hyperplasias have not been induced *in vitro* by chemical carcinogens, except for the nodulelike lesions described by Banerjee *et al.*[112]

The possible pathways of chemical carcinogen action on the progression of preneoplastic lesions is still a matter of debate. It is conceivable that the effects of chemical carcinogens on the immune and hormonal systems of the host enhance the progression of preneoplastic mammary lesions in the absence of a direct effect on the target cells. Several experiments have examined these possibilities,[113,114] since it has been well documented that chemical carcinogens can alter the immune response[115–117] and the functions of reproductive organs.[118,119]

Pituitary and serum PRL and serum progesterone levels were measured in chemical-carcinogen-treated BALB/c mice by polyacrylamide gel electrophoresis and by specific radioimmunoassay.[114] Serum and pituitary glands were collected from chemical-carcinogen-treated BALB/c female mice that were 15–17 or 44 weeks old. Neither MCA (1.5 mg) nor DMBA (1.5–6.0 mg) significantly altered pituitary content or serum PRL concentration in the 15- to 17-week-old mice, but DMBA slightly increased (33%) the total amount of pituitary PRL in the 44-week-old mice. The elevation of pituitary PRL content in 44-week-old mice was not correlated with the incidence of mammary tumors in the group nor with the presence of a mammary tumor in an individual mouse.

Serum progesterone levels were increased approximately 22% in MCA-treated mice by 50 days after the last carcinogen treatment. This increase could be attributed to higher serum levels during diestrus and proestrus. Progesterone levels were unaltered by ovariectomy, but were reduced approximately 60% by adrenalectomy. DMBA had no significant effect on serum progesterone levels in mice assayed at 44 weeks of age.

These results[114] provided little support for the concept that MCA and DMBA promote murine mammary tumorigenesis by leading to a sustained increase in pituitary PRL content or in serum PRL concentrations either shortly (15- to 17-week-old mice) after carcinogen treatment or at times during mammary tumor formation and growth (44-week-old mice).

The significance of chemical-carcinogen-induced immunosuppression in carcinogen-induced mammary tumorigenesis was investigated in BALB/c mice.[113] The immune response of mice treated with a carcinogenic dose of MCA and DMBA was measured between 1 and 100 days after carcinogen treatment. Cell-mediated immunity was measured by rejection of skin and heart homografts, and humoral immunity was measured by the direct hemolysin plaque assay.

Both MCA and DMBA depressed cell-mediated rejection of skin and heart homografts for up to 50 days after carcinogen treatment. The survival time of homografts in carcinogen-treated mice was 50–100% greater than in control mice.

DMBA, but not MCA, induced a profound suppression of the primary humoral response that lasted up to 100 days after carcinogen treatment. This effect was strain-specific, since MCA induced suppres-

Table VI
Tumor Potential of Nodule Outgrowth Lines in Mice Pretreated with Chemical Carcinogens[a]

	Recipient			Transplants			
Nodule line	Age transplanted (weeks)	Age carcinogen given (weeks)	Carcinogen	Tumors/ transplants	%	Mean latent period (weeks)	TE_{50} (weeks)
D1	14	—	—	1/32	3	44.0	—
D2	14	—	—	5/30	17	33.5	—
D1	14	8, 9, 10	MCA	1/32	3	40.0	—
D1	14	8, 9, 10	DMBA	1/22	5	36.0	—
D2	14	8, 9, 10	MCA	7/31	23	34.0	—
D2	14	8, 9, 10	DMBA	8/31	26	32.5	—
D1	14	17, 18, 19	MCA	6/28	21	23.5	—
D1	14	17, 18, 19	DMBA	24/35	68	25.0	27.5
D2	14	17, 18, 19	MCA	35/40	88	17.7	19.0
D2	14	17, 18, 19	DMBA	34/40	85	15.8	15.0

[a] All experiments were terminated at 44 weeks after transplantation of nodule tissues into recipient mice.

sion of the primary immune response in C3H but not in BALB/c mice. Finally, in an assay system that involves both cell-mediated immunity and a counteracting humoral blocking factor, namely, hypersensitivity to methylated bovine serum albumin (BSA), MCA did not alter the balanced immune response of the host.

In an attempt to answer directly whether carcinogen-induced immune suppression was a significant factor in mammary tumorigenesis in preneoplastic mammary tissue, samples of DMBA- and MCA-treated nodule outgrowth lines D1 and D2 were transplanted into the mammary gland free fat pads of mice already treated with 1.5 mg DMBA and MCA, respectively. The tumor potential of these outgrowths was the same as that of outgrowths transplanted into untreated mice and significantly less than that of outgrowths exposed directly to DMBA or MCA (Table VI).

These experiments support the contention that there is no simple correlation between carcinogen-induced immunosuppression and tumorigenesis in mammary tumors arising in preneoplastic outgrowths, and that the immunosuppressive function of DMBA and MCA is not essential for the carcinogenic function.

4.1.4. *Serial Transplantation of Carcinogen-Treated Nodule Outgrowth Lines*

To examine the stability of carcinogen-induced alterations in preneoplasias, carcinogen-treated nodule outgrowths were serially transplanted under a variety of protocols. Earlier results had shown that D1 nodule outgrowth lines exposed to MCA, DMBA, or irradiation failed to produce a significant number of tumors if transplanted as small pieces into untreated syngeneic mice,[4,120] but expressed their altered tumor potential if whole glands were transplanted onto the abdomens of untreated virgin BALB/c mice[120] or as small pieces into mice bearing pituitary isografts.[120] The relationship held even for nodule outgrowths that were treated for 4 consecutive generations with MCA.[120] The expression of the MCA-induced increase in tumor potential and the increased response to pituitary isograft stimulation was unstable, since the effects disappeared after serial transplantation of the treated outgrowths into untreated, unstimulated mice.[120]

In an attempt to understand the basis for this effect, a second series of long-term experiments was set up using a more potent carcinogen, DMBA. The results are illustrated in Fig. 5. The results were similar to the previous experiments in two respects and different in one other. Like MCA, a cumulative effect of DMBA was not seen if one considered the tumor incidence in DMBA treated outgrowths over 4 generations. Each generation produced 70–80% tumors with a similar latent period,

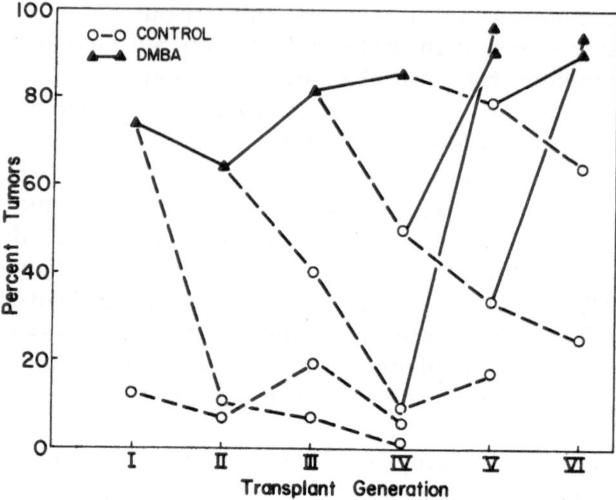

Fig. 5. Serial transplantation of DMBA-treated nodule line D1.

even though the cumulative dose of DMBA increased progressively from 1.5–3.0 to 4.5–6.0 mg. On transplantation of samples of the treated populations into untreated mice, however, the expression of the increased tumor potential was proportional to the total amount of carcinogen. Thus, outgrowths that had been treated with a cumulative dose of 3.0, 4.5, and 6.0 mg DMBA exhibited an increased expression of tumorigenesis. The tumor potential decayed, however, on successive transplantation. The explanation for these results is puzzling. It is clear that an increasing number of DMBA-altered cells are recruited with each successive dose of DMBA; however, some unknown effect of transplantation influenced their expression. The expression of altered cells was not influenced by the length of time between treatment and subsequent transplantation.[120] Outgrowths transplanted from 1 to 30 days after treatment still behaved the same way. Initially, it was proposed that these observations could be explained on the basis of a quantitative effect; i.e., the number of cells altered was small and the probability of picking out and transplanting altered cells was small when transplanting only 20–40% of the exposed cells. In the second series of experiments, however, at least 80% of the nodule outgrowth tissue was transplanted into recipient hosts; therefore, this explanation appeared untenable. An alternative explanation of most of the observed data is as follows: A nodule population is a mixed population of normal cells, unaltered nodule cells, and DMBA-altered nodule cells that is regulated by cell–cell interactions. Carcinogens may act either directly on nodule cells to alter them and increase their tumorigenicity or indirectly by disrupting the microenvironment and consequently disrupting normal cell–nodule cell interaction, or both directly and indirectly. The result in either case would be an increased probability of expression of potentially tumorigenic nodule cells. On transplantation of the DMBA-treated population, a minimum period of time (approximately 3 weeks) is necessary for the entire population to vascularize and adjust itself to the new surroundings. This period of "transplantation shock" might act to counterbalance any selective growth advantage of nodule cells. With increasing DMBA exposure, more cells are recruited, which increases their probability of surviving the initial transplantation as an "expressible" unit. However, each successive transplantation of a given population into normal mice counteracts the accumulative DMBA effect, i.e., allows normal cells to repair or regain control. The phenomenon of "transplantation shock" was demonstrated in experiments by Mehta *et al.*[(121)] in which nodule cells could grow in organ cultures only if they had been transplanted into the mammary fat pads for 3 weeks before explanation into organ culture. This time was needed to overcome the effects of *in vivo* transplantation before the cells were capable of starting their growth spurt *in vitro*. Instead of "transplantation shock," one could invoke the expression of potential reversibility of these DMBA-altered nodule cells as demonstrated in *in vitro* carcinogenesis experiments by Rabinowitz and Sachs.[(122)] The potential interactions of normal and nodule cells will be examined below.

Any hypothesis would have to explain why a cumulative effect was not seen in the treated hosts. A possible explanation exists based on the results of another experiment. When D1 outgrowths were treated with a high dose of urethane (100 mg), 83% of the treated outgrowths gave rise to tumors. A subsequent dose in the second generation gave the same result; thus, no apparent cumulative effect was demonstrable; i.e., a maximum threshold was reached by 100 mg. However, if a low dose of urethane (20 mg) was given each generation, the tumor incidence in the first generation was 43% and in the second generation, 70%.[(4)] A similar "threshold" may have been reached using 1.5 mg DMBA. Experiments using 0.1 mg DMBA per generation are now in progress to test this hypothesis.

The original aim of the serial transplantation experiments was to investigate the response of a target tissue to cumulative doses of a given carcinogen and to sequential doses of several different carcinogens. The results demonstrated that the effects of viral and chemical carcinogens in this system were additive.[(4,123)] More important, they suggested the possibility that the effects of a carcinogen, i.e., "increased tumorigenicity," were not permanent or irreversible, but could be modulated by the experimental and artifactual technique of transplantation. Moreover, the early events associated with transplanted cell populations might influence the expression and regulation of a preneoplastic or high-risk population.

4.2. Normal–Preneoplastic Cell Interactions

4.2.1. Evidence of Local Growth Regulation

The spatial pattern of mammary ducts in the rodent mammary gland is regulated by local growth regulators produced by normal mammary ducts.[63] Experiments have indicated that the minimum distance between ducts is 0.25 mm[63]; however, normal alveoli arising in the mammary glands of pregnant mice are not subject to such exact regulation. HANs, but not mammary tumors arising from HANs, are subject to the same growth regulation as local host mammary ducts. Nodules transplanted into intact fat pads do not overgrow the host ducts[4,63] as do tumors. Nodules transplanted into intact fat pads containing growing host mammary ducts stop growing at the interface of the growing host tissue.[4,63] In such cases, ducts can be seen to reverse this growth path by turning 180° and to continue growing in unoccupied space at the edge of the fat pad.[4] This ability to override local growth-regulatory factors remains the most important distinction between preneoplastic and neoplastic mammary tissues that has yet been described in experiments. The nature of the growth regulators operating in this system remains unknown, although this factor (or factors) is (are) not easily demonstrable by other means.[124]

In an elegant experiment by Mintz and Slemmer,[125] it was demonstrated that normal cells can influence the growth characteristics of a preneoplastic population when in cell-to-cell contact. Using C3H $\rightleftarrows$ C57BL allophenic mice in which each tissue had cells from both parents, these workers induced primary HANs. The HANs transplanted into hybrid (C3H × C57BL)F_1 female mice produced lobuloalveolar outgrowth in 100% of the mammary fat pad. Outgrowths transplanted into parental C57BL female mice produced 100% ductal outgrowths, indicating that the C57BL phenotype was a normal cell. Outgrowths transplanted into parental C3H female mice produced HAN outgrowths that filled only 30% of the fat pad, indicating that the C3H phenotype was the transformed cell type. These latter outgrowths produced tumors subsequently. More interestingly, they demonstrated the synergistic effect of normal cells on HAN cells and suggested that normal cells either were indistinguishable from HAN when in cell-to-cell contact or were modulated to the HAN phenotype.[125] The cells from the respective phenotypes were identified by their histocompatible antigens and their isoenzyme patterns, which were unique for each strain.

4.2.2. *Expression of Hyperplastic Alveolar Nodule Transformation*

The induction and appearance of HANs have been studied under a variety of stimuli. Factors such as virus infection, virus oncogenicity, hormonal milieu, immunological status, diet, and exogenous chemical carcinogens influence the appearance of HANs. HANs appear as foci of lobuloalveolar hyperplasia, visible by semigross observation, starting at 9–10 months of age in virgin BALB/cfC3H female mice. Hormone stimulation, by pregnancy or pituitary isografts, decreased the latent period to 4 months. While agents such as hormones clearly influence the expression of HAN phenotype, it is not clear whether this phenomenon is directly related to induction. In a recent series of experiments, Miyamoto and co-workers[(126,127)] demonstrated that the expression of the HAN phenotype is dissociable from the induction of the HAN phenotype. Enzymatically dissociated normal virgin mammary gland from MTV-positive BALB/cfC3H mice gave rise to HAN outgrowths on transplantation into the mammary fat pads of syngeneic mice. HAN cells could be detected in the mammary glands of 3-month-old virgin BALB/cfC3H mice, but not in 2-month-old BALB/cfC3H mice or in 10-month-old BALB/c or C57BL female mice.[(126)] The number of transplants yielding lobuloalveolar outgrowth rose from 30 to 65% in 3-month-old to 6-month-old mice. The lobuloalveolar phenotype was stable on serial transplantation of the outgrowths and was also highly tumorigenic.[(126)] A similar procedure was used to select HAN cells from pregnancy-dependent tumors (plaques) in GR mice, indicating the heterogeneous nature of the plaque population.[(127)]

Similar experiments have been performed using BALB/c nodule outgrowth lines as the source of dissociated cells. In these experiments, enzymatic dissociation led to an increase in tumor potential in nine different experiments using four different nodule outgrowth lines with differing tumor potentials.[(128)] In all experiments, the cells injected as single cells were compared with the same outgrowth line transplanted as small pieces (1 mm^3). In nodule line D1, the tumor potential increased from 19% (21/109) to 50% (44/88). In nodule line D2, the tumor potential increased from 54% (46/85), with the 50% tumor end point reached at 11 months, to 74% (32/43), with the 50% tumor end point reached at 7 months. A similar increase in tumor potential was seen with nodule lines C4 and CD-3. It seemed unlikely that the enhanced tumor potential could be explained on the basis of a larger number of cells transplanted in the cell suspension than in a small piece. Two facts argued against this possibility: First, cells transplanted as a cell suspension filled the fat pad at a slower rate than cells transplanted as a piece; i.e., by 8 weeks after transplantation, 10^5 cells transplanted as a suspension filled 60% of the fat pads as compared with 90% for cells transplanted as pieces. Second, an estimate of the number of epithelial cells in the "piece"-size transplant gave a minimum figure of 75,000 cells/piece for D1 and 83,000 cells/piece for D2, which was not significantly different from 10^5 cells. The estimate was made on the basis of experiments in which mammary fat pads containing nodule outgrowths were dissected into 100 standard-size transplantation pieces (1 mm^3), the pieces were pooled and enzymatically dissociated, and the cells were counted. The total number of cells divided by 100 yielded a minimum number of cells in one average-size piece. The estimate was considered minimum, since cells were lost during the dissociation procedure.

The basis for the enzymatically induced increase in tumor potential is unclear, although the hypothesis that the enzyme effect indirectly led to the selective expression of existing altered cells is reasonable and testable. It is known that broadly active proteases (i.e., trypsin, pronase) can induce a variety of changes in normal cells, including increased lectin agglutinability; changes in plasma membrane proteins, glycosaminoglycans, and cyclic nucleotide levels; an enhanced rate of glucose uptake; and stimulation of cell multiplication.[129] The similarity of responses to different environmental stimuli was termed a "coordinate response" by Rubin and Koide.[130] These responses are seen in neoplastic cells induced by chemical or viral carcinogens. Enzymes can induce changes that mimic the responses of neoplastic cells. In the nodule experiments, it is not clear what properties of this coordinate response the enzymatically dissociated cells would also exhibit. However, the significance of broadly active proteases in the enhancement of tumor potential was examined directly in a further series of experiments. Nodule outgrowth lines D1 and D2 were dissociated with just collagenase plus BSA plus hyaluronidase. In these experiments, the cell yield, percentage viability, and percentage single cells were less than with the three-enzyme procedure; under favorable conditions, however, the experiments could be carried out. Negative results would have been difficult to interpret due to the greater frequency of small clumps of cells (5–10 cells) transplanted along with the single cells. However, both nodule lines dissociated with just collagenase, hyaluronidase, and BSA exhibited a markedly enhanced tumor potential. Although collagenase preparations are thought to have contaminating broadly active proteases, BSA protects cells against their actions.[131] This was confirmed in our experiments by examining concanavalin A (Con A) agglutination of normal and neoplastic mammary cells treated with collagenase in the presence or absence of BSA. Collagenase alone increased mammary-cell susceptibility to Con A agglutination, but BSA counteracted the collagenase effect. The presence or absence of hyaluronidase in the dissociation procedure had no effect on Con A agglutinability of *in vivo* dissociated cells. While these results support the hypothesis that enzymatic digestion altered the expression of transformed cells, conclusive experiments would require the preparation of a single cell suspension by mechanical nonenzymatic methods, which has been difficult to attain thus far.

The stability of the altered tumor potentials seen in enzymatically dissociated nodule cells was investigated by serial transplantation of the dissociated cells into syngeneic mice. Nodule line D1 dissociated and transplanted at 10^5 cells per fat pad produced 50% tumors, compared with 19% for control implants. If these outgrowths were then transplanted serially as pieces, however, the tumor potential decreased to 32 and 15% in each transplant generation, respectively. Therefore, the increase in tumor potential was not stable, and these cell populations behaved similarly to the DMBA-treated nodule-cell populations.

4.2.3. *Interactions of Separate Cell Types*

Analysis of potential interactions between separate cell types of the mammary gland in normal and abnormal development was attempted in a few experiments.[125,132] Slemmer[132] postulated the existence of three cell types in normal mammary gland: ductal epithelial, alveolar epithelial, and myoepithelial. In his model, each cell type can be transformed to a specific class of preneoplastic lesion that can progress to neoplasia. Alveolar epithelial cells produce alveolar preneoplasia, ductal epithelial cells produce ductal hyperplasia, and myoepithelial cells produce epidermoid–mesenchymal metaplasia.[132] Experiments examining the interaction between normal and transformed cells of each type have been performed in allophenic mice and their parental strains. By inducing the lesion in allophenic mice, maintaining them in histocompatible hybrids, and analyzing them in the parental strains, Slemmer demonstrated that the normal cell type synergized and stimulated the growth of preneoplastic cells. However, progression to malignancy was accompanied by a loss of dependence on normal cells for optimal and malignant growth.[132] By reassociating normal cells with pure populations of neoplastic cells, Slemmer noted that the genotype of the normal component was not a significant factor.

Slemmer[132] studied the effects of reassociating normal with premalignant cells under conditions that examined the progression of premalignant growth to malignant growth. In many cases, it was clear that his premalignant lesions were not equivalent morphologically or biologically to the preneoplastic nodule lines described above. Experiments using the reassociation method with normal cells and preneoplastic cells, under defined conditions, suggested that normal cells can influence the neoplastic transformation in preneoplastic nodule lines.[128] In these experiments, normal cells were mixed in ratios of 1 : 1, 2 : 1, and 3 : 1 with a single cell suspension of nodule cells. Normal cells prepared from virgin, pregnant, or lactating mammary glands inhibited the rate of tumor formation in nodule line D2 (i.e., 75 to 37%), whereas a similar ratio of normal cells had no effect on tumor-cell growth. This experiment was also repeated with the D1 nodule line. Interestingly, in a reassociation between normal cells and nodule cells, morphological structures resembling normal ducts could not be seen in stained whole mounts, indicating the intimate interaction between the two cell populations. The further elucidation of events occurring at the cellular level in these experiments requires the development and application of markers that will distinguish between normal and preneoplastic cells.

5. Cell Biology of Preneoplasias

5.1. *In Vitro* Characteristics

5.1.1. *Growth Patterns*

The conditions necessary for the establishment of successful cell cultures of normal, preneoplastic, and neoplastic mammary gland have not been defined extensively. The majority of such information has resulted from experiments on neoplastic cells that clearly show that criteria developed for transformed fibroblasts *in vitro* do not extend to neoplastic mammary cells.[103,133–135] The growth characteristics of mammary glands can be summarized briefly. Unlike normal fibroblasts, the growth of which can be described by a sigmoid-shaped curve, reaching a common saturation density independent of the plating density, normal and neoplastic mammary epithelial cells reach a sharp density plateau that is dependent on the number of cells plated initially.[133,134] Furthermore, normal, preneoplastic, and neoplastic mammary cells reach a common saturation density if plated at similar initial concentrations.[133,134] This is unlike fibroblast systems, in which transformed fibroblasts show higher saturation densities than normal fibroblasts.

A variety of criteria based on growth parameters have been established to distinguish between normal and transformed fibroblast cells.

These criteria include plating efficiency, saturation density, growth in suspension, and growth in agar. In a recent study, using several cloned lines each from five different mammary tumors of differing etiologies, Butel *et al.*[103] demonstrated that none of these characteristics was strictly correlated with *in vivo* tumorigenicity.

Organ cultures of preneoplastic mammary tissues have been more successful than cell cultures. In one series of experiments, Mehta *et al.*[121] described the successful growth of preneoplastic nodule outgrowth tissue under conditions in which the whole fat pad was explanted *in vitro.* The amount of growth was remarkable, since 50% of the fat pad was filled with nodule outgrowth after only 12 days in culture under the influence of insulin, PRL, and aldosterone.[121] This growth rate was faster than generally seen *in vivo.* These experiments are important, since they provide a possible tool to examine the growth potential of dysplasias from human breasts that were recently described.[48–50]

In another series of experiments, Banerjee *et al.*[112] described the induction by chemical carcinogens of nodulelike lesions *in vitro* under organ culture conditions. These lesions are alveolar dysplasias and persist after the hormonal stimulus for lobuloalveolar differentiation is withdrawn. The tumor potential of these lesions has not yet been established. The frequency of these nodulelike lesions is correlated with the dose of DMBA and the potency of the carcinogen, i.e., DMBA > DBA > BA > A (personal communication). This latter result suggests that these lesions may be analogous to *in vivo* induced HAN.

5.1.2. Morphology

Normal mammary epithelial cells seeded at high (5×10^5 cells/cm^2) and medium ($1-2 \times 10^5$ cells/cm^2) cell densities have very similar morphology, growth patterns, DNA synthesis patterns, and saturation densities.[133–135] Epithelial cells form islands from which cells migrate outward and flatten to give the typical epithelial pavement of tightly fitting polygonal cells. Initial cultures, and up to 9 days in primary cultures, are primarily diploid, although cells seeded at low density (1×10^4 cells/cm^2) show a high incidence of multinuclearity and polyploidy by 10–12 days in culture. Tumor cells plated at low density exhibit 30% polyploidy by 4 days, whereas normal cells take 16 days to achieve this degree of polyploidy.

A unique feature of many epithelial cells *in vitro* is the formation of multicellular blisters or "domes." These multicellular structures are raised hemispheric areas of turgid epithelial cells, whose frequency is dependent on plating density.[135,136] Domes can be formed by both normal and neoplastic mammary cells, although neoplastic cells lose this tendency to form domes with increasing passage number.[135,136]

The development of epithelial cell junctions was examined in normal, preneoplastic, and neoplastic mammary cells in thin sections and freeze–fractured replicas.[137,138] Normal mammary gland is characterized by desmosomes, by tight junctions and gap junctions in duct cells of virgin and pregnant gland, and by tight junctions in the alveolar cells of lactating gland. Preneoplastic nodule tissues and neoplastic cells have irregular tight junctions, variable numbers of desmosomes, and gap junctions similar to normal.[137,138] There appeared to be more variation in the junctional complexes between cells found in the different developmental stages of normal mammary gland than between normal and preneoplastic and neoplastic cells. The cell junctions of the latter resemble normal ductal and pregnant mammary tissues. Junctional complexes in murine mammary gland survive neoplastic transformation and appear to be a stable property of the cell that is expressed normally under the appropriate conditions.

Very little information is available from scanning electron microscopy. The limited information on normal, preneoplastic, and neoplastic mammary cells grown in identical conditions *in vitro* suggests that no major differences in surface attachment patterns, microvilli, number and degree of surface distortions, and general three-dimensional topography can be easily seen (Medina and Brinkley, unpublished). More detailed comparisons utilizing different culture conditions may provide additional information.

5.2. Cytoskeletal Characteristics

The primary components of the cytoskeleton in eukaryotic cells are microfilaments and microtubules that appear to connect either directly or indirectly to macromolecules in the plasma membrane (reviewed in ref. 139). Several lines of evidence indicate that cytoplasmic microtubules, microfilaments, or both participate in regulating a number of membrane-associated cellular events, including cell morphology, cell motility, adhesion of a cell to its substrate, maintenance of contact inhibition of cells, mobility of cell surface receptors for immunoglobulins and lectins, and binding of hormones to cell-surface receptors.[139] Transformed cells show alterations in many of these properties, and such alterations may underlie the aberrant responses and interactions of cancer cells with their environment and with neighboring cells. These modifications may be due to changes in cytoskeletal elements, since recent electron-microscopic[140] and immunofluorescence[141,142] studies revealed that microtubules and microfilaments are often diminished in transformed cells.

The majority of such studies have utilized either early-passage cell cultures or established cell lines of mesenchymal origin usually derived from embryonic or neonatal tissues.[140–142] Relatively little work has been done on normal adult epithelial cells and their neoplastic derivatives. For instance, cytoplasmic microtubules were examined in cells of established lines from rat hepatomas, mouse mammary and renal carcinomas, and human cervical carcinoma (HeLa).[142] However, a corresponding normal epithelial cell of each type was not available for study.

We have begun studying the relationship of cell-surface properties and the cytoskeleton system to the progression of neoplasias in mouse mammary cells. Although a tremendous wealth of data exists on the growth, hormonal, morphological, and tumorigenic properties of normal, preneoplastic, and neoplastic murine mammary tissues (for a review, see ref. 4), there are no cellular markers known that will distinguish among the three cellular phenotypes. Since epithelial cells do not behave like embryonic or fibroblast cells after transformation *in vitro,* it has been extremely difficult to analyze the preneoplastic and neoplastic transformations on a cellular basis.

5.2.1. *Microfilaments and Microtubules*

The staining patterns of normal mammary cells, cells derived from D2 HAN outgrowths, and cells isolated from D2 tumors were examined after exposure to actin and tubulin antibodies.[143] Figures 6A–C show examples of these cells following treatment with antiactin antibodies. Most, but not all, of the normal epithelial cells (Fig. 6A), as well as fibroblast cells in the culture, displayed an elaborate network of actin cables resembling those found in normal fibroblasts of other systems. Some epithelial cells derived from the D2 HAN outgrowths were devoid of staining (Fig. 6B), suggesting a reduction or loss of actin structured in the form of microfilaments. In contrast, fibroblast cells in the same culture stained readily. Microfilaments were also diminished in some of the epithelial cells cultured from D2 mammary carcinomas (Fig. 6C), although normal fibroblast cells possessed striking arrays of actin cables.

Examples of the same three cell types after staining with tubulin antibodies are presented in Fig. 7A–C. The normal epithelial cells (Fig. 7A) usually contained a full complex of microtubules. Such cells have extensive fluorescent strands. Long delicate filaments extended to the cell periphery and either terminated or bent and continued along the cell

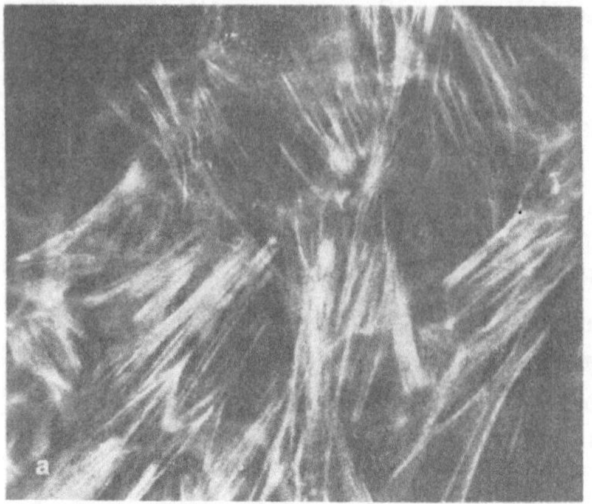

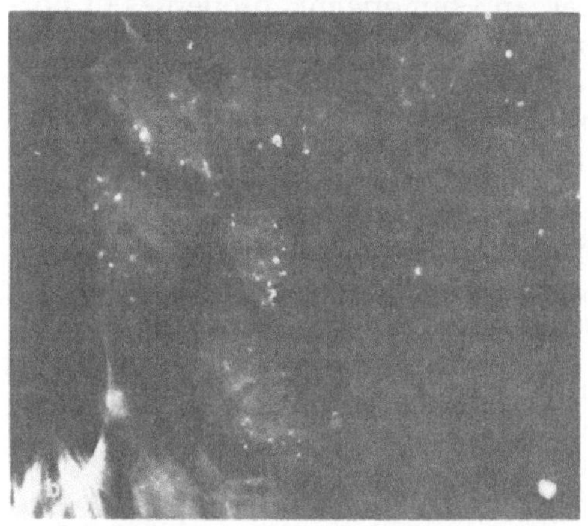

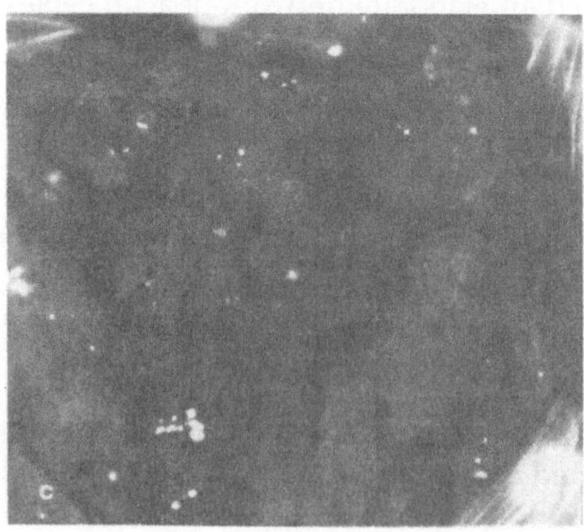

Fig. 6. Immunofluorescent localization of microfilaments in mammary cells grown *in vitro* using an anti-actin antibody. (a) Normal mammary cells; (b) preneoplastic HAN cell line D2; (c) D2 mammary tumor. Normal cells contain an abundant and well-developed network of microfilaments, whereas some preneoplastic and neoplastic cells contain a diminished content of microfilaments. ×600.

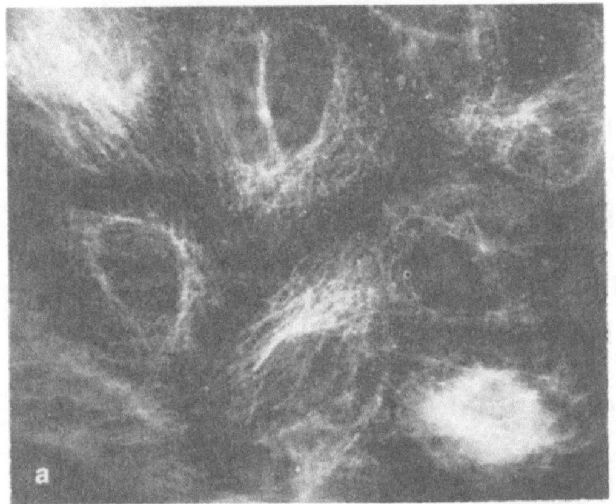

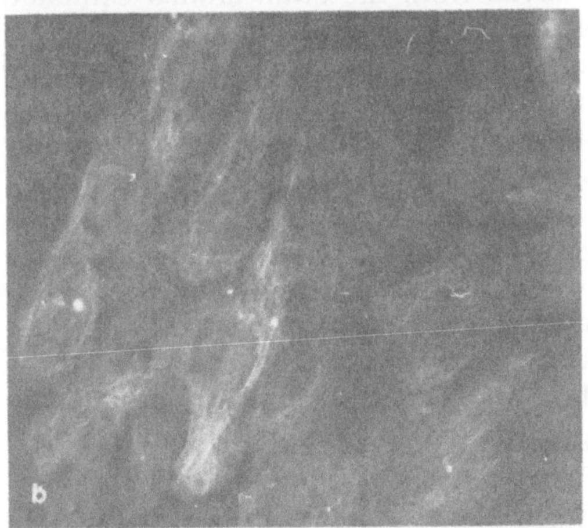

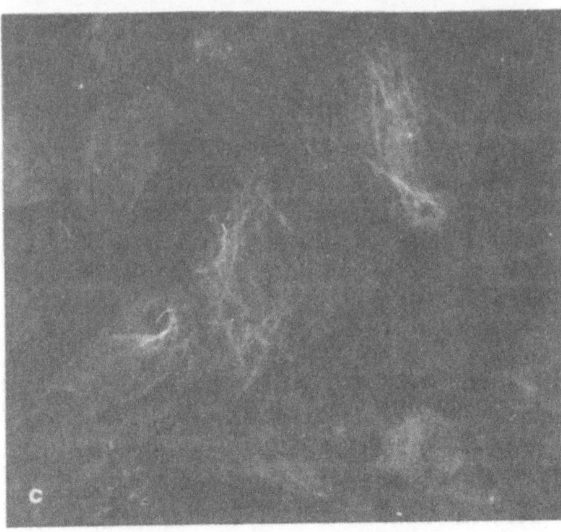

Fig. 7. Immunofluorescent localization of microtubules in mammary cells grown *in vitro* using an antitubulin antibody. (a) Normal mammary cells; (b) preneoplastic HAN cell line D2; (c) D2 mammary tumor. Normal cells contain a complete network of microtubules, whereas some preneoplastic and neoplastic cells contain a diminished network. ×600.

surface. The number of microtubules in cells from D2 HAN outgrowths (Fig. 7B) and tumors (Fig. 7C) was reduced and their organization was more disoriented.

The occurrence of cytoskeletal alterations in the cytoskeletons was also tested in HAN line D1 and in three other tumors, those developing from D1, C3, and C4 HAN outgrowths. The staining patterns of all four of these cell types were similar to those seen in cells of the D2 HAN and tumors. The reaction with antiactin antibodies of cultures established from normal and transformed populations revealed that many colonies of epithelial cells were mixed populations consisting of cells with a full complement of actin cables and cells that had few or no visible microfilaments.

The data demonstrated that despite many similarities, mammary epithelial cells only shortly removed from the adult animal exhibited no differences in their cytoskeletons regardless of whether their origin was from normal, HAN, or tumor tissue. These findings suggest that alterations in cytoplasmic microtubules and microfilaments could not offer a valuable marker for distinguishing populations of preneoplastic and neoplastic cells from normal mammary cells in culture, since individual cells in these transformed populations are capable of expressing a full complement of microtubules and microfilaments. The factors that influence this modulation of cytoskeletal expression pose an interesting problem to be elucidated. The presence of modulation in microtubules and microfilaments in cells of all the stages of mammary tumorigenesis suggests that these changes could not represent early manifestations of neoplastic progression. At the same time, it also seems probable that other changes in surface-related properties may be important for the neoplastic transformation and for the full expression of the neoplastic phenotype.

5.2.2. *Lectin Agglutination*

A second measure of the structural characteristics of cells that has been used to distinguish between normal and neoplastic cells is the property of lectin agglutination to the cell surface.[139] Although not all systems will show significant differences between normal and neoplastic cells, a majority of neoplastic cells, in comparison with their controls, will show enhanced agglutination by Con A or wheat germ agglutinin (WGA).[139] A recent report by Voyles and McGrath[144] suggested that mammary preneoplastic and neoplastic cells are highly agglutinable by Con A, in comparison with normal mammary cells. All three cell types were grown as primary cultures and assayed by the indirect hemagglutination technique.[145] The normal and neoplastic cells, when transplanted back into the mammary fat pads of syngeneic mice, developed into normal and neoplastic tissues, respectively.

At the time this report was published, we had been doing a series of experiments asking similar questions. In our experiments, agglutination of the three cell populations was examined in three different protocols.[146] Initially, cells were dissociated with collagenase plus BSA, then mixed in suspension with several concentrations of Con A, and the agglutination frequency determined semiquantitatively by visual inspection.[131] While this technique had its limitations (namely, background spontaneous clumping), neither the threshold dose, the half-maximal dose, nor the maximal dose of Con A or WGA was different for the three populations. To avoid misinterpretation of the results due to background spontaneous clumping, another series of experiments examined the agglutinability of mammary cells grown on cover slips *in vitro*, either directly or indirectly by a hemagglutination assay.[145] Both tests showed the same results, namely, that normal, preneoplastic, and neoplastic mammary cells grown under identical conditions *in vivo* show only a very low degree of agglutination by Con A (1+ at 50 μg Con A/ml). These experiments were repeated over a dozen times under different conditions. The use of fluorescein-labeled Con A demonstrated that all three cell types bound Con A in similar qualitative patterns and, judged by the intensity of the fluorescein staining, at similar quantitative levels. Mammary epithelial cells could be induced to agglutinate to maximal agglutination by drugs that disrupt the microtubule or microfilament complexes, or both. Colchicine, cytochalasin B, and Dibucaine all enhanced agglutination. In addition, agents that presumably act on the cell-surface level, such as hyaluronidase, also enhanced agglutination. Collagenase, in the presence of BSA to inhibit contaminating protease activity, and trypsin did not enhance agglutination. In our experiments, cells from the three different types of cultures were transplanted back into the mammary fat pads of syngeneic mice and produced outgrowths similar to the original starting material (i.e., normal produced normal, preneoplastic produced preneoplastic, and tumor produced tumor).

The differences between Voyles and McGrath's experiments and ours have not been definitively explained. Similar tissues were used in both experiments, and a similar protocol with minor differences was used to detect agglutination. Two brands of Con A were used (Miles-Yeda and Pharmaceia), and in our hands, neither agglutinated mammary tumor cells but both did agglutinate SV40-transformed 3T3 cells. The major difference appears to be in the dissociation procedure. Voyles and McGrath used trypsin, whereas we used collagenase. The former enzyme is known to alter cell-surface proteins in a reversible manner[129] and to enhance agglutination. Further experiments should clarify the significant reasons behind the different results. From the results available so far, it would appear that lectin agglutination will be of limited value in distinguishing between normal and preneoplastic mammary cells.

5.3. Nuclear Magnetic Resonance Studies

In 1971, Damadian[147] used nuclear magnetic resonance (NMR) techniques to study the spin–lattice (T_1) and spin–spin (T_2) relaxation times of water protons in normal and neoplastic tissues. His results, which were confirmed by numerous other investigators, demonstrated that neoplastic cells have elevated T_1 and T_2 relaxation times of water protons.[148–152] Hazlewood *et al.*[148,151] demonstrated a progressive increase in the relaxation times of water protons for preneoplastic nodule outgrowth lines D1 and D2 and mammary tumors, respectively. By this technique, preneoplastic mammary tissues could be distinguished from both normal and neoplastic tissues. Further studies indicated that elevated T_1 and T_2 relaxation times were not correlated with the individual tumor potential of the preneoplastic line, but with the preneoplastic phenotype *per se,* which indicated that the intracellular changes responsible for elevated T_1 and T_2 relaxation times were not sufficient for full expression of the neoplastic phenotype. The elevated relaxation times of neoplastic in comparison with preneoplastic tissues indicated that a progressive change occurred with neoplastic transformation.

Recently, Frey *et al.*[153] and others[151,154] demonstrated that elevated T_1 relaxation times occurred in uninvolved organs (i.e., spleen, kidney, liver, and hematopoietic) of tumor-bearing mice. This "systemic effect" extended to the sera of tumor-bearing animals[155] and to human cancer patients.[156] Beall *et al.*[157] demonstrated that the increase in water proton relaxation times in sera occurred in mice bearing small benign ductal mammary papillomas as well as ductal carcinomas, but was not present in mice bearing the preneoplastic stages that preceded these tumors. The elevation of T_1 in serum did not correlate with stress, serum iron levels, or serum protein concentration. These results indicated that the changes of water proton relaxation times in tissues occurred earlier than in sera and were a more sensitive indicator of an early change in the direction of neoplasia.

It is not clear at present whether the NMR techniques are discriminating enough to distinguish neoplastic states from nonneoplastic diseased states or suspected high-risk lesions in the breast. One study demonstrated that fibroadenomas and carcinomas could be distinguished from fibrocystic disease in the human breast,[158] but serum values were not done. At present, more work is needed in this area to develop NMR as a potential diagnostic aid for indicating recurrence of a breast neoplasm and distinguishing early proliferative changes in the human breast. This effort requires application of more advanced methodology, as demonstrated by the recent results of Damadian *et al.*(159)

6. Biochemical Aspects

6.1. Metabolic Pathways

Since much of this area was reviewed in an excellent article by Abraham and Bartley,(160) only the principle and general conclusions will be stated here.

Biochemically, HAN outgrowth cells resemble normal alveolar cells with respect to levels of enzyme activities involved in glucose utilization via the Embden–Meyerhof and pentose phosphate pathways.(161,162) The levels of enzyme activities of hexokinase, phosphoglucomutase, phosphoglucose isomerase, phosphofructokinase, glucose-6-phosphate dehydrogenase (GPDH), 6-phosphogluconate dehydrogenase, and enzymes responsible for ribose-5-phosphate breakdown are similar for nodule outgrowth cells and normal alveolar cells. The enzyme activities of aldolase, α-glycerophosphate dehydrogenase, and lactate dehydrogenase (LDH) were decreased in HANs as compared with normal alveolar cells. These enzyme activities, however, are not rate-limiting in any pathway in the mammary cell. All enzyme activities in nodule and normal cells, however, except phosphoglucomutase, differed from levels in adenocarcinomas derived from these nodule outgrowths.

Three sets of isoenzyme patterns investigated in HAN and tumor vs. normal alveolar cells demonstrated differences among the three cell types. Both tumor and normal had type I and II but no type III hexokinase enzyme. The five isoenzymes of LDH were examined in normal, HAN, and tumor cells. LDH-3 was greatly elevated in tumor and moderately elevated in HAN cells. The greatest differences were seen with GPDH, where a distinct isoenzyme, GPDH-1, was found in HAN and tumor cells, whereas normal lactating cells showed predominantly GPDH-3 and normal virgin cells a mixture of GPDH-2 and GPDH-3. The authors suggested that the appearance of GPDH-1 could be used to diagnose the presence of HANs.(163)

The levels and distribution of enzyme activities involved in utilization of di- and tricarboxylic acids were similar in HAN and normal alveolar cells, but different from those found in mammary adenocarcinomas. Enzymes measured were aconitase, isocitrate dehydrogenase, malate dehydrogenase, malic enzyme, and citrate cleavage enzyme.[162]

This analysis was extended by Bartley *et al.*[164] with the measurement of additional parameters, which included the uptake of labeled glucose, lipogenesis, and pentose phosphate cycle activity. HAN cells were similar in these parameters to normal, lactating unsuckled mammary gland cells. The only significant change seen in HAN cells was in their pattern of fatty acid synthesis. Although normal unsuckled and suckled lactating gland cells produced fatty acids of 10–12 carbons, nodules and mammary tumors produced fatty acids of 16 carbons. This was one of the first indications of a significant metabolic pattern in nodules distinct from normal, albeit the change was quantitative, not qualitative. The rate of protein synthesis and phosphatidylcholine synthesis in HANs and tumors was in a normal range, between that of pregnant and lactating mammary gland, although interestingly, the physiological changes from pregnancy to lactation did not alter the rate of protein synthesis in HANs and tumor as it did with the rates seen in normal mammary tissues.[163]

In general, HAN and tumor cells had a lower rate of lactate utilization.[163] This lower rate was due to a lower absolute number of mitochondria per gram tissue weight in HAN and tumor compared with normal, rather than a defect in the organelle itself. In isolated mitochondria, no major differences in substrate oxidation, phosphorylation, or phosphate ion transport were observed.[163]

Some enzymes active in nucleic acid metabolism were studied by Nahas *et al.*[165] In MTV-positive BALB/c mice, dihydrofolate reductase, thymidylate kinase, and thymidine kinase increased in activity as normal cells progressed to the nodule and tumor states, but no change was found in uridine kinase. In MTV-negative BALB/c mice, dihydrofolate reductase activity increased in nodule and tumor cells. Xanthine oxidase activity was the same in normal and HAN; both were significantly lower than in mammary tumors.[163] Interestingly, the xanthine oxidase activity responded to changes in the physiological state of the host. Activities of tRNA methylases were similar in HAN and lactating gland, although both were threefold lower than in mammary tumors.[163] Gantt *et al.*[166] reached the opposite conclusion based on differences in extent of methylation. They concluded that lactating mammary glands had a greater capacity to methylate RNA than did tumor extracts.

Whereas some of the enzymatic activation in HANs (i.e. xanthine oxidase and fatty acid syntheses) responded to the physiological change of the host in a manner similar to normal mammary gland, DNA and RNA exhibited little difference in rates of syntheses with such a change.[163] However, a dramatic drop in the level of pituitary hormones (in hypophysectomized mice) led to a dramatic decrease in DNA synthesis without a concomitant decrease in RNA synthesis.[160] Bodell[167] recently reported no difference in DNA repair rates between normal, HAN, and tumor cells.

In conclusion, the metabolic activity of HAN cells in a wide variety of pathways is similar to the normal pregnant and unsuckled lactating mammary gland in terms of levels of enzymatic activity. In a few instances, such as GPDH isoenzyme patterns, a significant qualitative difference could be found that might allow distinction between normal and HAN cells. Perhaps more significant was the general finding that HAN cells were much less responsive to shifts in the physiological state of the host (from virgin to pregnancy to lactation), thus placing the HAN between a fully responsive state like mammary gland and a generally unresponsive state like mammary tumors.

6.2. Hormonal Control of Molecular Events

A recent excellent review[24] covered the general topic of the responses of mammary cells *in vivo* and *in vitro* to hormones. Like other hormone-dependent organs, the hormonal regulation of normal mammary gland growth is mediated by hormone (estrogen and progesterone) stimulation of DNA, RNA, and protein synthesis, including hormone-induced synthesis of DNA polymerase. In a series of extensive experiments, Banerjee *et al.*[168,169] demonstrated that DNA polymerase activity and DNA synthesis in MTV-positive and MTV-free nodule outgrowth lines (D1, D1-MTV, D8, CH33) were independent of ovarian hormones. The HAN outgrowth lines, irrespective of tumor potential, exhibited the same degree of lack of responsiveness to normal regulation of DNA synthetic and polymerase activities.[169] HAN lines D1 and D8 also actively synthesized ribosomal precursor RNA and heterogenous nuclear RNA, the rate being unaffected by ovariectomy, adrenalectomy, prolonged treatment with estradiol-17β and progesterone, or the physiological states of pregnancy and lactation. The loss of hormonal (ovarian) control of DNA polymerase and DNA synthetic activities and rapidly labeled RNA synthesis correlated with loss of hormonal control of cell proliferation, lobuloalveolar structure, and tumorigenesis. This loss of hormonal control of cell proliferation along with an indefinite division potential represents two of the major fundamental changes in the preneoplastic mammary population that are often attributed only to neoplastic populations.

Lysine-rich histones from midpregnant mammary glands, preneoplastic nodule outgrowth lines (D1, D2a), and tumors derived from them were fractionated by chromatography. The subfractions from preneoplastic and neoplastic tissues were qualitatively similar to those obtained from normal mammary tissues.[170] Hormone-dependent induction of casein in normal mammary tissue *in vitro* was preceded by specific changes in the pattern of amino acid incorporation into lysine-rich histone subfractions.[171] These same hormones had no effect on casein induction and lysine-rich histone synthesis in one of two preneoplastic lines (D2) and tumors derived from either preneoplastic line. Line D1 responded in a qualitatively similar manner in casein induction and lysine-rich histone synthesis as did normal mammary tissues.[171]

6.3. Chromatin Biochemistry

A recent series of studies of the biochemical modification of chromatins in normal, preneoplastic, and neoplastic mammary tissues provided some information on the intranuclear organization, conformation, composition, and functional activity of chromatin subfractions.[172,173] Basically, the gross mass compositions of nuclei and of chromatin were not statistically distinguishable among the three types of tissues. Neoplastic cell chromatin, however, contained an elevated content of nonhistone protein relative to normal, which suggested that the elevated nonhistone protein content was a result of retention of nuclear nonhistone protein during an incomplete conversion of neoplastic cell nuclei to chromatin.[173] Analysis of the chromatin-associated nonhistone proteins by one-dimensional polyacrylamide gel systems, while precluding conclusions of absolute quantitative differences, suggested that neoplastic and preneoplastic chromatin showed extensive quantitative reduction and qualitative simplification in their heterogeneity pattern as compared with normal. There were many "apparent" qualitative differences in neoplastic chromatin. The differences between normal and tumor-cell chromatin reflected a unique intranuclear organization of tumor cell chromatin that prevented the removal of certain nuclear components during the usual solvent extraction protocols. In this respect, preneoplastic and lactating chromatin were indistinguishable. Further elucidation of the differences between tumor and preneoplastic chromatin await two-dimensional gel and template activity analysis.

7. Conclusions and Perspectives

The mammary gland offers one of the few systems in which defined preneoplastic lesions can be studied during the progression to neoplasia. A tremendous wealth of data exists on the biological characteristics of mammary preneoplasias that documents the significant role these preneoplastic lesions play in the genesis of mammary tumors. Table VII summarizes the important characteristics of mammary preneoplasias and compares them with neoplasms and normal tissues. Several features are worthwhile to consider further. While there is significant information on the hormonal sensitivities of preneoplastic HANs, very little information is available on hormone-induced molecule changes with the notable exception of the studies by Banerjee and co-workers. For instance, do preneoplasias induced by various etiological agents synthesize cell products (i.e., casein) characteristic of differentiated

Table VII
Characteristics of Normal, Hyperplastic Alveolar Nodule, and Neoplastic Mammary Tissues

Criteria	Normal[a]	HAN[a]	Neoplastic[a]
Morphological and physiological			
Myoepithelium	+	+	−
Apical alkaline phosphatase	−	−	+
Mg^{2+}-dependent ATPase	+	+	+
Virus particles (A and B)	+	++	+++
Polyploid DNA	−	−	+
NMR			
T_1, T_2 water protons (tissue)	Low	Intermediate	High
T_1, T_2 water protons (serum)	Low	Low	High
Microtubules	+	+	+
Microfilaments	+	+	+
Con A agglutination	−	−	−
Cell-surface junctional complexes	+	+	+
In vitro growth			
Saturation density	+	+	+
Contact Inhibition	+	+	+
Scanning electron microscopy:			
Shape	Flat	Flat	Flat
Microvilli	|→++	|→++	|→++
Growth in Methocel	N.D.	N.D.	→++
Transplantation			
Survival	+	+	+
Growth inhibition by cortisol	N.D.	−	+
Progressive growth	−	−	+
Serial transplantation	−	+	+
Mammary-fat-pad-dependent	+	+	−
Subject to local growth regulation	+	+	−
Hormonal dependency			
Ovariectomy			
Growth	+	+/−	−[b]
Alveolarity	+	+/−	−

(Continued)

Table VII (*Continued*)

Criteria	Normal[a]	HAN[a]	Neoplastic[a]
Transplantation (*continued*)			
Hypophysectomy			
Growth	+	+	−
Alveolarity	+	+	−
Growth inhibition by hormones/antihormones			
Testosterone	+	+	−
CB-154 (antiprolactin)	+	+/−	−
Nafoxidine (antiestrogens)	+	+/−	−
E_2 receptors	+	+/−	−[b]
Immunogenicity	−	+	+
Response to chemotherapeutic drugs			
Melphalan	N.A.	↑/↑	↓
5-FU	N.A.	↑/↓	↓
CTX	N.A.	↑/↓[c]	↓
VC	N.A.	↑/−	−
MTX	N.A.	↓/↓	↓
Prednisone	N.A.	↓/↓	−

[a] (+) Yes or positive; (−) no or negative; (+/−) yes for some lines, no for other lines; (N.D.) not determined; (N.A.) not applicable; (↑) enhances; (↓) inhibits; (↑/↓) D2/C4 HAN lines.
[b] The exceptions to the general rule are the tumors in GR mice that arise for plaques and the DMBA-induced tumors in (C57BL × DBAf)F_1 mice that arise for DHs.
[c] In line C4, low dose inhibits tumor formation, whereas higher doses enhance tumor formation.

mammary epithelial cells? If so, is the casein produced by the preneoplasias the same as that produced by normal cells? How are casein production and hormone receptors regulated in altered cells? Is the lack of casein production correlated positively with the evolution of hormone independence? Another related area, the molecular analysis of chromatin and DNA transcription changes, has been untouched.

A second area in which information is lacking is the cell biology of preneoplasias. Very little attention has been paid to the growth of these lesions *in vitro*, the state and regulation of the cAMP and cGMP systems during neoplastic progression, cell cycle kinetics during progression, cell-surface characteristics, and, most important, the development of stage-specific markers to analyze the cellular nature of neoplastic progression similar to studies currently being attempted on the rat preneoplastic hepatic nodules.

A third area that needs to be reinvestigated is the pathogenesis of virus- and chemical-carcinogen-induced preneoplasias. What are the basic cells types in normal mammary epithelia, and what are their cytochemical and ultrastructural characteristics? After treatment with chemical carcinogens or mammary tumor viruses, do unique and different cell types respond to each carcinogen? Do preneoplastic cells persist for long periods of time before they are expressed as large foci of hyperplastic cells? What are the factors that preneoplastic cells are still responsive to in the mammary fat pad? What is special about the mammary fat pad as compared with the gonadal fat depot? Finally, what cell types comprise a HAN population, which cell type gives rise to the mammary tumor, and is the progression to neoplasia the result of a qualitative change in a cell or due to selection of preexisting neoplastic cells?

Excellent model systems have been available over the past decade to study some of these problems, and with the current technological sophistication, it would seem that the time is ripe to exploit this system to answer some very critical questions concerning the progression to neoplasia. The mammary tumor system still remains the *only* system that combines the presence of a discrete unique preneoplastic lesion with the ability to analyze suspected alterations *in vitro* and *in vivo* and to be continually monitored *in vivo* in syngeneic, untraumatized, normal hosts. No other tumor system, whether it be skin, liver, or hematopoietic, has these capabilities and attributes. A tremendous effort has been spent developing this system to the point where it now offers a new and exploitable approach to understand both mammary carcinogenesis and carcinogenesis in general.

ACKNOWLEDGMENTS: The original results discussed in this chapter were supported by research grants CA-11944, CA-17074 and contracts NO1-CB-43907 and NO1-CM-57018 from the National Institutes of Health. I wish to acknowledge the expert technical assistance of Frances Shepherd, Barbara Click, Frances Miller, Tim Gropp, Jewel Brown, Raymond Windle, and Roy Parsons, and the professional collaboration of Dr. Bonnie Asch, who is a postdoctoral fellow examining the cytoskeletal and cell surface characteristics of cultured mammary cells, and Barbara Ruppert, who is a graduate student examining the immunological characteristics of mammary preneoplasias.

8. References

1. K. B. DeOme, L. J. Faulkin, Jr., H. A. Bern, and P. B. Blair, Development of mammary tumors from hyperplastic alveolar nodules transplanted into gland-free mammary fat pads of female C3H mice, *Cancer Res.* **19,** 515–520 (1959).
2. K. B. DeOme, The mouse mammary tumor system, in: *Proceedings of the Fifth Berkeley Symposium on Mathematical Statistics and Probability* (J. Neyman, ed.), pp. 649–655, University of California Press, Berkeley (1967).

3. D. Medina and K. B. DeOme, Influence of mammary tumor virus on the tumor-producing capabilities of nodule outgrowth free of mammary tumor virus, *J. Natl. Cancer Inst.* **40,** 1303–1308 (1968).
4. D. Medina, Preneoplastic lesions in mouse mammary tumorigenesis, in: *Methods in Cancer Research* (H. Busch, ed.), Vol. 7, pp. 3–53, Academic Press, New York (1973).
5. M. Haaland, Spontaneous tumors in mice, in: *Fourth Scientific Report,* Imperial Cancer Research Fund, pp. 1–111 (1911).
6. P. Rous and J. G. Kidd, Conditional neoplasms and subthreshold neoplastic states: A study of the tar tumors of rabbits, *J. Exp. Med.* **73,** 365–389 (1941).
7. I. MacKenzie and P. Rous, The experimental disclosure of latent neoplastic changes in tarred skin, *J. Exp. Med.* **73,** 391–415 (1941).
8. W. F. Friedewald and P. Rous, The pathogenesis of deferred cancer: A study of the after effects of methylcholanthrene upon rabbit skin, *J. Exp. Med.* **91,** 459–484 (1950).
9. J. C. Mottram, A developing factor in experimental blastogenesis, *J. Pathol. Bacteriol.* **56,** 391–402 (1944).
10. J. C. Mottram, A sensitizing factor in experimental blastogenesis, *J. Pathol. Bacteriol.* **56,** 391–402 (1944).
11. I. Berenblum and P. Shubik, The role of croton oil applications, associated with a single painting of a carcinogen, in tumor induction of the mouse's skin, *Br. J. Cancer* **1,** 379–382 (1947).
12. I. Berenblum and P. Shubik, The persistence of latent tumor cells induced in the mouse's skin by a single application of 9 : 10-dimethylbenzanthracene, *Br. J. Cancer* **3,** 384–386 (1949).
13. I. Berenblum, Sequential aspects of chemical carcinogenesis: Skin, in: *Cancer: A Comprehensive Treatise* (F. Becker, ed.), Vol. 1, pp. 323–324, Plenum Press, New York (1975).
14. I. Berenblum, The two-stage mechanism of carcinogenesis in biochemical terms, in: *The Physiopathology of Cancer* (F. Homburger, ed.), Vol. 1, pp. 393–402, S. Karger, Basel (1974).
15. R. K. Boutwell, Some biological aspects of skin carcinogenesis, *Prog. Exp. Tumor Res.* **4,** 207–250 (1964).
16. E. Farber, On the concept of minimal deviation hepatoma in the study of the biochemistry of cancer, *Cancer Res.* **28,** 1210–1211 (1968).
17. E. Farber, Hyperplastic liver nodules, in: *Methods in Cancer Research* (H. Busch, ed.), Vol. 7, pp. 345–375, Academic Press, New York (1973).
18. S. Epstein, N. Ito, L. Merkow, and E. Farber, Cellular analysis of liver carcinogensis: The induction of large hyperplastic nodules in the liver with 2-fluorenylacetamide or ethionine and some aspects of their morphology and glycogen metabolism, *Cancer Res.* **27,** 1702–1711 (1967).
19. L. P. Merkow, S. M. Epstein, B. J. Caito, and B. Bartus, The cellular analysis of liver carcinogensis: Ultrastructural alterations within hyperplastic liver nodules induced by 2-fluorenylacetamide, *Cancer Res.* **27,** 1712–1721 (1967).
20. E. Farber and M. B. Sporn (eds.), Early lesions and the development of epithelial cancer, *Cancer Res.* **36,** 2475–2706 (1976).
21. R. A. Willis, *The Pathology of Tumors,* Butterworths, London (1967).
22. R. D. Cardiff, S. R. Wellings, and L. J. Faulkin, Jr., Biology of breast preneoplasia, *Cancer,* **39,** 2734–2746 (1977).
23. S. Nandi and C. S. McGrath, Mammary neoplasia in mice, in: *Advances in Cancer Research* (G. Klein and S. Weinhouse, eds.), Vol. 17, pp. 353–414, Academic Press, New York (1973).
24. M. R. Banerjee, Responses of mammary cells to hormones, *Int. Rev. Cytol.* **47,** 1–97 (1976).

25. D. R. Pitelka, K. B. DeOme, and H. A. Bern, Virus-like particles in precancerous hyperplastic mammary tissues of C3H and C3Hf mice, *J. Natl. Cancer Inst.* **25,** 753–777 (1960).
26. K. B. DeOme, The mammary tumor system in mice: A brief review, in: *Viruses-Inducing Cancer* (W. J. Burdette, ed.), pp. 127–137, University of Utah Press, Salt Lake City (1966).
27. O. Mühlbock, Note on a new inbred mouse-strain GR/A, *Eur. J. Cancer* **1,** 123–124 (1965).
28. D. R. Pitelka, H. A. Bern, S. Nandi, and K. B. DeOme, On the significance of virus-like particles in mammary tissues of C3Hf mice, *J. Natl. Cancer Inst.* **33,** 867–885 (1964).
29. S. Nandi, New method for detection of mouse mammary tumor virus. I. Influence of foster nursing on incidence of hyperplastic mammary nodules in BALB/cCrgl mice, *J. Natl. Cancer Inst.* **31,** 57–73 (1963).
30. S. Nandi, New method for detection of mouse mammary tumor virus. II. Effect of administration of lactating mammary tissue extracts on incidence of hyperplastic mammary nodules in BALB/cCrgl mice, *J. Natl. Cancer Inst.* **31,** 75–89 (1963).
31. S. Nandi, K. B. DeOme, and M. Hardin, Mammary tumor virus activity in blood and mammary tissues of C3H and BALB/cfC3H strains of mice, *J. Natl. Cancer Inst.* **35,** 309–318 (1965).
32. S. Nandi, D. Knox, K. B. DeOme, M. Hardin, V. V. Finster, and P. B. Pickett, Mammary tumor virus activity in red blood cells of BALB/cfC3H mice, *J. Natl. Cancer Inst.* **36,** 809–815 (1966).
33. D. Medina, J. Vaage, R. Setlacek, Mammary noduligenesis and tumorigenesis in pathogen-free C3Hf mice, *J. Natl. Cancer Inst.* **51,** 961–965 (1973).
34. P. B. Blair and K. B. DeOme, Mammary tumor development in transplanted hyperplastic alveolar nodules of the mouse, *Proc. Soc. Exp. Biol. Med.* **108,** 289–291 (1961).
35. D. Medina, K. B. DeOme, and L. Young, Tumor-producing capabilities of hyperplastic alveolar nodules in virgin and hormone-stimulated BALB/cfC3H and C3Hf mice, *J. Natl. Cancer Inst.* **44,** 164–174 (1970).
36. L. J. Faulkin, Jr., Hyperplastic lesions of mouse mammary glands after treatment with 3-methylcholanthrene, *J. Natl. Cancer Inst.* **36,** 289–298 (1966).
37. D. Medina and M. Warner, Mammary tumorigenesis in chemical carcinogen-treated mice. IV. Induction of mammary ductal hyperplasis, *J. Natl. Cancer Inst.* **57,** 331–337 (1976).
38. D. Medina, Mammary tumorigenesis in chemical carcinogen-treated mice. VI. Tumor-producing capabilities of mammary dysplasias in BALB/cCrgl mice, *J. Natl. Cancer Inst.* **57,** 1185–1189 (1976).
39. D. Medina and K. B. DeOme, Effects of various oncogenic agents on tumor-producing capabilities of D series BALB/c mammary nodule outgrowth lines, *J. Natl. Cancer Inst.* **45,** 353–363 (1970).
40. A. Dux and O. Mühlbock, Enhancement by hypophyseal hormones on the malignant transformation of transplanted hyperplastic nodules of the mouse mammary gland, *Eur. J. Cancer* **5,** 191–194 (1969).
41. D. Medina, Preneoplastic lesions in murine mammary cancer, *Cancer Res.* **36,** 2589–2595 (1976).
42. A. Kirshbaum, W. L. Williams, and J. J. Bittner, Induction of mammary cancer with methylcholanthrene; histogenesis of induced neoplasms, *Cancer Res.* **6,** 354–362 (1946).

43. A. G. Liebelt and R. A. Liebelt, The "nodule–cancer" complex of mammary tissue in low cancer strains of mice, *Lav. Ist. Anat. Istol. Patol. Univ. Studi Perugia* **34,** 146 (1974).
44. L. J. Beuving, L. J. Faulkin, Jr., K. B. DeOme, and V. V. Bergs, Hyperplastic lesions in the mammary glands of Sprague–Dawley rats after 7,12-dimethylbenzanthracene treatment, *J. Natl. Cancer Inst.* **39,** 423–429 (1967).
45. L. J. Beuving, H. A. Bern, and K. B. DeOme, Occurrence and transplantation of carcinogen-induced hyperplastic nodules in Fischer rats, *J. Natl. Cancer Inst.* **39,** 431–447 (1967).
46. A. M. Cameron and L. J. Faulkin, Jr., Hyperplastic and inflammatory nodules in the canine mammary gland, *J. Natl. Cancer Inst.* **47,** 1277–1287 (1971).
47. M. Warner, Mammary gland morphology of female beagle dogs: Studies *in vivo* and *in vitro,* Ph.D. thesis, Department of Anatomy, University of California, Davis (1972).
48. S. R. Wellings, H. M. Jensen, and R. G. Marcum, An atlas of subgross pathology of the human breast with special reference to possible precancerous lesions, *J. Natl. Cancer Inst.* **55,** 231–274 (1975).
49. H. M. Jensen and S. R. Wellings, Preneoplastic lesions of the human mammary gland transplanted into the nude athymic mouse, *Cancer Res.* **36,** 2605–2610 (1976).
50. H. M. Jensen, J. R. Rice, and S. R. Wellings, Preneoplastic lesions in the human breast, *Science* **191,** 295–297 (1976).
51. L. J. Beuving, Mammary tumor formation within outgrowths of transplanted hyperplastic nodules from carcinogen-treated rats, *J. Natl. Cancer Inst.* **40,** 1287–1291 (1968).
52. D. Sinha and T. L. Dao, Hyperplastic alveolar nodules of the rat mammary gland: Tumor-producing capability *in vivo* and *in vitro, Cancer Lett.* **2,** 153–160 (1977).
53. S. Z. Haslam, Influence of age of treatment with DMBA on ovary-dependent and independent mammary tumor development in rats, *Proc. Am. Assoc. Cancer Res.* **67,** 366 (1976).
54. E. M. Rivera, M. Walbridge, and S. D. Hill, Tumor development in transplants of rat mammary hyperplastic alveolar nodules, *Proc. Am. Assoc. Cancer Res.* **68,** 810 (1977).
55. L. Beuving, Biological characteristics of preneoplastic lesions in the mammary glands of carcinogen-treated rats, Ph.D. thesis, University of California, Berkeley (1968).
56. I. Russo, J. Saby, and J. Russo, Pathogenesis of rat mammary carcinomas induced by DMBA, *Proc. Am. Assoc. Cancer Res.* **16,** 164 (1975).
57. J. Russo, I. H. Russo, M. Ireland, and J. Saby, Increased resistance of multiparous rat mammary gland to neoplastic transformation by 7,12-dimethylbenzanthracene, *Proc. Am. Assoc. Cancer Res.* **68,** 149 (1977).
58. C. W. Daniel, K. B. DeOme, L. J. T. Young, P. B. Blair, and L. J. Faulkin, Jr., The *in vivo* life span of normal and preneoplastic mouse mammary glands: A serial transplantation study, *Proc. Natl. Acad. Sci. U.S.A.* **61,** 53–60 (1968).
59. L. J. T. Young, D. Medina, K. B. DeOme, and C. W. Daniel, The influence of host and tissue age on life span and growth rate of serially transplanted mouse mammary gland, *Exp. Gerontol.* **6,** 49–56 (1971).
60. C. W. Daniel, L. J. T. Young, and D. Medina, The influence of mammogenic hormones on serially transplanted mouse mammary gland, *Exp. Gerontol.* **6,** 95–101 (1971).
61. C. W. Daniel, B. D. Aidells, D. Medina, and L. J. Faulkin, Jr., Limited division potential of precancerous mouse mammary cells after spontaneous or carcinogen-induced transformation, *Fed. Proc. Fed. Am. Soc. Exp. Biol.* **34,** 64–67 (1975).
62. R. A. Liebelt, C. B. Bordelon, and A. G. Liebelt, The adipose tissue system and food intake, *Prog. Physiol. Psychol.* **5,** 211–252 (1973).

63. L. J. Faulkin, Jr., and K. B. DeOme, Regulation of growth and spacing of gland elements in the mammary fat pad of the C3H mouse, *J. Natl. Cancer Inst.* **24,** 953–969 (1960).
64. P. B. Blair, K. B. DeOme, and S. Nandi, The preneoplastic state in mouse mammary carcinogenesis in: *Biological Interactions in Normal and Neoplastic Cells* (M. Brennan, ed.), pp. 371–389, Little, Brown, Boston (1962).
65. H. A. Bern and S. Nandi, Recent studies of the hormonal influence in mouse mammary tumorigenesis, *Prog. Exp. Tumor Res.* **2,** 91–145 (1961).
66. R. Yanai and H. Nagasawa, Enhancement by pituitary isografts of mammary hyperplastic nodules in adreno-ovariectomized mice, *J. Natl. Cancer Inst.* **46,** 1251–1255 (1971).
67. C. W. Welsch and C. Gribler, Prophylaxis of spontaneously developing mammary carcinoma in C3H/HeJ female mice by suppression of prolactin, *Cancer Res.* **33,** 2939–2946 (1973).
68. C. W. Welsch, Prophylaxis of early preneoplastic lesions of the mammary gland, *Cancer Res.* **36,** 2621–2625 (1976).
69. D. Medina, Tumor formation in preneoplastic mammary nodule lines in mice treated with nafoxidine, testosterone, and 2-bromo-α-ergocryptine, *J. Natl. Cancer Inst.* **58,** 1107–1110 (1977).
70. C. Watson, C. Medina, and J. H. Clark, Estrogen receptor characterization in a transplantable mouse mammary tumor, *Cancer Res.* **37,** 3344–3348 (1977).
71. G. Shyamala, Estradiol receptors in mouse mammary tumors: Absence of the transfer of bound estradiol from the cytoplasm to the nucleus, *Biochem. Biophys. Res. Commun.* **46,** 1623–1630 (1972).
72. J. R. Richards, G. Shyamala, and S. Nandi, Estrogen receptors in normal and neoplastic mouse mammary tissues, *Cancer Res.* **34,** 2764–2772 (1974).
73. D. W. Weiss, L. J. Faulkin, Jr., and K. B. DeOme, Acquisition of heightened resistance and susceptibility to spontaneous mouse mammary carcinomas in the original host, *Cancer Res.* **24,** 732–741 (1964).
74. D. W. Weiss, D. H. Lavrin, M. Dezfulian, J. Vaage, and P. B. Blair, Studies on the immunology of spontaneous mammary cancer in mice, in: *Viruses-Inducing Cancer* (W. J. Burdette, ed.), pp. 138–168, University of Utah Press, Salt Lake City (1966).
75. D. L. Morton, L. Goldman, and D. A. Wood, Acquired immunological tolerance and carcinogenesis by the mammary tumor virus. II. Immune responses influencing growth of spontaneous mammary adenocarcinomas. *J. Natl. Cancer Inst.* **42,** 321–329 (1969).
76. A. M. Attia, K. B. DeOme, and D. W. Weiss, Immunology of spontaneous mammary carcinomas in mice. II. Resistance to rapidly and slowly developing tumors, *Cancer Res.* **25,** 451–457 (1965).
77. D. S. Burton, P. B. Blair, and D. W. Weiss, Protection against mammary tumors in mice by immunization with purified mammary tumor virus preparations, *Cancer Res.* **29,** 971–973 (1969).
78. J. Vaage, Nonvirus-associated antigens in virus-induced mouse mammary tumors, *Cancer Res.* **28,** 2477–2483 (1968).
79. J. Vaage, Non-cross-reacting resistance to virus-induced mouse mammary tumors in virus infected C3H mice, *Nature (London)* **218,** 101–102 (1968).
80. J. Vaage, T. Kalinovsky, and R. Olson, Antigenic differences among virus-induced mouse mammary tumors arising spontaneously in the same C3H/Crgl host, *Cancer Res.* **29,** 1452–1456 (1969).
81. G. H. Heppner and G. Pierce, *In vitro* demonstration of tumor-specific antigens in spontaneous mammary tumors of mice, *Int. J. Cancer* **4,** 212–218 (1969).

82. D. W. Weiss, A. Sultizeanu, L. Young, M. Adelberg, and Y. Segev, Studies on the immunogenicity of preneoplastic and neoplastic mammary tissues of BALB/c mice free of the mammary tumor virus *Isr. J. Med. Sci.* **7,** 187–201 (1971).
83. Z. T. Halpin, J. Vaage, and P. B. Blair, Lack of antigenicity of mammary tumors induced by carcinogens in a nonantigenic preneoplastic lesion, *Cancer Res.* **32,** 2197–2200 (1972).
84. G. H. Heppner, J. S. Kopp, and D. Medina, Microcytotoxicity assay of immune responses to non-mammary tumor virus-induced, preneoplastic and neoplastic mammary lesions in BALB/c mice, *Cancer Res.* **36,** 753–758 (1976).
85. G. Slemmer, Host response to premalignant mammary tissues, *Natl. Cancer Inst. Monogr.* **35,** 57–71 (1972).
86. R. T. Prehn, Tumor progression and homeostasis, *Adv. Cancer Res.* **23,** 203–236 (1976).
87. D. H. Lavrin, P. B. Blair, and D. W. Weiss, Immunology of spontaneous mammary carcinomas in mice. III. Immunology of C3H preneoplastic hyperplastic alveolar nodules in C3Hf hosts, *Cancer Res.* **26,** 293–304 (1966).
88. D. H. Lavrin, P. B. Blair, and D. W. Weiss, Immunology of spontaneous mammary carcinomas in mice. IV. Association of the mammary tumor virus with the immunogenicity of C3H nodules and tumors, *Cancer Res.* **26,** 929–934 (1966).
89. D. Medina and G. H. Heppner, Cell-mediated "immunostimulation" induced by mammary tumor virus-free BALB/c mammary tumors, *Nature (London)* **242,** 329–330 (1973).
90. R. L. Stolfi, R. A. Fugmann, L. M. Stolfe, and D. S. Martin, Synergism between host anti-tumor immunity and combined modality therapy against murine breast cancer, *Int. J. Cancer* **13,** 389–403 (1974).
91. D. S. Martin, R. A. Fugmann, R. L. Stolfi, and P. E. Hayworth, Solid tumor animal model therapeutically predictive for human breast cancer, *Cancer Chemother. Rep.* **5,** 89–109 (1975).
92. B. Fisher, N. Wolmack, E. Saffer, and E. R. Fisher, Inhibitory effects of prolonged *Corynebacterium Parvum* and cyclophosphamide administration on the growth of established tumors, *Cancer* **35,** 134–143 (1975).
93. A. E. Bogden and D. J. Taylor, Predictive mammary tumor test systems for experimental chemotherapy, in: *Breast Cancer* (J. C. Heuson, W. H. Mattheiem, and M. Rozencweig, eds.), Vol. 2, pp. 95–110, Raven Press, New York (1976).
94. D. Medina and F. Shepherd, Enhancement and inhibition of mammary tumor formation and growth by cytostatic drugs, *Cancer Res.* **37,** 3571–3577 (1977).
95. A. Haddow, R. J. C. Harris, G. A. R. Kon, and E. M. F. Roe, The growth inhibitory and carcinogenic properties of 4-amino-stilbene and derivatives, *Philos. Trans. R. Soc. London Ser. A* **241,** 147–196 (1949).
96. C. Bertazzoli, T. Chiali, and E. Solcia, Different incidence of breast carcinomas or fibroadenomas in daunomycin or adriamycin treated rats, *Experientia* **27,** 1209–1210 (1971).
97. J. H. Weisburger, D. P. Griswold, Jr., J. D. Prejean, A. E. Casey, H. B. Wood, and E. K. Weisburger, The carcinogenic properties of some of the principle drugs used in clinical cancer chemotherapy, *Recent Results Cancer Res.* **52,** 1–17 (1975).
98. S. M. Sieber and R. H. Adamson, Toxicity of antineoplastic agents in man, chromosomal aberrations, antifertility effects, congenital malformations, and carcinogenic potential, *Adv. Cancer Res.* **22,** 57–155 (1975).
99. K. B. DeOme, D. Medina, and L. Young, Interference between the nodule-inducing virus and the mammary tumor virus at the level of the neoplastic transformation, in: *Immunity and Tolerance in Oncogenesis* (L. Severi, ed.), pp. 541–549, Division of Cancer Research, University of Perugia, Perugia (1970).

100. D. Medina, K. B. DeOme, D. R. Pitelka, and V. B. Colley, Appearance of virus particles in BALB/c mammary nodule outgrowth lines transplanted into BALB/cfC3H and (C3Hf × BALB/c) F_1 mice, *J. Natl. Cancer Inst.* **46,** 1153–1160 (1971).
101. A. Vaidya and E. Y. Lasfargues, Murine mammary tumor virus infection of mouse mammary epithelial cells *in vitro, Proc. Am. Assoc. Cancer Res.* **68,** 960 (1977).
102. D. Medina, unpublished observations.
103. J. S. Butel, J. P. Dudley, and D. Medina, Comparison of the growth properties *in vitro* and transplantability of continuous mouse mammary tumor cell lines and clonal derivatives, *Cancer Res.* **37,** 1892–1900 (1977).
104. C. M. McGrath, E. J. Marineau, and B. A. Voyles, Levels of MMTV sequences in DNA of "virus-" and "hormone-induced" malignant mammary epithelial cells of the BALB/c mouse. *Proc. Am. Assoc. Cancer Res.* **68,** 979 (1977).
105. W. Drohan, R. Kettman, D. Colcher, and J. Schlom, Isolation of the mouse mammary tumor virus sequences not transmitted as germinal provirus in the C3H and RIII mouse strains, *J. Virol.* **21,** 986–995 (1977).
106. E. Huberman and L. Sachs, Cell susceptibility to transformation and cytotoxicity by the carcinogenic hydrocarbon benzopyrene, *Proc. Natl. Acad. Sci. U.S.A.* **56,** 1123–1129 (1966).
107. E. Huberman and L. Sachs, Susceptibility of cells transformed by Polyoma virus and SV-40 to the cytotoxic effect of the carcinogen hydrocarbon benzopyrene, *J. Natl. Cancer Inst.* **40,** 329–336 (1968).
108. L. Sachs, An analysis of the mechanism of neoplastic cell transformation by Polyoma hydrocarbons. and X-irradiation, *Curr. Top. Dev. Biol.* **2,** 129–150 (1967).
109. T. T. Chen and C. Heidelberger, Quantitative studies on the malignant transformation of mouse prostate cells by carcinogenic hydrocarbons *in vitro, Int. J. Cancer* **4,** 166–178 (1969).
110. J. A. DiPaolo, K. Takano, and N. C. Popescu, Quantitation of chemically-induced neoplastic transformation of BALB/3T3 cloned cell lines, *Cancer Res.* **32,** 2686–2695 (1972).
111. M. J. Brennan, W. H. Grace, and J. A. Singly, Carcinogenesis in the rat mammary gland after exposure *in vitro* to DMBA, *Proc. Am. Assoc. Cancer Res.* **57,** 9 (1966).
112. M. R. Banerjee, B. G. Wood, and L. L. Washburn, Chemical carcinogen-induced alveolar nodules in organ culture of mouse mammary gland, *J. Natl. Cancer Inst.* **53,** 1387–1394 (1974).
113. D. Medina, G. Stockman, and D. Griswold, Significance of chemical carcinogen-induced immunosuppression in mammary tumorigenesis in BALB/c mice, *Cancer Res.* **34.** 2663–2668 (1974).
114. D. Medina, S. B. O'Bryan, M. R. Warner, Y. N. Sinha, W. P. Vander Laan, S. McCormack, and P. Hahn, Prolactin and progesterone levels in chemical carcinogen treated BALB/c mice, *J. Natl. Cancer Inst.* **59,** 213–219 (1977).
115. J. Stjernsward, Immunodepressive effect of 3-methylcholanthrene: Antibody formation at the cellular level and reaction against weak antigenic homografts, *J. Natl. Cancer Inst.* **35,** 885–892 (1965).
116. J. Stjernsward, Effect of non-carcinogenic and carcinogenic hydrocarbons on antibody forming cells measured at the cellular level *in vitro, J. Natl. Cancer Inst.* **36,** 1189–1195 (1966).
117. O. Stutman, Immunological aspects of resistance to the oncogenic effect of 3-methylcholanthrene in mice, *Isr. J. Med. Sci.* **9,** 217–228 (1973).
118. J. W. Jull, Hormonal mechanisms in carcinogenesis, *Can. Cancer Conf.* **6,** 109–123 (1968).

119. T. Krarup, Effect of 9,10-dimethyl-1,2-benzanthracene on the mouse ovary, *Br. J. Cancer* **24,** 168–186 (1970).
120. D. Medina, Serial transplantation of methylcholanthrene treated mammary nodule outgrowth line D1, *J. Natl. Cancer Inst.* **48,** 1363–1370 (1972).
121. R. G. Mehta, L. L. Washburn, P. N. Young, M. R. Banerjee, and H. A. Bern, Proliferation of preneoplastic mammary nodule outgrowth in mammary fat pads of BALB/c mice in organ culture, *J. Natl. Cancer Inst.* **52,** 1013–1018 (1974).
122. Z. Rabinowitz and L. Sachs, The formation of variants with a reversion of properties of transformed cells. V. Reversion to a limited life span, *Int. J. Cancer* **6,** 388–398 (1970).
123. D. Medina, L. J. Faulkin, Jr., and K. B. DeOme, Combined effects of 3-methylcholanthrene, mammary tumor virus, nodule-inducing virus, and prolonged hormonal stimulation on the tumor-producing capabilities of the nodule outgrowth line D1, *J. Natl. Cancer Inst.* **44,** 159–165 (1970).
124. C. S. Nicoll, Growth autoregulation and the mammary gland, *J. Natl. Cancer Inst.* **34,** 131–140 (1965).
125. B. Mintz and G. Slemmer, Gene control of neoplasia. I. Genotypic mosaicism in normal and preneoplastic mammary glands of allophenic mice, *J. Nat. Cancer Inst.* **43,** 87–95 (1969).
126. M. J. Miyamoto, K. B. DeOme, and R. C. Osborn, Detection of inapparent preneoplastic-transformed cells by *in vivo* cultivation of dissociated mouse mammary glands, *Proc. Am. Assoc. Cancer Res.* **66,** 57 (1975).
127. M. J. Miyamoto, Occurrence and preneoplastic significance of hyperplastic alveolar nodules in the strain GR mouse, *Proc. Am. Assoc. Cancer Res.* **67,** 131 (1976).
128. D. Medina and F. Shepherd, Enhancement of the tumor potential of preneoplastic mammary cells by enzymatic dissociation, *Proc. Am. Assoc. Cancer Res.* **68,** 335 (1977).
129. R. Robbin, I. N. Chou, and P. H. Black, Proteolytic enzymes, cell surface changes and viral transportation, *Adv. Cancer Res.* **19,** 203–259 (1975).
130. H. Rubin and T. Koide, Early cellular responses to diverse growth stimuli independent of protein and RNA synthesis, *J. Cell. Physiol.* **86,** 47–58 (1975).
131. J. J. Starling, S. C. Capetillo, G. Neri, and E. F. Walborg, Jr., Surface properties of normal and neoplastic rat liver cells: Lectin-induced cytoagglutination and lectin receptor activity of cell surface glycopeptides, *Exp. Cell Res.* **104,** 177–190 (1977).
132. G. Slemmer, Interactions of separate types of cells during normal and neoplastic mammary gland growth, *J. Invest. Dermatol.* **63,** 27–47 (1974).
133. H. L. Hosick, A note on growth of epithelial tumor cells in primary culture, *Cancer Res.* **34,** 259–261 (1974).
134. H. L. Hosick and K. B. DeOme, Plating and maintenance of epithelial tumor cells in primary culture: Interacting roles of serum and insulin, *Exp. Cell Res.* **84,** 419–425 (1974).
135. N. K. Das, H. L. Hosick, and S. Nandi, Influence of seeding density on multicellular organization and nuclear events in cultures of normal and neoplastic mouse mammary epithelium, *J. Natl. Cancer Inst.* **52,** 849–861 (1974).
136. C. M. McGrath, Cell organization and responsiveness to hormones *in vitro:* Genesis of domes in mammary cell cultures, *Am. Zool.* **15,** 231–236 (1975).
137. D. R. Pitelka, S. T. Hamamoto, J. G. Duafala, and M. K. Nemanic, Cell contacts in the mouse mammary gland. I. Normal gland in postnatal development and the secretory cycle, *J. Cell Biol.* **56,** 797–818 (1973).
138. P. B. Pickett, D. R. Pitelka, S. T. Hamamoto, and D. S. Misfeldt, Occluding junctions and cell behavior in primary cultures of normal and neoplastic mammary gland cells, *J. Cell Biol.* **66,** 316–332 (1975).

139. G. L. Nicolson, Trans-membrane control of the receptors on normal and tumor cells. II. Surface changes associated with transformation and malignancy, *Biochim. Biophys. Acta* **458,** 1–72 (1976).
140. V. Fonte and K. R. Porter, Topographical changes associated with the viral transformation of normal cells to tumorigenicity, in: *Eighth Int. Cong. Elect. Micro.* (J. V. Sanders and D. J. Goodchild, eds.). Vol. 2, pp. 334–335, Australian Academy of Sciences, Canberra (1974).
141. R. Pollack, M. Osborne, and K. Weber, Patterns of organization of actin and myosin in normal and transformed cultured cells, *Proc. Natl. Acad. Sci. U.S.A.* **72,** 994–998 (1975).
142. B. R. Brinkley, G. M. Fuller, and D. P. Highfield, Cytoplasmic microtubules in normal and transformed cells in culture: Analysis by tubulin antibody immunofluorescence, *Proc. Natl. Acad. Sci. U.S.A.* **72,** 4981–4985 (1975).
143. B. Asch, D. Medina, and B. Brinkley, Cytoskeletal changes associated with neoplastic progression in mouse mammary epithelial cells (submitted) (1978).
144. B. A. Voyles and C. M. McGrath, Markers to distinguish normal and neoplastic mammary epithelial cells *in vitro:* Comparison of saturation density, morphology, and Concanavalin A reactivity, *Int. J. Cancer* **18,** 498–509 (1976).
145. P. Furmanski, P. G. Phillips, and M. Lubin, Cell surface interactions with concanavalin A: Determination by microhemadsorption, *Proc. Soc. Exp. Biol. (N.Y.)* **140,** 216–219 (1972).
146. B. Asch, Surface properties of mouse mammary cells, *Proc. Am. Assoc. Cancer Res.* **68,** 96 (1977).
147. R. Damadian, Tumor detection by nuclear magnetic resonance, *Science* **171,** 1151–1153 (1971).
148. C. F. Hazlewood, D. C. Chang, D. Medina, G. Cleveland, and B. L. Nichols, Distinction between the preneoplastic and neoplastic state of murine mammary glands, *Proc. Natl. Acad. Sci. U.S.A.* **69,** 1478–1480 (1972).
149. G. L. Cottam, A. Vasek, and D. Lusted, Water proton relaxation rates in various tissues, *Res. Commun. Chem. Pathol. Pharmacol.* **4,** 495–502 (1972).
150. R. A. Floyd, T. Yoshida, and J. S. Leigh, Changes in tissue water proton relaxation rates during early phases of chemical carcinogenesis, *Proc. Natl. Acad. Sci. U.S.A.* **72,** 56–58 (1975).
151. C. F. Hazlewood, G. Cleveland, and D. Medina, Relationship between hydration and proton nuclear magnetic resonance relaxation times in tissues of tumor-bearing and non-tumor bearing mice: Implications for cancer detection, *J. Natl. Cancer Inst.* **52,** 1849–1853 (1974).
152. I. Weisminn, L. Bennett, L. Maxwell, Mark W. Woods, and D. Burk, Recognition of cancer *in vivo* by nuclear magnetic resonance, *Science* **178,** 1288–1290 (1972).
153. H. E. Frey, R. R. Knispel, J. Kruuv, A. R. Sharp, R. T. Thompson, and M. M. Pintar, Proton spin-lattice relaxation studies of non-malignant tissues of tumorous mice, *J. Natl. Cancer Inst.* **49,** 903–906 (1972).
154. W. R. Inch, J. A. McCredie, R. R. Knispel, R. T. Thompson, and M. M. Pintar, Water content and proton spin-relaxation time for neoplastic and non-neoplastic tissues from mice and humans, *J. Natl. Cancer Inst.* **52,** 353–356 (1974).
155. R. A. Floyd, J. S. Leigh, B. Chance, and M. Miko, Time course of tissue water proton spin-lattice relaxation in mice developing ascites tumor, *Cancer Res.* **34,** 89–91 (1974).
156. J. S. Economou, L. C. Parks, L. A. Saryan, D. P. Hollis, J. L. Czeisler, and J. Eggleston, Detection of malignancy by nuclear magnetic resonance, *Surg. Forum* **24,** 127–129 (1973).

157. P. T. Beall, D. Medina, D. C. Chang, P. K. Seitz, and C. F. Hazlewood, A systemic effect of benign and malignant mammary cancer on the spin-lattice relaxation time, T_1, on water protons in mouse serum, *J. Natl. Cancer Inst.* **59,** 1431–1433 (1977).
158. D. Medina, C. F. Hazlewood, G. C. Cleveland, D. C. Chang, H. J. Spjut, and R. Moyers, Nuclear magnetic resonance studies on human breast dysplasias and neoplasms, *J. Natl. Cancer Inst.* **54,** 813–818 (1975).
159. R. Damadian, L. Minkoff, M. Goldsmith, M. Stanford, and J. Koutcher, Field focusing nuclear magnetic resonance (FONAR): Visualization of tumor in a live animal, *Science* **194,** 1430–1432 (1976).
160. S. Abraham and J. C. Bartley, Comparisons among metabolic characteristics of normal, preneoplastic, and neoplastic mammary tumors, in: *Hormones and Cancer* (K. McKerns, ed.), pp. 29–74, Academic Press, New York (1974).
161. L. Kopelovitch, S. Abraham, H. McGrath, K. B. DeOme, and I. L. Chaikoff, Metabolic characteristics of a naturally occurring preneoplastic tissue. I. Glycolytic enzyme activities of hyperplastic alveolar nodule outgrowths and adenocarcinomas of the mouse mammary gland, *Cancer Res.* **26,** 1534–1546 (1966).
162. L. Kopelovitch, S. Abraham, H. McGrath, K. B. DeOme, and I. L. Chaikoff, Metabolic characteristics of a naturally occurring preneoplastic tissue. II. Soluble Krebs cycle enzyme activities of hyperplastic alveolar nodule outgrowths and adenocarcinomas of the mouse mammary gland, *Cancer Res.* **35,** 800–805 (1967).
163. R. Hilf, R. Ickowicz, J. C. Bartley, and S. Abraham, Multiple molecular forms of glucose-6-phosphate dehydrogenase in normal, preneoplastic, and neoplastic mammary tissues of mice, *Cancer Res.* **35,** 2109–2116 (1975).
164. J. C. Bartley, H. McGrath, and S. Abraham, Glucose and acetate utilization by hyperplastic alveolar nodule outgrowths and adenocarcinomas of mouse mammary gland, *Cancer Res.* **31,** 527–537 (1971).
165. A. Nahas, S. Abraham, and T. C. Hall, Tumorigenesis in mice and changes in thymidylate biosynthetic enzymes, *Proc. Am. Assoc. Cancer Res.* **63,** 451 (1972).
166. R. Gantt, G. H. Smith, and B. T. Julian, Virion-associated and cellular RNA methylase activity in normal and neoplastic mammary tissue from mammary tumor virus-infected and uninfected mice, *Cancer Res.* **35,** 1847–1853 (1975).
167. W. J. Bodell, Distribution of DNA repair in chromatin, *Proc. Am. Assoc. Cancer Res.* **68,** 470 (1977).
168. M. R. Banerjee, R. C. Mehta, and J. E. Wagner, DNA polymerase activity and DNA synthesis in preneoplastic nodule outgrowths of BALB/c and C3H mouse mammary gland, *J. Natl. Cancer Inst.* **50,** 339–345 (1973).
169. D. N. Banerjee, M. R. Banerjee, and R. G. Mehta, Hormonal regulation of rapidly labeled RNA in normal, preneoplastic and neoplastic tissues of mouse mammary gland, *J. Natl. Cancer Inst.* **51,** 843–849 (1973).
170. P. Hohmann, R. D. Cole, and H. A. Bern, A comparison of lysine-rich histones in various normal and neoplastic mouse tissues, *J. Natl. Cancer Inst.* **47,** 337–341 (1971).
171. P. Hohmann, H. A. Bern, and R. D. Cole, Responsiveness of preneoplastic and neoplastic mouse mammary tissues to hormones: Casein and histone synthesis, *J. Natl. Cancer Inst.* **49,** 355–360 (1972).
172. M. E. McClure and D. Medina, Similarity of chromatin nonhistone protein contents during breast neoplasia in the mouse, *Proc. Am. Assoc. Cancer Res.* **66,** 814 (1975).
173. M. E. McClure and D. Medina, Chromatin nonhistone protein variations during neoplastic progression in mouse mammary tissues, *Proc. Am. Assoc. Cancer Res.* **67,** 562 (1976).

4

Human Breast Cancer in Tissue Culture: The Effects of Hormones

C. KENT OSBORNE AND MARC E. LIPPMAN

1. Introduction

The hormone-dependent nature of some human breast cancers has been appreciated by physicians for nearly a century.[1] Clinical responses in breast cancer patients to ablative and additive hormone therapies suggest that several hormones are important growth regulators of mammary cancer. Recent studies of the basic mechanisms by which hormones influence target tissues have led to important advances in our understanding of steroid hormone action and the clinical care of women with breast cancer.[2,3] It is now recognized that the first step in steroid hormone action is the binding of the hormone to a cytoplasmic receptor protein.[4] In the absence of this receptor, the steroid hormone is unable to elicit a response in the cell. Using this principle, investigators have now identified receptors for estrogen and other steroid hormones in some breast tumor samples, providing a basis for more rational therapeutic decisions.

The effect of peptide hormones on human breast cancer is less clear. Studies of normal rodent mammary glands and spontaneous or

---

C. KENT OSBORNE AND MARC E. LIPPMAN • Medicine Branch, Division of Cancer Treatment, National Cancer Institute, National Institutes of Health, Bethesda, Maryland 20014. Dr. Osborne's present address is Department of Medicine, University of Texas Health Center at San Antonio, San Antonio, Texas 78284.

carcinogen-induced mammary tumors, however, suggest that insulin, prolactin, and probably other peptides are important growth factors.[5-7]

Nonetheless, clinical and animal studies have incompletely defined the biochemical mechanisms whereby hormones influence the growth and metabolism of human breast cancer. This deficiency may be partially explained by the difficulty in obtaining human tissue for the study of normal mammary gland physiology and by the absence of an adequate *in vitro* system to study hormone action in malignant tissue. *In vivo* studies are difficult to interpret, since secondary effects of the hormone on the activities or concentrations of other factors, or hormonal effects on the adjacent supporting stroma or immune system rather than the malignant epithelial component, cannot be excluded.

In this chapter, we will review alternative approaches to the study of hormone action in human breast cancer. Short-term explant tissue culture of breast tumor specimens will be considered briefly, followed by a more detailed discussion of hormone action using long-term continuous tissue culture cell lines derived from human breast cancer. These tissue culture systems enable the investigator to study mechanisms of hormone action in a defined medium and controlled environment devoid of the potential effects of other trophic factors.

Before we turn to an analysis of these cell systems, a few general comments and caveats need to be considered. First, before one concludes that a given hormonal stimulus evokes a response *in vitro,* one must substantiate that the *in vitro* system is composed of cells that appear by as many criteria as possible to resemble those found in the *in vivo* tissue. These criteria could include, for example, morphological criteria, chromosomal analysis, the elaboration of chemical products characteristic of mammary epithelium, and the ability to form tumors in nude mice. Second, one must always bear in mind that the failure to observe a response to a trophic stimulus may not reflect an absolutely refractory target cell, but may be due to the absence of additional factors the presence of which is "permissive" for a given effect, or conversely may be due to the presence of stimulatory concentrations of the hormone in the incubation medium (usually as a serum component) masking effects of exogenous hormone. Supporting stromal elements or specific cell–cell orientation or density may be required for response as well. Third, although the cells *in vivo* may respond to a given agent, sufficient selection may have occurred over prolonged hormonoprival cell-culture conditions that an unresponsive population of cells has appeared. As we examine studies in several systems, it will be helpful to keep the problems adumbrated above in mind. Finally, no matter how impressive the *in vitro* experimental results are, the ultimate general validity of such systems rests on verification of hypotheses in intact animal systems.

2. Human Breast Cancer in Organ Culture

Since the pioneering work of Fell nearly 50 years ago,[7] the technique of organ or explant tissue culture has proved to be a valuable biological tool. Most of our current understanding of mammary gland physiology has been achieved with studies of rodent glands in organ culture.[5] In 1937, a human breast carcinoma explant was maintained in short-term organ culture for the first time.[8] Since then, numerous attempts have been made to improve this technique for the culture of human breast tumors, with the hope of developing an *in vitro* system to predict hormonal or drug responsiveness.[9–12] Human breast tumors continue to be extremely difficult to cultivate in organ culture, however, probably because the scirrhous character of the tissue results in overgrowth by fibroblasts, deprives the tumor cells of essential nutrients or growth factors, or impedes epithelial cell division by some other process.[11,13] This feature does not predominate in benign human breast tumors or in animal breast carcinomas, which grow well in culture.[13] Paradoxically, a potential advantage of the organ-culture method is that the tumor explant with its surrounding stroma more closely approximates *in vivo* conditions than do monolayer cells in continuous culture. Unfortunately, this scirrhous component severely limits the viability and survival of the explant, and has limited this potentially useful method.

2.1. Effects of Hormones on Nonmalignant Human Breast Tissue in Organ Culture

Despite the difficulties in maintaining human breast tissue in organ culture, several studies have examined the effects of steroid and peptide hormones on explants from both normal and neoplastic specimens. In cultures of normal mammary gland, the effects of steroid hormones remain controversial. By morphological criteria, Ceriani *et al.*[14] concluded that physiologic concentrations of estradiol or progesterone produced full maintenance and lobuloalveolar development of the explant. Higher concentrations of estradiol were "toxic." Unfortunately, these descriptive morphological data were not complemented by supporting biochemical studies. Other investigators have found either no effect or an inhibitory effect of estradiol and testosterone on macromolecular synthesis or growth.[15,16] However, suprapharmacologic concentrations of steroids (10^{-6}–10^{-4} M) were employed, making interpretation of the data difficult.

Observations of the effects of insulin and prolactin on organ cultures of nonmalignant human breast tissue have also been variable.

Insulin has been claimed to induce "full maintenance" of explants,[14] to produce hyperplasia of the duct epithelium,[17,18] and to increase the thymidine labeling and mitotic index.[19–21] Other studies failed to show any effect of insulin on cell survival or proliferation.[15,22] Pharmacologic concentrations of insulin were invariably employed in all these studies. Similarly, pharmacologic concentrations of prolactin with or without insulin were required to observe any effect on maintenance or growth of explants.[14,19,21,22] Proof of an effect of these hormones on nonmalignant human breast tissue awaits the demonstration that physiologic concentrations of insulin and prolactin influence the epithelial component of the tissue by both morphological and biochemical criteria.

2.2. Effects of Hormones on Malignant Human Breast Tissue in Organ Culture

Studies of the effects of hormones on malignant human breast explants in organ culture are more numerous, but still inconclusive. By morphological criteria or crude survival of the tissue, pharmacologic concentrations of estradiol, testosterone, and prolactin enhanced the viability of a small number of tumor explants.[23–25] Other studies demonstrated no effects of these hormones, glucocorticoids, or insulin.[15,26]

To improve on the crude survival method of determining hormone dependence, other investigators have employed a biochemical approach. When oxygen consumption,[27] glucose utilization and lactate production,[28] a variety of enzyme determinations,[23,28–30] or incorporation of labeled precursors into macromolecules[16,31–34] were used to determine hormone responsiveness, the majority of tumors were unaffected by hormones. Although a few tumor explants were stimulated by estradiol, the most consistent finding was inhibition by all steroid hormones at concentrations greater than 1 μM. Welsch *et al.*[31] showed that thymidine incorporation in tumor explants was consistently enhanced by insulin, whereas a minority were stimulated by prolactin. Both hormones were used at concentrations more than 1000 times higher than the physiologic range. More important, perhaps, there has been no study convincingly correlating hormonal effects on breast cancers in organ culture with the clinical response to hormone therapy.

An improved organ-culture technique in which breast cancers could be maintained for up to 2 weeks was recently described by Heuson *et al.*[11] and Pasteels *et al.*[13] Increased survival of scirrhous tumors was achieved by treatment with collagenase. Although no significant effect of insulin, prolactin, or hydrocortisone on survival was observed, physiologic concentrations of estradiol improved survival especially in the scirrhous group, and was accompanied by collagen digestion in the explant. The data suggested the presence of an estrogen-dependent collagenolytic activity in these breast-cancer specimens. Whether or not the phenomenon is an important *in vivo* component of hormone dependency in human breast cancer is open to speculation.

In conclusion, organ culture of human breast cancer is still in its infancy and has not yet contributed significantly to our understanding of the effects of hormones in this disease. With improved techniques, however, this method of studying hormone action remains a potentially valuable adjunct to steroid-receptor assays in defining hormone-dependent tumors.

3. Human Breast Cancer Cells in Long-Term Tissue Culture

Proponents of the organ-culture method might argue that human breast cancer cells maintained in long-term tissue culture do not resemble the tumors from which they were derived. These cells, of course, are devoid of the supporting stromal elements present *in vivo* or in organ culture, a feature that may be an important prerequisite for hormone responsiveness. In addition, adaptation to and maintenance in long-term culture might select mutant strains or cell types present in only small numbers in the original tumor.

Nevertheless, this technique offers distinct advantages over *in vivo* or organ-culture model systems for the study of hormone action in human breast cancer. First, the continuous tissue culture of a fully viable cell line allows for repeat or sequential experiments on the same tissue, which are not possible with short-term explants. Second, the use of a cloned cell line ensures that the observed effects represent hormone interaction with the malignant epithelial component of the tumor. This assurance is particularly important, since normal mammary epithelial tissue or adipose and fibrous tissue may be targets for the action of many hormones and are potent metabolizers of steroid hormones.[35] Third, the environment of cells in monolayer or suspension may be controlled and manipulated easily. In addition, nearly all the cells are exposed to the defined conditions or medium, whereas in organ culture, a central core of tissue may be deprived of essential factors. Fourth, cells in continuous culture provide the opportunity to develop mutant or variant hormone-independent lines, making possible genetic studies oriented toward dissection of critical steps in hormone action. Finally, one can add back individual stromal components in cocultivation experiments or add their diffusible products to reconstitute *in vitro* the situation extant in the host.

In summary, human breast cancer cells grown in continuous long-term tissue culture facilitate the study of the mechanisms of hormone action at the molecular and biochemical levels. This review will now focus on recent investigations in this direction. We hope to demonstrate the usefulness of this technique and its potential for advancing our understanding of hormone-dependent breast cancer.

3.1. Characteristics of Breast Cancer Cells in Continuous Culture

Continuous cultivation of a human breast cancer cell line was first described in 1958 by Lasfargues and Ozzello.[36] This cell line, BT-20, was derived from a primary breast tumor specimen. Despite this success, long-term tissue culture of human breast cancer remains a difficult task, and only a short list can be compiled from literature reports of lines sufficiently well characterized to be confident of their mammary nature (Table I).* It is evident that primary or solid tumor metastases provide the source of a minority of long-term cultures. Indeed, Dr. R. Cailleau of the M. D. Anderson Hospital and Tumor Institute failed on more than 200 attempts to establish cell lines from solid breast tumor specimens, but had considerably more success with cells isolated from pleural effusions.[42]

The cell lines shown in Table I also emphasize the importance of documenting the human and mammary characteristics of an established culture and of excluding contamination by non-breast-tumor cells. Several features support the human and mammary origin as shown for four lines studied in depth by our laboratory (Table II). First, all these lines were derived from malignant effusions from women with metastatic breast cancer and have been in continuous tissue culture for at least 2 years. Second, routine chromosomal analyses reveal a human karyotype with a modal number near triploid or in the hypotetraploid range, and the presence of easily identifiable marker chromosomes resembling the original tumor. Chromosome banding studies demonstrate that these cells are not HeLa contaminants, which is amplified by their Type B isoenzyme mobility for glucose-6-phosphate dehydrogenase. As Nelson-Rees and co-workers[52,53] recently emphasized, HeLa contamination is common in tissue culture laboratories, and several "breast cancer" cell lines have now been shown to be HeLa cells (see Table I). Third, epithelial morphology by light and electron microscopy supports the mammary nature of these cells. The cells possess microvilli, desmosomes, and usually rough endoplasmic reticulum and golgi, and some cell lines tend to arrange themselves in ductlike structures.[51,54] This type of morphology is particularly evident in the MCF-7 and ZR75-1 cell

*Information on these cell lines was kindly supplied by Dr. E. M. Jensen and can be found in the Cell Culture Bank Inventory of the EG+G/Mason Research Institute, 1530 East Jefferson Street, Rockville, Maryland 20852.

Table I
Some Human Breast Carcinoma Cell Lines in Continuous Tissue Culture

Cell line	References[a]	ER[b]
From primary		
BT-20	36–38	
BOT-2	39	
HBT-3[c]	37, 40	
SW-613	*	
HS 0578T	*	−
From metastasis[d]		
SH-3[c]	41	
MDA-MB-134	42	+
MDA-MB-157	37, 43	
MDA-MB-175	42	
MDA-MB-231	42	
MDA-MB-330	*	
MDA-MB-361[e]	*	
MDA-MB-415	*	
CaMa	44	
HBT-39[c]	45	−
AlAb[f]	37, 46	
T-47D	*	+
G-11[c]	47, 48	−
SK-BR-3	48	
MCF-7	49	+
EVSA-T	50	−
ZR75-1	51[g]	+
HBL-100[h]	*	
BT-474	*	
BT-483	*	

[a] An asterisk (*) means that information was obtained from the Cell Culture Bank Inventory, EG+G/Mason Research Institute.
[b] Presence (+) or absence (−) of estrogen receptor if known.
[c] Indicates HeLa cell culture on basis of chromosomal banding and glucose-6-phosphate dehydrogenase typing.(52,53)
[d] Malignant effusion unless otherwise noted.
[e] Isolated from solid brain metastasis.
[f] Isolated from solid lung metastasis.
[g] This line was established by Dr. N. Young and Ms. L. Engel of the National Cancer Institute.
[h] Isolated from nonmalignant breast secretions.

lines, whereas the MDA-MB-231 line tends to have longer spindle-shaped cells without striking acinarlike formation. When the MCF-7 cells are grown on artificial capillaries(51,55) or in sponge culture,(54) they bear a striking resemblance to the original tumor from which they were derived. Examples of the MCF-7 cells grown *in vitro* are shown in Figs. 1 and 2. These were kindly supplied by Jose Russo, M.D., of the Michigan

Table II
Characteristics of Human Breast Cancer Cell Lines

	Source			Chromosomes			Milk proteins[b]		Hormone receptors[c]						Hormone activity[c]					
Cell line	Age	MS[d]	Race	N[e]	Banding[f]	G-6-PD[a] type	Cas.	α-Lac.	E	P	G	A	I	Pr	E	P	G	A	I	Pr
MDA-MB-231	51	Po	W	60–70	NH	B	−	+	−	−	+	−	+	−	−	−	+	−	−	−
MCF-7	69	Po	W	85	NH	B	−	+	+	+	+	+	+	−	+	−	+	+	+	−
EVSA-T	55	Po	W	76	NH	B	−	+	−	−	−	−	+	−	−	−	−	−	−	−
ZR75-1	63	Po	W	71–72	NH	B	−	+	+	+	+	+	+	−	+	−	+	+	+	−

[a] (G-6-PD) Glucose-6-phosphate dehydrogenase.
[b] Milk proteins: (Cas.) casein; (α-Lac.) α-lactalbumin.
[c] (E) Estrogen; (P) progesterone; (G) glucocorticoid; (A) androgen; (I) insulin; (Pr) prolactin.
[d] (MS) Menopausal status.
[e] Chromosome modal number or range.
[f] (NH) Non-HeLa cell by banding studies.

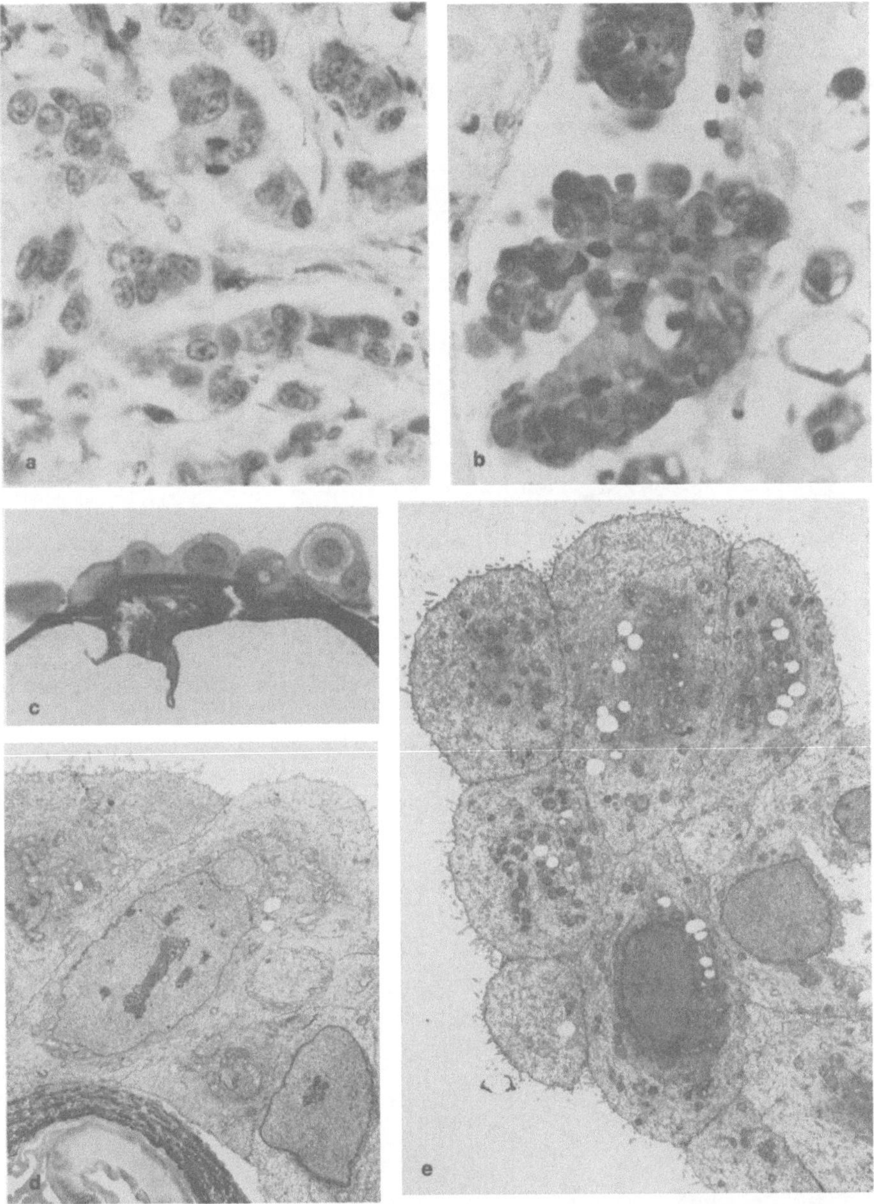

Fig. 1. (A) Antecedent primary breast tumor of MCF-7. Connective tissue surrounds clusters of epithelial cells. ×850. (B) Histological section of pleural effusion clot from which the MCF-7 cell line was derived. Seen are clusters of epithelial cells forming a lumenlike structure by necrosis of the central portion. Pyknotic cells are also observed. A fibrin clot surrounds the neoplastic component. ×850. (C) MCF-7 cells growing in monolayer on the collagen-coated cellulose sponge. ×600. (D) Transmission electron micrograph of a perpendicular section through a monolayer of cells growing on the collagen coat. ×4000. (E) Vertical section through a cluster of cells. ×2000. Supplied by J. Russo, Michigan Cancer Foundation.

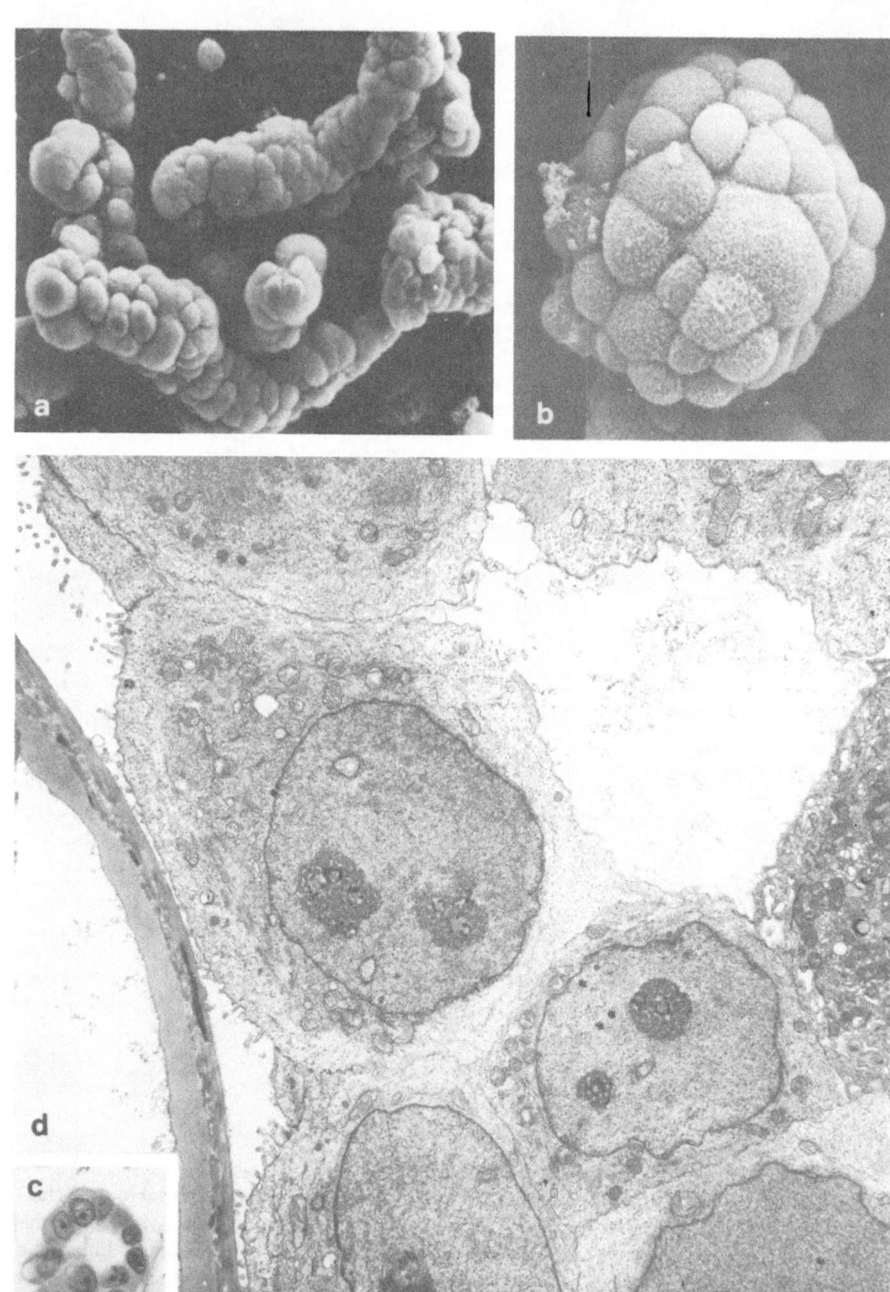
a
b
d
c

Cancer Foundation. Fourth, the mammary origin of a cell line may be suggested by the detection of milk proteins such as casein or α-lactalbumin. All four cell lines in Table II synthesize small amounts of α-lactalbumin by both enzymatic and radioimmunoassay.[47,50,56] Caution is necessary in interpreting these results, however, since the accumulation of this protein is apparently not hormonally controlled in these lines; none of the cell lines accumulates casein when measured by a sensitive radioimmunoassay,[57] and small amounts of α-lactalbumin have been found in several non-breast-cell lines including the HBT-39 and G-11 now known to be HeLa contaminants.[50] If malignancy is accompanied by derepression of genes, then the synthesis of many proteins by tumor cells is not surprising. This lack of cell specificity is supported by our demonstration that serum casein levels were elevated in about 10% of samples from breast cancer patients and 50% of samples from patients with colon cancer.[57] Thus, the presence of proteins thought to be specific for the differentiated mammary gland does not ensure the mammary nature of a cell line. Other markers such as carcinoembryonic antigen (CEA) that are occasionally secreted by the tumor cells *in vivo* may be secreted by the cells in culture, providing additional assurance of the cell of origin. The EVSA-T cell line, which was derived from malignant ascitic fluid from a patient with breast cancer and an elevated serum CEA, released this marker into the culture medium *in vitro*.[50]

Finally, the presence of receptors and biological responses to several hormones known to influence the growth and development of the breast strongly supports the mammary nature of an established cell line. A panoply of steroid and peptide hormones, and even iodothyronines,[58] have been shown to influence some breast cancer cell lines in culture. The four cell lines in Table II display a spectrum of hormone responsiveness. The EVSA-T line does not respond significantly to the hormones tested, whereas the MCF-7 cell line contains receptors for and responds to several hormones at physiologic concentrations. Furthermore, the mammary and epithelial nature of a cell line is more certain if it contains receptors for estrogen and progesterone, since adipose and fibrous tis-

Fig. 2. (A) Scanning electron microscopy of cells recovered from the supernatant medium by centrifugation onto a Millipore filter. Observe the long arrays of densely packed small round cells. Note the sharply defined cell limits. The surface is covered by short, thin microvilli. ×2000. (B) Cluster formed of small rounded cells that are covered by densely packed microvilli. ×2000. (C) MCF-7 cells forming a pseudoacinar structure by development of an eccentrically located lumen. This structure is attached to the collagen. ×600. (D) Cross section of a pseudoacinar structure in which the lumen contains cell detritus and a necrotic cell (far right). Microvilli and junctional complexes are located at the outer surface. Supplied by J. Russo, Michigan Cancer Foundation.

sue and strains of HeLa cells are devoid of these proteins. HeLa cell contaminants such as the G-11 do possess glucocorticoid receptors.[47] Hormone-receptor assays are becoming widely available and should provide additional markers for comparisons between established breast cancer cell lines and the original tumor from which they were derived.

In summary, although breast cancer cell lines in long-term culture are far from their *in vivo* environment, many maintain a striking resemblance to the original tumor by morphological, chromosomal, and biochemical criteria. With this in mind, we will now review in more detail the effects of hormones on these cell lines.

3.2. Effects of Estrogens and Antiestrogens

For nearly a century, the ovarian-dependent nature of breast cancer has been appreciated. As noted by Stoll,[1] Beatson first showed that removal of ovaries could lead to regression of breast cancer, yet the obscurity of the mechanism of this dependence has only recently begun to yield to scientific investigation. Many questions require a great deal more study, including why only some women respond to castration, why some tumors can be made to regress when pharmacologic concentrations of estrogens are administered, and what interactions androgens, progestins, and antiestrogens have with estrogen-responsive tumors.

Early studies in many systems suggested that binding of hormone to specific receptor proteins was a critical first step in steroid hormone action. The ability to measure estrogen receptor in the cytosols of human breast carcinoma specimens has represented a major advance in the treatment of this disease.[3] To further investigate and perhaps extend this observation, attempts were made to devise an *in vitro* model system to facilitate the study of hormone-receptor binding. Brooks *et al.*[59] first demonstrated estrogen receptor in an established breast tumor cell line, MCF-7. The binding protein had an equilibrium dissociation constant (K_d) of 2.5 nM, and the number of cytoplasmic binding sites approximated 60 femtomoles (fmol) [^{3}H]-17β-estradiol bound/mg cytosol protein. Identification of estrogen-receptor protein in the MCF-7 cell line was confirmed later by Horwitz *et al.*[60] and Lippman *et al.*[50] Receptor assays performed by sucrose density gradients,[2] protamine sulfate precipitation,[61] or dextran-coated charcoal[62] revealed receptor concentrations similar to that described by Brooks (60–100 fmol bound/mg protein), but with a lower K_d (0.06–0.7 nM). Sucrose density gradients run in low-ionic-strength buffer at 0°C showed an 8 S peak of [^{3}H]estradiol binding (Fig. 3) similar to that reported by Horwitz *et al.*[60] This binding peak was shifted to about 4 S when the gradients were performed in 0.4

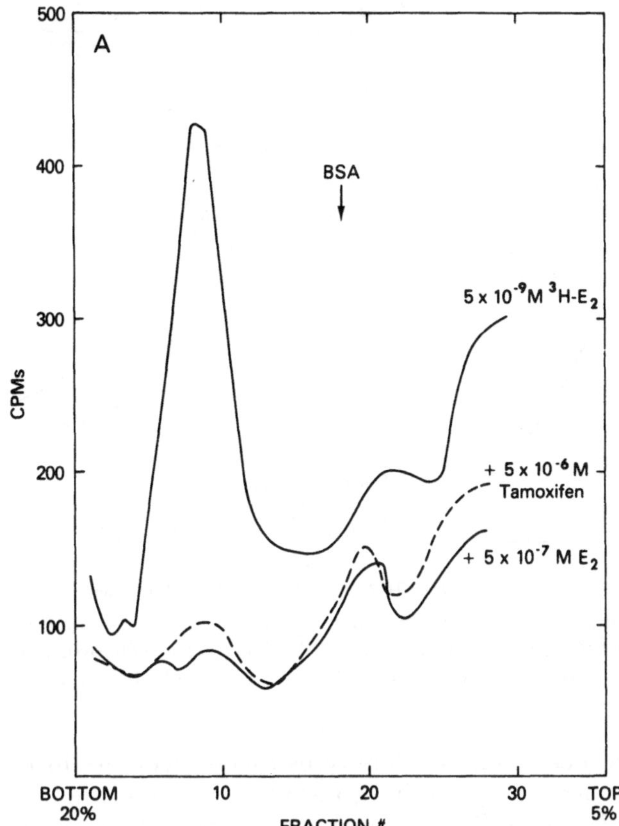

Fig. 3. Sucrose density gradients of cytoplasmic estrogen receptor from MCF-7 human breast cancer. The gradients shown in this figure were done under low-salt conditions. (BSA) A ^{14}C-labeled bovine serum albumin marker. Complete methodological details are supplied in Lippman *et al.*[50]

M KCl,[50] which may indicate receptor-subunit dissociation at high ionic strength. In any event, both binding peaks of [^3H]estradiol were completely inhibited by a 100-fold excess of unlabeled estradiol or by a 1000-fold excess of the antiestrogen tamoxifen (ICI 46474).

In addition to saturable, high-affinity binding, the estrogen receptors in the MCF-7 cells demonstrate remarkable steroid specificity (Fig. 4). Estrogens and antiestrogens [nafoxidine (U11,100A), tamoxifen, and Parke Davis CI628] can bind to the receptor and completely displace the labeled hormone, whereas progestins, androgens, and antiandrogens (R2956) have no effect at the concentrations shown. Similar results demonstrating limited binding specificity for estrogen and estrogen analogues were found by others.[59,60]

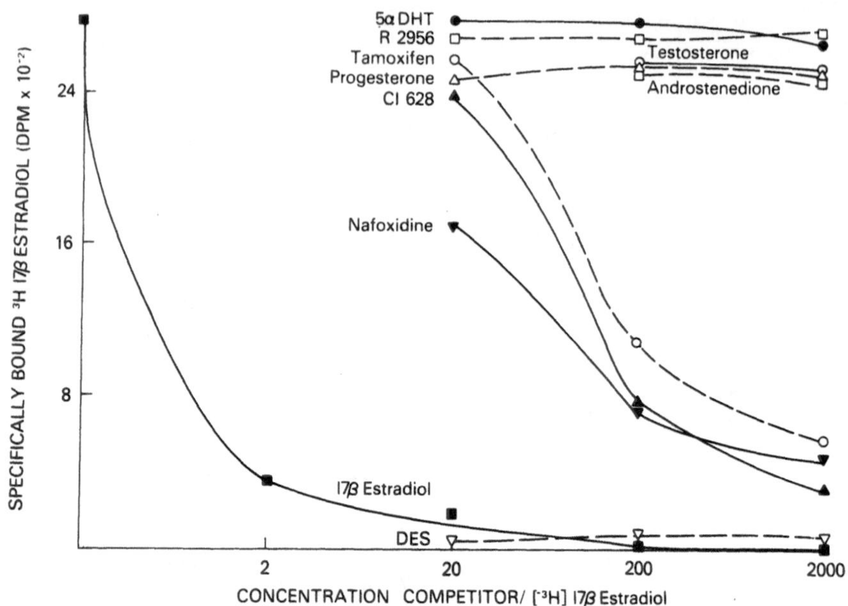

Fig. 4. Competition of various unlabeled compounds with [^{3}H]estradiol for cytoplasmic receptor sites in MCF-7 human breast cancer cells. (5α DHT) 5α-Dihydrotestosterone; (DES) diethylstibestrol. Complete methodological details are supplied in Lippman *et al.*[84]

We have been able to identify estrogen receptors in only two of about ten breast cancer cell lines tested, and interestingly, both these lines (MCF-7 and ZR75-1) also have progesterone, androgen, and glucocorticoid receptors (see Table II). In addition, we are aware of only two other cell lines that have been reported to contain estrogen receptor (see Table I). We do not know whether this paucity of estrogen-receptor-positive cell lines reflects: (1) differences in the intrinsic abilities of receptor-positive and -negative tumors to adapt to tissue culture; (2) selection of receptor-negative cells in tissue culture from a heterogeneous population in the original tumor; (3) that specimens for culture are most frequently obtained from effusions in patients with previously treated widely metastatic disease, which may select for fewer receptor positive tumors; or (4) that endogenous estrogen in the tissue culture medium may bind the available cytoplasmic receptors, or transport them to the nucleus, where they will not be detected by routine assays on cytoplasmic extracts, or both. The wider availability and application of estrogen-receptor assays should permit serial determinations of receptors on specimens derived from an original tumor, thus answering some of these queries.

Since the presence of estrogen and progesterone receptors predicts estrogen dependence in a high percentage of human tumors *in vivo*,[63] one might expect that cell lines possessing both receptors would be biologically responsive to estrogen. Indeed, this is the case for the MCF-7 and ZR75-1 cell lines. When the MCF-7 cells are grown in serum-free or charcoal-treated serum-containing medium with a physiologic concentration of estradiol, a significant growth effect is evident (Fig. 5). Estradiol treatment results in a 2-fold increase above controls in the number of cells per dish by the 8th day of culture. Similar results are obtained if one estimates growth by measuring the total protein per dish. On the other hand, significant inhibition of cell growth is observed with the addition of tamoxifen, a potent antiestrogen. When tamoxifen is added to cells maintained in totally serum-free conditions or when added at a concentration of 1.0 μM, a majority of the cells begin to detach from the dish and die after 48–72 hr. This appears to be a specific antiestrogen effect, because it does not occur in cells lacking

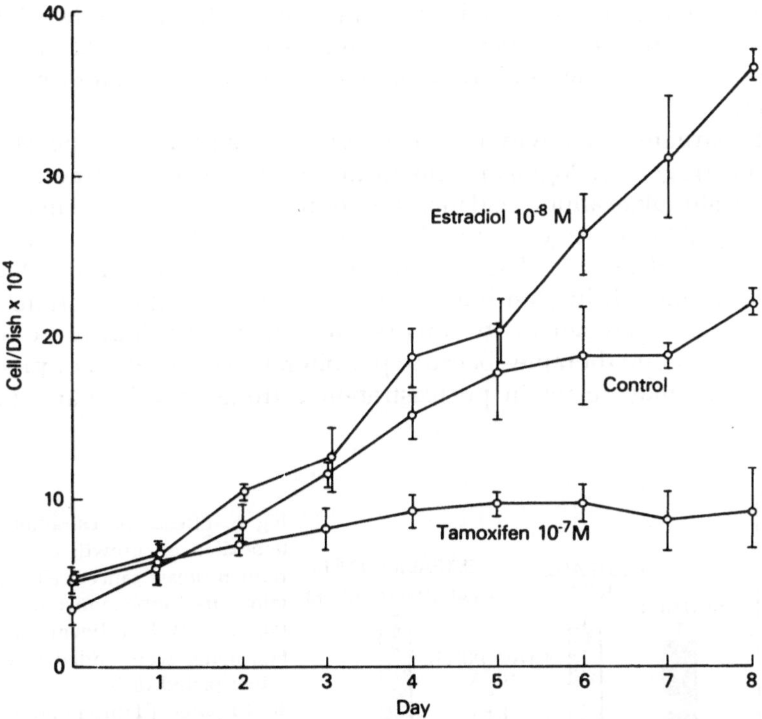

Fig. 5. Effects of estradiol and tamoxifen on growth of MCF-7 human breast cancer cells. Hormones were added at 0 time. Triplicate dishes of uniformly plated cells were harvested at various time intervals and counted. Results are shown ± 1 S.D. Complete methodological details are supplied in Lippman *et al.*[50]

estrogen receptor, is observed with several antiestrogens,[64] and can be reversed by the addition of 10-fold less estradiol as late as 48 hr after the addition of tamoxifen.[64]

Previous unsuccessful efforts to demonstrate an estrogen effect in human breast cancer cells *in vitro* may have resulted from endogenous estrogen inadvertently present when serum was used in the culture medium. This is demonstrated graphically in Fig. 6. Estradiol or tamoxifen or both were added to MCF-7 cells maintained in medium containing 10% fetal calf serum. Estradiol has no stimulatory effect on thymidine incorporation, whereas tamoxifen is inhibitory. This can be explained by the presence of estradiol in the medium sufficient to stimulate the cells nearly maximally, but insufficient to reverse the antiestrogen effect. On the other hand, the addition of estradiol with tamoxifen raised the total estrogen concentration sufficiently to abolish the antiestrogen effect. Estrogen levels frequently approach 1.0 nM in commercial sera preparations,[65] concentrations that when diluted 10-fold in medium still produce significant stimulation in these cells (see Fig. 7). It must be appreciated that for each 10-fold dilution of serum, there is only a 2-fold reduction in the free steroid concentration due to the equilibrium between free hormone and that associated with sex-steroid-binding globulin.

The striking sensitivity of macromolecular synthesis in the MCF-7 cells to estrogen analogues is shown in Fig. 7. Estrone, estradiol, and estriol all stimulate amino acid incorporation into acid-insoluble material. As little as 20 pM estradiol stimulates protein, RNA, and DNA synthesis in these cells.[51,66,67] The maximal fold of induction is similar for all three hormones. If human breast cancers *in vivo* also demonstrate this degree of estrogen sensitivity, then some patients may not respond to ablative hormone therapies because peripheral aromatization of precursor steroids may result in postcastration estrogen levels that remain

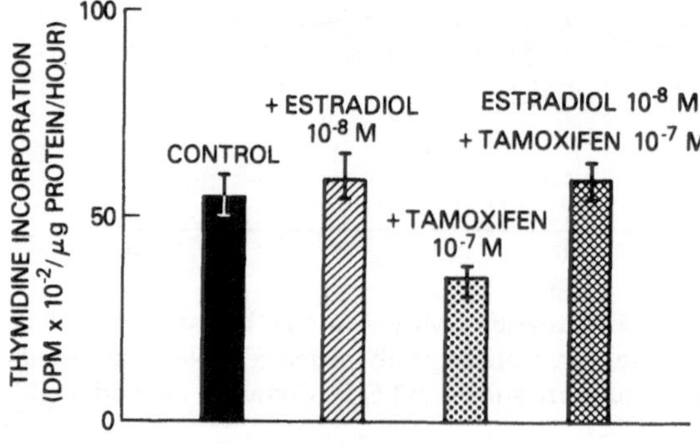

Fig. 6. Effects of estradiol and tamoxifen on growth of MCF-7 human breast cancer cells maintained in complete medium containing 10% fetal bovine serum. Hormones were added to replicately plated dishes of cells and 36 hr later [^{3}H]thymidine was added to each well. After 1 hr cells were harvested and incorporation measured as described in Lippman *et al.*[50]

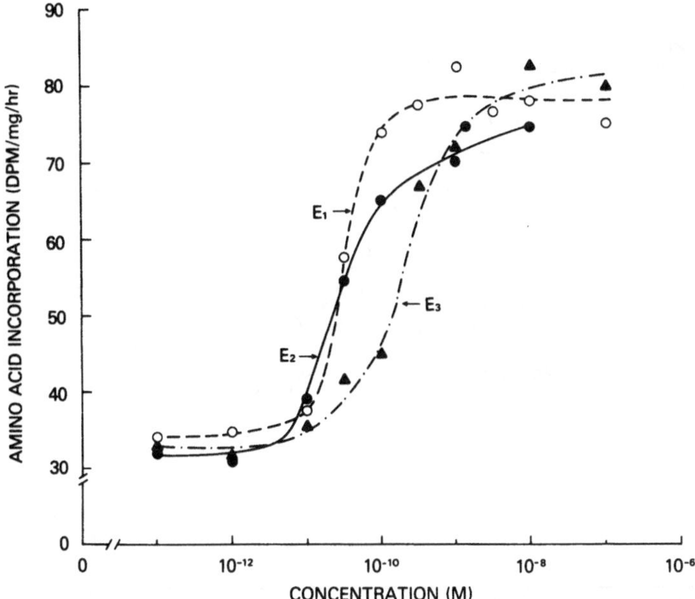

Fig. 7. Effects of various estrogens on amino acid incorporation in MCF-7 human breast cancer cells. (E_1) Estrone; (E_2) estradiol; (E_3) estriol. Complete methodological details are supplied in Lippman *et al.*[66]

sufficient to stimulate the tumor. Another interpretation for this striking estrogen sensitivity is also possible. The retention of even small amounts of estrogen by these cells throughout various preincubations prior to the addition of exogenous hormone would have the effect of both shifting the dose–response curve of the cells toward increased sensitivity and diminishing the apparent fold of induction.

The ability of these estrogens to compete for [^{3}H]estradiol-binding sites is proportional to their potency in stimulating thymidine incorporation (Fig. 8 and Table III). Estradiol is 4–5 times more potent than estrone and at least 5–10 times more potent than estriol in competing for the receptor and stimulating DNA synthesis. These data are supported by direct binding studies using tritiated estrogens. By Scatchard analysis, the affinity of the receptor for estradiol is about 10-fold higher than that for estrone and estriol.[66] Nevertheless, estrone and estriol are capable of maximally stimulating macromolecular synthesis. Since estriol is not converted in these cells to estradiol or estrone, we conclude that estriol has potent estrogenic activity in these human breast cancer cells.[66]

These data may have important clinical ramifications. Estriol has been considered an estrogen antagonist for many years.[68] In addition,

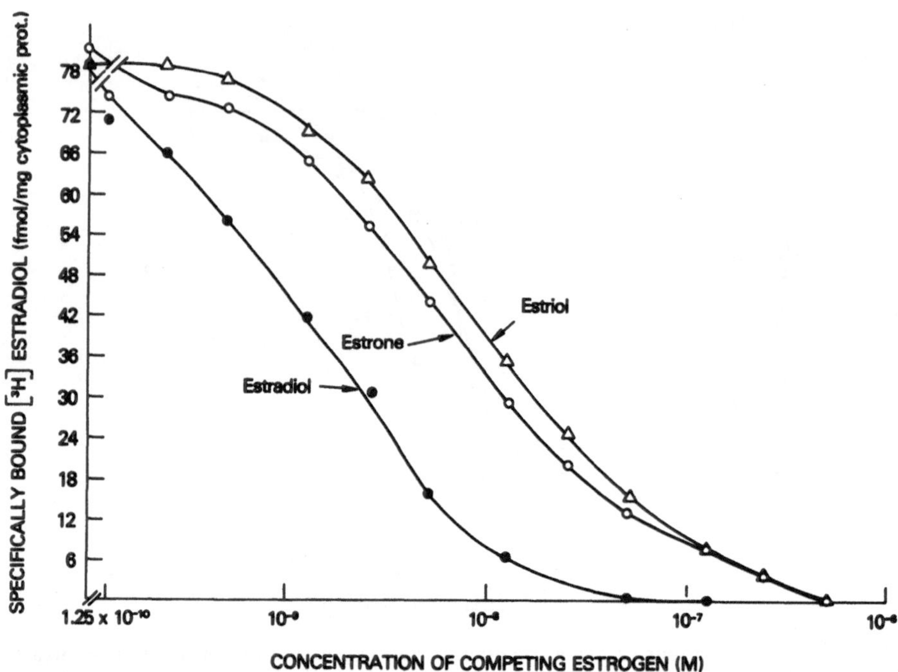

Fig. 8. Competition of various unlabeled estrogens with [^{3}H]estradiol for cytoplasmic binding sites in MCF-7 human breast cancer cells. Complete methodological details are supplied in Lippman *et al.*[66]

epidemiological studies showed that women with a high excretion of estriol relative to estrone and estradiol have a decreased incidence of breast cancer, suggesting a protective effect.[69] Some investigators have proposed that estriol should be considered for clinical trials in high-risk women in an attempt to reduce the incidence of breast cancer, since it prevented carcinogen-induced mammary carcinoma in rats.[70] Our studies indicate, however, that estriol is not an estrogen antagonist but a potent agonist in human breast cancer cells in culture. Recent data from Clark *et al.*[71] provide a basis for partially resolving these disparate results. They demonstrated that estriol acts as an estrogen antagonist when injected *in vivo* as a single bolus because of the short nuclear retention time of nuclear receptor–hormone complexes. When estriol is present continuously, however, it is a potent estrogen, since the nuclear level of receptor–hormone complexes is maintained. We therefore do not support the proposed use of estriol in clinical trials without further documentation of its "antiestrogenic" activity.

Another puzzling feature of the effect of estrogens on breast cancer in women is the observation that ablative therapies that lower estrogen

Table III
Correlation between Inhibition of [^{3}H]Estradiol Binding and Stimulation of Thymidine Incorporation by Estrogens in MCF-7 Cells[a]

Estrogen	Concentration of estrogen (M) causing one-half maximal: Inhibition of binding	Stimulation of thymidine incorporation
Estrone	7×10^{-9}	1.5×10^{-10}
Estradiol	2×10^{-9}	0.3×10^{-10}
Estriol	10×10^{-9}	6.0×10^{-10}

[a] Results are means of triplicate determinations.

levels or the administration of pharmacologic doses of estrogen will both induce tumor regression in about one third of cases.[72] The mechanism of this latter effect is not known, although like the former, it is correlated with the presence of estrogen receptor.[3] We therefore examined the effect of pharmacologic concentrations of estrogens on human breast cancer cells in culture in an attempt to define potential mechanisms (Table IV). Cells were exposed to physiologic and pharmacologic concentrations of 17β-estradiol or the much less active isomer, 17α-estradiol, which has a much lower affinity for the estrogen receptor in the MCF-7 cells. Stimulation of macromolecular synthesis is seen when physiologic concentrations of 17β-estradiol are incubated with cells containing receptor. Pharmacologic concentrations of either hormone are inhibitory and ultimately cause cell death. This inhibition appears to be

Table IV
Effects of Physiologic and Pharmacologic Estrogen on Human Cancer Cells in Culture

		17β-Estradiol[a]		17α-Estradiol[a]	
Cell line	ER[b]	1.0 nM	10 μM	1.0 nM	10 μM
MCF-7	+	↑	↓	–	↓
ZR75-1	+	↑	↓	–	↓
MDA-MB-231	–	–	↓	–	↓
EVSA-T	–	–	↓	–	↓
HeLa	–	–	↓	–	↓

[a] Stimulation (↑), inhibition (↓), or no effect (–) on macromolecular synthesis or growth.
[b] Presence (+) or absence (–) of estrogen receptor.

nonspecific, however, since: (1) it does not require the presence of estrogen receptor; (2) it is observed in non-breast-cell lines including HeLa cells; and (3) it is observed with a relatively inactive estrogen. Kiang and Kennedy[13] recently suggested in a preliminary presentation that high concentrations of estrogen can block nuclear translocation of cytoplasmic estrogen. More information will allow proper assessment of this observation. The mechanism of additive estrogen therapy remains a mystery and requires further study.

Antiestrogens induce tumor regressions in a significant number of estrogen-receptor-positive patients.[74] As illustrated above, we and others have also shown that antiestrogens inhibit macromolecular synthesis and growth in receptor-positive breast cancer cells in culture.[50,64,67,74] Antiestrogens can compete with estrogen for the estrogen receptor; they also inhibit the stimulatory effect of estrogens, and this effect can be reversed with estradiol.[64,67,75] Interestingly, in the apparent absence of estrogen, antiestrogens inhibit DNA synthesis in these cells below control levels, suggesting that they are capable not only of blocking estrogen effects, but also of inhibiting cellular processes themselves. Since antiestrogens can bind to estrogen receptor and translocate it to nuclear sites,[76–78] one might speculate that these complexes inhibit transcription of key mRNA molecules required for the synthesis of growth regulatory proteins in the basal state. Finally, if "free," biologically active nuclear estrogen receptor exists in these cells as suggested by Zava *et al.*,[75] then perhaps antiestrogens can bind to this nuclear receptor and deactivate it, thus inhibiting transcription.

Although physiologic concentrations of estradiol stimulate DNA synthesis and growth in the MCF-7 cells, this hormone is not absolutely necessary for the growth or maintenance of this cell line. In our hands, the cells grow well, although not optimally, in medium supplemented with insulin and serum treated with charcoal to remove steroid and probably many peptide hormones. Zava *et al.*[75] extended this observation and showed that the antiestrogen nafoxidine markedly reduced thymidine incorporation, whereas estradiol showed only 30–40% stimulation above control, suggesting that the growth of these cells is not dependent on estrogen. Their results may also be explained, however, by the fact that they performed their experiments in medium containing insulin, which also stimulates these cells. We have observed (Fig. 9) that insulin and estradiol effects are not additive, and when stimulation of RNA or DNA synthesis by estradiol is determined in the presence of insulin, a relatively small (30%) fold of induction above that seen with insulin alone is apparent. In the absence of insulin, a 2-fold or greater increase above controls is observed. This observation suggests that some final pathway may be stimulated by both these trophic hormones.

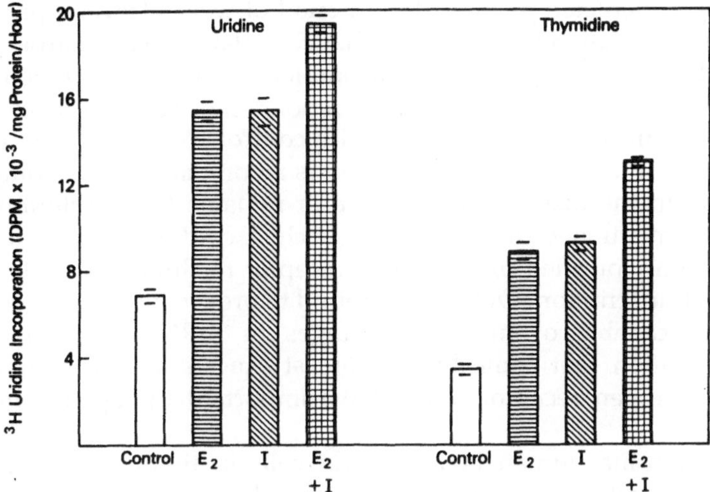

Fig. 9. Effects of estradiol and insulin on nucleoside incorporation in MCF-7 human breast cancer cells. (E_2) estradiol (10^{-8}M); (I) insulin (0.1 unit/ml). Methods are similar to those supplied in Lippman *et al.*[50]

On the other hand, we observe incomplete reversal of tamoxifen inhibition by insulin (as opposed to estradiol). We have found the fold of stimulation by estrogen in these cells variable, although usually in excess of 50%[50] and occasionally up to 6-fold. We have also observed that whereas antiestrogen inhibition is virtually always observed, this effect too may vary from 50 to 95% inhibition of cells as compared with control. A plausible explanation for this phenomenon was offered by Zava *et al.*[75] When the distribution of estrogen receptor was examined, most was located in the nucleus even when the cells were maintained in the absence of estrogen. In addition, this receptor was "free," i.e. not bound to endogenous estradiol, and the residual cytoplasmic receptor was capable of translocating estradiol to the nucleus. The authors suggested that stimulation of growth of these cells by unbound nuclear receptor might explain these results. The slight estrogen stimulation would result from the binding of estrogen to the few residual cytoplasmic receptors and their translocation to the nucleus, whereas blocking of the unbound nuclear receptor by antiestrogen would result in significant inhibition of transcription. The variability of stimulation might be due to the relative preponderance of nuclear compared with cytoplasmic receptor. Unfortunately, the effects of the other hormones in their systems (insulin, prolactin, and glucocorticoid) on cellular localization of estrogen receptor are not known. A preliminary observation in our laboratory is not easily explained by this interpretation. If the MCF-7 cells are maintained in estrogen-depleted medium for a period of weeks (several passages), estrogen stimulation is markedly enhanced relative to antiestrogen inhibition. This result suggests that incompletely removed estrogen in the earlier experiments may explain the preponderance of antiestrogenic vs. estrogenic effects when compared with control. Whether the enhanced stimulation in cells devoid of estrogen is accompanied by a compatible alteration in the distribution of receptor remains to be determined. It should be mentioned that in the case of chick oviduct at least, the addition of highly purified progesterone receptor to chromatin is devoid of effect on transcription. Only when bound to progestins are progesterone receptors capable of inducing changes in mRNA transcription.[79] Documentation that some human breast cancers *in vivo* contain free nuclear estrogen receptor might have important therapeutic implications.[75]

Thus, some human breast cancer cells in tissue culture possess high-affinity estrogen receptors and respond to estrogen with increased macromolecular synthesis and growth. One might now ask what the mechanisms of the estrogen effect are distal to receptor binding and nuclear translocation. Recently, we examined the effect of estradiol on the activity of cytoplasmic thymidine kinase (TK), a potentially rate-limiting enzyme of the salvage pathway of deoxynucleotide biosynthesis.[80] The rate of [^{3}H]thymidine incorporation, TK activity, and receptor binding are a function of the estrogen concentration in the MCF-7 cells (Fig. 10). The plots for thymidine incorporation and TK activity are similar; significant stimulation of both is evident with 50 pM 17β-estradiol, and in the experiment shown, a maximal 2-fold stimulation above control is observed with 1.0 nM estradiol. The plot of estrogen binding to receptor is shifted to the right about 10-fold, as we have previously shown.[50] This binding assay was performed on whole cells in conditions similar to those for the biological response studies. Scatchard analysis of the whole-cell binding data reveals a K_d similar to that seen with cytosol extracts at 0°C (about 0.6–0.8 nM).[47] Since total cellular receptor is measured with this assay, the shift in the binding curve cannot be explained by alteration in receptor distribution within the cell. These findings suggest that near-maximal stimulation of TK activity and thymidine incorporation is observed when only 50% of estrogen binding sites are filled. The significance of these "spare" receptors is not known.

As we have already noted, however, a small residual contamination of these cells by endogenous estrogen in serum would tend to shift the dose–response of these cells to the left, away from the binding curve. Similarly, if residual estrogen acted as a competitive inhibitor of binding of radiolabeled estradiol, it would tend to shift the binding curve to the

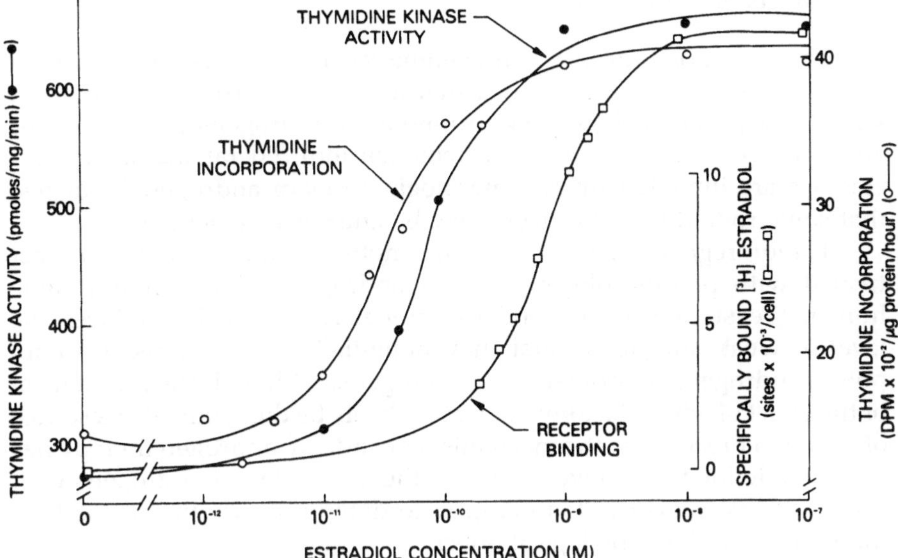

Fig. 10. Comparison of estradiol binding to whole MCF-7 human breast cancer cells with induction of thymidine kinase activity and increases in thymidine incorporation. Methods may be found in Monaco *et al.*[80]

right toward lower apparent affinity. The presence of unbound but biologically active nuclear receptor would have a similar effect. Thus, the exact mechanism of the observed disparity between binding and response is still unsettled.

As expected, antiestrogens inhibit TK activity in these cells.[80] In addition, TK activity does not respond to estrogen or antiestrogen in cells lacking estrogen receptor. Finally, time-course studies show that stimulation of TK activity by estrogen follows a 4- to 6-hr lag period and precedes estrogen stimulation of thymidine incorporation into DNA. Actinomycin D and cycloheximide, when added simultaneously with estradiol, block this induction. These data suggest that one mechanism by which estrogen stimulates DNA synthesis and growth in the MCF-7 cells may be mediated by estrogen interacting with its receptor, translocation of this complex to the nucleus, and then stimulation of transcription of mRNA molecules necessary for the synthesis or activation of TK. We cannot exclude, however, the possibility that the increased TK activity is merely the indirect result of estrogen stimulation of growth through some other mechanism. We hope that these estrogen-responsive tissue-culture cell lines will provide a system in which to study in further detail these and other aspects of estrogen action.

3.3. Effects of Androgens

The effects of androgens on mammary carcinoma are less clear than those of estrogens. The findings that a mouse breast tumor model is stimulated by physiologic concentrations of androgens and possesses androgen receptor[81] and that carcinogen-induced rat mammary carcinomas are inhibited by pharmacologic doses of androgen[82] suggest that some rodent breast cancers may be androgen-dependent.

Tumor regression is occasionally noted in human breast cancer treated with pharmacologic doses of androgens. Furthermore, some human breast cancers respond to adrenalectomy, which reduces the level of weak androgens that may potentially be converted to more potent androgens or aromatized to estrogens. Although the mechanism of the effect of adrenalectomy may more likely be the removal of a source of estrogen precursors, the mechanism of additive androgen therapy has not been defined. We have examined the interaction of androgens with human breast cancer cells in culture to determine whether these hormones exert a direct biological effect.

We have demonstrated that growth of the MCF-7 cells is enhanced, not inhibited, by pharmacologic concentrations (up to 1.0 μM) of 5α-dihydrotestosterone (DHT).[83,84] This is illustrated in Tables V and VI. 5α-DHT stimulates thymidine incorporation by more than 100% in the experiment shown, whereas the less active isomer 5β-DHT is ineffective. Two antiandrogens, R2956 (17β-hydroxy-2,2,17α-trimethylestra-4,9,11-triene-3-one) and cyproterone acetate are inhibitory (Table V). Increased DNA synthesis is accompanied by a 50% increase above controls in the number of cells after exposure to DHT for 8 days (Table VI). Again, R2956 inhibits cell growth and causes cell death when used at a concentration of 1.0 μM. Similar data are obtained when net protein

Table V
Relative Effects of Androgens and Antiandrogens on Thymidine Incorporation in MCF-7 Cells[a]

Hormone[b]	Thymidine incorporation (% change)
Control	0
5α-DHT	+ 128
5β-DHT	+ 9
R2956	− 46
Cyproterone acetate	− 36

[a] See Lippman *et al.*[84] for details.
[b] All hormones used at a concentration of 1.0 μM.

Table VI
Effect of Androgens and Antiandrogens on the Growth of MCF-7 Breast Cancer

Day	Cell number × (10^{-3})[a]		
	Control	5α-DHT[b]	R2956[b]
0	55	55	55
2	62	72	61
4	65	73	58
8	62	85	42

[a] Values represent the mean of replicately plated dishes.
[b] Concentration of 0.1 μM added on day 0.

synthesis is measured. Growth stimulation by androgens at these concentrations is not as striking as that induced by estrogens (see Fig. 5). This difference becomes even more impressive when one considers that DHT effects are not evident at concentrations below 50 nM, whereas maximal effects of estradiol are achieved by 0.5 nM estradiol.

DHT dose–response curves for thymidine and leucine incorporation in the MCF-7 cells are biphasic (Fig. 11), similar to that seen with estradiol. The curves are shifted far to the right, however, compared with that for estradiol. Maximal stimulation is observed at about 0.5 μM. A rapid fall-off of precursor incorporation is seen with concentrations greater than 1.0 μM, and cell death ensues. As is the case for estrogens, however, the inhibitory effect of androgens is nonspecific, since it is observed in androgen-receptor-negative cells and with 5β-DHT, an analogue that does not bind to the receptor in these cells (see Fig. 13). Thus, the mechanism of the response to additive hormone therapies with both estrogens and androgens remains to be defined.

As one might expect from the biological response studies, MCF-7 cells contain an androgen receptor, as shown by Horwitz *et al.*[60] and Lippman *et al.*[84] Sucrose density gradients of cytoplasmic extracts incubated with labeled DHT demonstrate an 8 S binding peak that is inhibited with cold DHT or antiandrogens (cyproterone acetate), but not with antiestrogens (tamoxifen) (Fig. 12). With the sucrose density gradient technique, Horwitz *et al.*[60] demonstrated that progestins were also capable of competing with DHT for binding, whereas estrogens had only a minimal effect. These data were confirmed with a detailed specificity study performed by the dextran-coated charcoal technique (Fig. 13). Competition for [^{3}H]-DHT binding is observed with androgens and antiandrogens and to a lesser extent with progesterone and 17β-estradiol.

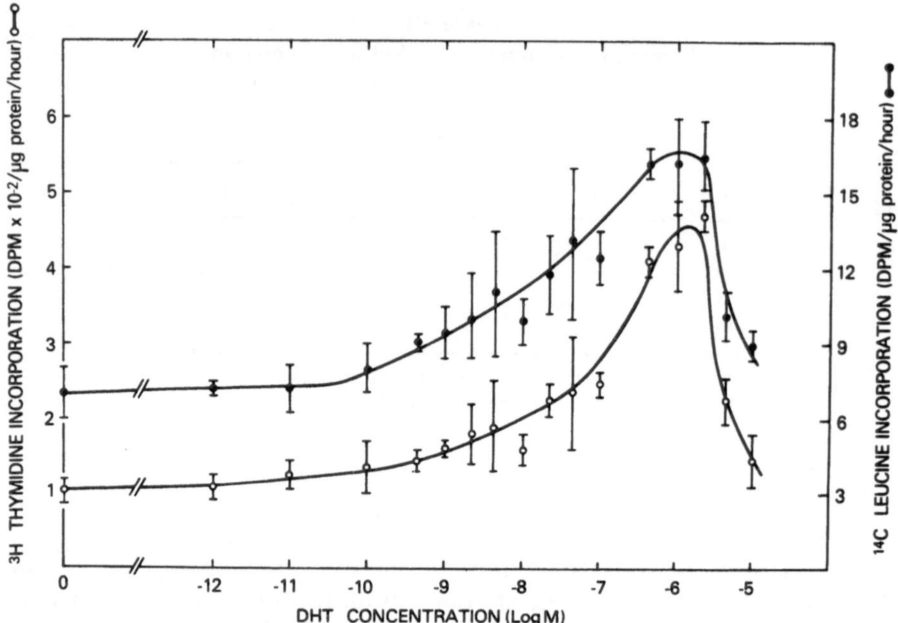

Fig. 11. Effects of increasing concentrations of 5α-dihydrotestosterone (DHT) on [^{3}H]thymidine and [^{14}C]leucine incorporation in MCF-7 human breast cancer cells. Methodological details may be found in Lippman *et al.*[84]

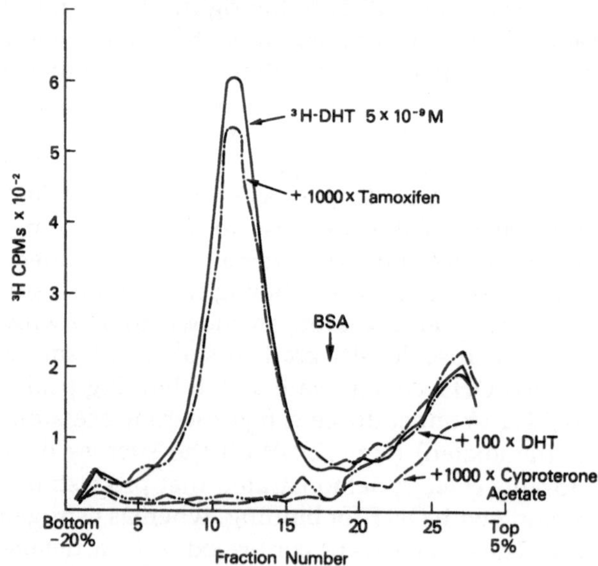

Fig. 12. Sucrose density gradients of cytoplasmic androgen receptor from MCF-7 human breast cancer cells. (BSA) A [^{14}C]bovine serum albumin marker. Complete methodological details may be found in Lippman *et al.*[84]

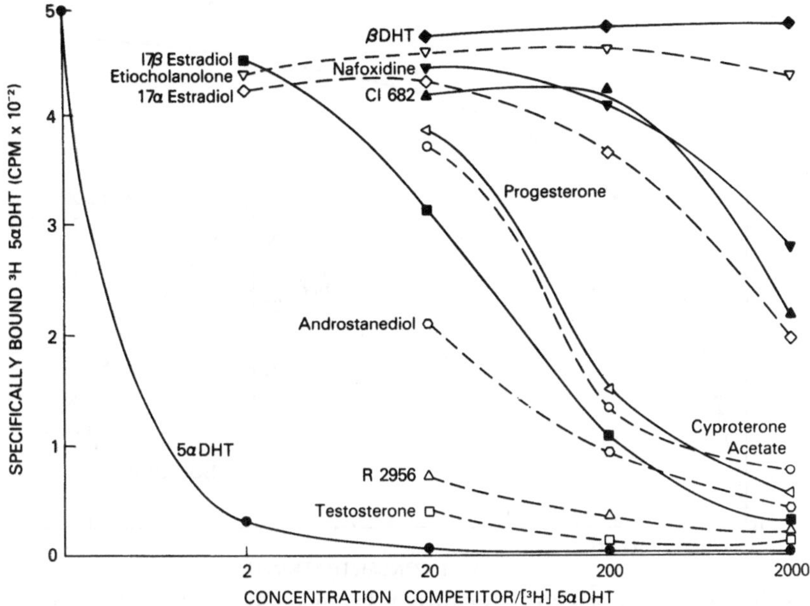

Fig. 13. Competition of various unlabeled compounds with [^{3}H]-5α-dihydrotestosterone (5α-DHT) for binding to cytoplasmic androgen receptor from MCF-7 human breast cancer cells. Details of methodology may be found in Lippman *et al.*[84]

Antiestrogens compete only at extremely high concentrations. The antiandrogen cyproterone acetate competes less well for receptor than R2956, which is also a more potent inhibitor of growth in these cells. These specificity data clearly distinguish the androgen receptor from the estrogen receptor in the MCF-7 cells (refer to Fig. 4 and Horwitz *et al.*[60]).

A saturation curve and Scatchard plot of DHT binding in MCF-7 cells reveals a limited-capacity, high-affinity receptor ($K_d = 8.7 \times 10^{-10}$ M) (Fig. 14). This dissociation constant approximates that found by Horwitz *et al.*[60] These data, however, immediately unearth an inconsistency with regard to the interaction of DHT with these cells. The concentration of DHT required for half-maximal stimulation of macromolecular synthesis is about 1000-fold greater than that which half-maximally saturates the receptor. This phenomenon might be explained by several potential mechanisms. First, one might argue that this discrepancy is artifactual due to the differing conditions in which the binding assay (cytoplasmic extract at 0°C) and biological response assay (whole cells at 37°C) were performed. When binding studies were repeated using these latter conditions, however, the dissociation constant was not altered significantly.[84] Second, metabolism of DHT to inactive analogues might

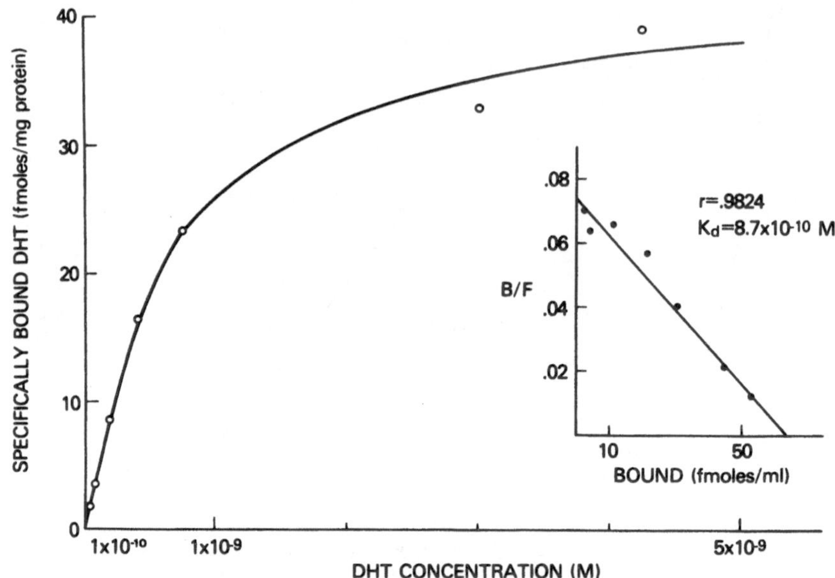

Fig. 14. Binding of [^{3}H]-5α-dihydrotestosterone to cytoplasmic receptor sites from MCF-7 human breast cancer cells. The same data are replotted in the inset using the Scatchard technique prepared by computer-assisted methods.[117]

also explain these findings. Conversion of DHT to inactive metabolites would be expected to shift the dose–response curve for biological activity to the right. We examined the metabolism of androgen in these cells using thin-layer chromatography to quantify the radioactive metabolites produced at various time intervals after the addition of [^{3}H]-5α-DHT or [^{3}H]testosterone.[84] These studies demonstrated a significant and extremely rapid conversion of DHT to androstanediol and other more polar metabolites. Only 30% of the DHT remained intact after an incubation period of 8 hr. In addition, the cells were capable of converting testosterone to DHT, indicating 5α-reductase activity. One might argue, however, that significantly more DHT would have to be converted to inactive products to account for the 1000-fold discrepancy. Furthermore, the major metabolite of DHT androstanediol, although less potent, is capable of competing for receptor binding (see Fig. 13), suggesting that it might be an active androgen in these cells.

A third alternative explanation for the discrepancy between DHT binding and biological response was recently offered in a preliminary report by Zava and McGuire.[85] In contrast to our results shown in Fig. 4, they showed in a previous study that 100-fold excess DHT inhibited [^{3}H]estradiol binding by 10%. They wondered whether this low-affinity DHT binding to the estrogen receptor might be responsible for its biological effects. Several lines of reasoning support this interpretation.[85] First, using several different methods, they repeated their initial observation that DHT can inhibit estradiol binding to estrogen receptor, an effect not observed with antiandrogens. Second, physiologic concentrations of DHT can bind to androgen receptor and translocate it to the nucleus, but no biological activity results. On the other hand, pharmacologic doses of DHT can translocate estrogen receptor to the nucleus, an effect that is accompanied by cell growth. Third, large doses of DHT can reverse antiestrogen inhibition of cell growth. Furthermore, this DHT "rescue" is not blocked by antiandrogens, suggesting that the effect is mediated via the estrogen receptor. Although these preliminary data await confirmation, they suggest that the effect of androgens on the MCF-7 cells is mediated by estrogen receptor. This interpretation cannot explain our observation that the MCF-7 cells are inhibited by antiandrogens, since neither R2956 nor cyproterone acetate has any affinity for the estrogen receptor. Furthermore, we showed that DHT can reverse most of the antiandrogen inhibition.[84] Resolution of these apparent disparities awaits further investigation.

3.4. Effects of Glucocorticoids and Progesterone

The mammary gland is thought to be a target tissue for glucocorticoids and progestins.[5] In addition, it has been recognized for many years that pharmacologic administration of these hormones induces objective tumor regression in 10–15% of women with breast cancer.[1] Although the response seen with glucocorticoids has frequently been attributed to its "medical adrenalectomy" effect, it is possible that a direct antitumor effect also occurs. Teulings *et al.*[86] detected significant levels of glucocorticoid receptor in about one third of human breast tumor samples. We found that about 50% of human breast cancer biopsy specimens contain glucocorticoid receptor (in prep). Some caution is required in the interpretation of these data in that most of the supporting stroma (fibroblasts, leukocytes, and adipocytes) are glucocorticoid target tissues and contain glucocorticoid receptor. Recent studies suggest that progesterone receptor in breast-cancer specimens provides a marker for estrogen dependence and helps to predict responses to endocrine therapy.[63] With this in mind, we investigated the interaction of glucocorticoids and progestins with human breast cancer cells in tissue culture.[87,88] Because of the apparent close relationship of the interaction of both hormones with these cells, we have chosen to discuss them together.

The effect of various steroid hormones on [^{3}H]thymidine incorporation in the ZR75-1 cell line is shown in Fig. 15. This cell line contains

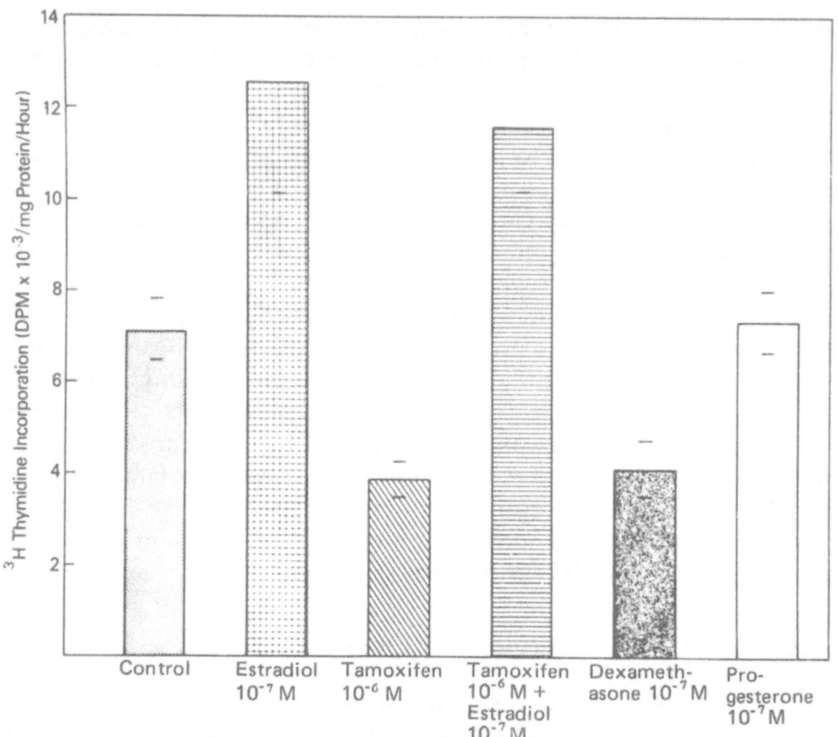

Fig. 15. Effects of various hormones on [^{3}H]thymidine incorporation in ZR75-1 human breast cancer cells. Detailed methods are described in Lippman *et al.*[64]

receptors for four classes of steroid hormones (see Table II). The response to estrogen and antiestrogen is similar to that described previously for the MCF-7 cells. Dexamethasone (DEX) inhibits significantly the rate of thymidine incorporation. Progesterone at the concentration shown (0.1 μM) has no effect. A more detailed dose–response curve of the effect of several steroid hormones on the rate of [^{3}H]thymidine incorporation in the MCF-7 cells is shown in Fig. 16. The potent synthetic glucocorticoid DEX inhibits thymidine incorporation, an effect that is half-maximal at about 10 nM. Cortisol, cortexolone, and R5020 (a progestational agent) also inhibit these cells, whereas 11α-cortisol, an inactive glucocorticoid that does not bind to the MCF-7 cells (see below), and the metabolite tetrahydrocortisol have no effect. Slight inhibition is observed with progesterone at a concentration of 1.0 μM. Leucine incorporation into protein and net protein synthesis are relatively unaffected for up to 48 hr in DEX.[87] Inhibition of DNA synthesis by DEX is manifested by a decrease in the cell growth rate, as we pre-

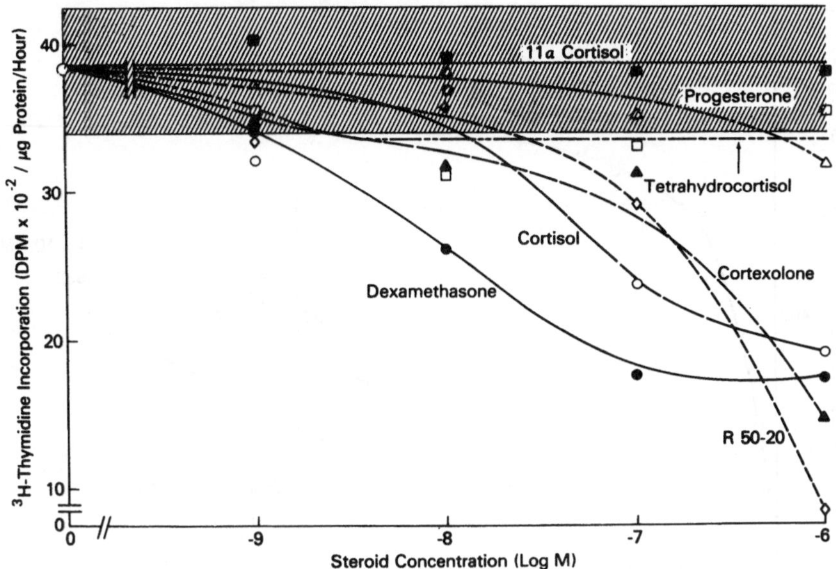

Fig. 16. Effects of various steroids on [^{3}H]thymidine incorporation in MCF-7 human breast cancer. Methodological details are supplied in Lippman *et al.*[87] The shaded area represents thymidine incorporation in control cells ± 2 S.D.

viously showed.[87] Furthermore, the EVSA-T cell line does not contain glucocorticoid receptor and accordingly fails to respond (see Table II). Thus, glucocorticoids effectively inhibit receptor-positive human breast cancer cells *in vitro.*

With the knowledge that glucocorticoids influence the growth of these cell lines, the presence of high-affinity receptors for these hormones was expected. By sucrose density gradients[60,87] or the dextran-coated charcoal assay,[87] glucocorticoid receptors were demonstrated in the cytosol from MCF-7 and other cell lines as shown in Table II. A typical DEX binding curve and Scatchard analysis with the MCF-7 cell line demonstrates a single class of receptors with a uniform K_d of about 2.4 nM, similar to that reported by Horwitz *et al.*[60] (Fig. 17). Binding-specificity studies demonstrated that the ability of various glucocorticoids to compete with [^{3}H]-DEX binding (Table VII) closely parallels their ability to inhibit thymidine incorporation as shown in Fig. 16. DEX is the most potent competitor, followed by cortisol and cortexolone. As expected, the inactive 11α-cortisol and tetrahydrocortisol do not compete for [^{3}H]-DEX binding. Of interest is the observation that progestins compete significantly with [^{3}H]-DEX for the receptor. Since R5020 exhibits significant glucocorticoidlike activity (see Fig. 16), one is not surprised to find that it also competes for this receptor. Progesterone, how-

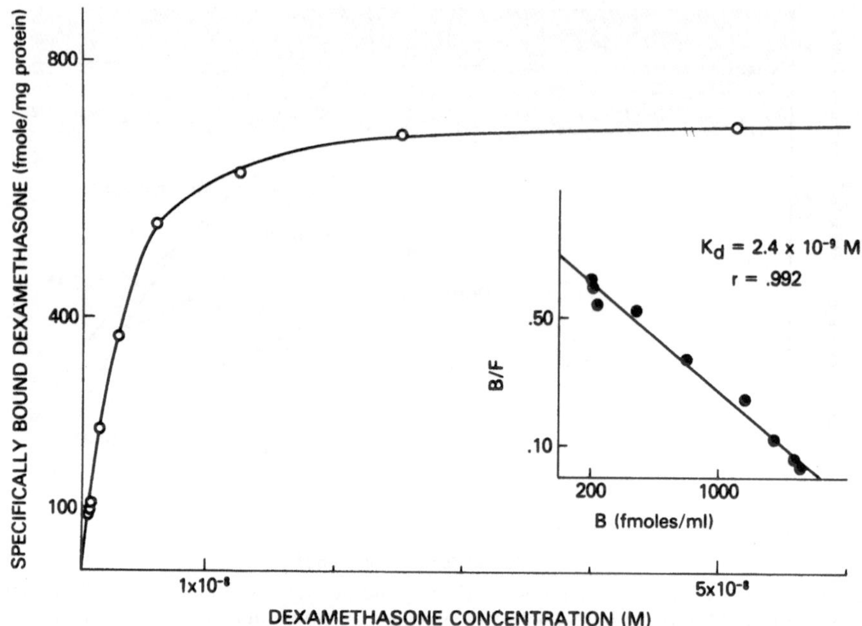

Fig. 17. Binding of [^{3}H]dexamethasone to glucocorticoid receptor from MCF-7 human breast cancer. The data shown are replotted in the inset using the Scatchard technique prepared by computer-assisted methods.[117] Methods are to be found in Lippman *et al.*[64] Other methods are to be found in Lippman *et al.*[87]

Table VII
Ability of Various Steroids to Inhibit the Binding of [^{3}H]Dexamethasone or [^{3}H]Progesterone[a]

Hormone	Relative inhibition of binding	
	[^{3}H]Dexamethasone	[^{3}H]Progesterone
Glucocorticoids		
DEX	++++	+
Cortisol	+++	−
Cortexolone	++	+
11α-Cortisol	−	−
Tetrahydrocortisol	−	−
Progestins		
R5020	+++	++++
Progesterone	+++	++++
Androgens (DHT)	+	++

[a] Data adapted from previously published binding specificity studies.[86,87]

ever, does not induce a detectable biological response in the MCF-7 cells, yet it too competes for [^{3}H]-DEX binding nearly as well as cortisol. Double reciprocal plots of specific binding of [^{3}H]-DEX to receptor with various concentrations of progesterone or R5020 (inhibitor) indicate competitive inhibition by these steroids for a common binding site(88) (see Figs. 20 and 21). This failure of progesterone to display glucocorticoid activity despite binding to the glucocorticoid receptor has been observed in other systems.(89) Thus, progestational agents appear to bind to glucocorticoid receptors in human breast cancer *in vitro*.

Characterization of the binding of [^{3}H]progesterone to receptor in the MCF-7 cells is more complex. When binding was initially examined using relatively low concentrations of progesterone, a single saturable high-affinity site was observed.(87) Binding-specificity studies (Table VII) showed that [^{3}H]progesterone binding was inhibited readily by R5020 and progesterone and to a lesser extent by DHT. The slight inhibition of binding by glucocorticoids may in fact result from their competition with [^{3}H]progesterone for the glucocorticoid receptor (see below). The results of the specificity studies reviewed in Table VII are nearly identical with those reported previously.(60)

When studies of [^{3}H]progesterone binding are performed using a broader range of concentrations, a different result is obtained (Fig. 18). When the binding data are plotted by the method of Scatchard and quantified as recommended by Buller *et al.*,(90) a two-component curve is

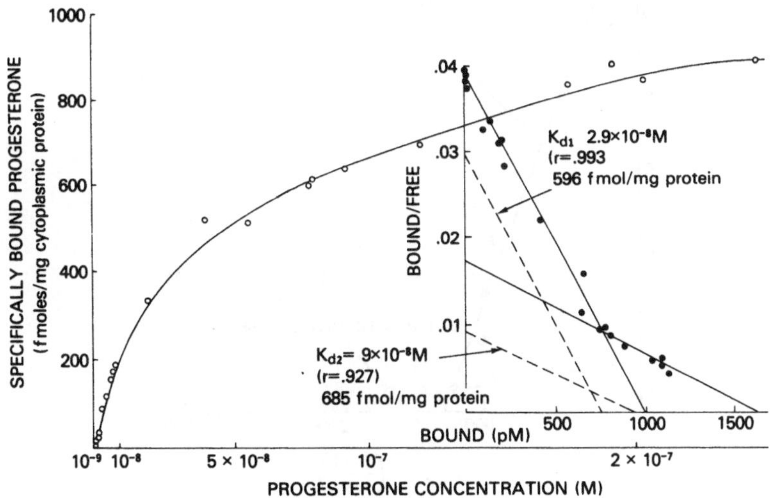

Fig. 18. Binding of [^{3}H]progesterone to receptors in MCF-7 human breast cancer cells. The binding data are replotted in the inset using the Scatchard technique.(117) Methods for quantification of these binding components are described in Buller *et al.*(90)

obtained. The first component, a high-affinity, low-capacity site, presumably represents progesterone binding to progesterone receptor. The second component, a low-affinity, high-capacity site, presumably reflects progesterone binding to glucocorticoid receptor, as suggested by the receptor-specificity studies shown in Table VII. Similar results were obtained when R5020 was used as the labeled progestin (Fig. 19).[88] Further illustration of the ability of progestins to bind to the glucocorticoid receptor and the failure of glucocorticoids to bind to the progesterone receptor is shown in Figs. 20 and 21. In Fig. 20, competitive inhibition by R5020 of [^{3}H]-DEX binding to receptor in the MCF-7 cell line is demonstrated by double-reciprocal analysis. In contrast, a similar plot of the data for [^{3}H]progesterone binding in the presence of cortisol reveals noncompetitive inhibition (Fig. 21).

In summary, some human breast cancer cell lines in tissue culture respond to glucocorticoids and the progestational agent R5020 with a decreased rate of DNA synthesis and growth. Other progestins are inactive. Furthermore, these studies support the hypothesis that in the

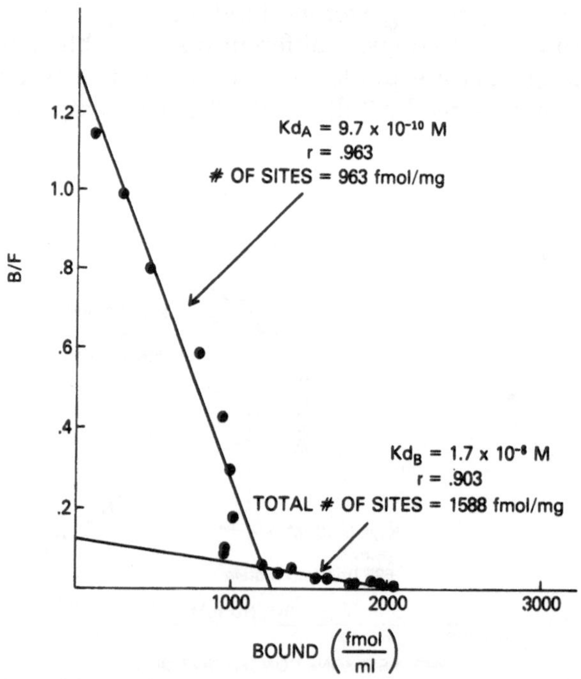

Fig. 19. Binding of [^{3}H]-R5020 to receptor sites in cytoplasmic extracts from MCF-7 human breast cancer cells. Detailed methods will be found in Lippman *et al.*[88,118]

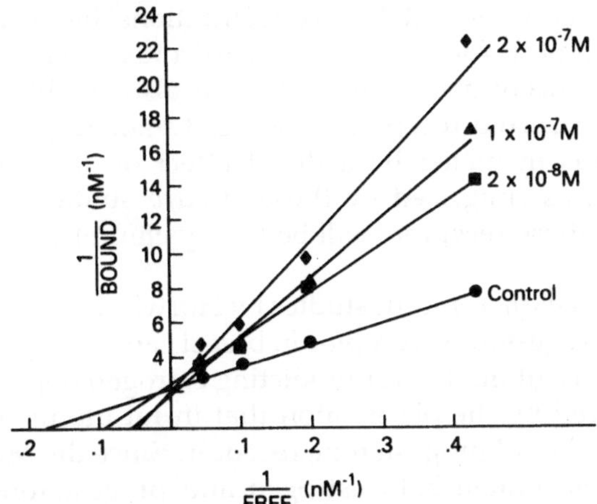

Fig. 20. Double-reciprocal analysis of specific binding of [^{3}H]dexamethasone to receptor from MCF-7 human breast cancer cells in the presence of various concentrations of R5020 (shown beside each curve). Specific binding was measured and analyzed as described by Lippman *et al.*(87) and Aitken and Lippman.(117)

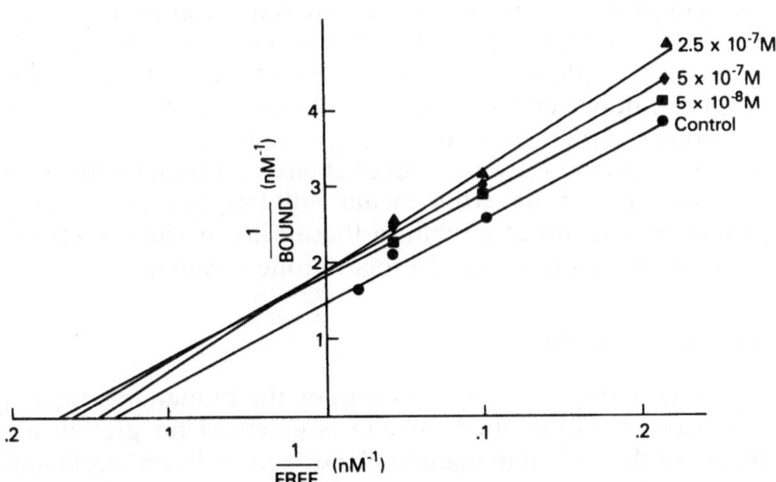

Fig. 21. Double-reciprocal analysis of [^{3}H]progesterone binding to specific receptor sites from MCF-7 human breast cancer cells in the presence of various concentrations of unlabeled cortisol (concentration shown beside each curve). Specific binding was measured and analyzed as described by Lippman *et al.*(87) and Aitken and Lippman.(117)

MCF-7 cells, progesterone at low concentrations binds to a distinct progesterone receptor for which glucocorticoids have little affinity, whereas at high concentrations, progestins are also capable of significant binding to glucocorticoid receptor. Whether tumor response *in vivo* to these hormones is mediated by a direct effect on the tumor cell via specific receptor, as suggested by these *in vitro* studies, and whether measurement of these receptors will be of any clinical value remains to be determined.

As mentioned earlier, recent studies by Horwitz *et al.*[63] suggest that the presence of progesterone receptor in breast cancer biopsy specimens might be an important marker for predicting estrogen dependence. This idea was prompted by the observation that in rat uterus, estrogen induces the appearance of progesterone receptor. Since the MCF-7 human breast cancer cells contain both estrogen and progesterone receptors, these investigators undertook to study in more detail regulation of progesterone receptor by estrogen in this system.[91] In a preliminary report, they showed that estradiol significantly increases progesterone receptor in these cells after an incubation period of 4 days. Surprisingly, the antiestrogen tamoxifen also increased progesterone-receptor levels. Both hormones bind to cytoplasmic estrogen receptor, and translocate it to the nucleus, where a phase they called "nuclear processing" occurs prior to the inductive effect. Progesterone-receptor levels fall to baseline on hormone withdrawal. Thus, tamoxifen has an estrogenic effect on induction of progesterone receptor, whereas it has antiestrogenic effects on growth (see Section 3.2). Delineation of the exact nature of this "nuclear processing" phase will require further study. These studies provide support, however, for the clinical observation that the presence of progesterone receptor aids in predicting estrogen dependence of human breast cancer, providing a more rational basis for therapeutic decisions. It is of some interest that while insulin will largely replace estrogen in the general promotion of growth in these cells, it will not obscure the estrogen-mediated induction of progesterone receptor.

3.5. Effects of Insulin

Although a requirement for insulin by the human mammary gland has not been clearly defined, insulin is essential for growth and development of the rodent mammary gland.[5] In addition, proliferation of certain animal mammary carcinomas has been shown to be insulin-dependent.[92,93] Heuson *et al.*[93] showed that administration of insulin to rats bearing 7,12-dimethylbenz(a)anthracene-induced mammary carcinoma resulted in tumor growth in a significant number of animals, whereas induction of diabetes with alloxan caused tumor regression.[92]

Similar results were obtained with explants of these tumors in organ culture.[93]

The effect of insulin on human breast cancer *in vivo* is not known. Interpretation of the effect of insulin on breast cancer in organ culture is difficult, as discussed in Section 2.2. Many studies showed no effect, whereas in others, extremely high concentrations of insulin were used to demonstrate a response. At such high concentrations, insulin could be eliciting an effect through a "growth factor" receptor such as that for nonsuppressible insulinlike activity (NSILA), rather than its own receptor.[94] Furthermore, organ cultures are heterogeneous and may contain fibroblasts and adipocytes, which may also be insulin-responsive. For similar reasons, a recent preliminary report of the demonstration of insulin binding to human breast cancer specimens is suspect.[95] Since human breast cancer cells in long-term tissue culture provide a homogeneous population of epithelial cells, study of the interaction of insulin with these cell lines has recently been intitiated.

Physiologic concentrations of insulin are mitogenic for the MCF-7 cells (Table VIII).[96] A 50% increase above controls in the number of cells is observed after 3 days with only 0.1 nM insulin. With 10 nM, a nearly 2-fold increase is evident by 3 days. Similar results were obtained with the ZR75-1 cell line, whereas the MDA-MB-231 and EVSA-T lines did not respond to insulin (see Table II).[97,98] To investigate the insulin effect on these cells in more detail, we examined the rates of DNA, RNA, protein, and fatty acid synthesis as a function of insulin concentration (Table IX).[96,99,100] The MCF-7 cells are extremely sensitive to insulin. As little as 0.05 nM insulin significantly stimulates all four parameters. This degree of insulin responsiveness was reproducible in multiple experi-

Table VIII
Effect of Insulin on the Growth of MCF-7 Human Breast Cancer[a]

	Cell count ($\times 10^{-4}$)		
		Insulin	
Day	Control	0.1 nM	10 nM
0	40	40	40
1	55	80	85
3	140	220	265

[a] Cells were replicately plated in medium with 10% fetal calf serum. They were changed to serum-free medium 24 hr prior to addition of insulin as described by Osborne.[96]

Table IX
Effect of Insulin on Macromolecular and Fatty Acid Synthesis in MCF-7 Human Breast Cancer[a]

Insulin concentration (M)	[^{3}H]Precursor incorporation (% above control)			
	Thymidine	Uridine	Leucine	Acetate
5×10^{-11}	48	36	22	12
1×10^{-10}	75	42	39	22
5×10^{-10}	85	45	40	34
1×10^{-9}	95	66	46	41
1×10^{-8}	120	122	50	37

[a] Insulin was added to cells maintained in serum-free medium. Labeled thymidine, uridine, and leucine incorporation into acid-insoluble material, and acetate incorporation into hexane-extractable material were measured as described previously.[96,100]

ments, and in addition, the ZR75-1 cell line responds similarly. The sensitivity of the MCF-7 cells to insulin was recently confirmed by Rillema and Linebaugh.[101] Thus, some human breast cancers at least *in vitro* respond to physiologic concentrations of insulin with enhanced macromolecular synthesis and growth and the more differentiated function of fat synthesis.

The effect of insulin on metabolism in the MCF-7 cells proceeds in an ordered manner. Protein and fatty acid synthesis are stimulated within 1 hr of the addition of insulin.[96,100,101] The effect on RNA synthesis is also rapid, whereas a significant lag period exists for insulin stimulation of DNA synthesis (Fig. 22). Uridine incorporation is stimulated by 3 hr and near maximal by 10 hr incubation with insulin. Increased thymidine incorporation is not evident until 10–15 hr and is maximal at 24 hr. This effect is preceded by an increase in the thymidine acid-soluble pool that is evident by 8 hr (Fig. 23). These results are nearly identical to those reported subsequently by Rillema and Linebaugh.[101] Preliminary observations in our laboratory suggest that the increased thymidine acid-soluble pool size might result from both enhanced thymidine transport into the cell and increased thymidine kinase activity (in preparation).

Stimulation of cell metabolism by insulin is thought to be primarily a posttranscriptional effect.[102] Results obtained on the effect of insulin on protein and fatty acid synthesis in the MCF-7 cells support this notion (Table X).[96,100] With concentrations of actinomycin D sufficient to inhibit new RNA synthesis by 90%, no inhibiting effect is seen on insulin stimulation of leucine incorporation into protein or acetate incorpora-

Table X
Effect of Actinomycin D and Cycloheximide on Insulin-Induced Leucine and Acetate Incorporation in MCF-7 Human Breast Cancer

	Leucine incorporation[a]	Acetate incorporation[a]
Insulin (alone)	50	100
+ Act D[b]	48	100
+ Cycloheximide[c]	0	120

[a] Values given are percentages above control for [^{3}H]leucine and [^{14}C]-acetate incorporation after 8 hr in insulin. Details can be found in Osborne *et al.*[96] and Monaco and Lippman.[100]
[b] Actinomycin D, 1 μg/ml, added 2 hr before insulin. This concentration inhibited RNA synthesis by more than 90%.
[c] Cycloheximide, 20 μg/ml, added 8 hr before insulin.

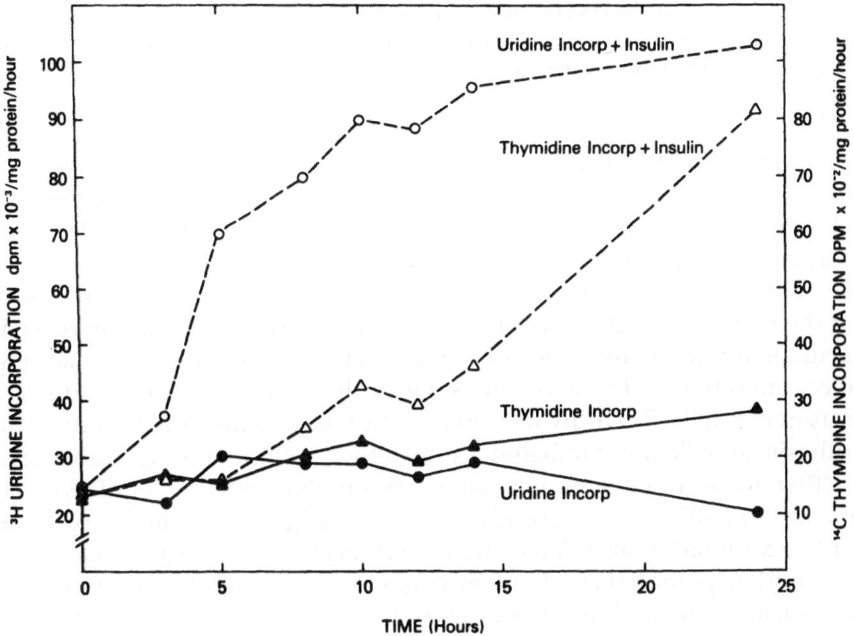

Fig. 22. Time course of the effect of insulin on thymidine and uridine incorporation into TCA-insoluble material in the MCF-7 cells. Insulin (10 nM) was added at time 0 after a 24-hr preincubation in serum-free medium. Cells were pulsed for 1 hr with [^{14}C]thymidine and [^{3}H]uridine and harvested at the time shown.

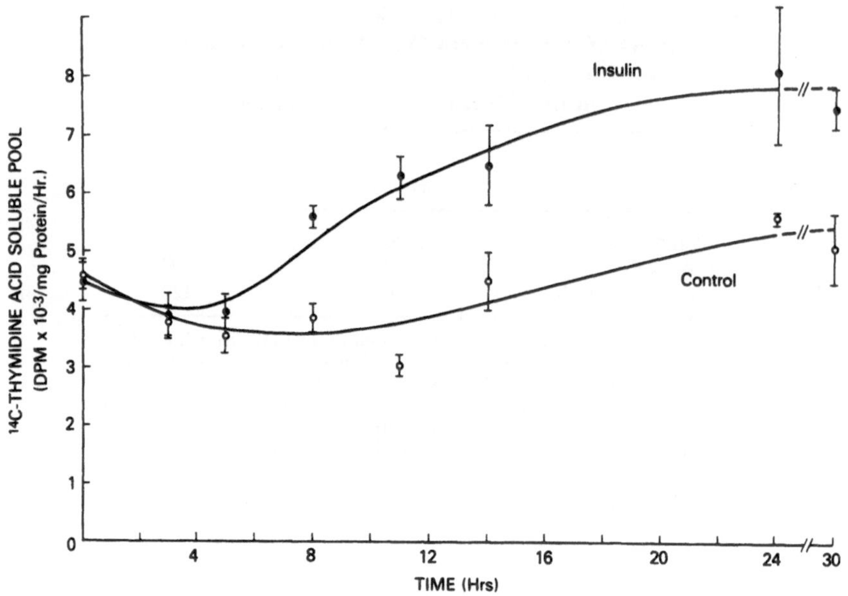

Fig. 23. Time course of the effect of insulin on the [^{14}C]thymidine TCA-soluble pool size in MCF-7 cells. The procedure was similar to that described in Fig. 22, except that the acid-soluble radioactivity was measured.

tion into fatty acids. Furthermore, when the effect on leucine incorporation is inhibited significantly with cycloheximide, the effect of insulin on acetate incorporation is again unchanged (Table X). Finally, [^{3}H]water incorporation into fats is also enhanced by insulin (Fig. 24), although the fold of stimulation is less than with [^{14}C]acetate. This finding suggests that the primary effect of insulin on fatty acid synthesis is not increased precursor transport into the cell, since water is freely diffusible across the cell membrane. These results suggest that enhanced fatty acid synthesis in the MCF-7 cells by insulin does not require new RNA or protein synthesis and is not mediated by an effect on membrane transport. Whether insulin activates key enzymes involved in the regulation of fatty acid synthesis is currently being investigated. Preliminary data point to an insulin-stimulated increase in acetyl-CoA carboxylase.

Another potential mechanism whereby insulin might stimulate cell metabolism is through increased glucose availability or utilization. This possibility was eliminated by the results obtained when macromolecular and fatty acid synthesis in response to insulin were determined in glucose-free tissue-culture medium.[96,100] Basal levels of precursor incorporation in cells maintained glucose-free were lower than in cells with glucose; however, the ability of insulin to stimulate the cells was unaltered.

The effect of insulin on fatty acid synthesis in these cells is specific, and does not result from an overall stimulation of cell proliferation (Fig. 25). Other hormones, estradiol (E_2) and 5α-dihydrotestosterone (DHT), previously shown to stimulate growth of the MCF-7 cells, do not influence acetate incorporation into fats. Furthermore, iodothyronines (T_3), dexamethasone (DEX), human placental lactogen (hPL), and progesterone (Po) have no effect. The combination of hPL and DEX, hormones known to enhance the lactogenic effect of insulin in rodents, does not stimulate fatty acid synthesis above that seen with insulin alone.

Table II illustrates that only two of the four human breast cancer cell lines tested (MCF-7 and ZR75-1) are insulin-responsive. To determine whether this failure to respond by the MDA-MB-231 and EVSA-T lines represents a defect in the initial interaction of the hormone with the cells, we studied insulin binding and degradation in detail in all four cell lines. All four lines possess insulin receptors that are qualitatively and quantitatively similar to receptors characterized in other tissues.[51,97,98,103] Furthermore, the differences in biological responsiveness cannot be explained on the basis of relative insulin-binding affinities or receptor concentration, since these parameters were similar among the cell lines. All four lines degrade insulin, but to a variable degree. The most insulin-responsive cell line, MCF-7, is also the most active insulin degrader. To demonstrate binding adequately in these cells, bacitracin, an

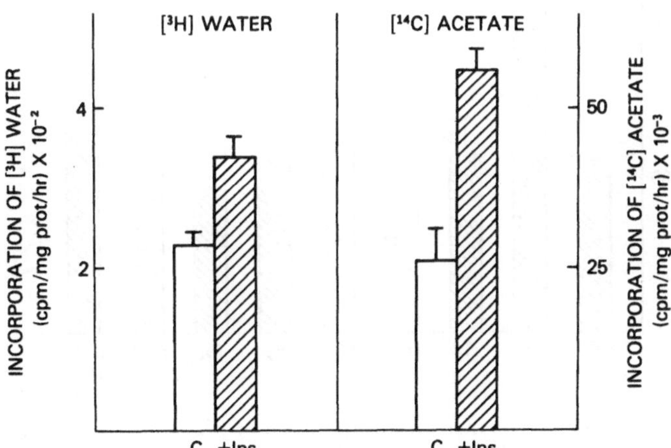

Fig. 24. Effect of insulin on incorporation of either [^{3}H]water or [^{14}C]acetate into hexane-extractable fatty acids in MCF-7 human breast cancer cells. Detailed methods are provided in Monaco and Lippman.[99,100]

antibiotic that is also a protease inhibitor, is necessary to inhibit degradation. A representative receptor-specificity study in the MCF-7 cells using bacitracin to block degradation is shown in Fig. 26. Insulin, insulin analogues, and other unrelated peptide hormones compete for [^{125}I]insulin binding to a degree roughly proportional to their capacity to stimulate glucose oxidation in adipocytes.[104] Porcine insulin and chicken insulin are equipotent, whereas proinsulin and guinea pig insulin are 50–100 times less potent. Epidermal growth factor does not compete for insulin binding (not shown). Multiplication-stimulating activity,[105] a growth factor derived from conditional medium of buffalo rat liver cells that displays similarities to human somatomedins, inhibits labeled-insulin binding, but with about 1% the potency of insulin. Half-maximal inhibition of binding is observed with about 6 ng/ml (1.0 nM) porcine insulin, a concentration similar to that causing half-maximal stimulation of macromolecular synthesis (see Table IX). These data suggest that human breast cancer cells in long-term tissue culture contain insulin receptors, and that growth stimulation by insulin is mediated through an "insulin" receptor rather than another "growth" receptor such as that for NSILA or somatomedin. The ability to investigate possible dif-

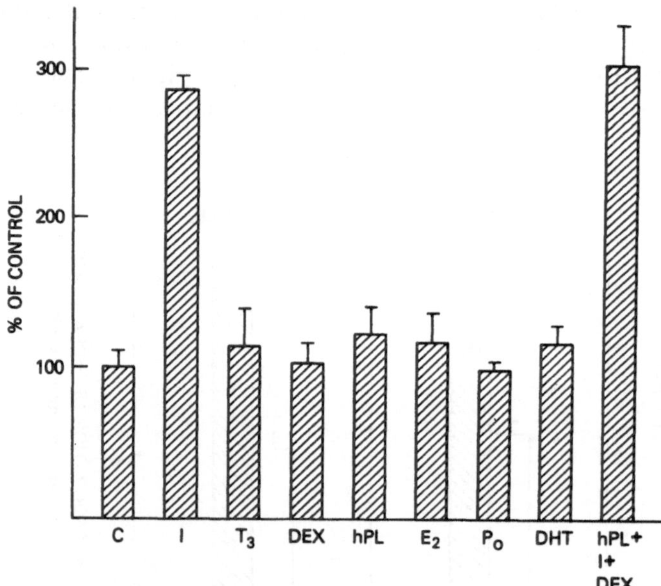

Fig. 25. Effects of various hormones on incorporation of [^{14}C]acetate into fatty acids in MCF-7 human breast cancer cells. (C) Control; (I) insulin, 0.1 U/ml; (T_3) triiodothyronine, 10^{-8} M; (DEX) dexamethasone, 10^{-7} M; (hPL) human placental lactogen, 10 μg/ml; (E_2) estradiol, 10^{-8}M; (P_0) progesterone, 10^{-7}M; (DHT) 5α-dihydrotestosterone, 10^{-6} M. Detailed methods are provided in Monaco and Lippman.[99,100]

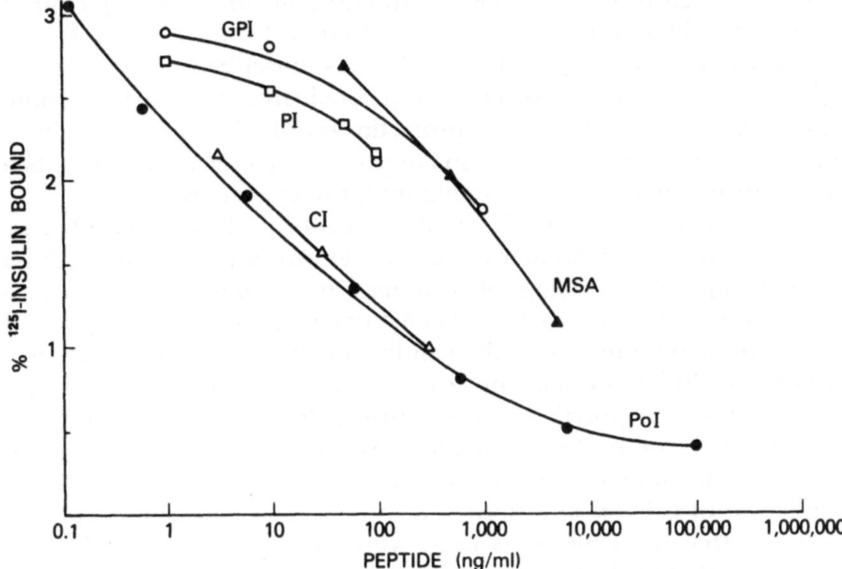

Fig. 26. Specificity of [^{125}I]insulin binding. Cells (3 × 10^6 ml) were incubated with [^{125}I]insulin (120 pg/ml) for 4 hr at 21°C in the presence of the indicated concentrations of polypeptide hormones: porcine insulin (PoI, ●), chicken insulin (CI, △), porcine proinsulin (PI, □), guinea pig insulin (GPI, ○), and multiplication-stimulating activity (MSA, ▲). Bacitracin (70 U/ml) was used to inhibit degradation. "Bound" and "free" hormone were separated by centrifugation of the cells through cold buffer, and the radioactivity in the "bound" fraction was determined.

ferences between the insulin responsive and unresponsive cells may be useful in studying the complex mechanisms of insulin action. The striking insulin sensitivity of the cells and the ability to monitor metabolic and growth parameters and insulin binding make this a unique model system.

3.6. Other Hormones

3.6.1. Prolactin

Whether prolactin has a role in growth regulation of human breast cancer is still controversial.[106] Prolactin does have, however, a major role in growth and development of rodent mammary glands[5] and breast carcinomas.[107] On the other hand, tumor response in patients undergoing hypophysectomy for metastatic breast cancer does not correlate well with prolactin blood level. In fact, tumor regressions have been noted in women after hypophysectomy or pituitary stalk section despite elevated prolactin levels.[108] In addition, pharmacologic inhibition of prolactin secretion in patients with breast cancer does not result in significant tumor regressions.[106,109] As described in Section 2.2, studies of the effect of prolactin are inconclusive. Kleinberg[110] noted stimulation of α-lactalbumin by prolactin in only 2 of 19 human breast cancer specimens studied in organ culture. These effects may have been due to normal breast cells contaminating the specimen.

In an attempt to better define the effect of prolactin and other lactogenic hormones on human breast cancer, we searched for both lactogenic receptor and biological responses in several breast cancer cell lines *in vitro*. To date, we have been unable to demonstrate prolactin receptor or activity in any of the cell lines using a variety of lactogenic hormones including human and ovine prolactin, human placental lactogen, and human growth hormone (unpublished observations). This could suggest true prolactin independence in the human tumors from which the cell lines were derived, or, alternatively, it might reflect dedifferentiation of cells in long-term culture. Last, we cannot exclude the possibility that the conditions used were not optimal for observing a response or that prolactin might influence a paramater not measured in our assays.

Shafie and Brooks[111] recently reported an effect of prolactin on growth of the MCF-7 cells. With high concentrations of human or ovine prolactin alone, no stimulation of thymidine incorporation or increase in DNA content was observed. In contrast, insulin increased both parameters significantly, as we described earlier. However, when dibutyryl cyclic AMP (cAMP) was administered with prolactin, stimulation of DNA synthesis was evident. Paradoxically, theophylline, which potentiates the action of endogenous cAMP by inhibiting its degradation, had the opposite effect. These puzzling results remain unexplained, and further investigation is required to clarify a possible interaction between prolactin and cyclic nucleotides.

These authors also presented data demonstrating that high concentrations of prolactin increased the estrogen-receptor level in the MCF-7 cells. Serious methodological difficulties preclude full interpretation of their data.

3.6.2. *Epidermal Growth Factor*

Epidermal growth factor (EGF) is a polypeptide hormone that stimulates growth in fibroblasts and rodent mammary epithelial cells in culture.[112–114] Although an effect of EGF on rodent or human breast cancer has not been reported, Stoker *et al.*[114] recently provided clear evidence that this growth factor in low concentrations stimulates DNA synthesis in benign human breast tumors in organ culture. Insulin and hydrocortisone were ineffective. In recent years, several other growth factors have been isolated from human and animal sera.[115] Whether these factors are important in the growth regulation of human tissues *in vivo* and, particularly, human breast cancer remains to be determined. Meanwhile, human breast cancer cells in culture should provide an interesting system in which to study their effects *in vitro*.

3.6.3. *Thyroid Hormones*

The mammary gland has also been thought to be a target of thyroid hormone action based on its growth-promoting and lactogenic effects in rodent mammary glands.[116] In a preliminary report, Burke *et al.*[58] studied the effect of iodothyronines on the MCF-7 breast cancer cell line. They found that physiologic concentrations of triiodothyronine (T_3) increased the growth rate of these cells. Furthermore, similar to the results described earlier for prolactin, T_3 increased the level of estrogen receptor. The authors also reported that nuclei from MCF-7 cells contain specific high-affinity thyroid-hormone receptors. The importance of these hormones in the development and/or growth regulation of breast cancer *in vivo* requires further study.

4. Conclusion

The various experimental results reviewed in this chapter make clear that human breast cancer cell lines are an extremely promising system in which to study hormone action. Clearly, these cell lines have already provided new information concerning estrogen and insulin action, and as discussed later, more progress is anticipated. While experiments on androgen effects have also been carried out, it cannot be stated with assuredness that true androgen-dependent cells are available. While cell lines with progesterone receptor have been defined, we are not aware of any results in which specific progesterone effects have been observed. Finally, thyroid-hormone receptor and response studies are in a relatively preliminary stage of development. Unequivocal prolactin, growth hormone, or other growth factor effects have not yet been identified. Next, we briefly identify a few areas that to us are most promising for future studies.

There are a variety of experiments for which human breast cancer cells may be useful in the future. Some of these new areas may be explored using existing cell lines; others will undoubtedly require cell lines initially developed under conditions that avoid preselection of hormone-independent cell lines.

The most fundamental general area in which the unique properties of these cells can be exploited is in the general field of mechanisms of hormone action. In the case of steroid action, little information is available for human cells on the events following initial binding of steroid to receptor. Cloned populations of cells such as these provide a useful starting point for studies of receptor processing and induction of specific RNA products. The potential ability to select against hormone-responsive cells (e.g., by the use of antiestrogens) could allow for the development of sublines defective in various steps in hormone action. Such genetic approaches have been critical in unraveling control mechanisms in prokaryotes and could be of similar value in eukaryotes.[119] Clearly, increased awareness of the various potential hormones and factors that may alter growth or differentiation (such as prostaglandins, growth peptides, thyronines, transferrin, retinoids) may allow the initial development of cell lines with previously unsuspected dependency. These new lines in turn can be used to study mechanisms of action of these additional factors. Eventually, this should lead to propagation of cells under entirely defined conditions.

Second, these cell lines may be employed for identification and purification of specific protein products of interest. For example, examination of the supernatant medium after incubation with cell lines might permit identification of new biological markers of interest as well as allow a greater understanding of regulation of production rates of these materials. Alternatively, these cells might provide a useful starting point for purification of human steroid receptors. While several cell lines containing estrogen receptor are available, it is possible that variants could be developed that have vastly increased levels of receptor. A general method of selection that might be suitable was recently developed.[120] Similarly, these or newly derived cell lines should be suitable for more detailed attempts to identify specific oncogenic viruses or viral products.

Finally, it should always be borne in mind that a tumor *in vivo* is a complex organ made up of numerous elements in addition to tumor cells themselves. It is critical to study the mechanisms by which tumor and host interact. Thus, angiogenesis factors, collagen-promoting activity, and other biochemical concomitants of the morphology of tumors need to be explored. These studies might well lead to alternative strategies limiting tumor growth independent of direct tumor-cell kill itself.

5. References

1. B. A. Stoll, Castration and oestrogen therapy, in: *Endocrine Therapy in Malignant Disease,* (B. A. Stoll, ed.), pp. 139–163, W. B. Saunders, London (1972).
2. E. V. Jensen, T. Suzuhi, T. Kawashima, W. E. Stumpf, P. W. Jungblut, and E. R. De Sombre, A two step mechanism for the interaction of estradiol with rat uterus, *Proc. Natl. Acad. Sci. U.S.A.* **59,** 632–638 (1968).

3. W. L. McGuire, P. P. Carbone, M. E. Sears, and G. C. Escher, Estrogen receptors in human breast cancer: An overview, in: *Estrogen Receptors in Human Breast Cancer* (W. L. McGuire, P. P. Carbone, and E. P. Volmer, eds.), pp. 1–7, Raven Press, New York (1975).
4. R. J. B. King and W. I. P. Mainwaring, *Steroid–Cell Interactions*, University Park Press, Baltimore (1974).
5. R. W. Turkington, Multiple hormonal interactions: The mammary gland, in: *Biochemical Actions of Hormones* (G. Litwack, ed.), pp. 55–77, Academic Press, New York (1972).
6. J. C. Heuson, N. Legros, J. A. Heuson-Stiennon, G. Leclercq, and J. L. Pasteels, Hormone dependency of rat mammary tumors, in: *Breast Cancer: Trends in Research and Treatment* (J. C. Heuson, ed.), pp. 81–93, Raven Press, New York (1976).
7. H. Fell and R. Robison, The growth, development and phosphatase activity of embryonic avian femora and limb-buds cultivated *in vitro*, *Biochem. J.* **23,** 767–784 (1929).
8. G. Cameron and R. Chambers, Neoplasm studies: Organization of cells of human tumors in tissue culture, *Am. J. Cancer* **30,** 115–129 (1937).
9. J. A. Dickson, Tissue-culture approach to the treatment of cancer, *Br. Med. J.* **1,** 817–823 (1966).
10. W. C. Gewant and I. S. Goldenberg, Techniques of human breast neoplasm cell culture, *Eur. Surg. Res.* **2,** 392–400 (1970).
11. J. C. Heuson, J. L. Pasteels, N. Legros, J. Heuson-Stiennon, and G. Leclercq, Estradiol-dependent collagenolytic enzyme activity in long-term organ culture of human breast cancer, *Cancer Res.* **35,** 2039–2048 (1975).
12. R. Tchao, G. C. Easty, E. J. Ambrose, R. W. Raven, and H. J. G. Bloom, Effect of chemotherapeutic agents and hormones on organ cultures of human tumours, *Eur. J. Cancer* **4,** 39–44 (1968).
13. J. L. Pasteels, J. Heuson-Stiennon, N. Legros, G. Leclercq, and J. C. Heuson, Organ culture of human breast cancer, in: *Breast Cancer: Trends in Research and Treatment* (J. C. Heuson, ed.), pp. 141–150, Raven Press, New York (1976).
14. R. L. Ceriani, G. P. Contesso, and B. M. Notaf, Hormone requirement for growth and differentiation of the human mammary gland in organ culture, *Cancer Res.* **32,** 2190–2196 (1972).
15. S. R. Wellings and V. L. Jentoft, Organ cultures of normal, dysplastic, hyperplastic, and neoplastic human mammary tissues, *J. Natl. Cancer Inst.* **49,** 329–338 (1972).
16. M. Finkelstein, A. Geier, H. Horn, I. S. Levij, and P. Ever-Hadani, Effect of testosterone and estradiol-17β on synthesis of DNA, RNA and protein in human breast in organ culture, *Int. J. Cancer* **15,** 78–90 (1975).
17. B. E. Barker, H. Fanger, and P. Farnes, Human mammary slices in organ culture. I. Method of culture and preliminary observations on the effect of insulin, *Exp. Cell Res.* **35,** 437–448 (1964).
18. J. J. Elias and R. C. Armstrong, Hyperplastic and metaplastic responses of human mammary fibroadenomas and dysplasias in organ culture, *J. Natl. Cancer Inst.* **51,** 1341–1342 (1973).
19. W. G. Dilley and S. J. Kister, *In vitro* stimulation of human breast tissue by human prolactin, *J. Natl. Cancer Inst.* **55,** 35–36 (1975).
20. L. J. van Bogaert, Glucose uptake by normal human breast tissue in organ culture, *Cell Tissue Res.* **171,** 535–541 (1976).
21. B. A. Flaxman and E. Y. Lasfargues, Hormone-independent DNA synthesis by epithelial cells of adult human mammary gland in organ culture, *Proc. Soc. Exp. Biol. Med.* **14,** 371–374 (1973).
22. B. A. Flaxman, J. Dyckman, and A. Feldman, Effect of prolactin on maintenance of prelactating human mammary gland *in vitro*, *In Vitro* **12,** 467–471 (1976).

23. H. Salih, H. Flax, and J. R. Hobbs, *In vitro* oestrogen sensitivity of breast cancer tissue as a possible screening method for hormonal treatment, *Lancet* **1,** 1198–1202 (1972).
24. M. D. Lagios, Hormonally enhanced proliferation of human breast cancer in organ culture, *Oncology* **29,** 22–23 (1974).
25. J. R. Hobbs, H. Salih, H. Flax, and W. Brander, Prolactin dependence in human breast cancer, *Proc. R. Soc. Med.* **66,** 16 (1973).
26. R. A. Sellwood and J. E. Castro, The effect of hormones on organ cultures of human mammary carcinoma, *J. Pathol.* **113,** 223–225 (1974).
27. K. G. Rienits, The effects of estrone and testosterone on respiration of human mammary cancer *in vitro, Cancer* **12,** 958–961 (1959).
28. J. R. Barker and C. Richmond, Human breast carcinoma culture: The effects of hormone, *Br. J. Surg.* **58,** 732–734 (1971).
29. J. R. W. Masters, K. Sangster, I. I. Smith, and A. P. M. Forrest, Human breast carcinomata in organ culture: The effect of hormones, *Br. J. Cancer* **33,** 564–566 (1976).
30. B. A. Stoll, Investigation of organ culture as an aid to the hormonal management of breast cancer, *Cancer* **25,** 1228–1233 (1970).
31. C. W. Welsch, G. C. De Iturri, and M. J. Brennan, DNA synthesis of human, mouse, and rat mammary carcinomas *in vitro:* Influence of insulin and prolactin, *Cancer* **38,** 1272–1281 (1976).
32. K. Aspegren, Hormone effects on human mammary cancer in organ cultures, *Am. J. Surg.* **131,** 575–580 (1976).
33. K. Aspegren and L. Hakansson, Human mammary carcinoma studied for hormone responsiveness in short term incubations, *Acta Clin. Scand.* **140,** 95–99 (1974).
34. J. C. Heuson and N. Legros, *In vitro* effect of testosterone and 17β-estradiol on h-leucine-^{14}C incorporation into human breast cancer tissue, *Cancer* **16,** 404–407 (1963).
35. A. Nimrod and K. J. Ryan, Aromatization of androgens by human abdominal and breast fat tissue, *J. Clin. Endocrinol. Metab.* **40,** 367–372 (1975).
36. E. Y. Lasfargues and L. Ozzello, Cultivation of human breast carcinomas, *J. Natl. Cancer Inst.* **21,** 1131–1147 (1958).
37. G. C. Buehring and A. J. Hackett, Human breast tumor cell lines: Identity evaluation by ultrastructure, *J. Natl. Cancer Inst.* **53,** 621–629 (1974).
38. L. Ozzello, Ultrastructure of human mammary carcinoma cells *in vivo* and *in vitro, J. Natl. Cancer Inst.* **48,** 1043–1050 (1972).
39. R. E. Nordquist, D. R. Ishmael, C. A. Lovig, D. M. Hyder, and A. F. Hogl, The tissue culture and morphology of human breast tumor cell line BOT-2, *Cancer Res.* **35,** 3100–3105 (1975).
40. R. H. Bassin, E. J. Plata, B. I. Gerwin, C. F. Mattern, D. K. Haapola, and E. W. Chu, Isolation of a continuous epithelioid cell line, HBT-3, from a human breast carcinoma, *Proc. Soc. Exp. Biol. Med.* **141,** 673–680 (1972).
41. G. Seman, S. J. Hunter, R. C. Miller, and L. Dmochowski, Characterization of an established cell line (SH-3) derived from pleural effusion of patient with breast cancer, *Cancer* **37,** 1814–1824 (1976).
42. R. Cailleau, R. Young, M. Olive, and W. J. Reeves, Jr., Breast tumor cell lines from pleural effusions, *J. Natl. Cancer Inst.* **53,** 661–674 (1974).
43. R. K. Young, R. M. Cailleau, B. Mackay, and W. J. Reeves, Jr., Establishment of epithelial cell line MDA-MB-157 from metastatic pleural effusion of human breast carcinoma, *In Vitro* **9,** 239–245 (1974).
44. Y. V. Dobrynin, Establishment and characteristics of cell strains from some epithelial tumors of human origin, *J. Natl. Cancer Inst.* **31,** 1173–1195 (1963).

45. E. J. Plata, T. Aoki, D. D. Robertson, E. W. Chu, and B. I. Gerwin, An established cultured cell line (HBT-39) from human breast carcinoma, *J. Natl. Cancer Inst.* **50,** 849–862 (1973).
46. M. V. Reed and G. O. Gey, Cultivation of normal and malignant human lung tissue, *Lab. Invest.* **11,** 638–652 (1962).
47. M. Lippman, Hormone responsive human breast cancer in continuous tissue culture, in: *Breast Cancer: Trends in Research and Treatment* (J. C. Heuson, ed.), pp. 111–140, Raven Press, New York (1976).
48. J. Fogh and G. Trempe, New human tumor cell lines, in: *Human Tumor Cells in Vitro* (J. Fogh, ed.), pp. 115–159, Plenum Press, New York (1975).
49. H. D. Soule, J. Vazquez, A. Long, S. Albert, and M. Brennan, A human cell line from a pleural effusion derived from a breast carcinoma, *J. Natl. Cancer Inst.* **51,** 1409–1416 (1973).
50. M. Lippman, G. Bolan, and K. Huff, The effects of estrogens and antiestrogens on hormone-responsive human breast cancer in long-term tissue culture, *Cancer Res.* **36,** 4595–4601 (1976).
51. M. E. Lippman, C. K. Osborne, R. Knazek, and N. Young, *In vitro* model systems for the study of hormone-dependent human breast cancer, *N. Engl. J Med.* **296,** 154–159 (1977).
52. W. A. Nelson-Rees, R. R. Flandermeyer, and P. K. Hawthorne, Banded marker chromosomes as indicators of intraspecies cellular contamination, *Science* **184,** 1093–1096 (1974).
53. W. A. Nelson-Rees and R. R. Flandermeyer, HeLa cultures defined, *Science* **191,** 96–98 (1976).
54. J. Russo, H. D. Soule, C. McGrath, and M. A. Rich, Reexpression of the original tumor pattern by a human breast carcinoma cell line (MCF-7) in sponge culture, *J. Natl. Cancer Inst.* **56,** 279–282 (1976).
55. R. Knazek, M. E. Lippman, and H. Chopra, Formation of solid human mammary carcinoma *in vitro*, *J. Natl. Cancer Inst.* **58,** 419–423 (1977).
56. H. N. Rose and C. M. McGrath, α-Lactalbumin production in human mammary carcinoma, *Science* **190,** 673–675 (1975).
57. M. E. Monaco, D. A. Bronzert, D. C. Tormey, P. Waalkes, and M. E. Lippman, Casein production by human breast cancer, *Cancer Res.* **37,** 749–754 (1977).
58. R. E. Burke, D. T. Zava, and W. L. McGuire, Human breast cancer cells contain thyroid hormone receptors, *Clin. Res.* **25,** 461A (1977).
59. S. C. Brooks, E. R. Locke, and H. D. Soule, Estrogen receptor in a human cell line (MCF-7) from breast carcinoma, *J. Biol. Chem.* **248,** 6251–6253 (1973).
60. K. B. Horwitz, M. E. Costlow, and W. L. McGuire, MCF-7: A human breast cancer cell line with estrogen, androgen, progesterone, and glucocorticoid receptors, *Steroids* **26,** 785–795 (1975).
61. G. C. Chamness, K. Huff, and W. L. McGuire, Solid phase ligand exchange assay for charged cytoplasmic estrogen receptor, *Fed. Proc. Fed. Am. Soc. Exp. Biol.* **33,** 1511 (1974).
62. W. L. McGuire and M. DeLaGarza, Improved sensitivity in the measurement of estrogen receptor in human breast cancer, *J. Clin. Endocrinol. Metab.* **36,** 548–552 (1973).
63. K. B. Horwitz, W. L. McGuire, O. H. Pearson, and A. Segaloff, Predicting response to endocrine therapy in human breast cancer: A hypothesis, *Science* **189,** 726–727 (1975).

64. M. Lippman, G. Bolan, and K. Huff, Interactions of antiestrogens with human breast cancer in long-term tissue culture, *Cancer Treatment Rep.* **60,** 1421–1429 (1976).
65. H. Esber, I. Payne, and A. Bogden, Variability of hormone concentrations and ratios in commercial sera used for tissue culture, *J. Natl. Cancer Inst.* **50,** 559–562 (1973).
66. M. Lippman, M. E. Monaco, and G. Bolan, Effects of estrone, estradiol, and estriol on hormone-responsive human breast cancer in long-term tissue culture, *Cancer Res.* **37,** 1901–1907 (1977).
67. M. Lippman and G. Bolan, Oestrogen-responsive human breast cancer in long-term tissue culture, *Nature (London)* **256,** 592–593 (1975).
68. C. Huggins and E. V. Jensen, The depression of estrone-induced uterine growth by phenolic estrogens with oxygenated functions at position 6 or 16: The impeded estrogens, *J. Exp. Med.* **102,** 335–346 (1955).
69. B. Mac Mahon, P. Cole, J. B. Brown, K. Aoki, T. M. Lin, R. W. Morgan, and N. C. Woo, Urine oestrogen profiles of Asian and North American women, *Int. J. Cancer* **14,** 161–167 (1974).
70. H. M. Lemon, Estriol prevention of mammary carcinoma induced by 7,12-dimethylbenzanthracene and procarbazine, *Cancer Res.* **35,** 1341–1353 (1975).
71. J. H. Clark, Z. Paszko, and E. J. Peck, Jr., Nuclear binding and retention of the receptor estrogen complex: Relation to the agonistic and antagonistic properties of estriol, *Endocrinology* **100,** 91–96 (1977).
72. J. C. Heuson, Hormones by administration, in: *The Treatment of Breast Cancer* (H. Atkins, ed.), pp. 113–164, University Park Press, Baltimore (1974).
73. D. T. Kiang and B. J. Kennedy, "Intranuclear" castration effect of high dose estrogens: Proceedings of the American Association of Cancer Research, *Cancer Res.* **17,** 194, Abstract 774 (1976).
74. J. C. Heuson, A. Coune, and M. Staquet, Clinical trial of nafoxidine, an oestrogen antagonist in advanced breast cancer, *Eur. J. Cancer* **8,** 387–389 (1972).
75. D. T. Zava, G. C. Chamness, K. B. Horwitz, and W. L. McGuire, Human breast cancer: Biologically active estrogen receptor in the absence of estrogen, *Science* **196,** 663–664 (1977).
76. J. H. Clark, E. J. Peck, Jr., and J. N. Anderson, Oestrogen receptors and antagonism of steroid hormone action, *Nature (London)* **251,** 446–448 (1974).
77. B. S. Katzenellenbogen and E. R. Ferguson, Antiestrogen action in the uterus: Biological ineffectiveness of nuclear bound estradiol after antiestrogen, *Endocrinology* **97,** 1–12 (1975).
78. J. N. Anderson, E. J. Peck, Jr., and J. H. Clark, Estrogen-induced uterine responses and growth: Relationship to receptor estrogen binding by uterine nuclei, *Endocrinology* **96,** 160–167 (1975).
79. R. J. Schwartz, C. Chang, W. T. Schrader, and B. W. O'Malley, Effect of progesterone receptors on transcription in biochemical actions of progesterone and progestins (E. Gurpide, ed.), *Ann. N.Y. Acad. Sci.* **286,** 147–160 (1977).
80. M. E. Lippman, M. E. Monaco, and L. Pinkus, Estradiol induces thymidine kinase activity in estrogen responsive human breast cancer in continuous tissue culture, in: *Proc. 59th Annual Meeting of the Endocrine Society,* p. 122 (1976) (abstract #132).
81. J. A. Smith and R. J. King, Effects of steroids on growth of an androgen-dependent mouse mammary carcinoma in cell culture, *Exp. Cell Res.* **73,** 351–359 (1972).
82. M. N. Teller, C. C. Stock, G. Stohr, P. C. Merker, R. J. Kaufman, G. C. Escher, and M. Bowie, Biologic characteristics and chemotherapy of 7,12-dimethylbenz[a]anthracene-induced tumors in rats, *Cancer Res.* **26,** 245–254 (1966).
83. M. E. Lippman, G. Bolan, and K. Huff, Human breast cancer responsive to androgen in long-term tissue culture, *Nature (London)* **258,** 339–341 (1975).

84. M. Lippman, G. Bolan, and K. Huff, The effects of androgens and antiandrogens on hormone-responsive human breast cancer in long-term tissue culture, *Cancer Res.* **36,** 4610–4618 (1976).
85. D. T. Zava and W. L. McGuire, Pharmacological effects of androgen in human breast cancer cells are mediated by estrogen receptor, *Proc. 59th Annual Meeting of the Endocrine Society,* p. 180 (1977) (abstract #247).
86. F. A. G. Teulings and H. A. van Gilse, Demonstration of glucocorticoid receptors in human mammary carcinomas, *Hormone Res.* **8,** 107–116 (1977).
87. M. Lippman, G. Bolan, and K. Huff, The effects of glucocorticoids and progesterone on hormone-responsive human breast cancer in long-term tissue culture, *Cancer Res.* **36,** 4602–4609 (1976).
88. M. Lippman, K. Huff, and G. Bolan, Progesterone and glucocorticoid interactions with receptor in breast cancer cells in long-term tissue culture, *Ann. N. Y. Acad. Sci.* **286,** 101–115 (1977).
89. G. G. Rousseau, J. D. Baxter, and G. M. Tomkins, Glucocorticoid receptors: Relations between steroid binding and biological effects, *J. Mol. Biol.* **67,** 99–116 (1972).
90. R. E. Buller, W. T. Schrader, and B. W. O'Malley, Steroids and the practical aspects of performing binding studies, *J. Steroid Biochem.* **7,** 321–326 (1976).
91. K. B. Horwitz and W. L. McGuire, Induction of progesterone receptor in a human breast cancer cell line, *Clin. Res.* **25,** 295A (1977).
92. J. C. Heuson and N. Legros, Influence of insulin deprivation on growth of the 7,12-dimethylbenz[a]anthracene-induced mammary carcinoma in rats subjected to alloxan diabetes and food restriction, *Cancer Res.* **32,** 226–232 (1972).
93. J. C. Heuson, N. Legros, and R. Heimann, Influence of insulin administration on growth of the 7,12-dimethylbenz[a]anthracene-induced mammary carcinoma in intact, oophorectomized, and hypophysectomized rats, *Cancer Res.* **32,** 233–238 (1972).
94. K. Megyesi, C. R. Kahn, J. Roth, D. M. Neville, Jr., S. P. Nissley, R. E. Humbel, and E. R. Froesch, The NSILA-s receptor in liver plasma membranes, *J. Biol. Chem.* **250,** 8990–8996 (1975).
95. I. M. Holdaway and I. Worsley, Specific binding of human prolactin and insulin to human mammary carcinomas, *Endocrinology (Suppl.)* **96,** 160 (220A) (1975).
96. C. K. Osborne, G. Bolan, M. E. Monaco, and M. E. Lippman, Hormone responsive human breast cancer in long-term tissue culture: Effect of insulin, *Proc. Natl. Acad. Sci. U.S.A.* **73,** 4536–4540 (1976).
97. C. K. Osborne, M. E. Monaco, and M. E. Lippman, Insulin receptors in human breast cancer: Relationship of binding, degradation and biological activity, *Diabetes* **25** *(Suppl. 1),* 380 (1976).
98. C. K. Osborne, M. E. Monaco, M. E. Lippman, and C. R. Kahn, Correlations among insulin binding, degradation, and biologic activity in human breast cancer cells in long-term tissue culture, *Cancer Res.* **38,** 94–102 (1978).
99. M. E. Monaco and M. E. Lippman, Insulin stimulation of fatty acid synthesis in human cancer cells, *J. Natl. Cancer Inst.* **58,** 1591–1593 (1977).
100. M. E. Monaco and M. E. Lippman, Insulin stimulation of fatty acid synthesis in human breast cancer in long-term tissue culture, *Endocrinology,* **101,** 1238–1246 (1977).
101. J. A. Rillema and B. E. Linebaugh, Characteristics of the insulin stimulation of DNA, RNA and protein metabolism in cultured human mammary carcinoma cells, *Biochim. Biophys. Acta* **475,** 74–80 (1977).
102. I. G. Wool, Relation of effects of insulin on amino acid transport and on protein synthesis, *Fed. Proc. Fed. Am. Soc. Exp. Biol.* **24,** 1060–1070 (1965).
103. C. R. Kahn, Membrane receptors for polypeptide hormones, in: *Methods in Membrane Biology* (E. Korn, ed.), Vol. 3, pp. 81–146, Plenum Press, New York (1975).

104. P. Freychet, J. Roth, and D. M. Neville, Jr., Insulin receptors in the liver: Specific binding of ^{125}I-insulin to the plasma membrane and its relation to insulin bioactivity, *Proc. Natl. Acad. Sci. U.S.A.* **68,** 1833–1837 (1971).
105. N. C. Dulak and H. M. Temin, A partially purified polypeptide fraction from rat liver cell conditioned medium with multiplication-stimulating activity for embryo fibroblasts, *J. Cell. Physiol.* **81,** 153–160 (1973).
106. F. Smithline, L. Sherman, and H. D. Kolodny, Prolactin and breast carcinoma, *N. Engl. J. Med.* **292,** 784–792 (1975).
107. F. Vignon and H. Rochefort, Regulation of estrogen receptors in ovarian-dependent rat mammary tumors. I. Effects of castration and prolactin, *Endocrinology* **98,** 722–729 (1976).
108. R. W. Turkington, L. E. Underwood, and J. J. Van Wyk, Elevated serum prolactin levels after pituitary-stalk section in man, *N. Engl. J. Med.* **285,** 707–710 (1971).
109. E. Engelsman, J. C. Heuson, J. Blonk-Van der Wijst, A. Drochmans, H. Maass, F. Cheix, L. G. Sobrinho, and H. Nowakowski, Controlled clinical trial of L-dopa and nafoxidine in advanced breast cancer: An E.O.R.T.C. study, *Br. Med. J.* **2,** 714–715 (1975).
110. D. L. Kleinberg, Human α-lactalbumin: Measurement in serum and in breast cancer organ cultures by radioimmunoassay, *Science* **190,** 276–278 (1975).
111. S. Shafie and S. C. Brooks, Effect of prolactin on growth and the estrogen receptor level of human breast cancer cells (MCF-7), *Cancer Res.* **37,** 792–799 (1977).
112. S. Cohen, G. Carpenter, and K. J. Lembach, Interaction of epidermal growth factor (EGF) with cultured fibroblasts, *Adv. Metab. Disord.* **8,** 265–284 (1975).
113. R. W. Turkington, The role of epithelial growth factor in mammary gland development *in vitro, Exp. Cell Res.* **57,** 79–85 (1969).
114. M. G. P. Stoker, D. Pigott, and J. Taylor-Papadimitriou, Response to epidermal growth factors of cultured human mammary epithelial cells from benign tumours, *Nature (London)* **264,** 764–767 (1976).
115. D. Gospodarowicz and J. S. Moran, Growth factors in mammalian cell culture, *Annu. Rev. Biochem.* **45,** 531–558 (1976).
116. B. K. Vonderhaar, Studies on the mechanism by which thyroid hormones enhance α-lactalbumin activity in explants from mouse mammary glands, *Endocrinology* **100,** 1423–1431 (1977).
117. S. C. Aitken and M. E. Lippman, A simple computer program for quantitation and Scatchard analysis of steroid receptor proteins, *J. Steroid Biochem.* **8,** 77–99 (1977).
118. M. E. Lippman, K. K. Huff, G. Bolan, and J. P. Neifeld, Interactions of R5020 with progesterone and glucocorticoid receptors in human breast cancer and peripheral blood lymphocytes *in vitro,* in: *Progesterone Receptors in Normal and Neoplastic Tissues* (W. L. McGuire, ed.), pp. 193–210, Raven Press, New York (1977).
119. S. Bourgeois and R. F. Newby, Diploid and haploid states of the glucocorticoid receptor gene of mouse lymphoid cell lines, *Cell* **11,** 423–430 (1977).
120. R. E. Kellems, F. W. Alt, and R. T. Schimke, Regulation of folate reductase synthesis in sensitive and methotrexate-resistant sarcoma 180 cells, *J. Biol. Chem.* **251,** 6987–6993 (1976).

5

Antiestrogens: Mechanism of Action and Effects in Breast Cancer

KATHRYN B. HORWITZ AND WILLIAM L. McGUIRE

1. Biological Activities of Antiestrogens

1.1. Introduction

Several of the nonsteroidal antiestrogens are in experimental use for the treatment of breast cancer. The treatment goal is to obtain, with minimal toxicity, specific control over cell growth by chemical means, thereby avoiding, on the one hand, pharmacologic doses of hormones and, on the other, major surgical ablative procedures.[1] One rationale for use of estrogen antagonists is an outgrowth of our current awareness of the role of estrogen receptors (ERs) and estrogen in breast cancer. Recent reports show that hormone dependence can be predicted by use of ER measurements. If estrogen antagonists block the action of estrogen at its receptor, it would be possible to obtain, by noninvasive means, the same therapeutic end as ablative hormonal procedures. Despite considerable research to that end, however, our knowledge of the mechanism of antiestrogen action remains unclear. The purpose of this review is to summarize this research. We will review first some of the biological activities of antiestrogens in normal tissues, and the mechanisms that have been proposed for their effects, then the role of antiestrogens in experimental animal breast cancer, and the data currently available on use of antiestrogens in humans and human breast cancer. Finally, we

KATHRYN B. HORWITZ AND WILLIAM L. McGUIRE • Department of Medicine, University of Texas Health Science Center at San Antonio, San Antonio, Texas 78284.

will describe some of our work involving the mechanisms of antiestrogen action in human breast cancer cells in tissue culture.

1.2. Estrogenic and Antiestrogenic Properties of Estrogen Antagonists

Many compounds can interfere with the biological effectiveness of estrogens. These compounds include the natural steroids (progesterone, androgens, and estriol), the synthetic steroids, and the synthetic nonsteroidal compounds. Among the latter are derivatives of triphenylethylenes (Fig. 1), of which the most common are ethamoxy-

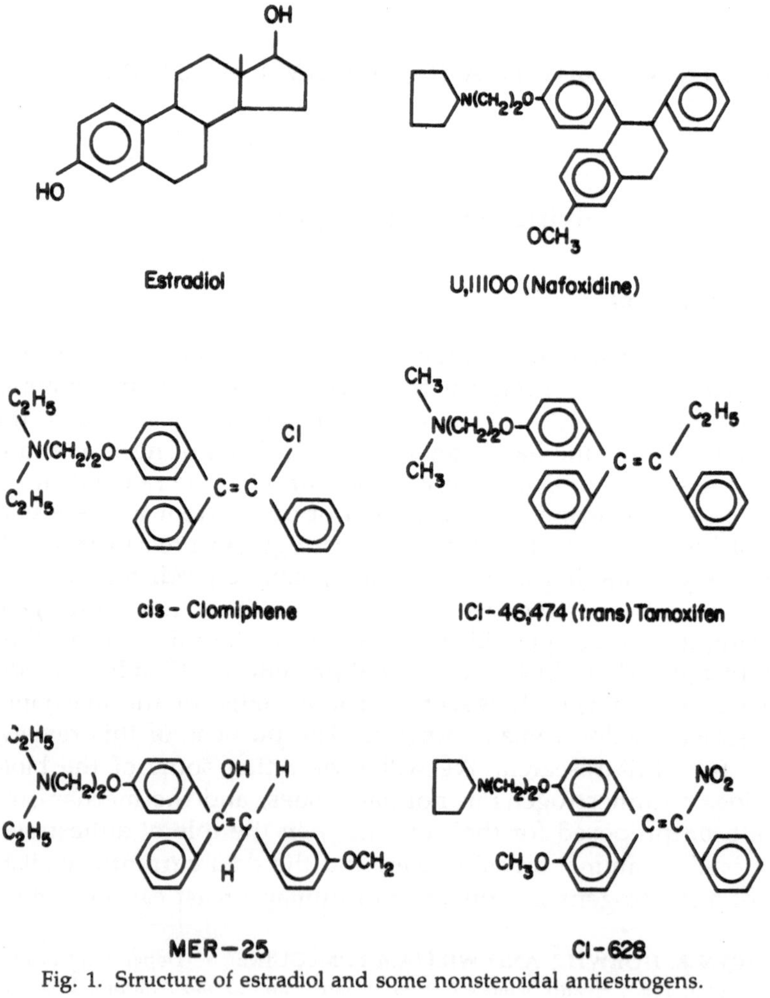

Estradiol

U,11100 (Nafoxidine)

cis - Clomiphene

ICI-46,474 (trans) Tamoxifen

MER-25

CI-628

Fig. 1. Structure of estradiol and some nonsteroidal antiestrogens.

triphetol (MER-25), nafoxidine (U-11, 100A), clomiphene, tamoxifen (ICI 46,474), and CI 628.

Because estrogen antagonists were originally developed as antifertility agents,[2] their biological activity has for the most part been measured and compared with estrogen-sensitive parameters in the female reproductive tract. Without exception, all antiestrogens have been shown to be estrogenic.[3-19] They induce vaginal cornification[7,11] and are uterotrophic. Antiestrogens can double uterine weights of ovariectomized rats or mice[2-10]; the uterine epithelial ultrastructure resembles that of the estrogen-stimulated tissue.[12] Antiestrogens increase [^{3}H]leucine incorporation and protein synthesis,[13,14] increase mitosis and DNA content,[9,10,15-17] increase RNA polymerase activity,[18] and increase glucose metabolism.[19] Why, then, are these compounds considered antagonists? Their antagonistic properties are defined in contrast to the effects of estradiol. In general, estrogens alone provoke more potent effects than do antiestrogens alone (in this respect, antiestrogens may simply be considered weak estrogens). In the presence of estradiol, however, antiestrogens prevent the full expression of the estrogenic response. This property characterizes true antagonists, though even in the presence of estrogens, antiestrogens usually do not prevent the estrogenic response, but rather reduce it to the level seen with antiestrogen alone.[2-4,15,20]

The distinctions between estrogenic and antiestrogenic effects of estrogen antagonists are most clearly seen in comparisons of single and multiple doses of these substances given alone or together. Such studies, in which changes in uterine weight and DNA and protein synthesis are used as markers of estrogenic responses, have been performed mainly in the laboratories of Clark,[10,16] Katzenellenbogen,[8,9,10,21,22] and Rochefort.[23] After single doses of antiestrogens, purely estrogenic responses are observed. In immature female rats, a single injection of nafoxidine[10,22,23] or other antiestrogens[7] results in a pronounced, slowly developing increase in uterine weight, exceeding the effect seen with a single dose of estradiol. There are also sustained (72-hr) increases in RNA polymerase activity[18] and DNA and protein synthesis,[9,16] in contrast to estradiol, with which these effects have receded by 72 hr. No antagonism is observed when the two compounds are injected together.[10] Curiously, low doses (5 μg) of nafoxidine are more potent estrogens than high doses (50 μg), an unexplained effect that has repeatedly been seen with antiestrogens in several responses studied.[4,11,22,24-26]

The antagonistic properties of antiestrogens appear when estrogens or antiestrogens are injected separately or together two or more times, 24 hr apart. With estradiol alone,[19] a second injection causes uterine weights to continue to rise considerably above weights evoked by a single injection. In contrast, with nafoxidine and other antiestrogens alone,[8,9] no increase occurs beyond that seen in the first 24 hr. If both estrogens and antagonists are given together, the heightened second estrogen response is prevented. These observations have been confirmed in studies of the effects of nafoxidine and tamoxifen on uterine ornithine decarboxylase activity.[24] Thus, antiestrogens are antagonistic because they fail to produce a second response when injected alone, and because they inhibit the second response to estradiol. Note, however, that the timing of injections may be quite critical. For instance, the second estrogen dose fails to enhance weight above that seen after the first dose, if the second is given too soon after the first.[27]

In sum, estrogens and antiestrogens are remarkably similar in the first 24–48 hr after a single injection, but differ markedly after long-term, multiple injections. Estrogen antagonists may be either estrogenic or antiestrogenic, depending on concentration, on timing and number of doses, and on the presence or absence of estradiol.

In addition, other factors may modify the response obtained with antiestrogens. For instance, hormones other than estrogens may alter antiestrogen effects. Progesterone serves as an example. When the effects of antiestrogens on uterine ultrastructure in ovariectomized rats are considered, they are estrogenic if given alone, have no effect if given together with progesterone, and are antiestrogenic in the presence of estradiol.[12,17] Since progesterone can modify the uterine response to estrogen,[28–30] its ability to modify responses to estrogen antagonists should not be surprising. To further complicate matters, progesteronelike effects have been described for nafoxidine.[31] Antiestrogens may also have androgenic activity[25] and interact with androgen receptors.[32] Furthermore, antiestrogens can have direct effects on ovaries and pituitaries.[33]

Given such complex interactions, it is quite likely that the effects of antiestrogens in intact cycling females may be different from those described in immature or ovariectomized adult animals normally used as models. For instance, in the mature uterus at proestrus, there is diminished capacity to respond to exogenous estrogens, unrelated to cytoplasmic ER levels (see below), suggesting that there exist in the cycling rat other factors the replenishment or reactivation of which is slower than that of the receptor.[34] Unfortunately, studies in mature intact animals are sparse. Until the biological effects of antiestrogens are studied in the adult cycling animal, it will be difficult to explain their variable effectiveness (as for instance when comparing pre- and postmenopausal women) in breast cancer.

The most important contributions that have been made toward understanding the mechanisms of action of estrogen antagonists have come from comparisons of their effect with those of active estrogens on cellular estrogen receptors. These studies are discussed in the following section.

2. Mechanisms of Antiestrogen Action

Antiestrogens are being extensively studied in the hope that this will help to explain not only their own mechanism but also the mechanism of normal estrogen action. Despite this effort, there is no single unifying concept to explain antiestrogen action. Instead, a wide range of mechanisms have been proposed that envision antiestrogen intervention at virtually every site that is susceptible to estrogen control. Therefore, before considering the possible sites of antiestrogen action, we will briefly summarize the general model of estrogen action.

2.1. Mechanisms of Estrogen Action

It is widely[35] though not universally[36] believed that estrogens enter the cell by passive diffusion, and that these steroids and others accumulate in the cell because they are bound there by specific receptor proteins. Binding is to a limited number of high-affinity sites.[37] On sucrose gradients, the receptor is an 8 S molecule in low-salt buffers, or a 4 S molecule in hypertonic buffers. Since sedimentation properties depend on experimental conditions,[38] the sedimentation velocity of the native cytoplasmic form remains unknown. When steroid binds with the receptor, a change in the conformation of the molecule appears to be induced,[39] which in hypertonic buffers *in vitro* is temperature-dependent, and is accompanied by an increase in sedimentation rate to 5 S.[40] The 4 S to 5 S transformation may occur while the receptor is still in the cytoplasm,[40] although there is also evidence that transformation occurs only after steroid–receptor complex is in the nucleus.[41] In any case, estrogen, bound to a 5 S receptor essentially indistinguishable from the cytoplasmic receptor, can be extracted from the nucleus.[42] According to the basic "two-step" model,[43] derived for the most part from studies with the rat uterus, unfilled receptors exist only in the cytoplasm, and filled receptors are rapidly translocated to the nucleus. This may prove to be an oversimplification, since we[44,45] and others[46] have demonstrated unfilled ER sites in the nuclei of tumor cells in culture and *in vivo*.[47,48] These free receptors bind estrogen directly, and may possess biological activity even in the absence of estrogen.[44]

The fate of the ER–hormone complex in the nucleus is unclear, but several nuclear components, including DNA,[49,50] nuclear proteins,[51] ribonucleoproteins,[52] and nuclear membranes,[53] have been proposed as acceptor sites. It is possible that more than one site is involved in these interactions.[54] While in the case of progesterone receptor sequential binding to two sites may precede nuclear activation, for ER it has been proposed that one, relatively weaker, binding site releases estradiol quickly and a second, tighter binding site is responsible for long-term retention of the receptor–hormone complex[55] required for true uterine growth. It is not known whether the primary response is transitory, after which the receptor–hormone complex is inactivated, or whether sustained receptor–acceptor interactions are required for prolonged effects. Once estrogen is bound to acceptor, the earliest effect is regulation of gene transcription following a rise in polymerase activity.[18,56] Most nuclear receptors disappear within 6 hr of estrogen treatment, preceding cytoplasmic receptor replenishment. Nuclear loss and cytoplasmic replenishment are not linked events, since cycloheximide inhibits the latter without affecting the former.[57]

Any or all of these steps are potential sites of action of an antagonist, which in the broadest definition of the term is a compound that should interfere either with the availability of estrogen to the cell or with the cellular responses to estrogen.

2.2. Antiestrogen Effects on Estradiol Levels

Although there are reports that antiestrogens inhibit ovarian hormone secretion,[58] there are considerable data showing that antiestrogens do not decrease circulating estradiol levels. Tamoxifen induces rat mammary tumor regression without affecting circulating prolactin or estradiol.[59] Patients with breast cancer whose tumors are responding to tamoxifen have unchanged plasma estradiol and prolactin levels,[60] while menstrual cycles in premenopausal women are unchanged.[61] This finding suggests that a complete block of estrogen secretion is not required to obtain antiestrogen-induced tumor regressions. Though antiestrogens may have little direct effect on estradiol levels, they appear able to suppress estradiol-induced effects on the pituitary. Nafoxidine[31] and tamoxifen[62] have been shown to antagonize the estradiol-induced increase of prolactin secretion by a direct action on pituitary prolactin-secreting cells.[31]

Interestingly, antiestrogen treatment failures in humans may be associated with increased estradiol levels.[61,63] The possibility arises that the pituitary, the estrogen sensitivity of which is blocked by antiestrogens, reacts to an apparent decrease in estrogen activity by hypersecretion of gonadotropins, resulting in an actual increase in estrogen secretion from the ovaries and escape from antiestrogen suppression. Such a mechanism might explain the finding of high circulating prolactin levels after some antiestrogen treatments,[64] and could also explain why tamoxifen may be more effective at low doses than at high doses in promoting breast-tumor regression.[65,66]

Very little is known about the effect of antiestrogens on intracellular estradiol availability, though this is another possible means of reducing estrogen activity. One method would be by promoting estradiol metabolism to less active estrogens.[67] Antiestrogens can also stimulate microsomal enzymes and accelerate the metabolism of estrogens.[68,69]

2.3. Antiestrogen Metabolism

We now turn to consideration of intracellular estrogen antagonism. Since estrogen entry into the cell is generally assumed to be a passive process, little effect of antiestrogens can be expected at this point. There is, however, considerable interest in studies involving turnover rates of antiestrogens and the role of metabolism in controlling both extracellular and intracellular antiestrogen levels. Studies from Katzenellenbogen's laboratory[8,9,19] suggest that chemical alterations of a compound can markedly change its biological potency. Ethinyl estriol cyclopentyl ether is a more potent estrogen than ethinyl estriol. The higher estrogenicity of the former has been attributed to its gradual metabolism to produce a continual supply of ethinyl estriol, the active form.[19] Similarly, it is proposed that nafoxidine may be a long-acting antiestrogen because it is in fact a "prohormone" that is metabolized slowly to an active form; alternatively, it may simply have a slow rate of clearance or inactivation in the blood. All these effects could, by controlling concentration at the target cell, influence its potency.[8,9]

We have shown that the effectiveness of tamoxifen *in vitro* cannot be explained by its conversion to a more active metabolite. We incubated human breast cancer cells the growth of which is inhibited by tamoxifen with [^{3}H]tamoxifen, and then analyzed the nature of the products extracted from the cells. Figure 2 shows the LH-20 elution patterns of ethyl-acetate-extractable radioactivity from medium incubated with or without cells, and from cell cytoplasm and nuclei. The position of purified tamoxifen is indicated by the bars. Even in the absence of cells, tamoxifen in the medium breaks down at 37°C to an unidentified compound. Although extensive amounts of this breakdown product are present in the medium, only tamoxifen itself is recovered from the cell. We conclude, first, that *in vitro* only the intact tamoxifen molecule is taken up into cells, and, second, that the dual estrogenic/antiestrogenic properties of this compound therefore cannot be explained by metabolic formation of the true antiestrogen from a prohormonal (and estrogenic) precursor.

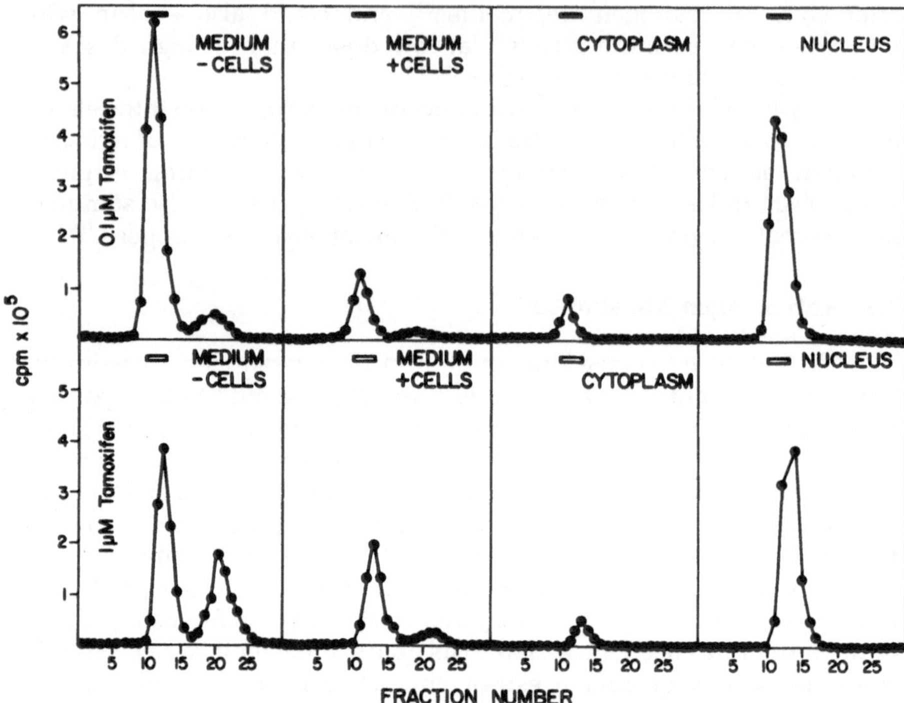

Fig. 2. LH-20 elution pattern of extractable radioactivity from media and cells treated with [^{3}H]tamoxifen. Elution of purified tamoxifen is shown by the bar.

The thesis that chemical modification of a prohormone sustains serum levels of antiestrogens and increases their potency is attractive, but requires experimental validation. It must be shown whether the compounds are only sequestered in the body and slowly released in unchanged form, or whether they are in fact chemically altered. If chemical modification occurs, is the conversion peripheral or within the target cell? Ultimately, we need to know the identity of the compounds actually present in cytoplasmic and nuclear binding sites.

2.4. Antiestrogens and Cytoplasmic Estrogen Receptor Binding

The interaction of antiestrogens with ERs has been studied by *in vitro* and *in vivo* methods, the assumption being that *in vitro* analyses could be extrapolated to *in vivo* situations. As is discussed below, such conclusions are likely to be erroneous.

The ability of many antiestrogens to inhibit binding of [^{3}H]estradiol to its receptor by *in vitro* competition studies at equilibrium has been repeatedly demonstrated,[70–77] providing many data on comparative binding affinities and hypothetical structure–activity relationships. For the most part, the conclusions have not been sustained[8,72,78] because *in vivo* estrogenic potency, i.e., the true biological effectiveness of a compound, depends in part on its metabolic fate. For instance, compounds with widely varying *in vitro* affinities may have equal biological potencies because less active compounds can, *in vivo,* be metabolized to more active derivatives, or because compounds that are active *in vitro* are much more rapidly inactivated *in vivo.*

Antiestrogens were originally thought to be antagonistic to estrogens because they competed for ER binding sites.[70,71] For example, in both rat and human mammary tumors, competition for specific receptors and depression of 8 S binding peaks have been demonstrated.[79,80] From competitive inhibition studies, it was concluded that antiestrogens bind to the same site as estradiol.[34] Others, however, have proposed that antiestrogens are noncompetitive inhibitors. Hahnel *et al.,*[77] after subjecting competitive binding data to varied kinetic analyses, concludes that while the weak estrogen estriol competes for a primary estrogen-binding site, the inhibition by antiestrogens is caused by allosteric changes, indicating that there may be separate binding sites for estradiol and the antiestrogens.

Some of these questions are currently being reexamined because of the recent availability of radioactively labeled antiestrogens. According to Rochefort and Capony,[81] the affinity of antiestrogens for ER is much higher when determined directly than when evaluated by competitive experiments with estradiol. They concluded that estradiol changes the ER into a form with less affinity for antiestrogen and that this phenomenon is in fact an "*in vitro*" demonstration of ER transformation.

It is still not clear whether all antiestrogens bind to the 8 S form of ER. The antiestrogen [^{3}H]dimethylstilbestrol has an 8 S sedimentation constant when complexed to ER,[82] but studies with [^{3}H]tamoxifen fail to show such a peak, giving rise to the theory that the receptor–antiestrogen complex is an impaired form.[15] Like the binding of tamoxifen, that of [^{3}H]cortexolone (an antiglucocorticoid) is only to a 4 S form of the glucocorticoid receptor, while binding of [^{3}H]triamcinolone acetonide is to 8 S receptor.[83]

Using crude cytosol preparations, the complete answers to the relationships of estrogens and antiestrogens with the binding protein, and the characterization of the functional units of receptor, will not be attainable. Attempts are now under way to purify and characterize receptors. With these receptors, it should be possible to demonstrate whether distinct binding subunits exist, and whether their function with respect to the behavior of estrogens and their antagonists can be contrasted.

2.5. Transformation and Translocation

After association of hormones with the receptor protein, the complex undergoes transformation characterized by an increase in sedimentation rate from 4 S to 5 S and acquisition of nuclear binding capacity.[84] There is currently little, if any, evidence that antiestrogens can modify ER action at this stage. In early studies with antiestrogens, it was envisioned that the cytoplasmic receptor exists in equilibrium between an active and inactive form. The latter was stabilized by estrogen and was able to translocate, while the former, a nontranslocatable form, was stabilized by antiestrogens.[85] This model is derived from an allosteric receptor system proposed for glucocorticoid regulation of the tyrosine aminotransferase enzyme,[86] in which the postulated conformational change provoked by an inducer enhances receptor affinity for the nuclear acceptor sites, while an inhibitor such as progesterone stabilizes an untransformed form.[87] Similarly, Jordan[15] postulated that the tamoxifen–ER complex may undergo an imperfect transformation.

According to Ruh and Ruh,[88] the antiestrogen effect on ER translocation is dose-dependent. While translocation of antiestrogen-receptor complex occurs at low doses, an inhibiting effect appears at high doses, so that both antiestrogen and estradiol-receptor complex translocation fails. They also suggest that only the ER complex and not antiestrogen-receptor complex is translocated to the nucleus when antiestrogens are in direct competition with estradiol. Though similar effects have also been observed by others,[76] that estradiol itself demonstrates a biphasic dose–response curve suggests that this effect may be due to an artifact engendered by endogenous unlabeled hormone at high doses depressing the binding of label in the receptor assay.

There are several possible fates for antiestrogen-bound cytoplasmic receptor (Rc). First is the possibility that antiestrogens are not bound at all; this is untenable in light of considerable *in vivo* and *in vitro* evidence to the contrary. Second is the possibility that receptors are bound and not translocated. This can be discounted on the basis of demonstrations of nuclear ERs after antiestrogen treatment. Third, it is possible that though all Rc may be bound at sufficiently high doses, only part of the bound receptor is translocated. Katzenellenbogen and Katzenellenbogen[21] have data showing that severalfold higher concentrations of antiestrogens than would be expected from *in vitro* studies are needed to inhibit nuclear binding of subsequent [^{3}H]estradiol. One explanation for this is that antiestrogen-filled sites are less readily translocated, leaving untranslocated receptor to which [^{3}H]estradiol binds. Our data and those of others,[8,16,56] however, show that most cytoplasmic ER is potentially translocatable, if the antiestrogen dose is sufficiently high or administered continuously, and adequate time is allowed for the effect. The same may be true for weak estrogens such as estriol.[56,89] The requirement of high dose and extended time may be explained by either slowed antiestrogen entry into cells, lowered affinity of antiestrogens for Rc so that higher doses are required to saturate the receptor, or slower transfer of antiestrogen-receptor complex into the nucleus. It would seem, however, that given adequate exposure of receptor to the compound and enough time, complete translocation can be achieved, so that an explanation for estrogen antagonism cannot be found at this level.

2.6. Binding to Nuclear Receptors and Nuclear Residence Time

In vivo time-course studies using nuclear and cytosol exchange techniques to measure the movements of ER by antiestrogen clearly show that cytoplasmic depletion of Rc is paralleled by appearance of approximately equal amounts of nuclear receptor bound to the antiestrogen.[16,22,90,91] In dealing with nuclear residence of receptor, we must distinguish between two questions: the length of time the receptor remains in the nucleus and the length of time hormones remain bound to the receptor. The nuclear exchange assays, all based on the method of Anderson *et al.*,[92] employ elevated temperatures to effect exchange, and do not distinguish between hormone-filled and unfilled sites. Our data show that after exposure to tamoxifen or nafoxidine, most of the nuclear receptor is initially bound to antiestrogen. In this respect, antiestrogens probably behave like estradiol.[93] Until [^{3}H]antiestrogens are used for more direct studies, however, it will be difficult to demonstrate absolutely whether, and for how long, the nuclear receptor remains occupied by antiestrogens. The reason is that estradiol, used in exchange assays, can partially displace certain lower-affinity antiestrogens from the receptor even at 0–4°C.

Though in the rat uterus high levels of nuclear ER are achieved within 1 hr of estradiol injection, this maximum is followed by a phase (1–6 hr after estradiol) of rapid disappearance of ER from the nuclear fraction.[55,94] Antiestrogens have a markedly different effect on nuclear receptor occupancy. A single injection of nafoxidine may cause retention of the ER by uterine nuclei for as long as 19 days.[16] This atypical, long-term nuclear retention of ER after antiestrogen treatment has been repeatedly confirmed for nafoxidine[8,9,23,91] and for a variety of other antiestrogens, including other triphenylethylene derivatives[8,9,23,90,91,95] and clomiphenes,[91] as well as for long-acting derivatives of estriol.[9,19]

According to Capony and Rochefort,[23] in immature rat uteri, the majority of nafoxidine-translocated nuclear sites are hormone-bound between 4 and 48 hr, but by the 5th day, they are unoccupied. It is not known whether the presence of bound vs. free receptors is associated with a corresponding difference in their biological effects. What is clear is that in the first 24–48 hr after a single injection, presumably a time when they are still receptor-bound, antiestrogens have both early and late responses similar to those of estradiol. Responses to estradiol are usually classified as early or late (short-term or long-term) on the basis of events occurring after a single injection of the hormone. Early responses, seen within 1–4 hr, are water imbibition[96,97]; hyperemia[17,98]; amino acid and nucleotide uptake[13,14]; activation of RNA polymerases I and II[18]; RNA, lipid, and protein synthesis[56,78,99] (including stimulation of a specific "induced protein"[100]); and increased glucose metabolism.[27,56,94] Late responses measured 24 hr after hormone administration include true uterine growth (cellular hypertrophy and hyperplasia),[18,74] 2-deoxyglucose phosphorylation,[19] increased mitotic activity and DNA synthesis,[19,101] and sustained RNA polymerase I and II activity.[18]

Compounds that are capable of inducing late responses also induce early ones. The reverse, however, is not true,[18,19,89] giving rise to the classification of some estrogens as short-acting (also called partial agonists or weak estrogens). When injected together with estradiol, such weak estrogens, probably by virtue of their ability to compete with estradiol for Rc, act as partial estrogen antagonists.[89] This type of antagonism is distinctly different, however, from that brought about by antiestrogens. Clark and co-workers showed that estriol (E_3, a weak estrogen), in contrast to nafoxidine, fails to cause significant retention of nuclear receptor (Rn) and suggests that the short nuclear residence time of the RnE_3 complex accounts for its estrogen antagonism. Estriol also differs from both estradiol and nafoxidine by its failure to stimulate sustained RNA polymerase I and II activity, 24-hr RNA and protein synthesis, and increased uterine weight.[18] If, however, E_3 is administered so as to maintain continuously high blood levels,[19,89] nuclear receptor complexes remain elevated, and long-term estrogen effects are obtained.

These studies give strong support to the concept that true uterine growth requires the direct and prolonged influence of the nuclear ER complex (see also refs. 27, 56, and 102). It is therefore surprising that antiestrogens when given in multiple doses that provoke long-term retention of Rn fail to support sustained uterine growth in a manner similar to estradiol and, furthermore, antagonize the effects of estradiol when they are injected together (see Section 1.2).

To explain this paradox, it has been suggested that different nuclear binding sites exist for ER and antiestrogen-receptor complexes, or that binding of the antiestrogen-receptor complex to chromatin is somehow atypical. Several investigators[93,102–106] have observed that extraction of nuclei with 0.3–0.4 M KCl does not remove all estradiol-bound nuclear sites (RnE): most sites (80–90%) are low-affinity and extractable, while a few (10–20%) are nonextractable, high-affinity sites. It is believed that the long-term nuclear retention of the RnE complex is due to binding to the high-affinity, possibly "acceptor" sites. The existence in the nucleus of limited numbers of high-affinity binding sites has also been deduced from kinetic studies of estradiol dissociation rates[107] and exchange rates.[108]

In contrast, antiestrogens bound in the nucleus form only the weaker salt-extractable species of ER at all times studied (1–24 hr).[91,95] Furthermore, studies showing opposing effects of intercalating agents on ER and antiestrogen-receptor complex binding suggest that the two classes of compounds have different DNA site specificities.[105] Such a difference in binding sites may explain the impaired biological response or "nuclear paralysis" seen with estradiol-bound nuclear complexes following antiestrogen treatment.[22] Other studies similarly show that a complete estrogen response cannot be achieved if antiestrogen treatment precedes the estrogen injection.[8,9,15,20,24] This is in contrast to the direct effects of a single dose of antiestrogens. Recall that most antiestrogens can behave as complete estrogens after a single dose, exhibiting a full 24-hr response.[8–10] This ability does not correlate with the observation that antiestrogens form only a salt-extractable nuclear receptor complex and means that different nuclear binding sites, if they exist, will have to be demonstrated by more sophisticated techniques than salt extraction. Not all antiestrogens are potent estrogens after a single injection; MER-25 has negligible effects on uterine weight, at doses that bind and translocate Rc.[23] It is possible that the nuclear MER-25-receptor complex not only binds to inappropriate nuclear sites but also, unlike more potent antagonists, is somehow unable to regulate gene expression from these sites.

Taken together, the data lead to the conclusion that the biological ineffectiveness of some antiestrogen-receptor complexes is due in part to binding to abnormal or less responsive nuclear loci. However, until the existence of such loci is experimentally confirmed, and binding and genome activation is demonstrated under cell-free conditions, these conclusions remain only inferential. Certainly any mechanism proffered must explain why antiestrogens are first estrogenic, then subsequently become antagonists. One attempt at this explanation comes from studies on replenishment of cytoplasmic receptors.

2.7. Cytoplasmic Estrogen Receptor Replenishment

In addition to impaired nuclear binding, the antagonistic properties of antiestrogens may be caused by their effects on replenishment of Rc.

It was first thought that after antiestrogen treatment, Rc's were not restored, and that this prevented the second estrogenic effect 24 hr after the first.[8,10,76] More recent studies[8,9,23,90] show that one reason compounds may be effective estrogen antagonists is that they keep Rc levels depressed. Eventually, however, Rc is replenished, and this replenished receptor can be retranslocated. Thus, the antiestrogenic potency of a compound depends in part on the time required for Rc replenishment. This time varies with the injection regimen[9,89] and dosage[15,106] employed. Antiestrogenic potency may additionally depend on the amount of Rc replenished. Varying studies show replenishment to be greater than,[23,90] equal to,[106] or less than[56] controls. When injected together with estradiol, antagonists inhibit normal replenishment.[106] Although usually no detectable levels of Rc's are seen during the first 24 hr of antiestrogen treatment, and very little second estradiol response can be elicited,[8,9,22,27] administration of radioactive estradiol at this time does result in appearance of radioactivity in the nucleus.[22] This may mean either that there is replenished receptor in the cytoplasm, which is not measured in cytosol assays but which is translocatable, or alternatively that the nuclear bound antiestrogen can be directly displaced by estradiol. In the former case, total receptor in the nucleus after estradiol should exceed the sum of cytoplasmic and nuclear receptor measured at zero time, while if the radioactivity in the nucleus is due to exchange, total receptor in controls and after estrogen injection should remain the same. This question is not yet resolved. In any case, it appears that the new nuclear estradiol-receptor complex is biologically inactive, supporting the theory that the ER was placed at ineffective nuclear loci by the antiestrogen to begin with. Katzenellenbogen and Ferguson[22] concluded that since the Rc resynthesized after antiestrogen is biologically ineffective after secondary translocation by estradiol, there must exist nuclear responses that are inhibited much longer than Rc resynthesis.

2.8. Other Mechanisms

Estrogenic compounds may play a role in regulating cyclic AMP (cAMP) levels in cells. Mammary tumor regression can be induced by dibutyryl cAMP,[109] and antiestrogens may inhibit the actions of estrogens through an effect on cAMP levels.[110]

It is also possible that steroids at high concentrations may have nonspecific, toxic effects on cells[111] that need not involve a receptor mechanism. High doses of estrogens, in contrast to low doses, inhibit macromolecular synthesis in human breast cancer cell lines containing negligible ER levels[78] and induce solid tumor regression.[112] A similar mechanism may operate in preneoplastic mammary nodules, which are inhibited by nafoxidine but not by ovariectomy. These nodules apparently have no cytoplasmic or nuclear ER.[113]

Furthermore, if we define antiestrogens as compounds that antagonize the actions of estrogens, the natural steroids (progesterone, androgens, estriol) would be included. Progesterone and to a lesser extent testosterone prevent the full Rc restoration above control levels normally seen after estrogen treatment[29,30] and result in lowered uterine weight. This inhibition may be directed to only certain types of cells in a target tissue.[17,28,29,68] Estriol, a weak estrogen, has a short nuclear residence time. When given together with estradiol, estriol prevents long-term retention of nuclear ER.

Finally, the possibility that nuclear ER interacts with genes to provoke increase of all RNA species required for full development of the estradiol response does not exclude the further possibility that, initially, there is synthesis of only a small number of induced proteins acting as second messengers of estrogens. Such compounds might be potential sites of antiestrogen action.

It is not clear that a single unifying concept to explain antiestrogen action will be possible or even correct. The problem is compounded since most studies have been done in immature or castrated animals the tissues of which are atrophic, and with single or pulsed doses of estrogens or antiestrogens. Little is known of the effects of continuous exposure to these hormones in intact tissues, which represent the true physiological situation. It is clear that the inherent estrogenicity of a compound must be distinguished from its antagonistic properties: a compound that is an effective estrogen after one dose may fail to sustain or enhance estrogen effects in subsequent doses. Similarly, relative estrogenicity based on *in vitro* competition studies cannot be extrapolated to *in vivo* situations, in which clearance rates, tissue uptake, binding, and translocation rates of various compounds are undoubtedly quite different.

Control of receptor depletion and replenishment appears to be a more accurate indication of biological effectiveness. However, Clark's hypothesis that suppression of receptor replenishment is characteristic of all antiestrogens does not alone suffice to explain antiestrogen action; nuclear insensitivity appears to persist despite receptor replenishment.[22] Cidlowski and Muldoon[72] point out, for instance, that dimethylstilbestrol (DMS) is antiestrogenic even though it effectively elicits depletion and replenishment. It could be argued, however, that DMS exhibits all the properties of a weak estrogen and, like E_3, may not be a true antagonist.[82]

This quite naturally brings us to the question of the definition of an estrogen antagonist. In the broadest sense, we might classify compounds as antiestrogens if, by themselves, they fail to sustain long-term estrogenic effects and if, when given together with an estrogen, they inhibit the estrogen's ability to sustain these effects. In that case, weak estrogens such as estriol, other steroids such as progesterone, and the nonsteroidal antiestrogens may all be considered antagonists.

To narrow this definition, we might contrast the effects of weak estrogens such as estriol or nonestrogenic steroids such as progesterone or testosterone with the effects of the nonsteroidal antiestrogens. The former seem to act ultimately by reducing the amount of Rc later available to estrogen for translocation. The nonsteroidal antiestrogens also do this, but in addition appear to place the translocated receptor at atypical, biologically ineffective nuclear loci.

3. Antiestrogens and Experimental Breast Cancer

Since chronic treatment with antiestrogens suppresses the actions of estrogens and antagonizes estrogen-dependent responses, these agents may prove to be noninvasive tools for arresting or inhibiting estrogen-dependent tumor growth. Though this has been the tacit assumption behind use of antiestrogen therapy in humans, the experimental groundwork is fragmentary, at best. We describe below some experimental work with antiestrogens and animal breast tumor models.

3.1. Normal Mammary Gland

Little is known of the effects of antiestrogens on normal mammary tissue. Clomiphene may inhibit gland development produced by estradiol–progesterone combinations.[114] Chlormadinone acetate, a progestin with antiestrogenic properties, either inhibits or stimulates [^{3}H]estradiol uptake into normal glands in culture, depending on administration schedule.[115] In considering these results, it should be recalled that normal nonlactating mammary tissue contains little or no ER.

3.2. Antiestrogens and Mammary Tumor Induction

Estrogens have a dual, dose-dependent effect on mammary tumor induction and growth. Large doses inhibit tumor development[116] and suppress growth of established tumors.[117,118] However, they can also induce mammary tumors to develop *de novo*. Injections of lower, physiologic doses of estrogens stimulate tumor growth in ovariectomized animals.[119,120]

Antiestrogens have similarly been implicated in tumor induction and suppression. Perhaps the most dramatic demonstration of the carcinogenic potential of antiestrogens was recently reported by Clark and McCormack,[121] who showed that a single injection of clomiphene or nafoxidine in neonatal rats causes multiple abnormalities of the reproductive tract in the adult female. Similarly, in humans, exposure as fetuses to diethylstilbestrol may result in genital tract abnormalities in young women 20 years later.[122]

In the mammary gland, most studies show that antiestrogens inhibit tumor formation.[31,113,123–127] Even compounds such as estriol that are not usually considered antiestrogens but that nevertheless inhibit estrogen action also suppress tumor formation.[127] Nafoxidine and a structurally related antiestrogen, U23,469, inhibit both formation of preneoplastic nodules and their transformation into tumors.[31,126] Since the 7,12-dimethylbenz(a)anthracene (DMBA)-induced mammary tumor is considered to be estrogen- and prolactin-dependent,[31,120,128] the antitumor effects of nafoxidine have been ascribed to inhibition of prolactin secretion.[31] Interestingly, nafoxidine inhibits tumorigenesis in mammary preneoplastic nodules that are not suppressed by ovariectomy,[113] suggesting perhaps the existence in these nodules of free, biologically active estrogen receptors that can be antagonized by antiestrogens. Such a mechanism was proposed by Zava *et al.*[44] to explain growth suppression by antiestrogens of breast tumor cells in culture, in the absence of estradiol. Alternatively, this may simply be a toxic effect of the compound at high dose.

Inhibition of prolactin secretion cannot be the sole mechanism of antiestrogen-induced suppressive effects. Tamoxifen, after only 2 days of administration, produces prolonged tumor suppression despite the presence of normal estrus cycles and only weak suppression of estrogen-induced prolactin secretion.[15,123] Another antiestrogen, RU16117 (11α-methoxyethinyl estradiol), completely prevents appearance of DMBA tumors at doses that normally stimulate plasma prolactin.[124,125]

3.3. Antiestrogens and Mammary Tumor Growth

Just as they inhibit tumor induction, antiestrogens suppress growth of established mammary tumors. Understanding this effect of antiestrogens is complicated by the fact that the role of estrogens themselves is far from clear. For example, though most DMBA tumors are hormone-dependent, it is unknown whether the essential hormone(s) required for growth is prolactin, estradiol, or both.

Prolactin has been implicated as the major growth-promoting hormone,[128–130] since for short duration it supports mammary tumor growth in the absence of a pituitary, ovaries, and adrenals.[128,130] On the other hand, estradiol may in turn regulate mammary responsiveness to prolactin,[120] so that estrogens appear to be essential, but not sufficient for tumor growth (for a review, see McGuire *et al.*[131]). In endocrine-ablated rats, some tumors, nonresponsive to prolactin alone, resume growth when given prolactin–estrogen combinations.[120] Of interest is the fact that antiestrogens can similarly rescue such tumors. Nafoxidine–prolactin combinations can restore growth of prolactin-insensitive tumors,[120] suggesting an estrogenlike effect for nafoxidine in this case. Manni *et al.*[132] showed that tamoxifen fails to block prolactin-induced growth and argued that this demonstrates the predominant role of prolactin on growth. Their data are equally consistent, however, with an estrogenic effect of tamoxifen that, like nafoxidine, is potentiating the effects of prolactin.

In general, chronic administration of antiestrogen causes regression of many but not all DMBA tumors. Responses may vary depending on treatment schedules, doses, and compounds used. Nafoxidine significantly inhibits tumor growth when given chronically to intact rats.[31,133] On the other hand, not all tumors regress.[120] Furthermore, nafoxidine may have growth-stimulating properties in ovariectomized rats,[134] which may be an expression of its estrogenic activity. When U23,469 is used at very high doses (250 μg/day), it elicits regression in almost all tumors. The time course of regression is similar to that obtained with ovariectomy.[126] Tamoxifen[15,123,135,136] and CI 628[137,138] also have variable effects on tumors in intact rats: some grow, some remain static, and others regress, depending in part on the dose of hormone used, though even with large doses, variable responses persist.[137,138] Such doses either have no effect on[138] or enhance[135] prolactin levels. High doses of RU16117 suppress tumor growth in the face of markedly elevated serum prolactin. On the other hand, low doses, which have little effect on prolactin, enhance tumor growth.[124,139] Thus, it would seem that the growth-suppressive effects of this antiestrogen, as well as others, may not require parallel suppression of prolactin.

De Sombre and Arbogast[137] studied the effects of CI 628 on tumor regression and made several important observations regarding antiestrogen treatment. First, antiestrogens may be most effective in those tumors treated soon after their appearance; such tumors are more likely to be hormone-dependent, and are faster growing.[140,141] Second, the tendency of tumors to become autonomous is not prevented by chronic antiestrogen administration: the majority of tumors have incomplete remissions followed by regrowth. Other tumors seem to disappear completely, only to reappear despite continued antiestrogen treatment. The relationship of these observations to the problems of treatment regimen and recurrence of human tumors need not be belabored.

3.4. Estrogen Receptors

The mechanism by which antiestrogens inhibit mammary tumors is unknown. Since estrogens (at low doses) stimulate tumor growth, probably through ER, antiestrogens may act directly at the tumor to antagonize ER effects. The presence of ER in DMBA tumors is well documented,[119,120,126,140,142–145] though ER relationship to hormone dependence is not as clear.[119,143–145] In human tumors, the presence of ER correlates with both hormone dependence and tumor regression following endocrine therapy.[146]

At about the same time that the correlation of ER and response to hormone therapy was being described, it was first shown that a radioactive antiestrogen ([^{14}C]clomiphene) accumulates in mammary tumors,[142] and that the specific binding of [^{3}H]estradiol could be inhibited by nafoxidine and clomiphene.[133] The 8 S estrogen binding in tumor cytosols is suppressed *in vitro* by unlabeled tamoxifen,[15,127,147] even when the antiestrogen is present at relatively low doses (5 nM).[148] *In vivo* binding of [^{3}H]estradiol is also suppressed by tamoxifen pretreatment.

Antiestrogens translocate ER to tumor cell nuclei. Rc levels virtually disappear after chronic antiestrogen treatment.[133,141] Doses of RU16117 that stimulate tumor growth have no effect on Rc levels, while high doses that induce tumor regression lower Rc.[136,139] U23,469 also depletes cytoplasmic receptors and induces tumor regression despite the presence of very high (94% of total) levels of nuclear receptors. Of interest in this case is that regression of the tumor is accompanied by estrogenic stimulation of the uterus with maintainance of progesterone receptors.[126]

The time courses of receptor redistribution in DMBA-tumor cells after estradiol (5 μg) or tamoxifen (100 μg) injection are quite different.[149,150] At 30 min after estradiol treatment, Rc is depleted; replenishment is virtually complete within 4 hr. Note that this cycle is accelerated in tumor cells compared with normal target tissues such as the uterus, in which 18–24 hr is required for complete replenishment. In contrast, tamoxifen has little early effect on Rc. In fact, in the first 4 hr, a slight (10–20%) depletion is quickly followed by an increase in Rc above controls. Persistent depletion is not seen until 24 hr, and 48 hr is needed for 95% of Rc to disappear. Nuclear retention of ER is, unlike the case in rat uterus, prolonged for both estradiol and tamoxifen, but both return to control in 48 hr. Thus, if one looks at total cell receptor levels 48 hr after treatment with estradiol, receptors in the cytoplasm have returned to control levels. After tamoxifen treatment, total cellular ERs are extremely low. Receptors are absent from both cytoplasmic and nuclear compartments,[150] an effect that is distinctly different from that seen with U23,469.[126] These data, from a single laboratory, require confirmation and extension. They suggest that the dynamics of receptor movement in the tumor cell differ markedly from those in a normal cell and would have important implications. Furthermore, they may mean that after antiestrogen treatment, tumors are unresponsive to estrogen, not because of an intrinsic property of the antiestrogen-receptor complex, i.e., localization of Rn at nonfunctional sites, but because of total cellular depletion of receptor.

In sum, though it seems reasonable to suppose that antiestrogens suppress tumor growth through an effect on ER, the complete mechanism is unknown, as is the site of action. In intact animals, at least three sites (the tumor, the ovary, and the hypothalamic–pituitary axis) may be directly affected by antiestrogens and thereby lead to tumor regression. Apparently, tumor-inhibiting doses are those capable of suppressing Rc levels. Currently, the data are inconsistent regarding the levels of nuclear receptors at these doses. Other aspects of antiestrogen effects on tumors have been largely ignored. What is the behavior of ER during spontaneous or estrogen-induced escape from antiestrogen suppression described by De Sombre and Arbogast[137]? Although an autonomous cell line could arise during antiestrogen suppression, accounting for spontaneous escape, it is more difficult to envision a mechanism of estrogen-induced escape if cell receptors are completely depressed as in the studies of Nicholson *et al.*[149,150]

Total cell receptors, as well as receptor distribution, are different after ovariectomy-induced regression compared with antiestrogen-induced regression.[126] In ovariectomy-induced regression, total tumor ER levels are high, consistent with estrogen absence (see Section 5.3). After regression due to U23,469,[126] however, total receptor levels decrease markedly. We find such low, or processed, levels during periods of estrogenic stimulation (see also below). This may mean that the mechanism of regression in these two cases is quite different. In fact, it may well be that antiestrogen-induced regression is analogous to that induced by high-dose estrogens, though no studies contrasting these two methods of achieving tumor regression have been done.

4. Antiestrogens and Human Breast Cancer

Hormonal treatments, whether ablative or additive, prove successful in approximately one third of patients with breast cancer, and successful response to treatment is correlated with presence of ER in metastatic tumors.[146] Since circulating estrogens may be important factors in stimulating tumor growth, one rationale for use of estrogen antagonists has been to block estrogen binding to hormone receptors at the target tissue, thereby accomplishing, by noninvasive means, the same therapeutic ends as ablative procedures. As seen in experimental animal tumor models, however, there is no *a priori* reason to suppose that these compounds are merely antagonistic to estrogens in the sense of mimicking the effects of ovariectomy, i.e., estrogen withdrawal. Rather, in the intact animal (and human), their effects may be either estrogenic, estrogen antagonistic, or antigonadotropic.

4.1. General Endocrine Effects

Estrogens control the development and functions of the female reproductive tract and mammary gland directly through effects on the target cells and indirectly by regulation of the pituitary. Thus, multiple interaction sites will have to be considered for estrogen antagonists.

4.1.1. *Role of Reproductive Cycles: Age and Menopausal Status*

The uterine estrogenic effects of antagonists, though not manifested in adult intact animals and humans, are usually unmasked following ovariectomy. The extent of biological estrogenic and antiestrogenic activity on the pituitary may also differ in intact and oophorectomized women, and furthermore may be critically dose-dependent. With regard to the latter, estradiol can increase plasma luteinizing hormone (LH) levels at low doses, and has the opposite effect at high doses.[151] Similarly, antiestrogens have contradictory effects on gonadotropin secretion, depending on dosages. Clomiphene may stimulate gonadotropin secretion at low doses and inhibit it at high doses[152]; nafoxidine inhibits at low doses, and has no effect at high doses.[153] Thus, in sexually mature females, the role of antagonists may shift from antiestrogenic to estrogenic or antigonadotropic, depending on dose.[33] That sexual maturation may be an important determinant of the effectiveness of antiestrogens is shown by their differential effect in humans of different menopausal status. In postmenopausal women, prolactin and estrogen levels are not changed by tamoxifen,[60] while follicle-stimulating hormone (FSH) and LH are decreased.[60,63] In premenopausal women, tamoxifen has no effect on LH and only slight effect on FSH; however, estradiol levels are markedly elevated while prolactin levels are depressed.[154] Increased LH and FSH secretion is thought to explain the usefulness of tamoxifen in the treatment of oligospermia.[155]

Morgan *et al.*[156,164] found that tumor regressions in response to antiestrogen treatment are more likely to occur in women from the oldest age groups, while premenopausal women have the poorest responses. Among postmenopausal women of different ages, however, Lerner *et al.*[65] found little differences in response rate. Since more postmenopausal than premenopausal women have positive tumor ER,[146] this may also explain the former's improved response to antiestrogens, and might imply that the antiestrogens are acting directly at the tumor, through the ER system.

4.1.2. *Estrogenic Effects**

Nafoxidine therapy is successful in more than 35% of cases[155] in which it appears to be estrogenic; vaginal cytology in postmenopausal women is transformed from atrophic to estrogenic during treatment.[157,158] Doses of tamoxifen that cause tumor regression or prevent progression fail to interrupt the menstrual cycle of premenopausal women, so that complete block of estrogen action may not be required to suppress tumor growth.[61] Since tumors can be reduced further by oophorectomy, however, maximal suppression of estrogen action may be an important therapeutic goal. Initial estrogenic effects of tamoxifen may explain the flare-up of disease seen just after start of treatment.[32]

4.1.3. *Effects on Estradiol*

While Golden *et al.*[60] reported that plasma estradiol levels are unchanged by tamoxifen, Willis *et al.*[63] found that during tamoxifen therapy, estradiol levels rise, but only in treatment-failure groups. This may be due to escape from antiestrogen suppression, and may indicate differential sensitivity of target cells (pituitary?, tumor?) in the two response categories. According to Manni *et al.*,[61] estradiol levels increase slightly in all groups.

*See also Section 2.2.

4.1.4. *Effects on Pituitary**

Similarly, while Golden *et al.*[60] found prolactin levels unchanged, Willis *et al.*[63] found that tamoxifen distinguishes between two populations of patients. Prolactin levels in patients with normal basal prolactin are not affected by tamoxifen, while tamoxifen reduces prolactin in patients with initial hyperprolactinemia whose tumors regress, but fails to do this in treatment-failure groups. Since elevated prolactin is associated with failure, these authors suggest that prolactin suppression together with antiestrogen treatment would enhance response. It should be recalled, however, that regression is achieved in experimental animal tumors (Section 3.3), despite normal or elevated prolactin levels.

4.1.5. *Effect of Dose*

Though Ward[160] found relatively little difference in response between two doses of tamoxifen, studies from two other groups[33,65] show that tamoxifen may be more effective at low doses than at high doses. For unexplained reasons, estrogen antagonists are often more estrogenic at low that at high doses,[4,11,22,24–26] and they may be differentially antigonadotropic as well (see also Section 2.2). It is also possible that at high doses, there are interactions with other receptors (such as androgen receptors) or increased metabolite formation. If the latter, then at high doses, the circulating compound may consist largely of hydroxylated metabolites, the properties of which may differ considerably from those of the parent compound.[161] Cytoplasmic ER replenishment may also differ considerably at different doses.[106] It is clear that when considering antiestrogen therapy, the dosage used must be rationally selected and then perhaps modified, depending on initial changes in circulating hormones.

4.2. Estrogen Receptors

As is the case for all other endocrine therapies, effectiveness of antiestrogen therapy is correlated with presence of ERs in the tumors (Table I).

Terenius[133] showed in 1971 that some human breast cancer biopsy specimens bound estradiol specifically, and that nafoxidine and clomiphene inhibit this binding. He suggested that antiestrogen therapy

*See also Section 4.1.1.

Table I

Antiestrogen Treatment of Human Breast Cancer: Response Rate and Estrogen Receptor Correlation

			Total responders[a]		Menopausal status: responders				Estrogen receptor: responders			
					Pre		Post[b]					
Investigator	Antiestrogen	Dose	Number	%	Number	%	Number	%	ER+[c]	%	ER−	%
Hecker *et al.*[163]	Clomiphene	200 or 300 mg/day	19/50	38	0		19/50	38				
E.O.R.T.C. Breast Cancer Group[157]	Nafoxidine	60 mg × 3/day	8/23	35	0		8/23	35				
Bloom and Boesen[158]	Nafoxidine	60–90 mg × 3/day	18/48[d]	38	0/1		17/46	37				
Sasaki *et al.*[159]	Nafoxidine	180–240 mg/day	8/23	35					7/10	70	0/8	0
Ward[160]	Tamoxifen	10 mg × 2/day	12/33	36	0		12/33	36				
		20 mg × 2/day	14/35	40	0		14/35	40				
Cole *et al.*[165]	Tamoxifen	10 mg × 1 or 2/day	9/39	23	0		9/39	23				
Golder *et al.*[60]	Tamoxifen	20 mg × 2/day	10/24	42	0		10/42	42				
Willis *et al.*[63]	Tamoxifen	20 mg × 2/day	18/45	40	0		18/45	40				
Morgan *et al.*[156]	Tamoxifen	20 mg × 2/day	24/72	33	1/7	14	23/65	35	11/25[e]	44	0/6	0
Lerner *et al.*[65]	Tamoxifen	10 mg × 2/day	16/30	53	1/2		15/28	54	9/13	69	0/4	0
		≈ 25 mg × 2/day	19/44	43	1/1		18/43	42				
Manni *et al.*[61]	Tamoxifen	20 mg × 2/day	19/39	49	1/2		18/37	49	6/11	58	0/4	0
Tormey *et al.*[66]	Tamoxifen	< 12 mg/m^2 × 2/day	2/4	50					6/18[f]	30	1/2	33
		> 12 mg/m^2 × 2/day	3/14	21								
Heuson[167]	Tamoxifen	20 mg × 2/day	2/10	20	2/10	20	0					
TOTALS			201/533	38	6/23	26	181/486	37	39/77	51	1/25	4

[a] Objective criteria as defined by the E.O.R.T.C. Breast Cancer Group[157,163,167] or similar criteria[61,63,65,66,156]; subjective criteria only.[165,166]
[b] Postmenopausal: natural or ablative.
[c] (ER+) > 1 fm/mg protein[65]; (ER+) > 3 fm/mg protein[61,66]; others not stated.
[d] 1 menopausal status unknown.
[e] Includes a few patients lacking objective criteria for regression.
[f] Includes some patients with combination tamoxifen–fluoxymesterone.

might be useful for patients whose tumors had high levels of ER. In human tumor cytosols derived from biopsy specimens[80] or cells in tissue culture,[78] the 8 S binding of estradiol is suppressed by tamoxifen. Garola *et al.*[162] showed that in a tumor that contains appreciable Rc before treatment, chronic clomiphene-induced remission is accompanied by complete loss of cytoplasmic ER. Thus, as in experimental animal tumors, in humans, there is some evidence that the ER system mediates antiestrogenic effects. As a corollary, this would imply that the suppressive effects are directed against the tumor itself. As we have alluded to before, however, there is at present no reason for eliminating antiestrogen action at sites other than the tumor.

4.3. Tumor Response to Antiestrogen Therapy

Despite these potentially modifying influences, antiestrogens have been almost uniformly found to be efficacious in the treatment of breast cancer. The cumulative response rates from a number of studies are summarized in Table I. Their effectiveness equals that of other additive or ablative endocrine procedures.[146] It must be kept in mind, however, that with few exceptions, all the patients on antiestrogen treatment have been postmenopausal. The response rate in the small series of premenopausal patients has been low (Table 1 and refs. 156 and 159), and emphasizes again that the patient's endocrine status may critically determine which biological activity an antagonist may display.

Although antiestrogens are undoubtedly useful in treatment of breast cancer, it is difficult, if not impossible, to derive clues to the mechanisms involved from a review of the endocrine effects. The following section is a brief review of our studies on estrogen and antiestrogen action in cultured human breast cancer cells with the eventual goal of understanding some of these mechanisms.

5. Antiestrogens and Human Breast Cancer in Long-Term Tissue Culture

5.1. Introduction

Though antiestrogens are widely used for treatment of a variety of malignancies[168] and endocrine disorders,[156,169] their mechanism remains unclear. The major reason for this failure has been the lack of a suitable model. In the first place, the responses of the sexually immature or ovariectomized adult animal may not resemble those of the intact cycling adult. Second, most studies have been done using single or pulsed doses of hormones. This procedure fails to reproduce conditions as they actually occur in the normal animal. If our long-term goal is to explain the actions of steroids and their antagonists in humans with a view to their rational deployment in disease, then such conditions will have to be approached. Third, antiestrogen research has been hampered by inadequate knowledge of the role of peripheral and intracellular metabolism of antiestrogens. This is one explanation for the failure of *in vivo* studies to uphold findings and predictions made from *in vitro* (cell-free) work and for dose-dependent differences in responses often observed. Fourth, for the most part, studies on the mechanism of estrogen antagonists have used the uterus as a model system. There is now considerable evidence that within a specific organ, important tissue differences exist.(17,28,53,72) There is therefore little reason to expect breast cells to respond in ways analogous to uterine cells, so that for the purposes of breast-cancer-related problems, studies should be carried out using breast cells. Fifth, nuclear receptor assays have been critically deficient. Under the conditions used (exchange at high temperatures in intact nuclei), the receptors may undergo further processing or inactivation (see below), so that their precise quantitation is difficult. Thus, receptor content in the nuclei at the time of homogenization and after receptor assay at elevated temperatures may be different.

In our attempts to circumvent some or all of these problems, we have employed both a different model system and a new nuclear exchange assay. The human breast cancer cell line, MCF-7,(170) has been maintained in long-term tissue culture, contains receptors for all four major steroids known to influence breast cells,(171,172) and is estrogen-responsive.(173) These cells are ideally suited to study the influences of estrogen and its antagonists under conditions in which metabolite formation, hormone concentration, and treatment time are carefully controlled, and in which the direct actions of these compounds at the target cell can be studied independently of their indirect effect on trophic hormones. The new nuclear exchange assay employs the principle that protamine precipitates estrogen receptor.(174) The free precipitated Rc binds [^{3}H]estradiol directly(175); salt-extracted free and bound Rn's are measured directly (free, 4°C binding) or by exchange (bound, 30°C binding).(44,45,93) With this method, receptors can first be removed from nuclear inactivation or further processing sites, and then assayed under cell-free conditions.

5.2. Subcellular Distribution of Estrogen Receptors in MCF-7

We have found that estrogen receptors in MCF-7 cells have an unusual subcellular distribution.(44,45) These cells contain unfilled (free) nuclear estrogen receptors (Rn's) (Fig. 3), in addition to the usual cytoplas-

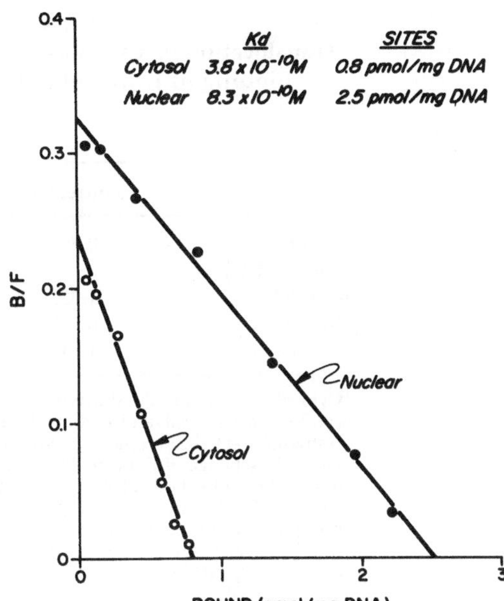

Fig. 3. Scatchard plots of specific [^{3}H]estradiol binding to unfilled cytoplasmic and salt extracted nuclear estrogen receptors from MCF-7 cells.

mic sites (Rc's). The presence of Rn's in MCF-7 cells was first indirectly suggested by the experiments of Brooks *et al.*,[171] who found that after cells were incubated at 0°C with [^{3}H]estradiol for 1 hr, the radiolabel was located in the nucleus. While they believed that estradiol had translocated Rc at 0°C, according to our data, an alternative explanation is that [^{3}H]estradiol binds directly to unfilled sites present in the nuclei of untreated cells. Unfilled nuclear ER sites can also be demonstrated in solid human breast tumors[47,48] and in other tissue culture cell lines.[46] However, receptors for other steroid hormones, though present in the cytoplasm, are not found free in nuclei of MCF-7 cells.

5.3. Effects of Estrogen Treatment

5.3.1. *Receptor Redistribution*

Rc can bind estradiol and translocate it into the nucleus simultaneous with direct binding of estradiol to Rn. Thus, within 1 hr after exposure to estradiol (Table II), Rc disappears from the cytoplasm and reappears in the nucleus as part of RnE. Rn binds E directly to become RnE, so that virtually no unfilled sites (Rc or Rn) remain within a short time, and total cellular receptors are in the nucleus in bound form.

Table II
Translocation of Cytoplasmic Receptor and Direct Binding of Estradiol to Nuclear Receptor[a]

	pmol/mg DNA	
	Untreated	Estradiol-treated
Rc (4°C)	0.391	0.021
RcE (30–4°C)	0	0.028
Rn (4°C)	1.322	0.057
RnE (37–4°C)	0	1.492
TOTALS	1.713	1.598

[a] Cytosol and nuclear salt extracts were prepared from untreated intact confluent MCF-7 cells and after 1 h of exposure to radioinert estradiol (estradiol-treated). Receptor content was determined by direct binding and exchange of [^{3}H]estradiol using the single-dose protamine assay. Unoccupied cytoplasmic (Rc) and nuclear (Rn) sites were determined by direct uptake of [^{3}H]estradiol into receptor–protamine precipitates at 4°C. Estradiol-occupied cytoplasmic (RcE) and nuclear (RnE) sites were assessed by the difference in total sites (30 or 37°C) and unoccupied (Rc or Rn) sites. From Zava and McGuire.[45]

5.3.2. *Processing of Nuclear Bound Estrogen Receptors*

If the cells are exposed to estradiol for longer periods of time (from 5 hr to several days), further changes are manifested by decreases in total nuclear receptor levels. Table III shows the effect of different, continuous concentrations of estradiol on the compartmentalization and total levels of ER. In this study, the numbers of unfilled cytoplasmic and nuclear receptors in untreated cells are approximately equal. With in-

Table III
Effect of Estradiol Dose on Estrogen Receptor Distribution and Progesterone Receptor Synthesis[a]

	Rc	Rn	RnE	Total	Processed	PgR
Control	1.79	1.80	0.22	3.81	—	0.25
10^{-12}M E_2	1.61	1.57	0.09	3.27	0.54	0.56
10^{-11}M E_2	0.99	1.44	0.18	2.61	1.20	0.69
10^{-10}M E_2	0.28	0.49	0.69	1.46	2.35	1.17
10^{-9}M E_2	0.02	0.21	0.97	1.20	2.61	1.25
10^{-8}M E_2	0	0.17	0.98	1.15	2.66	1.10
10^{-7}M E_2	0	0	1.13	1.13	2.68	1.09

[a] Cytosol and nuclear extracts were prepared from MCF-7 cells treated with varying doses of estradiol for 4 days. ER receptor content was determined by direct binding and exchange of [^{3}H]estradiol using the single-saturating-dose protamine assay. Unoccupied cytoplasmic (Rc) and nuclear receptor (Rn); occupied nuclear receptors (RnE). Progesterone receptor (PgR) measured by dextran-coated charcoal assay.

creasing doses of estradiol, there is progressive depletion of Rc and Rn. At 10^{-10} M, only 15% of cytoplasmic sites remain unfilled, and virtually complete depletion occurs at higher doses. Rc does not remain in the cytoplasm in bound form (RcE, 30°C incubation; not shown) or in cytoplasmic organelles (0.6 M KCl extract of high-speed pellets). We conclude that Rc translocates to the nucleus while simultaneously Rn sites fill, so that all receptor is in the nucleus in bound form (RnE).

The receptor is then processed in a dose-dependent fashion. At the lower doses, though Rc and Rn decrease, there is little accumulation of occupied nuclear receptors (RnE). Instead, total cellular receptors (Rc + Rn + RnE) are progressively lower. At the higher estradiol doses, free cytoplasmic and nuclear receptors are entirely depleted, but total receptor levels (as RnE) are only 30% of total cell receptor present in controls.

One interpretation of these effects is that at low doses, all RnE formed from both Rc and Rn is rapidly utilized or processed in a subsequent step to a steady-state level that is dependent on dose. The number of RnE sites that can be processed may be limited, however, so that RnE formed at higher doses remain unprocessed.

5.3.3. *Kinetics of Estrogen Receptor Processing*

The processed receptor levels seen in Fig. 4 were measured 4 days after start of treatment. If processing is an essential step in genome

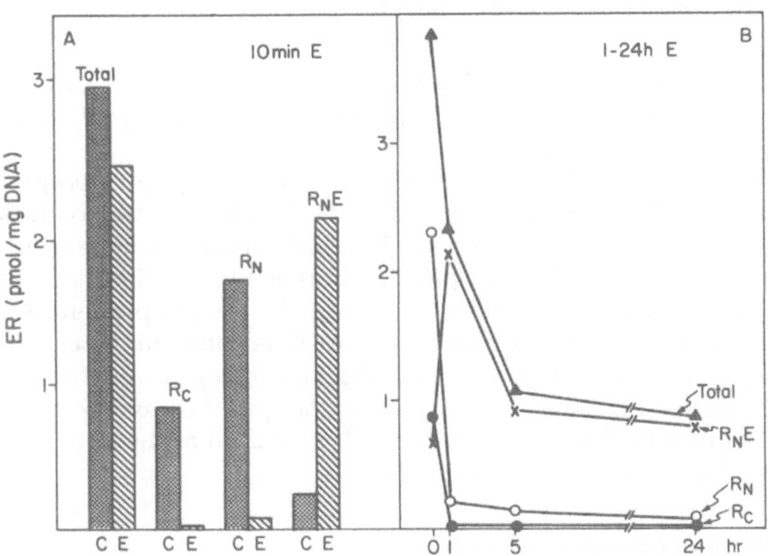

Fig. 4. Estrogen-receptor levels in MCF-7 cells treated 10 min or 1–24 hr with unlabeled estradiol. Receptors measured by single-dose protamine sulfate exchange assay. (R_C) unfilled cytoplasmic receptor, (R_N) unfilled nuclear receptor; (R_NE) filled nuclear receptor; (Total) R_C + R_N + R_NE.

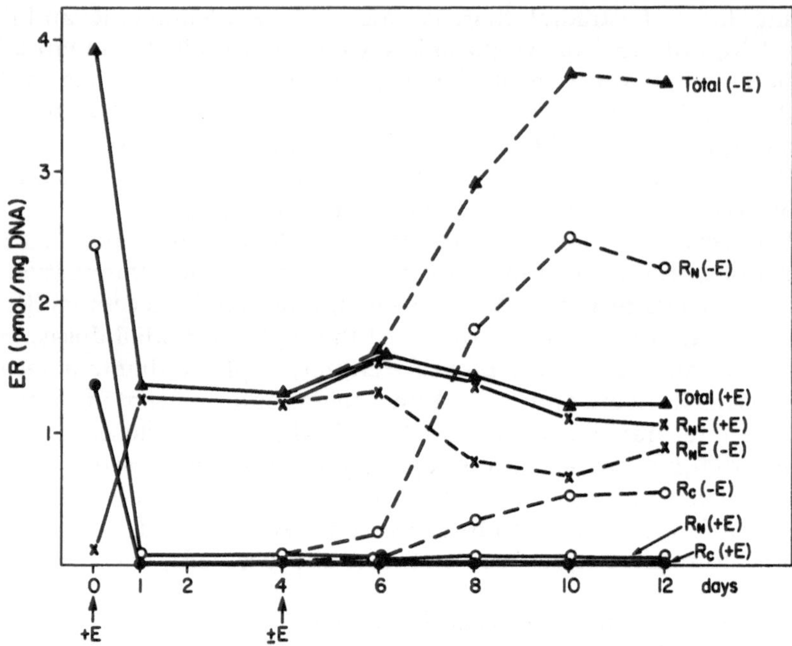

Fig. 5. Estrogen-receptor distribution in MCF-7 cells after continuous 12-day estrogen treatment (+E) or withdrawal of estrogen from days 4 to 12 (−E). Receptors measured by protamine sulfate exchange assay. (R_C) unfilled cytoplasmic receptor; (R_N) unfilled nuclear receptor; (R_NE) filled nuclear receptor; (Total) $R_C + R_N + R_NE$.

activation one would expect it to be an early event in estrogen action. Figures 4A and B show the effect of brief (10-min) or more prolonged (1- to 24-hr) 10 nM estrogen treatment.

The untreated cells shown in Fig. 4A have 70% of free receptor in the nucleus and 30% in the cytoplasm. After 10 min on estradiol, Rc and Rn are no longer measurable and all cellular receptors appear in the nucleus bound to estradiol. Despite the short incubation time (cells were treated, harvested, and cooled to 4°C within 30 min), receptor processing has started as shown by the decrease in total receptors in the estradiol-treated group. Figure 4B shows again that processing is well under way by 1 hr, so that maximal RnE buildup is not seen. Processing is essentially complete by 5 hr; thereafter, RnE is stabilized at the new steady-state level (Fig. 4B, 24 hr, and Fig. 5).

5.3.4. *What Is Nuclear Processing?*

The nature of processing is unclear. It may be an active state in which a new equilibrium between receptor degradation and synthesis is achieved,[57] or a redistribution of receptor within nuclear binding sites of differing affinities[108] or specificities,[54] or sequestration of receptor to sites inaccessible to salt extraction.[91,102]

The ER processing seen in breast cancer cells may or may not be the same phenomenon observed in the rat uterus, where bound nuclear receptors are maximal after 1 hr of estradiol treatment, with loss of 70–80% of sites by 6 hr and complete loss by 24 hr as cytoplasmic sites replenish.[102] This may mean that without continuous stimulation, once activation occurs RnE function ceases so that it is degraded[56] and Rc resynthesized. It has also been proposed, however, that continued binding and action of estrogens is required to elicit a sustained hormone response.[27,56] Similarly, the ability to stimulate uterine weight by antiestrogens or weak estrogens correlates with the time of nuclear-receptor occupancy.[9,56] This would suggest that RnE processing is not simply a mechanism to terminate RnE action. Our studies show that during continuous estrogen treatment, processing suffices only to stabilize RnE at a new steady-state level and that continued stimulation may be required to activate the genome and maintain synthetic function.

Both Rc and Rn appear to be involved in estrogen action, since the amount of receptor lost in processing is often greater than can be accounted for by total loss of receptor from one of these compartments alone. Similarly, the restoration of receptors in both compartments on estrogen withdrawal (see below) suggests that both receptors participate in estrogen action.

Most studies designed to show the effects of estrogens on subcellular ER distribution involve a single or pulsed dose of hormone,[9,55,57,102] and show shifting receptor distributions during recovery from estrogen treatment. *In vivo* cells are almost never absolutely deprived of estrogen; instead, they are under continuous, albeit fluctuating, stimulation. Under such steady-state conditions, cytoplasmic and nuclear receptor levels represent the sum of receptor synthesis, translocation, and processing. This may be much lower than receptor levels that are potentially present in the unstimulated cell or that the cell is capable of synthesizing. It is such processed levels of receptor that are being measured in biopsied human tissues.

5.3.5. *Estrogen Receptor Distribution during Replenishment*

Processed receptor levels seen during estradiol treatment return to control values when the hormone is withdrawn, showing that the cells are capable of new receptor synthesis.

The compartmentalization of ER following 4-day estrogen treatment followed by estrogen withdrawal is shown in Fig. 5. After estrogen treatment (+E), Rc and Rn disappear, and total receptor levels fall approximately 70% and are found in the nucleus as RnE. Receptor levels

Table IV
Cytosol and Nuclear Receptor Distribution in Untreated, Processed, and Receptor-Replenished Cells[a]

	Receptor (pmol/mg DNA)		
	Untreated	Processed	Replenished
Rc	1.37 (35%)	0	0.55 (15%)
Rn	2.43 (62%)	0.07 (6%)	2.25 (61%)
RnE	0.12 (3%)	1.10 (94%)	0.88 (24%)
TOTALS	3.92	1.17	3.68

[a] Cytosol and nuclear extracts were prepared from MCF-7 cells untreated with estradiol, treated 12 days with continuous 10 nM estradiol (processed), or treated 4 days with 10 nM estradiol, then 8 days without it (replenished). Receptor content was determined by direct binding and exchange of [^{3}H]-estradiol using the single-saturating-dose protamine assay. Unoccupied cytoplasmic (Rc) and nuclear receptors (Rn) measured by 4°C binding. Occupied nuclear receptors (RnE) measured by the difference in total sites (30°C) and unoccupied (Rn, 4°C) sites.

then remain unchanged during the entire 12-day course of estrogen treatment. In cells from which estradiol has been removed (−E), several effects are seen. The binding of estradiol to Rn (RnE) is remarkably prolonged. Though there is loss of RnE on days 4–12, at least some estrogen always remains bound to nuclear receptor. Thus, restoration of cell ER cannot be explained by loss of E from the nuclear receptor followed by redistribution of the newly emptied sites. Instead, both cytoplasmic receptors (Rc) and nuclear receptors (Rn) are clearly being synthesized *de novo,* and this synthesis is reflected in the restoration of total cellular ER. Since hormone withdrawal serves as a trigger of Rn reappearance, it seems unlikely that its nuclear localization is a result of translocation from the cytoplasm after ligand binding. The possibility remains that after synthesis of new receptor in the cytoplasm, some of the molecules, by an unknown mechanism, move to the nucleus.

Table IV summarizes receptor distribution in untreated, processed, and replenished cells. In the last, final Rc levels are below control while Rn levels have returned to control. Totals are also the same as controls, and the difference is in the high levels of RnE remaining in cells despite 8 days without estradiol.

5.4. Progesterone Receptor Synthesis: A Biological Response to Estrogen Action

One way to demonstrate that these shifts in ER distribution and ER levels have biological significance is by demonstrating parallel, estrogen-induced responses. Progesterone receptors (PgRs) are specific products synthesized in the uterus under the control of estrogen.[176] We have measured PgRs in breast cancer to serve as an indicator of sustained estrogenic stimulation and of the integrity of the ER system.[177] Table III shows that like the uterus, MCF-7 cells are capable of responding to estradiol treatment with increased PgR synthesis. The data also suggest that estrogen stimulation of PgR involves ER because, first, the extent of PgR induction parallels closely both the binding and translocation of Rc and the binding of Rn, and, second, PgR induction is correlated with ER processing during estradiol stimulation. When processing ceases (not shown) PgR levels fall and ER levels are restored.

5.5. Effects of Antiestrogen Treatment

After studying the effects of estradiol on ER redistribution, processing, and PgR synthesis, we contrasted these effects with the effects of antiestrogens. The following section describes results obtained using tamoxifen and nafoxidine to study MCF-7 cell growth, PgR synthesis, and ER distribution and processing.

We measured MCF-7 cell growth rate and PgR synthesis to distinguish estrogenic from antiestrogenic properties of estrogen antagonists. Leavitt *et al.*[178] showed that a variety of antiestrogens promote synthesis of uterine PgR as well as uterine weight gain, and suggested that of the two, the PgR response is a more sensitive endpoint of estrogen action.

5.5.1. Tamoxifen and Growth

We found that the effect of tamoxifen on growth is dose-dependent. Figure 6 compares the growth rate of cells given two doses of tamoxifen. Though these cells are estrogen-responsive, they are not estrogen-dependent for growth; in this study, growth of preconfluent cells is unaffected by 10 nM estradiol (not shown) or the antiestrogen tamoxifen at 0.1 μM. At 1 μM, however, tamoxifen is a potent growth inhibitor. To show whether the antiestrogen is acting through the ER system, cells were coincubated with tamoxifen and estradiol. We found that the growth suppression by 1 μM tamoxifen can be prevented by simultaneous incubation of cells with estradiol (Fig. 7); estrogen doses 10- to 100-fold lower than tamoxifen completely reverse the inhibition. The antiestrogenic suppression of growth is therefore mediated through the ER system, and is not just a nonspecific, toxic effect of the compound. Furthermore, since growth can be either suppressed or stimulated by manipulation with estrogens, the ER system is, at least in part, involved in growth regulation.

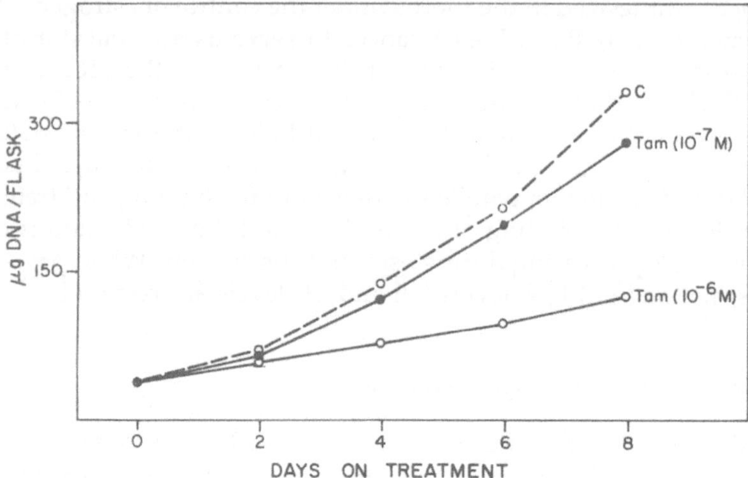

Fig. 6. Growth of cells treated with insulin, hydrocortisone, or prolactin (C) alone, or together with tamoxifen (Tam) at 10^{-6} or 10^{-7}M.

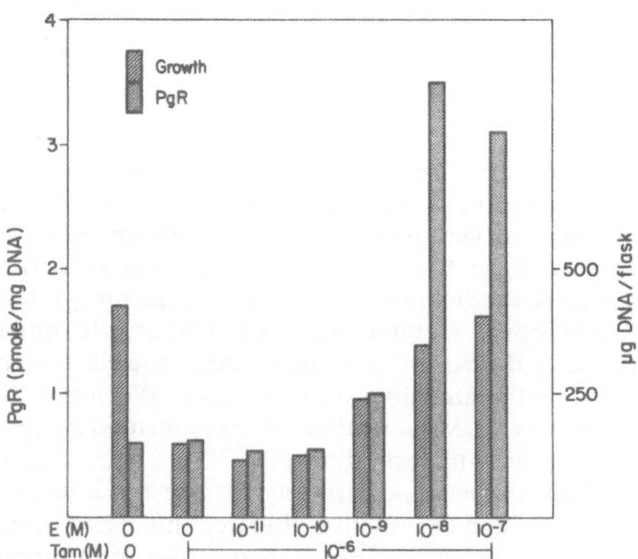

Fig. 7. Rescue of tamoxifen-inhibited growth and PgR synthesis by varying doses of estradiol.

5.5.2. *Tamoxifen and PgR*

We showed that physiologic doses of estradiol induce PgR in these cells (see Table III). Though tamoxifen at high doses (1 μM) prevents PgR induction by 0.1 nM estradiol, higher doses of estradiol can overcome this inhibition (Fig. 7). This would be expected if estradiol acts by displacing tamoxifen from ER binding site and suggests that the antiestrogen effect on PgR is also mediated by the estrogen receptor.

In view of the dose-dependent growth response of these cells to tamoxifen and estradiol, we assayed PgR and ER after varying doses of tamoxifen. We expected to completely suppress PgR so that subsequently an enhanced estradiol effect could be demonstrated. Consequently, we studied the effects of tamoxifen on PgR and ER in cells that were incubated 4 days with doses ranging from 0.1 nM to 1 μM of the antiestrogen (Fig. 8). As might be expected, with increasing doses, Rc's are progressively lost from the cytoplasm and translocated into the nucleus. The effects on PgR were, however, surprising. The low doses had little effect on PgR, but at 10 nM tamoxifen, PgR induction equaled that obtained with the same dose of estradiol used as a control. A re-

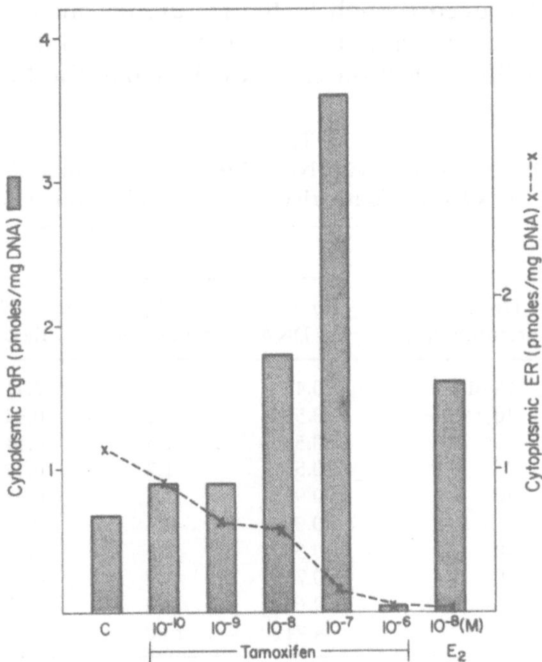

Fig. 8. Effects of 4-day treatment with estradiol or varying doses of tamoxifen on cytoplasmic ER and PgR in MCF-7 cells.

markable superinduction occurred at a higher dose (0.1 μM), and only when doses were raised further (1 μM) did true antiestrogenic properties emerge, with suppressed basal PgR levels. Tamoxifen thus shows remarkable dual properties; it is estrogenic at low doses, antiestrogenic at high doses.

To confirm that the PgR induced by tamoxifen is potentially functional, we studied its sedimentation behavior, binding affinity, and ability to be translocated to the nucleus. We found that the PgR induced by tamoxifen and by estrogen comigrate on sucrose density gradients at 7–8 S. Cytoplasmic PgR in tamoxifen treated cells has a K_d of 1.7 nM (4°C) and 0.87 nM (15°C). The receptor can be translocated to the nucleus by progesterone; 50–60% of the original cytoplasmic sites can be recovered in the nucleus in bound form.

5.5.3. *Nafoxidine*

The dual estrogenic/antiestrogenic properties seen with tamoxifen are not a general property of antiestrogens or an anomalous response of these cells to antiestrogens. Nafoxidine, another widely used antiestrogen, has very little effect on PgRs when tested over a large concentration range (Table V), even though in the rat uterus, increased PgRs have been demonstrated[178] with this compound. At lower doses (0.1–10 nM), nafoxidine has little effect on cell growth, while the higher doses are

Table V
Effect of 6-Day Treatment with Nafoxidine on Growth and Progesterone Receptor in MCF-7: Comparison with Estradiol and Tamoxifen[a]

		PgR		Growth	
Hormone	Hormone concentration (M)	pmole/ mg DNA	Fold stimulation	mg DNA/ flask	Percentage of control
None	Control	0.42	—	0.41	—
Nafoxidine	10^{-10}	0.55	1.2	0.44	107
	10^{-9}	0.55	1.3	0.44	100
	5×10^{-9}	0.52	1.2	0.42	102
	10^{-8}	0.92	2.2	0.37	90
	5×10^{-8}	0.98	2.3	0.35	85
	10^{-7}	0.90	2.1	0.27	66
	10^{-6}	0.79	1.9	0.22	54
Estradiol	10^{-8}	2.62	6.2	0.40	98
Tamoxifen	10^{-7}	4.21	10.0	0.42	102

[a] Cytosols were prepared from MCF-7 cells in each treatment group and assayed for PgR by dextran-coated charcoal assay using [^{3}H]-R5020 with and without excess unlabeled hormone. Growth was determined from the DNA content.[179]

progressively more inhibitory. However, effects on PgRs at all doses are minimal. This is in contrast to the effects of estradiol, which stimulates PgRs 6-fold, and of tamoxifen, with which PgR induction is 10-fold.

In sum, these studies show that certain estrogenic and antiestrogenic properties can be distinguished by use of growth and PgRs as indicators. When the cells are exposed to continual high levels of antiestrogens, both nafoxidine and tamoxifen inhibit cell growth, while PgRs are maintained at or below basal levels. At lower doses, however, tamoxifen but not nafoxidine can be a potent estrogen when induction of PgRs is considered as an end point. We recently showed that in the rat uterus, tamoxifen can also induce PgRs,[(90)] and it is likely that the effect is similar in mammary tumors. For instance, in DMBA tumors, PgRs are under estrogen control,[(119)] and fall precipitously in tumors regressing after ovariectomy. When regression is in response to antiestrogen treatment, however, PgR loss fails to occur,[(126,131)] suggesting that the antiestrogen is capable of maintaining induced PgR levels. The other estrogenic and antiestrogenic effects of these compounds have already been amply documented.

Our data show that antiestrogens can have dose-related biphasic actions, and that there seems to be a critical dose range below which a compound may be estrogenic and above which inhibitory effects become apparent. We do not know the mechanisms involved. Incomplete antiestrogenic effects are often seen in studies in which metabolism or injection regimen may affect hormone levels at the target organ, and in fact, dose-dependent stimulation, followed by inhibitory effects of antiestrogens, have been described for uterotrophic activity, for autoinduction of ER sites in the rooster liver, and for translocation of estrogen receptors.[(4,11,22,24–26)]

We have now described an estrogen-responsive system in which estrogenic and antiestrogenic effects can be compared, and in which ER binding, translocation, and nuclear processing mediate estrogen-induced protein synthesis. Below we contrast the actions of estrogen antagonists in this system.

5.5.4. Antiestrogen Binding to Estrogen Receptors

Using competition studies, we first determined whether antiestrogens bind to the free cytoplasmic and nuclear estrogen receptors (data not shown). Free receptors were incubated with 2×10^{-9} M (Rc) or 7×10^{-9} M (Rn) [^{3}H]estradiol. At these concentrations, (approximately 10-fold K_d), receptor sites are maximally bound while nonspecific binding is minimal. At equimolar concentrations of unlabeled estradiol, binding is reduced by 50%. When increasing does of tamoxifen or nafoxidine are added to Rc, a 50% decrease of [^{3}H]estradiol binding is achieved at 1–2 × 10^{-8} M nafoxidine and 5–6 × 10^{-8} M tamoxifen. Thus, the *in vitro* affinity of nafoxidine is 10-fold lower, and of tamoxifen is 20-fold lower, than E for Rc. By analogy, nafoxidine has 3-fold and tamoxifen 8- to 10-fold lower affinity than E for Rn.

The *in vitro* data show that the antiestrogens bind both cytoplasmic and nuclear forms of ER. *In vivo,* we see progressive depletion of Rc with increasing antiestrogen doses, as would be expected from a direct effect of hormone on receptor. To show whether Rn is bound *in vivo,* we precharged cells for 1 hr with antiestrogens, then extracted the receptor and measured the binding rate of [^{3}H]estradiol at 4°C. We reasoned that if the antiestrogens were bound to nuclear receptor, the binding of [^{3}H]estradiol would require exchange and be reduced compared with binding of free receptor. Total receptor was measured by incubation with [^{3}H]estradiol at 30°C to permit complete exchange. Table VI showed the time course of binding. Free Rn binds rapidly and is two thirds complete at 6 hr. RnE does not exchange at 4°C over an 18-hr period. Binding of [^{3}H]estradiol by RnNaf and RnTam is intermediate; the hormones exchange at rates corresponding to their affinity, suggesting that they bind Rn *in vivo.* Nafoxidine, which has a higher affinity based on competition studies, has a slower exchange rate at 4°C. Total

Table VI
Exchange Rate of Estrogens and Antiestrogen-Bound Nuclear Receptors (pmol/mg DNA)[a]

Time	Rn	RnE	RnNaf	RnTam
15 min (4°C)	0.41	0.07	0.08	0.08
30 min	0.57	0.02	0.11	0.13
60 min	0.95	0.01	0.19	0.19
90 min	1.12	0.00	0.16	0.28
2 hr	1.48	0.06	0.15	0.35
3 hr	1.84	0.06	0.19	0.47
6 hr	2.25	0.04	0.33	0.78
18 hr	3.11	0.09	0.77	2.06
Total				
5 hr (30°C)	3.80	3.27	4.63	4.12

[a] MCF-7 cells were incubated with estradiol (10^{-8} M), tamoxifen (10^{-6} M), or nafoxidine (10^{-6} M) for 1 hr at 37°C. Cells were harvested, washed, homogenized, and crude nuclei prepared by centrifugation at 800*g*. After being washed, nuclei were extracted with 0.4 M KCl, and protamine precipitates of the extracts were incubated with [^{3}H]estradiol (10 nM) alone or together with 1 μM diethylstilbestrol for the times and at the temperatures indicated. (Rn) Free nuclear receptor; (RnE) estradiol-bound nuclear receptor; (RnNaf) nafoxidine-bound nuclear receptor; (RnTam) tamoxifen-bound nuclear receptor.

nuclear receptor in the antiestrogen-treated cells is greater than in untreated because Rc is translocated and added to Rn. However, subsequent processing of nuclear receptor results in considerable decrease of total RnE after 1 hr.

5.5.5. Processing of Nuclear Estrogen Receptors with Antiestrogens

Though both estradiol and antiestrogens bind to Rc and Rn, the response of the ER is quite distinctive. All the compounds deplete Rc. This depletion is much faster with estradiol than with the antiestrogens. Similarly, appearance of bound receptor in the nucleus is accelerated with estradiol. The nuclear receptor then undergoes rapid processing; however, the steady-state level achieved for each hormone is quite different. Estradiol-bound nuclear receptor decreases to 30% of its maximal value by 5 hr (Fig. 9). Tamoxifen-bound receptor is partially processed, decreasing to 70% of its maximal value. Interestingly, nafoxidine-bound nuclear receptor (not shown) is not processed at all. Processed receptor levels remain as long as estrogen or tamoxifen are present. When these are removed, total receptor levels return to control.

Tamoxifen-induced PgR synthesis (Fig. 10) and decay coincide with ER processing. As long as ER levels are low, PgR synthesis continues even during the interval (48 hr) following tamoxifen withdrawal and

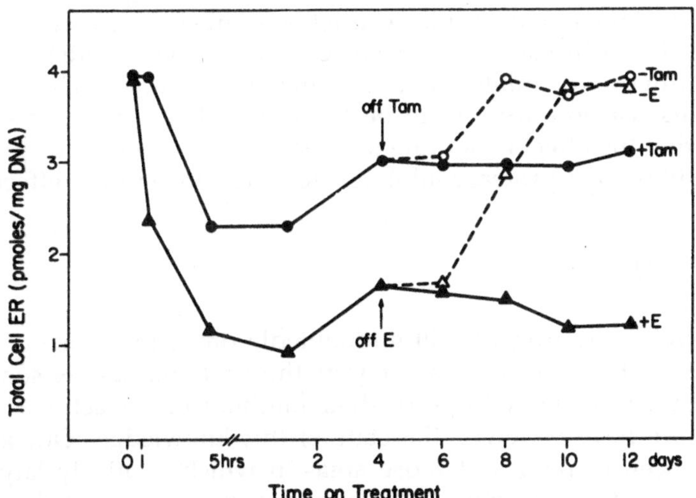

Fig. 9. Effects of continuous 12-day treatment with estradiol (10^{-8}, +E) or tamoxifen (10^{-7}M, +Tam), or withdrawal of these hormones from days 4 to 12 (−E, −Tam), on total cell ER levels. Methodological details and original data to be published elsewhere (manuscript in preparation).

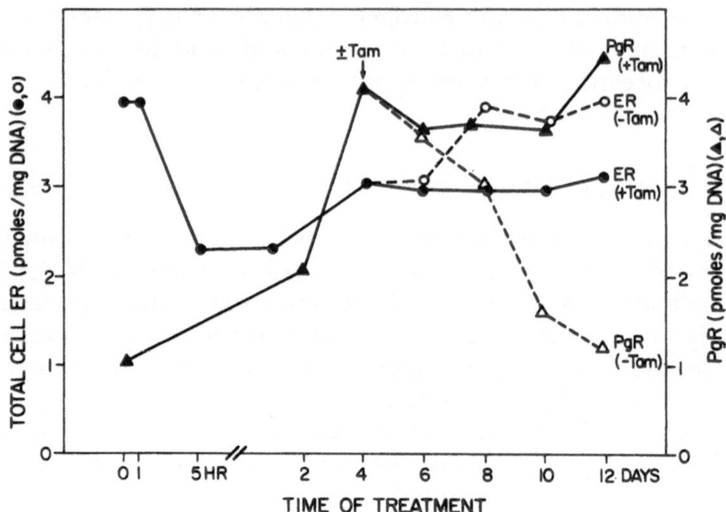

Fig. 10. Effects of continuous 12-day tamoxifen treatment (+Tam) or tamoxifen withdrawal from days 4 to 12 (−Tam) on ER and PgR levels in MCF-7 cells. Methodological details and original data to be published elsewhere (manuscript submitted).

preceding ER restoration. On restoration of ER, PgR induction ceases. We believe that processing of ER is involved in estrogenic stimulation, whether by estradiol or tamoxifen. The failure of processing seen with nafoxidine may be related to its inability to stimulate PgR synthesis. That processing is not a nonspecific destruction of nuclear receptor is shown by the fact that in some circumstances (with some ligands), it fails to occur; furthermore, unfilled Rn's also fail to be processed. Thus, triggering of this reaction appears to be hormone-related. Preliminary data suggest that the effect is not due to release of an endogenous protease, since addition of a protease inhibitor does not prevent the effect.

6. Summary

Though antiestrogens will undoubtedly prove to be useful for the treatment of breast cancer, at present their rational use is somewhat limited by scanty knowledge of their mechanisms of actions. In this chapter, we have reviewed the state of this knowledge. Our aim has partially been to point out those areas in which relatively large gaps persist—gaps that will have to be closed if the promise of these compounds is to be completely fulfilled. We feel that, in particular, areas that require extensive further studies are, first, development of model systems that resemble the intact, adult human; second, studies of prohormonal forms and metabolic fates of estrogen antagonists, both extra- and intracellularly; third, distinguishing between effects of antiestrogens at target tissues (i.e., the uterus or breast tumor) and indirect effects on other endocrine glands; fourth, the role of the ER system and particularly the nuclear sites of action of the antiestrogen–receptor complex, including the means by which genomic stimulation occurs, and the mechanisms that truly distinguish antiestrogen from estrogen actions at these sites.

ACKNOWLEDGMENTS: We thank Dr. J. P. Raynaud (Roussel Uclaf) for the R-5020, Dr. H. Soule (Michigan Cancer Foundation) for the MCF-7 cells, and L. Trench (ICI chemicals) for the tamoxifen. Studies from the authors' laboratory are supported in part by grants from the NIH (CA 11378, CB 23862), the American Cancer Society (BC-23G), and the Robert A. Welch foundation.

7. References

1. P. P. Carbone, Editorial: Antiestrogens and breast cancer treatment, *Ann. Intern. Med.* **83**(5), 730–731 (1975).
2. G. W. Duncan, S. C. Lyster, J. J. Clark, and D. Lednicer, Antifertility activities of two diphenyl-dihydronaphthalene derivatives, *Proc. Soc. Exp. Biol. Med.* **112,** 439–442 (1963).
3. G. DiPasquale, C. L. Rassaert, E. McDougall, and L. Tripp, Action of an estradiol-17β antagonist in intact, ovariectomized, hypophysectomized and hypophysectomized–ovariectomized rats, *Contraception* **5,** 39–51 (1972).
4. L. Terenius, Structure–activity relationships of anti-estrogens with regard to interaction with 17β-oestradiol in the mouse uterus and vagina, *Acta Endocrinol.* **66,** 431–447 (1971).
5. L. Terenius, Hexoestrol analogues as probes of oestrogen receptors. II. Importance of hydrogen-bonding groups for binding to uterine tissue and for uterotrophic activity, *Acta Pharmacol. Toxicol.* **31,** 449–455 (1972).
6. V. C. Jordan, Prolonged antioestrogenic activity of ICI 46,474 in the ovariectomized mouse, *J. Reprod. Fertil.* **42,** 251–258 (1975).
7. K. D. Schulz, and S. August, Female endocrine control mechanisms during the neonatal period, *Acta Endocrinol.* **74,** 144–156 (1973).
8. B. S. Katzenellenbogen, E. R. Ferguson, and N. C. Lan, Fundamental differences in the action of estrogens and antiestrogens on the uterus: Comparison between compounds with similar duration of action, *Endocrinology* **100,** 1252–1259 (1977).
9. E. R. Ferguson and B. S. Katzenellenbogen, A comparative study of antiestrogen action: Temporal patterns of antagonism of estrogen stimulated uterine growth and effects on estrogen receptor levels, *Endocrinology* **100,** 1242–1251 (1977).
10. J. H. Clark, J. N. Anderson, and E. J. Peck, Jr., Oestrogen receptors and antagonism of steroid hormone action, *Nature (London)* **251,** 246–248 (1974).
11. C. W. Emmens and L. Martin, Biological activities of U-11100A, *J. Reprod. Fertil.* **9,** 269–275 (1965).

12. L. Terenius and I. Ljungkvist, Aspects on the mode of action of antiestrogens and antiprogestogens, *Gynecol. Invest.* **3,** 96–107 (1972).
13. K. D. Schulz, S. August, K. Gosde, and G. Kramer, Studies on the anti-oestrogen-like action of clomiphene citrate in animal experiments, *Gynecol. Invest.* **3,** 135–141 (1972).
14. B. S. Katzenellenbogen and J. A. Katzenellenbogen, Antiestrogens: Studies using an *in vitro* estrogen-responsive uterine system, *Biochem. Biophys. Res. Commun.* **50,** 1152–1159 (1973).
15. V. C. Jordan, Antiestrogenic and antitumor properties of tamoxifen in laboratory animals, *Cancer Treatment Rep.* **60,** 1409–1419 (1976).
16. J. H. Clark, J. N. Anderson, and E. J. Peck, Jr., Estrogen receptor anti-estrogen complex: A typical binding by uterine nuclei and effects on uterine growths, *Steroids* **22,** 707–718 (1973).
17. I. Ljungkvist and L. Terenius, MER 25 and U11-100A, two antiestrogens with tissue selective and incomplete estrogenic activity, on rat uterus, *Contraception* **10,** 395–404 (1974).
18. J. W. Hardin, J. H. Clark, S. R. Glasser, and E. J. Peck, Jr., RNA polymerase activity and uterine growth: Differential stimulation by estradiol, estriol, nafoxidine, *Biochemistry* **15,** 1370–1374 (1976).
19. N. C. Lan and B. S. Katzenellenbogen, Temporal relationships between hormone receptor binding and biological responses in the uterus: Studies with short- and long-acting derivatives of estriol, *Endocrinology* **98,** 220–227 (1976).
20. B. R. Komisaruk and C. Beyer, Differential antagonism, by MER-25, of behavioral and morphological effects of estradiol benzoate in rats, *Horm. Behav.* **3,** 63–70 (1972).
21. B. S. Katzenellenbogen and J. A. Katzenellenbogen, Antiestrogens: Studies using an *in vitro* estrogen-responsive uterine system, *Biochem. Biophys. Res. Commun.* **50,** 1152–1159 (1973).
22. B. S. Katzenellenbogen and E. R. Ferguson, Antiestrogen action in the uterus: Biological ineffectiveness of nuclear bound estradiol after antiestrogen, *Endocrinology* **97,** 1–12 (1975).
23. F. Capony and H. Rochefort, *In vivo* effect of anti-estrogen on the localization and replenishment of estrogen receptor, *Mol. Cell. Endocrinol.* **3,** 233–251, 1975.
24. W. H. Bulger and D. Kupfer, Induction of uterine ornithine decarboxylase (ODC) by antiestrogens—Inhibition of estradiol-mediated induction of ODC: A possible mechanism of action of antiestrogens, *Endocrinol. Res. Commun.* **3,** 209–218 (1976).
25. M. K. Harper and A. L. Walpole, A new derivative of triphenylethylene: Effect on implantation and mode of action in rats, *J. Reprod. Fertil.* **13,** 101–119 (1967).
26. M. Geschwendt, The effect on antiestrogens on egg yolk protein synthesis and estrogen-binding to chromatin in the rooster liver, *Biochem. Biophys. Acta* **399,** 395–402 (1975).
27. J. N. Anderson, E. J. Peck, Jr., and J. H. Clark, Nuclear receptor–estradiol complex: A requirement for uterotrophic responses, *Endocrinology* **95,** 174–178 (1974).
28. J. Mester, D. Martel, A. Psychoyos, and E. E. Baulieu, Hormonal control of oestrogen receptor in uterus and receptivity for ovoimplantation in the rat, *Nature (London)* **250,** 776–778 (1974).
29. A. J. W. Hsueh, E. J. Peck, Jr., and J. H. Clark, Progesterone antagonism of the oestrogen receptor and oestrogen-induced uterine growth, *Nature (London)* **254,** 337–339 (1975).
30. A. J. W. Hseuh, E. J. Peck, Jr., and J. H. Clark, Control of uterine estrogen receptor levels by progesterone, *Endocrinology* **98,** 438–444 (1976).
31. J. C. Heuson, C. Waelbroeck, N. Legros, G. Gallex, C. Robyn, and M. L. Hermite, Inhibition of DMBA-induced mammary carcinogenesis in the rat by 2-

Br-α-ergocryptine (CB 154), an inhibitor of prolactin secretion, and by nafoxidine (U-11,100A), an estrogen antagonist, *Gynecol. Invest.* **2,** 130–137 (1971/1972).
32. D. C. Tormey, R. M. Simon, M. E. Lippman, J. M. Bull, and C. E. Myers, Evaluation of tamoxifen dose in advanced breast cancer: A progress report, *Cancer Treatment Rep.* **60,** 1451–1459 (1976).
33. K. D. Schulz, S. August, K. Gasde, and G. Kramer, Studies on the anti-oestrogenic and oestrogen-like action of clomiphene citrate in animal experiments, *Gynecol. Invest.* **3,** 135–141 (1972).
34. B. S. Katzenellenbogen, Synthesis and inducibility of the uterine estrogen-induced protein, IP, during the rat estrous cycle: Clues to uterine estrogen sensitivity, *Endocrinology* **96,** 289–297 (1975).
35. E. J. Peck, Jr., J. Burgner, and J. H. Clark, Estrophilic binding sites of the uterus: Relation to uptake and retention of estradiol *in vitro, Biochemistry* **12,** 4596–4603 (1973).
36. E. Milgrom, M. Atger, and E. E. Baulieu, Studies on estrogen entry into uterine cells and on estradiol receptor complex attachment to the nucleus—Is the entry of estrogen into uterine cells a protein-mediated process?, *Biochem. Biophys. Acta* **320,** 267–283 (1973).
37. R. J. B. King and W. I. P. Mainwaring, *Steroid–Cell Interactions,* pp. 190–262, University Park Press, Baltimore (1974).
38. G. C. Chamness and W. L. McGuire, Estrogen receptor in the rat uterus: Physiological forms and artifacts, *Biochemistry* **11,** 2466–2472 (1972).
39. A. C. Notides and S. Nielsen, The molecular mechanism of the *in vitro* 4 S to 5 S transformation of the uterine estrogen receptor, *J. Biol. Chem.* **249,** 1866–1873 (1974).
40. E. V. Jensen and E. R. DeSombre, Mechanism of action of the female sex hormones, *Annu. Rev. Biochem.* **41,** 203–230 (1972).
41. K. R. Yamamoto, On the specificity of the binding of the estradiol receptor protein to deoxyribonucleic acid, *J. Biol. Chem.* **249,** 7068–7075 (1974).
42. G. A. Puca and F. Bresciani, Receptor molecules for oestrogens from rat uterus, *Nature (London)* **218,** 967–969 (1968).
43. E. V. Jensen, T. Suzuki, T. Kawashima, W. E. Stumpf, P. W. Jungblut, and E. R. DeSombre, A two step mechanism for the interaction of estradiol with rat uterus, *Proc. Natl. Acad. Sci. U.S.A.* **59,** 632–638 (1968).
44. D. T. Zava, G. C. Chamness, K. B. Horwitz, and W. L. McGuire, Human breast cancer: Biologically active estrogen receptor in the absence of estrogen?, *Science* **197,** 663–664 (1977).
45. D. T. Zava and W. L. McGuire, Unoccupied sites in nuclei of a breast tumor cell line, *J. Biol. Chem.* **252,** 3703–3708 (1977).
46. G. Sonnenschein, A. M. Soto, J. Cologiore, and R. Farookhi, Estrogen target cells: Establishment of a cell line derived from the rat pituitary tumor MtT/F_4, *Exp. Cell Res.* **101,** 15–22 (1976).
47. R. Garola and W. L. McGuire, An improved assay for nuclear estrogen receptor in experimental and human breast cancer, *Cancer Res.* **37,** 3333–3337 (1977).
48. R. Garola and W. L. McGuire, Estrogen receptor and proteolytic activity in human breast tumor nuclei, *Cancer Res.* **37,** 3329–3332 (1977).
49. D. O. Toft, The interaction of uterine estrogen receptors with DNA, *J. Steroid Biochem.* **3,** 515–522 (1972).
50. R. J. B. King and J. Gordon, Involvement of DNA in the receptor mechanism for uterine estradiol receptors, *Nature (London) New Biol.* **240,** 185–186 (1972).
51. T. C. Spelsberg, A. W. Steggles, and B. W. O'Malley, Progesterone binding components of chick oviduct. III. Chromatin acceptor sites, *J. Biol. Chem.* **246,** 4188–4197 (1971).

52. T. Liang and S. Liao, Association of the uterine 17β-estradiol receptor complex with ribonucleoprotein *in vitro* and *in vivo*, *J. Biol. Chem.* **249,** 4671–4678 (1974).
53. V. Jackson and G. R. Chalkley, The binding of estradiol 17β to the bovine endometrial nuclear membrane, *J. Biol. Chem.* **249,** 1615–1627 (1974).
54. W. T. Schrader, D. O. Toft, and B. W. O'Malley, Progesterone-binding protein of chick oviduct. VI. Interaction of purified progesterone-receptor components with nuclear constituents, *J. Biol. Chem.* **247,** 2401–2407 (1972).
55. G. Giannopoulos and J. Gorski, Estrogen receptors: Quantitative studies on transfer of estradiol from cytoplasmic to nuclear binding sites, *J. Biol. Chem.* **246,** 2524–2529 (1971).
56. J. N. Anderson, E. J. Peck, Jr., and J. H. Clark, Estrogen-induced uterine responses and growth: Relationship to receptor estrogen binding by uterine nuclei, *Endocrinology* **96,** 160–167 (1975).
57. M. Sarff and J. Gorski, Control of estrogen binding protein concentration under basal conditions and after estrogen administration, *Biochemistry* **10,** 2557–2563 (1971).
58. A. Matsuzawa and T. Yamamoto, Inhibited growth *in vivo* of a mouse pregnancy-dependent mammary tumor (TPDMT-4) by an antiestrogen, 2 alpha, 3 alpha-epithio-5 alpha-androstan-17 beta-ol (10275-S), *Cancer Res.* **36,** 1598–1606 (1976).
59. R. I. Nicholson and M. P. Golder, The effect of synthetic anti-oestrogens on the growth and biochemistry of rat mammary tumours, *Eur. J. Cancer* **11,** 571–579 (1975).
60. M. P. Golder, M. E. A. Phillips, D. R. Fahmy, P. E. Preece, V. Jones, J. H. Henks, and K. Griffiths, Plasma hormones in patients with advanced breast cancer treated with tamoxifen, *Eur. J. Cancer* **12,** 719–723 (1976).
61. A. Manni, J. Trujillo, J. S. Marshall, and O. H. Pearson, Antiestrogen-induced remissions in stage IV breast cancer, *Cancer Treatment Rep.* **60,** 1445–1450 (1976).
62. V. C. Jordan and S. Koerner, Tamoxifen as an anti-tumor agent: Role of oestradiol and prolactin, *J. Endocrinol.* **68,** 305–311(1976).
63. K. J. Willis, D. R. London, H. W. Ward, W. R. Butt, S. S. Lynch, and B. T. Rudd, Recurrent breast cancer treated with the antiestrogen tamoxifen: Correlation between hormonal changes and clinical course, *Br. Med. J.* **1,** 425–428 (1977).
64. P. A. Kelly, J. Asselin, M. G. Caron, F. Labrie, and J. P. Raynaud, Potent inhibitory effect of a new antiestrogen (RU 16117) on the growth of 7,12-dimethylbenz(a)anthracene-induced rat mammary tumors, *J. Natl. Cancer Inst.* **58,** 623–628 (1977).
65. H. J. Lerner, P. R. Baud, L. Israel, and B. S. Leung, Phase II study of tamoxifen: Report of 74 patients with Stage IV breast cancer, *Cancer Treatment Rep.* **60,** 1431–1435 (1976).
66. D. C. Tormey, R. M. Simon, M. E. Lippman, J. H. Bull, and C. E. Myers, Evaluation of tamoxifen dose in advanced breast cancer: A progress report, *Cancer Treatment Rep.* **60,** 1451–1459 (1976).
67. S. G. Richardson and E. Killen, Metabolism of oestradiol by human mammary tumor 800 × *g* supernatants pretreated with dihydrolipoic acid, *Cancer Lett.* **2,** 299–304 (1977).
68. W. Levin, R. M. Welch, and A. H. Cormey, Decreased uterotrophic potency of oral contraceptives in rats pretreated with phenobarbital, *Endocrinology* **83,** 149–156, 1968.
69. W. Levin, R. M. Welch, and A. H. Cormey, Effect of phenobarbital and other drugs on the metabolism and uterotrophic action of estradiol 17-β and estrone, *J. Pharmacol. Exp. Ther.* **159,** 361–371 (1968).
70. S. G. Korenman, Relation between estrogen inhibitory activity and binding to cytosol of rabbit and human uterus, *Endocrinology* **87,** 1119–1123 (1970).
71. S. G. Korenman, Comparative binding affinity of estrogens and its relation to estrogenic potency, *Steroids* **13,** 163–177 (1969).

72. J. A. Cidlowski and T. G. Muldoon, Dissimilar effects of antiestrogens upon estrogen receptors in responsive tissues of male and female rats, *Biol. Reprod.* **15,** 381–389 (1976).
73. H. Rochefort and F. Capony, Binding properties of an anti-estrogen to the estradiol receptor of uterine cytosol, *FEBS Lett.* **20,** 11–15 (1972).
74. L. Terenius, Two modes of interaction between oestrogen and anti-oestrogen, *Acta Endocrinol.* **64,** 47–58 (1970).
75. C. Martucci and J. Fishman, Uterine estrogen receptor binding of catecholestrogens and of estetrol [1,3,5,(10)-estratriene-3,15α,16α,17β-tetrol], *Steroids* **27,** 325–333 (1976).
76. L. J. Black and R. J. Kraay, Evaluation of two types of estrogen inhibition with regard to effects on uptake and binding of ^{3}H-β-estradiol in the uterus, *J. Steroid Biochem.* **4,** 467–475 (1973).
77. R. Hahnel, E. Twaddle, and T. Ratajczak, The influence of synthetic antiestrogens on the binding of tritiated estradiol-17β by cytosols of human uterus and human breast carcinoma, *J. Steroid Biochem.* **4,** 687–695 (1973).
78. M. Lippman, G. Bolan, and K. Huff, Interactions of antiestrogens with human breast cancer in long term tissue culture, *Cancer Treatment Rep.* **60,** 1421–1429 (1976).
79. W. Powell-Jones, P. Davies, and K. Griffiths, Influence of antiestrogens on specific binding of ^{3}H-β-oestradiol *in vitro* by nuclei from rat mammary tumors, *J. Endocrinol.* **66,** 437–438 (1975).
80. V. C. Jordan and S. Koerner, Tamoxifen (ICI 46,474) and the human carcinoma 8 S oestrogen receptor, *Eur. J. Cancer* **11,** 205–206 (1975).
81. H. Rochefort and F. Capony, Estradiol dependent decrease of binding inhibition by antiestrogens (a possible test of receptor activation), *Biochem. Biophys. Res. Commun.* **75,** 277–285 (1977).
82. F. Capony and H. Rochefort, *In vitro* and *in vivo* interactions of ^{3}H-dimethylstilbestrol with the estrogen receptor, *Mol. Cell. Endocrinol.* **8,** 47–64 (1977).
83. R. W. Turnell, N. Kaiser, R. J. Milholland, and F. Rosen, Glucocorticoid receptors in rat thymocytes: Interactions with the antiglucocorticoid cortexolone and mechanism of its action, *J. Biol. Chem.* **249,** 1133–1138 (1974).
84. E. R. DeSombre, S. Mohla, and E. V. Jensen, Receptor transformation, they key to estrogen action, *J. Steroid Biochem.* **6,** 469–473 (1975).
85. H. Rochefort, F. Lignon, and F. Capony, Effect of antiestrogens on uterine estradiol receptors, *Gynecol. Invest.* **3,** 43–62 (1972).
86. H. H. Samuels and G. M. Tomkins, Relation of steriod structure to enzyme induction in hepatoma tissue culture cells, *J. Mol. Biol.* **52,** 57–74 (1970).
87. G. G. Rousseau, Interaction of steriods with hepatoma cells: Molecular mechanisms of glucocorticoid hormone action, *J. Steroid Biochem.* **6,** 75–89 (1975).
88. T. S. Ruh and M. F. Ruh, The effect of antiestrogens on the nuclear binding of the estrogen receptor, *Steroids* **24,** 209–224 (1974).
89. J. H. Clark, Z. Paszko, and E. J. Peck, Jr., Nuclear binding and retention of the receptor estrogen complex: Relation to the agonistic and antagonistic properties of estriol, *Endocrinology* **100,** 91–96 (1977).
90. Y. Koseki, D. T. Zava, G. C. Chamness, and W. L. McGuire, Estrogen receptor translocation and replenishment by the antiestrogen tamoxifen, *Endocrinology* **101,** 1104–1110 (1977).
91. T. S. Ruh and L. J. Baudendistel, Different nuclear binding sites for antiestrogen and estrogen receptor complexes, *Endocrinology* **100,** 420–426 (1977).
92. J. Anderson, J. H. Clark, and E. J. Peck, Jr., Oestrogen and nuclear binding sites: Determination of specific sites by ^{3}Hβ oestradiol exchange, *Biochem. J.* **126,** 561–567 (1972).

93. D. T. Zava, N. Y. Harrington, and W. L. McGuire, Nuclear estradiol receptor in the adult rat uterus: A new exchange assay, *Biochemistry* **15,** 4292–4297 (1976).
94. J. N. Anderson, E. J. Peck, Jr., and J. H. Clark, Nuclear receptor–estrogen complex: Relationship between concentration and early uterotrophic responses, *Endocrinology* **92,** 1488–1495 (1973).
95. J. S. Rinehart, T. S. Ruh, and M. S. Ruh, Antiestrogen action: Uterine nuclear retention of the CI-628 antiestrogen receptor complexes *in vitro, Acta Endocrinol.* **84,** 367–373 (1977).
96. C. M. Szego and S. Roberts, Steroid action and interaction in uterine metabolism, *Recent Prog. Horm. Res.* **8,** 419–469 (1953).
97. F. L. Hisaw, Comparative effectiveness of estrogens on fluid imbibition and growth of the rat's uterus, *Endocrinology* **64,** 276–289 (1959).
98. J. Gorski, Estrogen binding and control of gene expression in the uterus, in: *Handbook of Physiology, Endocrinology II, Part 1* (R.O. Greep, ed.), pp. 525–536, American Physiology Society, Washington D.C. (1973).
99. T. H. Hamilton, Isotopic studies on estrogen-induced accelerations of ribonucleic acid and protein synthesis, *Proc. Natl. Acad. Sci. U.S.A.* **49,** 373–379 (1963).
100. T. S. Ruh, B. S. Katzenellenbogen, J. A. Katzenellenbogen, and J. Gorski, Estrone interaction with the rat uterus: *In vitro* response and nuclear uptake, *Endocrinology* **92,** 125–134 (1973).
101. F. Stormshak, R. Leake, N. Wertz, and J. Gorski, Stimulatory and inhibitory effects of estrogen on uterine DNA synthesis, *Endocrinology* **99,** 1501–1511 (1976).
102. J. H. Clark and E. J. Peck, Jr., Nuclear retention of receptor–estrogen complex and nuclear acceptor sites, *Nature (London)* **260,** 635–637 (1976).
103. J. Mester and E. E. Baulieu, Dynamics of oestrogen-receptor distribution between the cytosol and nuclear fractions of immature rat uterus after oestradiol administration, *Biochem. J.* **146,** 617–623 (1975).
104. J. V. Juliano and G. H. Stancel, Estrogen receptors in the rat uterus: Retention of hormone–receptor complexes, *Biochemistry* **15,** 916–920 (1976).
105. L. J. Baudendistel and T. S. Ruh, Antiestrogen action: Differential nuclear retention and extractability of the estrogen receptor, *Steroids* **28,** 223–237 (1976).
106. V. C. Jordan, C. J. Dix, L. Rowsby, and G. Prestwich, Studies on the mechanism of action of the nonsteroidal antiestrogen tamoxifen (ICI 46,474) in the rat, *Mol. Cell. Endocrinol.* **7,** 177–192 (1977).
107. J. M. Sala-Trepat and E. Reti, Dissociation studies of different forms of the estradiol–receptor complex from calf uterus: Higher stability of the complex bound to chromatin, *Biochem. Biophys. Acta* **338,** 92–103 (1974).
108. R. De Hertogh, E. Ekka, I. Vanderheyden, and J. J. Hoet, Slowly exchangeable pool of estradiol in the rat uterus, *J. Steroid Biochem.* **4,** 313–320 (1973).
109. Y. S. Cho-Chung and B. H. Redler, Dibutyryl cyclic AMP mimics ovariectomy: Nuclear protein phosphorylation in mammary tumor regression, *Science* **197,** 272–275 (1977).
110. S. M. Paul and P. Skolnik, Catechol oestrogens inhibit oestrogen elicited accumulation of hypothalamic cyclic AMP suggesting role as endogenous antiestrogens, *Nature (London)* **266,** 559–561 (1977).
111. E. B. Thompson and M. E. Lippman, Mechanism of action of glucocorticoids, *Metab. Clin. Exp.* **23,** 159–202 (1974).
112. J. Meites, Relation of prolactin and estrogen to mammary tumorigenesis in the rat, *J. Natl. Cancer Inst.* **48,** 1217–1224 (1972).
113. D. Medina, Tumor formation in preneoplastic mammary nodule lines in mice treated with nafoxidine, testoterone, and 2-bromo-alpha-ergocryptine, *J. Natl. Cancer Inst.* **58,** 1107–1110 (1977).

114. J. Richards and D. R. Griffith, Effects of *cis*- and *trans*-clomiphene on mammary gland development in the rat, *Fertil. Steril.* **25,** 74–78 (1974).
115. G. Bedes, The effect of chlormadione acetate upon estradiol uptake by the rat mammary gland in organ culture, *Am. J. Obstet. Gynecol.* **188,** 1050–1053 (1974).
116. C. Huggins, R. C. Moon, and S. Morii, Extinction of experimental mammary cancer. I. Estradiol 17-β and progesterone, *Proc. Natl. Acad. Sci. U.S.A.* **48,** 379–386 (1962).
117. G. S. Kledzik, C. J. Bradley, S. Marshall, G. A. Campbell, and J. Meites, Effects of high doses of estrogen on prolactin-binding activity and growth of carcinogen-induced mammary cancers in rats, *Cancer Res.* **36,** 3265–3268 (1976).
118. J. Meites, E. Cassell, and J. Clark, Estrogen inhibition of mammary tumor growth in rats: Counteraction by prolactin, *Proc. Soc. Exp. Biol. Med.* **137,** 1225–1227 (1971).
119. K. B. Horwitz and W. L. McGuire, Progesterone and progesterone receptors in experimental breast cancer, *Cancer Res.* **37,** 1722–1738 (1977).
120. B. S. Leung and G. H. Sasaki, On the mechanism of prolactic and estrogen action in 7,12-dimethylbenz(a)anthracene-induced mammary carcinoma in the rat. II. *In vivo* tumor responses and estrogen receptors, *Endocrinology* **97,** 564–572 (1975).
121. J. H. Clark and S. McCormack, Clomid or nafoxidine administered to neonatal rats causes reproductive tract abnormalities, *Science* **197,** 164–165 (1977).
122. D. C. Poskanzer and A. L. Herbst, Epidemiology of vaginal adenosis and adenocarcinoma associated with exposure to stilbestrol *in utero, Cancer* **39,** 1892–1895 (1977).
123. V. C. Jordan, Effect of tamoxifen (ICI 46,474) on initiation and growth of DMBA-induced rat mammary carcinomata, *Eur. J. Cancer* **12,** 419–424 (1976).
124. F. Labrie, P. A. Kelly, J. Asselin, and J. P. Raynaud, Potent inhibitory activity of a new antiestrogen, RU 16,117, on the development and growth of DMBA-induced rat mammary adenocarcinoma, *Recent Results Cancer Res.* **57,** 109–120.
125. P. A. Kelly, J. Asselin, M. G. Caron, J. P. Raynaud, and F. Labrie, High inhibitory activity of a new antiestrogen, RU 16117 (11 alpha-methoxy ethinyl estradiol), on the development of dimethylbenz(a)anthracene-induced mammary tumors, *Cancer Res.* **37,** 76–81 (1977).
126. T. L. Tsai and B. S. Katzenellenbogen, Antagonism of development and growth of 7,12 dimethylbenz(a)anthracene-induced rat mammary tumors by the antiestrogen U23,469 and effects on estrogen and progesterone receptors, *Cancer Res.* **37,** 1537–1543 (1977).
127. L. Terenius, Effect of anti-oestrogens on initiation of mammary cancer in the female rat, *Cancer* **7,** 65–70 (1971).
128. O. H. Pearson, O. Llerena, L. Llerena, A. Molina, and T. Butler, Prolactin-dependent rat mammary cancer: A model for man?, *Trans. Assoc. Am. Physicians* **82,** 225–238 (1969).
129. A. Sterental, J. M. Dominguez, C. Weissman, and O. H. Pearson, Pituitary role in the estrogen dependency of experimental mammary cancer, *Cancer Res.* **23,** 481–485 (1963).
130. H. Nagasawa and R. Yanai, Effects of prolactin or growth hormone on growth of carcinogen-induced mammary tumors of adeno-ovariectomized rats, *Int. J. Cancer* **6,** 488–495 (1970).
131. W. L. McGuire, G. C. Chamness, M. E. Costlow, and K. B. Horwitz, in: *Hormone Receptors in Breast Cancer* (G. S. Levy, ed.), pp. 265–299, Marcel Dekker, New York (1976).
132. A. Manni, J. E. Trujillo, and O. H. Pearson, Predominant role of prolactin in stimulating the growth of 7,12-dimethylbenz(a)anthracene-induced rat mammary tumor, *Cancer Res.* **37,** 1216–1219 (1977).
133. L. Terenius, Anti-oestrogens and breast cancer, *Eur. J. Cancer* **7,** 57–64 (1971).

134. G. Gallez, J. C. Heuson, and C. Waelbroeck, Growth stimulating effect of nafoxidine on rat mammary tumor after ovariectomy, *Eur. J. Cancer* **9,** 699–700 (1973).
135. V. C. Jordan and S. Koerner, Tamoxifen as an antitumor agent: Role of oestradiol and prolactin, *J. Endocrinol.* **68,** 305–311 (1976).
136. V. C. Jordan and T. Jaspan, Tamoxifen as an anti-tumor agent: Oestrogen binding as a predictive test for tumor response, *J. Endocrinol.* **68,** 453–460 (1976).
137. E. R. DeSombre and L. Y. Arbogast, Effect of the antiestrogen CI 628 on the growth of rat mammary tumors, *Cancer Res.* **34,** 1971–1976 (1974).
138. R. I. Nicholson and M. P. Golder, The effect of synthetic anti-oestrogens on the growth and biochemistry of rat mammary tumor, *Eur. J. Cancer* **11,** 571–579 (1975).
139. P. A. Kelly, J. Asselin, M. C. Caron, F. Labrie, and J.-P. Raynaud, Potent inhibitory effect of a new antiestrogen (RU 16117) on the growth of 7,12-dimethylbenz(a)-anthracene induced rat mammary tumors, *J. Natl. Cancer Inst.* **58,** 623–628 (1977).
140. D. P. Griswold, Jr., and C. H. Green, Observation on the hormone sensitivity of 7,12-dimethylbenz(a)anthracene-induced mammary tumors in the Sprague–Dawley rat, *Cancer Res.* **30,** 819–826 (1970).
141. C. J. Bradley, G. S. Kledzik, and J. Meites, Prolactin and estrogen dependency of rat mammary cancers at early and late stages of development, *Cancer Res.* **36,** 319–324 (1976).
142. K. D. Schulz, B. Haselmayer, and F. Holzel, The influence of clomid and its isomers on dimethyl benzathracene-induced rat mammary tumors, in: *Basic Actions of Sex Steroids on Target Organs,* pp. 274–279, Karger, Basel (1971).
143. W. L. McGuire and J. Julian, Comparison of macromolecular binding of estradiol in hormone-dependent and hormone-independent rat mammary carcinomata, *Cancer Res.* **31,** 1440–1445 (1971).
144. E. R. DeSombre, G. Kledzik, S. Marshall, and J. Meites, Estrogen and prolactin receptor concentrations in rat mammary tumors and response to endocrine ablation, *Cancer Res.* **36,** 354–358 (1976).
145. I. M. Holdaway and H. G. Friesen, Correlation between hormone binding and growth response of rat mammary tumor, *Cancer Res.* **36,** 1562–1567 (1976).
146. W. L. McGuire, P. P. Carbone, M. E. Sears, and G. C. Escher, Estrogen receptors in human breast cancer: An overview, in: *Estrogen Receptors in Human Breast Cancer* (W. L. McGuire, P. P. Carbone, and E. P. Vollmer, eds.), pp. 1–7, Raven Press, New York (1975).
147. V. C. Jordan and L. J. Dowse, Tamoxifen as an anti-tumor agent: Effect on oestrogen binding, *J. Endocrinol.* **68,** 297–303 (1976).
148. W. Powell-Jones, D. A. Jenner, R. W. Blamey, P. Davies, and K. Griffiths, Influence of anti-oestrogens on the specific binding *in vitro* of ^{3}H-β-oestradiol by cytosol of rat mammary tumours, *Biochem. J.* **150,** 71–75 (1975).
149. R. I. Nicholson, M. P. Golder, P. Davies, and K. Griffiths, Effect of oestradiol-17β and tamoxifen on total and accessible cytoplasmic oestradiol-17β receptors in DMBA-induced rat mammary tumours, *Eur. J. Cancer* **12,** 711–717 (1976).
150. R. I. Nicholson, P. Davies, and K. Griffiths, Effects of oestradiol-17β and tamoxifen on nuclear oestradiol-17β receptors in DMBA-induced rat mammary tumors, *Eur. J. Cancer* **13,** 201–208 (1977).
151. M. Callantine, Nonsteroidal estrogen antagonists, *Clin. Obstet. Gynecol.* **10,** 74–87 (1967).
152. S. Roy, V. B. Manesh, and R. B. Greenblatt, Effects of clomiphene on the physiology of reproduction in the rat. I. Changes in the hypophyseal–gonadal axis, *Acta Endocrinol.* **47,** 645–656 (1964).
153. G. W. Duncan and A. D. Forbes, Blastocyst survival and nidation in rats treated with oestrogen antagonists, *J. Reprod. Fertil.* **10,** 161–167 (1965).

154. G. V. Groom and K. Griffiths, Effect of the antiestrogen tamoxifen on plasma levels of luteinizing hormone, follicle-stimulating hormone, prolactin, oestradiol, and progesterone in normal pre-menopausal women, *J. Endocrinol.* **70,** 421–428 (1976).
155. F. Comhaire, Treatment of oligospermia with tamoxifen, *Int. J. Fertil.* **21,** 232–238 (1976).
156. L. R. Morgan, Jr., P. S. Schein, P. V. Woolley, D. Hoth, J. Macdonald, M. Lippman, L. E. Posey, and R. W. Beazley, Therapeutic use of tamoxifen in advanced breast cancer: Correlation with biochemical parameters, *Cancer Treatment Rep.* **60,** 1437–1443 (1976).
157. E.O.R.T.C. Breast Cancer Group, Clinical trial of nafoxidine, an oestrogen antagonist in advanced breast cancer, *Eur. J. Cancer* **8,** 387–389 (1972).
158. H. J. G. Bloom and E. Boesen, Antiestrogens in treatment of breast cancer: Value of nafoxidine in 52 advanced cases, *Br. Med. J.* **2,** 7–10 (1974).
159. G. H. Sasaki, B. S. Leung, and W. S. Fletcher, Therapeutic use of nafoxidine in advanced breast cancer—A correlation with endocrine ablation and tumor estrogen response, *Proc. Am. Soc. Clin. Oncol.* **16,** 271 (1975).
160. H. W. C. Ward, Antiestrogen therapy for breast cancer: A trial of tamoxifen at two dose levels, *Br. Med. J.* **1,** 13–14 (1973).
161. J. M. Fromson, S. Pearson, and S. Bramali, The metabolism of tamoxifen (ICI 46,474). II. In female patients, *Xenobiotica* **3,** 711–714 (1973).
162. R. Garola, C. M. Levy, I. Vegh, C. Magin, J. C. Martinez, and E. Hecker, *In vivo* blockade of the estradiol-binding protein (EBP) by clomiphene citrate in human breast cancer, *Oncology* **30,** 105–112 (1974).
163. E. Hecker, I. Vegh, C. M. Levy, C. A. Magin, J. C. Martinez, J. Loureiro, and R. E. Garola, Clinical trial of clomiphene in advanced breast cancer, *Eur. J. Cancer* **10,** 747–749 (1974).
164. L. R. Morgan, P. S. Schein, D. Hoth, J. McDonal, L. E. Posey, R. W. Beazley, and L. Trench, Therapeutic use of tamoxifen in advanced breast cancer: Correlation with biochemical parameters, *Proc. Am. Soc. Cancer Res.* **17,** 126 (1976).
165. H. P. Cole, C. T. A. Jones, and I. D. H. Todd, A new anti-oestrogenic agent in late breast cancer: An early clinical appraisal of ICI 46,474, *Br. J. Cancer* **25,** 270–274 (1971).
166. M. J. O'Halloran and P. G. Maddock, ICI 46,474 in breast cancer, *J. Ir. Med. Assoc.* **67,** 38–39 (1974).
167. J. C. Heuson, Current overview of E.O.R.T.C. clinical trials with tamoxifen, *Cancer Treatment Rep.* **60,** 1463–1466 (1976).
168. S. Legha and F. M. Muggia, Antiestrogens in the treatment of cancer, *Ann. Intern Med.* **84,** 751 (1976).
169. M. C. Macnaughton, Treatment of female infertility, in: *Clinics in Endocrinology* (J. A. Loraine, ed.), pp. 545–560, W. B. Saunders, London (1973).
170. H. D. Soule, J. Vasquez, A. Long, S. Albert, and M. Brennan, A human cell line from a pleural effusion derived from a breast carcinoma, *J. Natl. Cancer Inst.* **51,** 1409–1416 (1973).
171. S. C. Brooks, E. R. Locke, and H. D. Soule, Estrogen receptor in a human cell line (MCF-7) from breast carcinoma, *J. Biol. Chem.* **248,** 6251–6253 (1973).
172. K. B. Horwitz, M. E. Costlow, and W. L. McGuire, MCF-7: A human breast cancer cell line with estrogen, androgen, progesterone and glucocorticoid receptors, *Steroids* **26,** 785–795 (1975).
173. M. E. Lippman and G. Bolan, Oestrogen-responsive human breast cancer in long term tissue culture, *Nature (London)* **256,** 592–593 (1975).
174. A. W. Steggles and R. J. B. King, The use of protamine to study 6,7-^{3}H oestradiol-17β binding in rat uterus, *Biochem. J.* **118,** 695–701 (1970).

175. G. C. Chamness, K. Huff, and W. L. McGuire, Protamine-precipitated estrogen receptor: A solid phase ligand exchange assay, *Steroids* **25,** 627–635 (1975).
176. E. Milgrom, L. Thi, M. Atger, and E.-E. Baulieu, Mechanisms regulating the concentration and the conformation of progesterone receptors in the uterus, *J. Biol. Chem.* **247,** 8000–8004 (1972).
177. K. B. Horwitz, W. L. McGuire, O. H. Pearson, and A. Segaloff, Predicting resonse to endocrine therapy in human breast cancer: A hypothesis, *Science* **189,** 726–727 (1975).
178. W. W. Leavitt, T. J. Chen, and T. C. Allen, Regulation of progesterone receptor formation by estrogen action, *Ann. N. Y. Acad. Sci.* **286,** 210–225 (1977).
179. K. Burton, A study of the conditions and mechanism of the diphenylamine reaction for the colorimetric estimation of deoxyribonucleic acid, *Biochem. J.* **62,** 315–323 (1956).

6

Biological Markers in Breast Cancer

RONALD B. HERBERMAN

1. Introduction

Many important problems related to the biology, diagnosis, and management of breast cancer require the accurate assessment of the presence of tumors and of their size and extent of growth. Biological markers are substances made by tumors, or substances or biological phenomena closely associated with the presence of tumors. These markers may be specific for breast cancer, or they may be quantitatively altered in tumor-bearing subjects, and thereby can aid in the identification of tumors and in the assessment of tumor burden. Although some useful markers can be measured chemically, most have been detected by immunological assays. In recent years, there has been growing interest in the clinical application of immunological assays for the evaluation of patients with cancer. This increasing interest has been due in part to recent advances in radioimmunoassays (RIAs) and other highly sensitive immunological techniques that allow measurement of picogram or nanogram quantities of antigens. In addition, immunological assays can be exquisitely specific, and in some instances can discriminate between molecules with differences in a single amino acid or sugar. Such procedures have provided the basis of discrimination of products of various types of neoplastic cells from normal cellular materials.

In this chapter, I will review the various clinical applications of biological markers in breast cancer, and will discuss the various issues

RONALD B. HERBERMAN • Laboratory of Immunodiagnosis, National Cancer Institute, Bethesda, Maryland 20014.

and problems related to each of the possible applications. Following this discussion, I will review and attempt to assess the current status of the wide variety of markers that have been reported to be associated with human breast cancer.

2. Potential Clinical Applications

A high level of effort has been expended and continues to be expended on research related to biological markers in breast cancer. It is important to note, however, that only a few markers have been definitely shown to have a place in clinical oncology. This is largely a reflection of the difficulties involved in satisfactory transfer of technology from the research laboratory to the bedside. Many of the problems are not unique to immunodiagnosis, but are also presented by other types of diagnostic tests, including some that have been available for many years and some that have been incorporated into widespread use without real validation or objective assessment of utility. It is important for laboratory investigators and for clinicians using tests for tumor markers to be aware of these difficulties and limitations, and to recognize the general criteria for useful markers:

(1) A first, obvious point is that the test must be able to detect some consistent difference between cancer and noncancer. It is desirable, but not necessary, that the difference be qualitative. The presence of a marker in cancer patients that is absent in nonneoplastic states, or the loss of a normal component in cancer patients, would provide strong bases for development of a useful diagnostic test. However, quantitative differences between cancer patients and controls could also be sufficient. Either increased levels of normal substances above the normal range, or decreases below the normal range, could provide useful information.

(2) A good diagnostic test should have a high degree of specificity; there should be very few false positives, i.e., subjects without cancer who have tests indicating cancer. In this sense, the percentage specificity of a test may be defined as

$$\left[1 - \text{Incidence of false positive tests}\left(\frac{\text{Number of positive tests among subjects without cancer}}{\text{Total noncancer subjects tested}}\right)\right] \times 100$$

(3) Also, the test should be very sensitive and have few false negative results, i.e., be able to detect cancer in a large proportion of cancer patients. The percentage sensitivity of a test may be defined as

$$\left[1 - \text{Incidence of false negative tests}\left(\frac{\text{Number of negative tests among cancer patients}}{\text{Total cancer patients tested}}\right)\right] \times 100$$

A particularly useful test for a tumor marker would be one that is positive even in cancer patients with localized tumors or small, metastatic deposits that are asymptomatic and undetectable by conventional diagnostic tests.

(4) Some tests for tumor markers give only qualitative results, i.e., are either positive or negative. A test is much more valuable, however, if it provides quantitative information on the levels of the marker. Tests that have a large quantitative range between clinically detectable tumors and absence of tumors are particularly useful, since they offer the possibility of closely monitoring changes with tumor burden and are most likely to provide indications of small amounts of tumor. Many tumor markers are produced by the tumor cells themselves, and their levels would therefore be expected to be dependent on the mass of tumor. The levels of tumor products may be influenced, however, by a variety of factors: (a) Number of tumor cells present. (b) Proportion of tumor cells synthesizing the marker and the synthetic rate per cell. Only certain cells within a tumor may make a marker, and production may vary with the phase in the cell cycle and with the stage of differentiation of the cell. (c) Location of marker within tumor cell and mechanism for release from cells and entry into circulation. Some tumor markers are cell-membrane constituents or secretory products and may be shed or released from viable cells. Other markers may be intracellular constituents that would be released only when the tumor cells lose viability. Some tumor markers released from solid tumors might enter the circulation in appreciable quantities only after invasion of blood vessels. With such markers, levels in the region of the tumor or in directly contiguous body fluids or excretions might be much higher than in the circulation, and testing of these levels might be more useful than tests on serum. (d) The half-life of the marker in the circulation can also vary considerably, depending on the size and nature of the substance. If the circulating marker is immunogenic to the host and antigen–antibody complexes are formed, it is likely to be cleared much more rapidly than a nonimmunogenic marker. With markers that are not produced directly by the tumor, but are produced in response to tumor growth, the relationship between marker levels and the mass and extent of spread of the tumor might be quite different, and it is not possible to set down any general principles for such reactive markers.

(5) It would be very helpful if a test to be used for initial detection or diagnosis could provide information about the tumor type and location. One of the main concerns that has been raised about detection of occult, clinically undetectable cancer is the difficulty in determining what type of cancer it is and where it is located. Clearly, at present, more information than a diagnosis of "cancer, type and site unknown" would be needed for rational therapy. Specificity of the immunological test for a particular organ site or histological type of cancer is therefore an important factor initially, but is not as essential for monitoring of previously diagnosed patients.

A further issue relates not so much to the marker itself as to the design of the studies to evaluate the usefulness of the marker. It is essential that the measurements and the data analyses be performed objectively without knowledge of the clinical diagnosis or status of the patient. To achieve this objectivity it is important that the laboratory receive coded specimens, without any identifiers as to source or type of donor. Further, it is necessary to design appropriate studies for the particular clinical application for which the tumor marker will be used. Adequate studies of this type have been performed with very few of the available markers, and even then for only some of the possible clinical applications. The oncologist can and should play a central role in designing good studies, to provide solid information on the clinical value of a tumor marker. Each of the types of clinical applications for tumor markers will be discussed in detail below. It should be noted that most tests for markers may be considered for their utility for several distinct clinical applications: (1) Some tumor markers might be useful in the detection of cancer cases by screening of general populations or of groups at high risk of developing cancer. (2) With patients who have come to a physician with signs or symptoms consistent with, or suggestive of, cancer, marker assays may aid in distinguishing between patients with cancer and those with benign diseases. In patients with known cancer, tumor markers may help in (3) the localization of tumor and (4) determining the stage of disease and prognosis. (5) Furthermore, after primary therapy, serial testing of patients may aid in the early detection of recurrences or metastases.

Another point to emphasize in regard to the usefulness of various assays for one or more of the applications discussed above is that a single test may not have sufficient specificity or sensitivity, but the simultaneous use of several tests may provide highly discriminatory data. It is possible that assays for two or more tumor markers would have additive or synergistic effects for improving the sensitivity or specificity of detection of tumor cells. The studies exploring the use of a combination of markers in breast cancer will be reviewed in Section 3.12.

2.1. Screening

In the past few years, much attention and effort have been directed toward screening women for breast cancer. Women have been encouraged to go to their physicians or to hospitals for physical and radiological examinations, and many centers have set up large-scale mammography screening programs. These efforts have been based mainly on evidence that early detection of localized breast tumors may result in a substantial improvement in the efficacy of therapy and consequently in prolonged survival.[1–3] Recently, however, there has been recognition of some of the problems or limitations of these screening programs: (1) The oncogenic risk of mammography, especially in women under age 55.[2] (2) The sensitivity and specificity of physical and mammographic screening procedures are substantially less than 100%. For example, it should be noted that the false positive rate from mammography may be 80%[1] or higher. This high rate results in a significant morbidity from the biopsies that need to be performed after a positive mammogram.

It is therefore very desirable and important to develop alternative, noninvasive screening procedures that will reduce the risks in screening and improve the accuracy of early detection of breast cancer. Unfortunately, the use of marker tests for screening for cancer is probably the most difficult of the various potential applications to bring to fruition. If a test has been shown objectively to discriminate between cancer and control groups, it then has to be evaluated for its use in screening by a study with an appropriate design. Despite the large number of breast-cancer-related markers, none has thus far been directly evaluated for its use in screening. Many factors can affect the feasibility of a particular test for screening purposes. Most of these factors need to be extensively considered before a study on the possible usefulness of screening can be initiated.

(1) It is particularly important that the assay be relatively simple and practical for applications to testing of very large numbers of specimens or subjects. The procedures must be sufficiently well developed and standardized that reproducible results can be obtained over time and in many laboratories. Further, a screening test, which will be given to a high proportion of normal subjects, should present little or no risk to the recipients.

(2) A suitable population must be available for study. For rapid identification of a useful test, it is very helpful to identify populations or families at high risk of developing breast cancer (see, for example, Anderson[4]). It is very important to have sufficient access to the population, to permit retesting as appropriate, and to perform extensive clinical evaluations, particularly of the test-positive subjects. Furthermore, most

of the subjects in the population must be available for clinical follow-up, over a period of several years, to determine which initially disease-free subjects, among both the test-positive and the test-negative subjects, subsequently develop cancer.

(3) Of fundamental concern is the specificity of the tumor marker assay. It is difficult to make general statements about the acceptable levels of specificity for screening tests. However, since physical examinations and mammography have already been shown to be effective screening procedures, it would be important for a new screening test to be shown to have better specificity than these techniques, or to lead to improved specificity when used in conjunction with these procedures.

(4) A further important issue is that the test should be very sensitive. To be useful as a screening procedure, the test should be able to detect asymptomatic subjects bearing small, localized tumors, at a time when the disease is treatable and has not yet metastasized. The longer the lead time (i.e., the interval between test positivity and clinical detection of disease) a test provides, the more likely it is that the test will contribute to better response to therapy and to survival. It should be noted that to accurately determine the lead time for a particular assay, tests must be performed repeatedly, to establish the point when the test first becomes positive. The amount of lead time a test might provide, and the likely clinical benefits to be accrued by early detection of disease, are likely to be determined in considerable part by the rate of tumor growth. Screening tests are more likely to be useful for slowly growing tumors with long latent periods than for explosively growing tumors that metastasize early. The previous experience with breast cancer screening programs has provided some information on these issues, particularly in regard to the lead time provided by physical and mammographic screening procedures. In the HIP Breast Cancer Screening Project, the lead time was estimated to be about one year.[5] This earlier detection of cases has been associated with an overall increased survival in the screened population and a decreased case fatality rate among the women with breast cancer diagnosed by mammography alone. The screening, however, was not sufficiently sensitive to detect many cases of breast cancer, which were detected by other means within 12 months of screening. These cases have been ascribed to false negatives from screening and to rapidly advancing breast cancer.[1] The mammography screening also detected a higher proportion of breast cancer cases with no extension of disease to the axillary lymph nodes[1]; even so, one of four patients had nodal involvement at the time of screening. To augment our current screening capabilities, it would be desirable to develop other screening tests with biological markers that could provide a longer lead time and detect cases prior to any metastases, regional or systemic.

2.2. Adjunct to Diagnosis of Patients with Signs or Symptoms

Some of the issues discussed above for screening also apply to the application of tumor markers as adjuncts to the diagnosis of cancer in patients with signs or symptoms, or both, suggestive of breast cancer. The concerns regarding the sensitivity and specificity of the assay, however, are somewhat different. In regard to specificity in this setting, one is not concerned with identifying the small proportion of subjects with cancer in a very large population of normal subjects. Rather, it is necessary to discriminate among a group of subjects with breast disease, to determine which have malignant vs. benign diseases. This presents particular problems, since most forms of cancer arise out of the background of an older population with a variety of underlying benign chronic diseases, often affecting the same organ system as the cancer. This is especially true in breast cancer, since there is an increased risk of cancer in subjects with chronic fibrocystic disease.[6] Therefore, for a test to be expected to be useful as an adjunct to the initial diagnosis of breast cancer, it should be able to discriminate between patients with cancer and patients with chronic fibrocystic disease or fibroadenomas. However, if a major cause for false positive results for a tumor marker is benign conditions involving organs other than the breast, e.g., the liver, it may be possible to perform the relevant test to rule out those other conditions.

In regard to sensitivity, the test would not have to be able to detect very small, asymptomatic lesions. To be useful as a diagnostic adjunct, however, it should still be capable of detecting most resectable or otherwise treatable localized tumors.

2.3. Aid to Histopathological Evaluation of Tumors

Recently, the value of evaluation of tumor specimens for the presence of various markers has begun to be increasingly appreciated. This analysis may provide assistance in the histopathological classification of the tumor. The presence or the amount of a marker may provide useful prognostic information, since this may reflect the state of differentiation, immunogenicity, or metastatic potential of the tumor.

Another very important aspect of marker evaluation in tumors is to provide needed information for the subsequent monitoring of the patient. Identification of one or more markers within the tumor would provide a solid basis for use of the assays for those markers for following the course of disease. Although similar information might be obtained by testing for the markers in a pretherapy serum sample, direct examination of the tumor is likely to be more sensitive and specific. In patients with small, localized tumors, the serum levels before therapy are often in the normal range despite active production of the markers by the tumor. Therefore, the failure of the patient to have an initial elevated marker level should not rule out the possible later use of that marker. In fact, with some of the markers discussed below, such a disparity between tumor and circulatory levels has been noted. In contrast, failure to detect a marker within the tumor would make it much less likely to be subsequently detectable. However, although there is little evidence for this, it remains possible that some primary tumors would be negative for a marker but the metastases might become positive. This could be envisioned if production of the marker were more likely in a less differentiated or more aggressively growing tumor cell. This possibility needs to be directly explored.

Very recently, improved methods have been developed for examination of markers within tumors. In addition to the usual studies of intact cells or tissue sections by immunofluorescence, it is now possible to look for the distribution of markers in fixed and stained tissue sections, using conventional light microscopy. This has been made possible by the development of immunoperoxidase staining techniques. Using this procedure, one can now accurately determine both the presence of the marker and its location within various cell types in the tumor.

2.4. Aid in Staging

The measurement of tumor markers in newly diagnosed breast cancer patients, on serum obtained before any therapy or after primary surgical removal of tumor, can be very useful as an aid in the assessment of stage of disease. There are two distinct aspects to the use of tumor markers in staging, which will be discussed separately below.

Since the circulating levels of some tumor markers have been found to be dependent on the overall tumor burden and on the extent of spread of the tumor, determination of marker levels on preoperative specimens can provide useful prognostic information. Elevated pretherapy levels, particularly when quite high, may suggest the presence of metastases and poor prognosis. They might also reflect the state of differentiation of the tumor and its inherent aggressiveness. Most studies on this aspect of marker utilization have been performed on groups of patients, to determine the overall relationship between marker level and extent of tumor, as determined at surgery and by subsequent clinical course. To supplement information obtainable by other means, the marker data should be able to make prognostic discriminations among patients with the same clinical or histopathological stage of disease. If this is possible, then the use of markers in conjunction with clinical and other laboratory information could provide the basis for improved staging of patients and for the administration of therapy appropriate to the assessed extent of disease. This type of application might be particularly useful in breast cancer, since adjuvant chemotherapy after mastectomy appears useful in treating residual micrometastases. It should be noted that the type of data provided in this area provides information on the probable prognosis of patients within a group, rather than on the clinical status of the individual patient. As with other staging criteria, the presence or absence of elevated levels of a marker could not be taken as definitive evidence for occult metastases vs. localized tumor.

One of the major challenges to the field of immunodiagnosis is to make the transition from population studies to a study in which statements can be made regarding the individual patient, on which therapy or other important clinical decisions may be based with some degree of certainty. One promising approach in this regard is the testing for levels of a circulating tumor cell product after primary therapy, especially surgery, to determine whether all the tumor was removed or eradicated. Persistence of elevated levels might provide strong evidence for residual tumor at the primary site or for regional or distant metastases. After mastectomy, this persistence could indicate the need for axillary lymph node dissection and removal, or for radiotherapy or chemotherapy. On the other hand, fall of elevated levels of a tumor-produced marker into the normal range might provide some assurance of curative surgery, and obviate the need for further therapeutic maneuvers. As will be discussed below, some of these markers are already being examined for this application. To base important clinical decisions on marker data, a number of factors need to be considered: The assays for markers should be quantitative, and it would be desirable for them to be very sensitive, providing the ability to observe changes in levels over a wide range—a range of, it is to be hoped, 10-fold or more differences in concentration—between the initial elevated values and the normal levels. It must be emphasized that markers should not be expected to disappear immediately after tumor removal. The level at a particular time after therapy would reflect the original level and the length of time that the markers would remain in circulation. Thus, when elevated markers are detected after surgery, their significance can be evaluated only be obtaining additional specimens over a period of time. Furthermore, if the marker was not produced by the tumor, but was a reactive type of marker, the disappearance after curative resection might not be expected. In fact, with some markers of this type, levels might increase after tumor removal.

2.5. Detection of Metastases

Markers produced by tumors could be very helpful in detecting the location of primary tumors and metastases, by at least two different approaches. The first, the determination of changes in levels of markers in the blood in various regions of the body, is currently practical for some markers. The other approach, by *in vivo* localization of marker by antibodies labeled with a radioisotope or heavy metal, is theoretically very appealing and potentially a powerful tool, but has not yet been sufficiently developed for practical application.

To detect the occult source of elevated tumor-produced marker levels, one can measure the differences in concentration in arterial and venous blood (A/V difference) in different regions in which the metastasis is likely to be found. The utility of this approach is dependent on several factors. The ease of detecting A/V differences in the region of tumor is: (1) Inversely proportional to the $T_{\frac{1}{2}}$ of the marker (or directly proportional to the fractional catabolic rate of the marker). If the marker has a long survival time in the circulation, the arterial level will be relatively high and the increment in the venous drainage may be quite small. (2) Inversely proportional to the fraction of blood volume drained by the venous return being sampled. For example, it would be easier to detect significant A/V differences with brain metastases than it would be to measure A/V differences of blood flow to the liver. (3) Inversely proportional to experimental uncertainty of the values; i.e., the lower the coefficient of variation (S.D./mean), the more reliable the assay and the greater the ease in detecting significant small A/V differences. Therefore, the more accurate the assay, the greater the ability to catheterize discrete areas of blood flow, and the more rapid the turnover of the marker, the more likely it is that this approach would be of value in localization of tumor deposits.

A related approach to localization would involve the measurement of marker levels in extravascular fluids in the region of the tumor. For example, this might be particularly of use to detect recurrent breast tumors by testing of pleural fluid, or cerebral metastases by testing of cerebrospinal fluid.

The possibility of localization of tumors by labeled antibodies to antigens on the tumors is a very attractive one. The rationale behind this approach is quite straightforward. If antibodies are specifically directed against antigens on a tumor, and these antigens are absent on normal tissues or are inaccessible to systemically inoculated antibodies, one would expect to have localization of the antibodies at the site of tumor growth. Radiolabeling of the antibodies, or labeling with heavy metals, would allow the accumulation of antibodies to be detected by isotopic or radiological scanning equipment. Despite these simple principles, little progress has been made in this area, and application at the clinical level is just now being attempted. Some of the practical problems to be faced include: (1) Location of marker in tumor cell. One would anticipate that this procedure would be more easily applied to markers that are on the tumor cell surface than those that are present only inside viable tumor cells. (2) Adequacy of blood flow to the tumor. (3) Levels of circulating marker. One would expect that inoculation of labeled antibodies into the circulation of a patient with high serum levels of marker would result in rapid formation of antigen–antibody complexes and nonspecific clearance of the label. (4) Proximity of the tumor to the major sites of nonspecific uptake of the labeled material. For example, it might be much more difficult to detect tumor within, or in the region of, the liver than it would to localize tumor elsewhere in the abdomen. (5) Proportion of specific antibody to nonspecific immunoglobulins in the labeled reagent. The more purified the antibody to the marker, and the less denatured or likely the preparation is to bind nonspecifically, the more likely it will be to detect small foci of tumors.

2.6. Serial Monitoring to Determine Efficacy of Therapy and to Detect Recurrence or Metastases

Some of the most important applications of tumor markers to clinical oncology are for following the response of patients to chemotherapy or other forms of therapy, and for monitoring patients without evidence of disease for early detection of recurrence or metastases. For these purposes, the requirements or desirable features for a tumor-produced marker* include the following: (1) The assay should be sensitive and quantitative, and levels should be proportional to tumor mass. As pointed out earlier, markers that vary over a very wide range of concentrations are more likely to be useful. (2) For monitoring of patients after primary surgical resection of tumor, levels of the marker should return to normal after tumor removal. This could then provide a good, low baseline for detection of rising levels. For this application, the repeated levels of the patient herself provide the best baseline, and progressive increases in levels, even when still within the range of normal, could provide a meaningful indication of tumor recurrence. (3) The amount of test-to-test fluctuation in the assay must be considered. A variety of non-tumor-related factors, including technical variation, therapy itself, and benign inflammatory or other transient diseases, could cause transient fluctuations or elevations in the assay. For almost all markers, there-

*Some but not all of these features also apply to markers not produced by the tumor.

fore, it would be important to obtain repeated specimens over a period of time and to observe persistent or progressive elevation in marker levels before the data should be taken as indicative of recurrence. (4) For accurate monitoring, it is important that the marker be produced by all or most of the tumor cells so that recurrence of nonproducing tumors would be infrequent. If this is likely to be a problem, the use of two or more markers to detect each of the major elements would be important. (5) The assay should be suitable for frequent, repeated testing. Assays requiring only small volumes of blood would obviously be advantageous. (6) For following the efficacy of therapy for metastatic or unresectable tumors, most of the principles enumerated above apply. The issues are somewhat different, however, in that one would be starting with elevated levels and usually looking for a decrease: (a) Here, it would be important to establish a baseline of elevated levels by repeat sampling prior to therapy. (b) Shortly after chemotherapy or radiotherapy, the levels might transiently rise rather than fall, due to destruction of tumor cells and consequent release of their markers into the circulation. Although this phase must be avoided during the usual monitoring, the detection of increase in levels immediately after therapy might actually be useful as an indication of some responsiveness of the tumor to the therapy. Failure to see a rise might rapidly predict the failure of the particular therapy being used. This approach appears useful for polyamines, but has not yet been sufficiently well investigated for immunological markers. (c) Significant and progressive decreases in marker levels following therapy could provide a reliable indication of response to therapy and could even provide quantitative information on the degree of decrease in tumor burden and the extent of residual tumor.

In making a decision as to whether to base therapy on elevated levels of markers in the absence of any other indications of tumor recurrence, one must consider the predictive accuracy of the marker and the possible benefits to be gained from the therapy vs. the risks to the patient from the additional therapy.

3. Biological Markers of Potential Value in Breast Cancer

In the past few years, a wide variety of substances and approaches have been evaluated for their possible use as markers for human breast cancer (Table I). Some are normally present in the tissues of the fetus and then either disappear or are much reduced in amount by the end of gestation or shortly after birth. These have been termed oncofetal antigens or embryonic antigens. Other tumor markers may be normally

Table I
Types of Markers of Potential Value in Breast Cancer

I. Oncofetal antigens
  - Carcinoembryonic antigen (CEA)
  - Gamma fetal proteins

II. Placental markers
  - Human chorionic gonadotropin (HCG)
  - Human placental lactogen (HPL)
  - Pregnancy α_2-glycoprotein

III. Breast- or milk-associated antigens
  - Casein
  - Gross cystic disease fluid protein (GCDFP)
  - Nonhistone nuclear antigen

IV. Other ectopic hormones
  - Calcitonin

V. Enzymes
  - Lactic dehydrogenase (LDH) isoenzymes
  - Sialyl transferase

VI. Normal body constituents
  - Ferritin
  - Blood group substances
  - Tissue polypeptide antigen (TPA)
  - Hydroxyproline
  - Dimethylguanosine
  - Acid glycoprotein

VII. Histopathological markers
  - Type of lymphoid cell infiltrates
  - Immunoglobulins
  - Placental markers

VIII. Alterations in immune function
  - Depressed levels of T cells and other lymphocytes
  - Depressed delayed hypersensitivity reactions
  - Depressed lymphoproliferative responses

IX. Immune responses to breast-cancer-associated antigens
  - Antibodies
  - Cell-mediated immunity

X. Antigen–antibody complexes

produced by the placenta. Some tumor markers may be characteristic of cancers of a particular tissue or organ, and some of these may be present in some normal adult tissues, but may be functionally or quantitatively altered in tumors or may be released in higher concentrations into the circulation of cancer patients. In breast cancer, particular attention has been directed toward the expression of milk proteins. Antigens of viruses associated with mammary tumors in experimental animals, or at least antigens cross-reactive with viral proteins, may also be detectable

Table II. Elevated Circulating Levels of

Diagnosis and clinical status	Number of elevated values/total					
	LoGerfo *et al.*[(7)a]	Reynoso *et al.*[(8)a]	Chu and Nemoto[(9)a]	Laurence *et al.*[(10)b]	Steward *et al.*[(11)b]	Wang *et al.*[(12)a]
Breast cancer	30/45 (67)			37/79 (47)		
Pretherapy						
Operable	9/24 (38)	1/10 (10)				— (33)
Localized (N_0)				12/39 (31)	2/10 (20)	
Axillary lymph node metastasis (N+)				9/20 (45)	4/12 (33)	
Metastatic	21/21 (100)	15/25 (60)	57/83 (68)	16/20 (80)	28/31 (90)	— (69)
Bone			14/22 (64)			
Liver			14/15 (93)			
Lung			20/24 (83)			
Soft tissues			9/22 (41)			
After surgery						
Remained tumor-free			8/15 (53)			16/54 (30)
Recurrence			4/6 (67)			18/27 (67)
Benign breast disease	0/13 (0)			6/74 (8)	1/17 (6)	
Normal	0/49 (0)	1/138 (1)		0/60 (0)		(35)

[a–g]Elevation defined as: [a]> 2.5 ng/ml; [b]> 12.5 ng/ml; [c]> 5 ng/ml; [d]> 10 ng/ml; [e]> 20 ng/ml; [f]> 2.7 for nonsmokers, > 5 ng/ml for smokers; [g]≥ 4 ng/ml.

in some human tumors. Most tumor markers are characterized only by their immunological or physiocochemical properties, but some have functional activities or are variants of normal functional products. These include hormones, enzymes, and metal-binding and secretory proteins. In addition to detection of circulating tumor markers, useful information can also be gained by examination of tumor tissues and regional lymph nodes for their content of tumor-associated substances and also for the types of infiltrating host lymphoid cells. A further large area that potentially can provide useful diagnostic and monitoring information is that concerned with the immunological functions of the patient with breast cancer. Although tests for alterations in immunological competence and for immunological reactivity against tumor-associated antigens are not usually thought of as biological markers, they should be included in this context. As will be discussed in detail later, some of these tests can detect tumor-associated reactions or alterations in immune function quite early in the course of disease.

Carcinoembryonic Antigen in Breast Cancer

(% elevated)

Henderson *et al.*[13][c]	Franchimont *et al.*[14][d]	Coombes *et al.*[15][e]	Tormey *et al.*[16][f]	Myers *et al.*[17][g]	Haagensen *et al.*[18][d]
	20/39 (51)	2/25 (8)	2/14 (14)		
			0/7 (0)		
			2/7 (28)		
74/172 (43)	14/25 (56)	13/16 (81)	83/117 (71)	206/342 (60)	65 /200 (33)
9/36 (25)			52/66 (79)		— (50)
24/28 (86)			22/27 (82)		—
41/80 (51)			— (61)		—
—			24/46 (52)		(15)
			3/39 (8)	105/482 (22)	
				88/447 (20)	24[c]/222 (11)
				17/35 (49)	(50)[c]
	0/55 (0)				
	0/935 (0)			4/81 (5)	

3.1. Oncofetal and Placental Markers

3.1.1. *Carcinoembryonic Antigen*

Carcinoembryonic antigen (CEA) was originally described as an antigen present only in gastrointestinal cancers and fetal endodermal tissues. However, with development of very sensitive RIAs for CEA and with extensive research by many investigators on this marker, it became apparent that CEA was not specific for gastrointestinal cancers and elevated circulating levels of CEA were described for some patients with cancer of a variety of histological types and for some patients with non-neoplastic diseases, particularly those of an inflammatory type (e.g., ulcerative colitis, hepatitis, cirrhosis). In the past few years, many laboratories have demonstrated elevated CEA levels in the plasma of some patients with breast cancer.[7–18] A summary of the findings of these studies is given in Table II.

Although most attention has been focused on the possible use of CEA for monitoring breast cancer patients for the detection and treatment of recurrent or metastatic disease, considerable information has also been provided on the CEA levels in untreated patients with breast disease. Most of the studies have shown that some patients with primary operable breast cancer have elevated CEA plasma levels, and that the incidence of elevations in patients with benign breast disease is very low and significantly different from that in primary breast cancer. An exception to these findings was the finding of Wang *et al.*[12] that CEA levels were elevated in about one third of both operable breast cancer and benign breast disease patients. The reason for this disparate finding is not clear. The study was performed with Hoffmann-LaRoche reagents, as used by the majority of the other groups. The authors raised the possibility that their finding might be due to the high incidence of heavy smokers in the population.[12] In a later study with other reagents,[14] however, the same laboratory found a major difference between the incidence of elevations in breast cancer and benign disease patients, with more of the latter group elevated. Such variations among studies in incidence of elevated CEA levels have been prevalent, with from 8 to 51% of operable cancer patients having elevated levels. It is unclear how much of this variation can be attributed to technical variations among the laboratories or to differences in patient selection.

There has been general agreement that elevated CEA levels in untreated breast cancer patients are found mainly in patients with metastatic disease. As with operable cases, however, the incidence of elevations has varied substantially among studies, from 33 to 100%. A high incidence of elevations has been found in patients with liver metastases, and, with the exception of one study,[13] also with bone metastases. Patients with metastases only to the skin and other soft tissues have usually had the lowest incidence of elevations. In patients with operable disease, there has not been a clear difference between N_0 and N+ cases, but rather only a trend in that direction. Because of the usual low incidence of elevations in operable patients, even those with lymph node involvement, it has been suggested that elevated CEA levels reflect occult systemic metastases, especially in the liver.[13] The low levels in most primary breast cancer patients do raise the question whether CEA is actually made by breast cancer cells, as opposed to its being a reactive marker, i.e., released from the liver or other sites on metastatic involvement. Some direct evidence for the presence of CEA in breast cancer cells has been provided, however, by immunofluorescent[19] and immunoperoxidase[20] studies of biopsy specimens and by measurement of

CEA levels in tumor extracts.[21,22] Some laboratories have failed to detect CEA in most breast cancers examined,[23,24] but this failure may be related to differences in the specificity of the antisera used and the sensitivity of the methods. Even in the positive studies, only low amounts of CEA were detected; it was estimated that breast cancer cells contain 50- to 75-fold lower concentrations of CEA than do colorectal cancer cells.[21]

Despite this evidence for the presence of CEA in breast cancer, the relationship between CEA levels and tumor burden or extent of disease in untreated breast cancer patients remains unclear. It might be expected that elevated CEA levels before surgery would reflect a poor prognosis, but no appreciable information on this point has yet been presented.

In contrast to the paucity of information on the prognostic significance of pretherapy CEA levels, several studies have concentrated on the usefulness of CEA levels in specimens obtained after surgery. In the first study on this aspect, Chu and Nemoto[9] found that CEA frequently rose transiently after mastectomy. In addition, 8 of 15 patients without detectable metastases had intermittent elevations in CEA. Sustained elevations were noted in only 2 of 6 patients developing recurrent disease. In contrast, Wang *et al.*[12] measured CEA levels at 10 days after surgery and found that patients with elevated values had a significantly poorer prognosis, with recurrence in 65% of the patients as compared with recurrence in only 20% of the patients with normal CEA levels at that point. In two recent larger studies,[17,18] the association of elevated postoperative CEA levels with poor prognosis was confirmed. In the study of Myers *et al.*,[17] CEA data provided stronger prognostic discrimination than pathological staging. Despite the significant differences between the groups of patients, however, the sensitivity and specificity of the discrimination have been somewhat limited. Many patients who remained tumor-free for long periods of follow-up had elevated postoperative CEA levels, and one third to one half of patients developing recurrences had normal values. Part of this may be accounted for by variations in the time at which the samples were collected after surgery, and the tendency to base the analyses on one CEA determination. One might anticipate that stronger discrimination might be found if repeated CEA determinations were performed during a standard period after surgery. It is hoped that more information in this area might provide a basis for better selection of high-risk patients for further therapy. There is also a need for more data on the value of serially monitoring patients after mastectomy, to determine whether progressively rising levels would provide further early evidence for occult metastases.

Repeated CEA measurements have been utilized mainly to monitor the response to therapy of patients with metastatic breast cancer. The overall conclusions have been that decreasing CEA levels correlate with response to therapy,[9,11,16,18] and that patients with elevated levels responded less well to therapy.[16] CEA determinations appeared to provide clinically useful data in some of these studies; Chu and Nemoto[9] concluded, however, that CEA was a poor marker for monitoring because clinical improvement tended to occur prior to a fall in CEA levels.

3.1.2. *Human Chorionic Gonadotropin*

Human chorionic gonadotropin (HCG) has been found to be associated with a variety of nontrophoblastic malignancies.[25] In the initial report, using a radioimmunoassay for the β subunit of HCG, Braunstein *et al.*[25] found elevated circulating HCG levels in 3 of 33 (9%) breast cancer patients. Shortly thereafter, Sheth *et al.*[26] found elevated βHCG levels in 9 of 65 (14%) breast cancer patients, and all the patients with elevated values had metastatic disease. In contrast, none of 22 patients with benign breast disease had elevated levels of HCG. Tormey *et al.*[27] found elevated βHCG levels in 9 of 20 (45%) postoperative N+ patients, as well as in 37 of 74 (50%) patients with metastatic disease. In further studies, Tormey *et al.*[28] found elevated levels in 5 of 14 (36%) patients before surgery, 9 of 33 (27%) N+ patients at 1–6 months postoperatively, and in 65 of 134 (49%) patients with metastatic disease. Elevated levels were particularly associated with liver metastases. No correlation was found between elevations in HCG, either before or after surgery, and subsequent recurrence of disease, and in only 4 of 10 patients with recurrence did the HCG elevations precede clinical detection. The main clinical usefulness of HCG in this study was in monitoring response to chemotherapy in patients with metastatic disease. Changes in HCG levels correlated with response to therapy, and patients with normal HCG levels before therapy had a higher response rate. Franchimont *et al.*[14] performed assays for both native HCG and the β subunit. They found elevated levels of HCG in 6 of 39 breast cancer patients before surgery (15%) and of βHCG in only 1 (2%). Surprisingly, none of 25 patients with metastatic disease had elevations in either assay.

Overall, only a low incidence of elevated HCG levels has been found in primary breast cancer, and even when the values were elevated, there was usually only a small increment above the normal range. These results and the failure of the elevations to correlate with prognosis[28] again raise the question whether the marker is produced by the tumor cells. Horne *et al.*,[29] however, in a study of tissue sections by the immunoperoxidase technique, found that 60% of breast cancers and none of 12 benign breast tissues contained HCG. It would appear, therefore, that although ectopic production of HCG is frequent in breast cancer, release from primary tumors is infrequent and quite variable, and not necessarily related to tumor burden. Thus, measurement of HCG levels in primary breast cancer patients will probably not be of substantial clinical value. There would appear to be more regular release of HCG from metastatic breast cancer lesions, and HCG assays may be useful for monitoring therapy in these patients.

3.1.3. *Human Placental Lactogen*

Sheth *et al.*[30] measured the levels of another placental hormone, human placental lactogen (HPL), in the sera of an appreciable number of patients with breast cancer. Of 72 patients with breast cancer, 10 (14%) had detectable levels, in contrast to none of 24 patients with benign breast diseases. The patients with elevated HPL levels were distributed among most clinical stages, except that none of 9 patients with stage I disease were positive. The findings are reminiscent of those with HCG. Although 82% of breast cancer were found to contain HPL,[29] few of these release detectable amounts of hormone into the circulation, and then only in quite low concentrations.

3.1.4. *Pregnancy-Associated α_2-Glycoprotein*

Pregnancy-associated α_2-glycoprotein has been found to be increased in the serum of some patients with breast cancer.[31] Cancer patients, however, have a wide range of values that overlap with those in persons without cancer, and Stimson *et al.*[31] therefore concluded that this marker could not be used as a diagnostic tool. Even so, they found that serial measurements on breast cancer patients, looking for deviations from the patients' own baseline levels, provided correlation with clinical course in 79% of 168 patients. All patients developing recurrent disease after mastectomy had elevated values before clinical detection of metastases, whereas patients who remained tumor-free did not show sustained increases. This marker, however, does not appear to be produced by the tumor, and levels have been found to rise progressively with age.[32] Considerably more information will therefore be needed before the clinical value of this marker can be properly assessed.

3.1.5. *Placental Alkaline Phosphatase*

Placental alkaline phosphatase, also known as the Regan isoenzyme, has been found to be elevated in the sera of 12% of patients with breast cancer.[33] It has been pointed out, however, that most elevations are only slight, in a range in which correlations with clinical status would be expected to be least accurate.[33]

3.1.6. *Gamma Fetal Proteins*

Edynak *et al.*[34,35] described two proteins with gamma mobility to be present in cancerous and fetal tissues. The sera of a very low proportion of patients with breast cancer or other types of cancer were found to have precipitating antibodies for gamma fetal protein, which was detected in extracts of 73 of 97 breast cancer and 11 of 13 fibroadenomas. In contrast, the antigen was not found in 30 breast tissues with fibrocystic disease.[34] A second antigen, termed gamma fetal protein-2, was also detected by precipitating antibodies in the sera of a few cancer patients.[35] Gamma fetal protein-2 was found in 17 of 42 breast cancers and in about half the specimens of grossly normal breast tissue obtained from breast cancer patients. Antigen was detected as far as 12 cm away from the tumor margin, yet was undetectable in normal tissues or in 35 breast tissues of patients with benign diseases. Although some of these findings with the gamma fetal proteins appear of interest for possible diagnostic applications, there have been no reports on the development of more sensitive procedures that might detect circulating levels of the markers.

3.2. Breast- or Milk-Associated Markers

3.2.1. *Casein*

Hendrick and Franchimont[36] set up an RIA for human casein and surveyed the levels in the sera of patients with breast cancer and other diseases. Of 11 untreated breast cancer patients, 8 (70%) had elevated levels, and of 41 treated stage I patients, 41% had elevated values. In contrast, elevations were found in 1 of 7 sera from benign breast diseases and in only 3% of normal women. Elevated casein levels were not specific for breast cancer, since a large proportion of specimens from patients with lung and gastrointestinal cancers were also positive. Based on these findings, Franchimont *et al.*[14] performed a more extensive study on the association of elevated casein levels with breast cancer. Only 6 of 39 patients (15%) with untreated primary breast cancer were found to have elevated values, whereas 10 of 25 (44%) patients with metastatic disease had elevations. Only 1 of 55 patients with benign breast diseases had elevated casein levels. Elevations in patients after mastectomy appeared to be related to stage of disease, with 7 of 39 stage II patients positive vs. 1 of 30 stage I patients. Monaco *et al.*[37] also developed an RIA for casein, but they failed to detect any elevated levels in the sera of breast cancer patients. Furthermore, only 8 of 47 (17%) tumors themselves and none of seven cultured cell lines from breast cancer had detectable amounts of casein. Similarly, Hurlimann *et al.*[38] detected casein synthesis in breast cancers, but the amounts were less than those found with dysplastic breast tissues. The reason for these major discrepancies between these studies and those of Franchimont[14,36] remains to be elucidated. It seems possible that the assay of the latter group detects antigenic specificities not recognized by the antisera of the other investigators. Franchimont *et al.*[39] attributed their reactivity to the kappa type of casein.

3.2.2. Gross Cystic Disease Fluid Protein

Haagensen *et al.*[40,41] recently developed a radioimmunoassay for a glycoprotein isolated from breast gross cystic disease fluid. This gross cystic disease fluid protein (GCDFP) is believed to be an epithelial cell secretory product, is present in milk and saliva, and is expressed in very high concentrations in the fluid of breast cysts. One third to one half of patients with either primary breast cancer or benign breast diseases have low elevations of GCDFP in their sera, and therefore the marker does not appear useful for initial diagnosis. However, high elevations (> 150 ng/ml) have been found only in patients with metastatic breast cancer or recurrent disease. About 25–33% of such patients have had values above 150 ng/ml, and monitoring for such elevations appears to be useful for early detection of recurrent disease and for monitoring response of metastatic patients to therapy. In the serial monitoring of metastatic patients, 28 of 87 had high levels of GCDFP, and changes in levels correlated with response to therapy. Frequently, the changes occurred earlier and were more reliable indications of change in disease status than were clinical examination, other laboratory studies, or X rays. Recurrence in 18 patients after surgery was first detected in 6 by high elevations in GCDFP.

3.2.3. *Lactalbumin*

α-Lactalbumin is another milk protein that has been shown to be produced by some breast cancers *in vivo* and *in vitro*. However, measurements of serum levels by RIA revealed no elevations of this protein in the sera of breast cancer patients as compared with the levels in normal sera.[15,42]

3.2.4. *Lactoferrin*

Lactoferrin is yet another milk protein that has been examined for its usefulness as a breast cancer marker. Although it is produced by tumor cells, the levels were lower in cancer cells than in dysplastic breast tissues.[38] Rümke *et al.*[43] developed an RIA for measurement of lactoferrin in plasma. Although this protein was present in the circulation of some breast cancer patients, there were many false positives, which were attributed to the release of lactoferrin from normal granulocytes.

3.2.5. *Nonhistone Nuclear Antigen*

Chiu *et al.*[44] presented evidence for tissue-specific antigens on nonhistone nuclear antigens in human breast and lung cancer. Antisera produced in rabbits against dehistonized chromatin of breast cancer reacted well in complement fixation with breast cancers, but not with chromatin from normal or benign breast disease tissue or with normal lung or placenta. These results appear promising, and efforts are being made to develop a more sensitive assay to look for circulating levels of the breast cancer chromatin antigen.

3.3. Other Ectopic Hormones

In addition to HCG and HPL, calcitonin has also been associated with breast cancer.[45,46] The presence of calcitonin in the sera of breast cancer patients has been found to be stage-related, with a high incidence in patients with metastatic disease. However, some patients without metastasis but with poor prognosis related to lymph-node histology or tumor grade had elevated calcitonin levels. Calcitonin was present in extracts of 7 of 8 breast cancers and not detectable in benign breast disease tissues.[47] Furthermore, its production by breast cancer was confirmed by demonstrating its synthesis by *in vitro* cell lines and by a breast cancer cell line passaged in nude mice.[45,46] As with HCG and HPL, it would appear that some tumors producing calcitonin do not regularly release it into the circulation. In only 2 of 5 cases with calcitonin in the tumor extract was it also detectable in the serum.[47] In a recent study,[15] only 2 of 17 patients with metastatic breast cancer had elevated levels. In studies of calcitonin in medullary carcinoma of the thyroid, it has been found that administration of various agents can stimulate release into the circulation. Similarly, calcium infusion led to rise in calcitonin in 2 patients, and oral ethanol ingestion led to elevated levels in 5 of 9 other patients with metastatic disease.[46] It would be of interest to further explore the value of calcitonin as a marker, particularly after stimulation by more potent agents, like pentagastrin.

Prolactin has also been examined as a hormone that might be ectopically produced or elevated in breast cancer. In most studies, however, the circulating levels of prolactin were not elevated in breast cancer (see, for example, Sheth *et al.*[48]).

3.4. Other Enzymes and Isoenzymes

3.4.1. *Lactic Dehydrogenase*

Hawrylewicz *et al.*[49,50] noted a markedly altered lactic dehydrogenase (LDH) isoenzyme pattern in some breast cancer tissues, with increases in the cationic isoenzymes, LDH-5 and LDH-4. Examination of sera from breast cancer patients has given similar results. Sera from preoperative patients undergoing mammography or breast biopsy had elevated LDH-5 or -4 in 63% from breast cancer patients, 26% from patients with benign breast diseases, and 4% from normal women. Of sera from patients with metastatic breast cancer, 83% had this isoenzyme pattern. In longitudinal studies after mastectomy,[50] recurrence was predicted by the LDH isoenzyme pattern in 69%, and a normal pattern was observed in 82% of women who remained without evidence of disease.

3.4.2. *Sialyl Transferase and Other Enzymes*

Bosmann and Hall[51] studied the activity of various enzymes in neoplastic and normal breast tissues and found that sialyl transferase was present in 7-fold higher concentrations in cancer tissues. They also noted a 5-fold increase in neuraminidase, a 3-fold increase in β-galactosidase, and also increases in some other glycosidases. Intermediate levels of some of these enzymes were found in fibrocystic disease tissues. Henderson and Kessel[52] measured serum levels of sialyl transferase and in 55 of 69 tests on sera of 31 patients, values above the normal range were found. It was associated mainly with metastatic disease. In 11 patients studied serially, a fairly good correlation was found between sialyl transferase levels and extent of disease and response to therapy. The mean enzyme level in cancer patients, however, was less than 2 times that of normal donors, and this narrow range reduces its potential as a marker. Another limitation is that elevated sialyl transferase levels may occur in some benign diseases, e.g., rheumatoid arthritis. Coombes *et al.*[15] found elevated levels in 9 of 16 (56%) patients with metastatic breast cancer, but in only 1 of 25 patients with localized disease.

3.5. Normal Body Constituents or Their Variants

3.5.1. *Ferritin*

Marcus and Zinberg[53] found that ferritin was present in high concentrations in extracts of some breast cancers. The ferritin isolated from the tumors was found to be an acidic type, which has been termed carcinofetal ferritin or $\alpha 2_H$ globulin. These investigators set up an RIA for ferritin and examined the levels in sera of breast cancer patients and controls.[54] Elevated levels were detected in 41% of sera from cancer patients prior to surgery and in 67% of patients with recurrent or metastatic disease. Elevated ferritin levels were also found in patients with hepatic inflammation, in 43% of patients with hepatitis or cirrhosis, and in 13% of patients with ulcerative colitis or peptic ulcers. No information was provided in this report regarding levels in patients with benign breast disease. Coombes *et al.*[15] detected elevated ferritin levels in 88% of patients with metastatic breast cancer, but none of 25 patients with localized disease had elevated values. No data have yet been provided on the possible clinical applications of this marker for the management of breast cancer patients, but a study of this type is in progress.

3.5.2. *Blood Group Substances*

Expression of blood group antigens on cancer tissues has frequently been found to be altered. Following the many studies in this area by Davidsohn,[55] Gupta and Schuster[56] studied 83 breast tissues for expression of A, B, and H blood group antigens, by a mixed cell agglutination technique. All 25 benign breast tissues contained these antigens in the epithelial cells. In striking contrast, no reactions were obtained with 45 primary and 15 metastatic cancer tissues. This major difference in antigen expression has yet to be exploited as a useful clinical marker.

Springer *et al.*[57] reported that T antigen, a precursor of MN blood group substances, was detectable in all of 15 breast cancers studied, including *in situ* lesions, but not in 5 benign breast tissues. As will be discussed later, this finding provided the basis for a study of the titers of anti-T antibodies in the sera of breast cancer patients.[58]

3.5.3. *Tissue Polypeptide Antigen*

Björklund[59] and Björklund *et al.*[60] studied the levels of tissue polypeptide antigen (TPA), a substance that may be related to tissue regeneration, in the sera and urine of cancer patients, using a hemagglutination inhibition assay. The clinical findings related to TPA were recently summarized by Holyoke and Chu.[61] Elevated levels were found in 67 of 94 (71%) breast cancer patients, as compared with 40 of 112 (36%) patients with a variety of benign diseases.[62] A major problem with this marker is that it is increased in conditions with increased metabolic activity, e.g., infections or injury. A possible solution to this problem has been to perform repeated measurements. Whereas stable or increasing levels have usually been seen in untreated cancer patients, the levels have tended to decline in the patients with benign diseases. In a study of the response of advanced breast cancer patients to chemotherapy or to hormones, more than 80% of the patients had elevated levels of TPA.[63] Decreases tended to occur in patients responding to therapy or in those with stable disease, whereas progressive increases were seen in patients with poor prognosis or recurrence. In another longitudinal study of patients with cancer, patients with repeatedly negative TPA tests survived significantly longer than those with positive tests.[61]

3.5.4. *Hydroxyproline*

Measurement of the hydroxyproline/creatinine ratio in urine was shown to be a useful index of bone breakdown, and provided early indications of bone metastases in breast cancer.[64,65] Guzzo *et al.*[64] found that increased urinary ratios often preceded clinical evidence of bone metastases by 1–7 months, and that persistently normal ratios were associated with an absence of bone metastases during a period of clinical follow-up of 6–19 months. The hydroxyproline/creatinine ratio appeared to be more useful than measurement of serum alkaline phosphatase. Powles *et al.*[65] evaluated the use of urinary hydroxyproline measurements for predicting response to therapy in breast cancer. Among 10 patients responding to therapy, all had decreases in their ratios. In contrast, all 20 nonresponders had increases in their ratios after therapy. Coombes *et al.*[15] found elevated ratios in 11 of 15 (73%) patients with metastatic breast cancer, but in only 1 of 25 patients with apparently localized disease.

3.5.5. *Dimethylguanosine*

Tormey *et al.*[27] measured 24-hr urinary N^2-dimethylguanosine and other nucleoside levels in breast cancer patients. Elevated levels of dimethylguanosine were detected in 57% of the breast cancer patients studied. In addition, 8 of 18 (44%) N+ patients after mastectomy had elevated values. These data appear promising but are difficult to evaluate, since no information on the specificity of the elevations was provided.

3.5.6. *Polyamines*

Although measurement of urinary polyamines appears to be useful in monitoring some forms of cancer, it does not seem very promising in breast cancer. Tormey *et al.*[27] found elevated levels in fewer than 15% of patients with metastatic disease and in none of 16 postoperative N+ patients. Similarly, Coombes *et al.*[15] detected elevated levels in only a small number of patients with metastatic disease.

3.6. Acute Phase Reactants

3.6.1. *Erythrocyte Sedimentation Rate*

A question often asked about a new potentially useful biological marker is whether it is better than the erythrocyte sedimentation rate (ESR). This is not an idle question, since the ESR may often be elevated in cancer patients, as well as in a variety of nonneoplastic conditions. Riley[66] recently reported on an extensive study of the ESR in patients with breast diseases. The results with blood from 385 patients with breast cancer and 544 patients with benign breast diseases were compared with those from over 3500 normal women. The normal controls and benign disease patients had virtually identical frequency distributions, whereas with breast cancer, there was a shift toward higher levels, which was highly significant. About 40% of cancer patients had elevated values, compared to about 20% of the controls.

3.6.2. *Acid Glycoprotein*

Serum α-1-acid glycoprotein is produced by the liver and is elevated in response to a variety of disease states.[67] Elevated levels have been noted in the sera of some breast cancer patients, including some with localized, small tumors. High values have been particularly associated, however, with large tumors, recurrent disease, and metastases. Elevations were found in 18 of 26 patients with positive bone scans and in 18 of 29 patients with other evidence of metastases. It has been suggested that normal acid glycoprotein levels may be a good prognostic sign.[67] Coombes *et al.*[15] found elevated levels in 12 of 16 (75%) patients with metastatic breast cancer, and in only 1 of 25 patients with localized disease.

3.7. Antigens of Oncogenic Viruses

A few recent studies have pointed to possible cross-reactions between human breast cancer and antigens of oncogenic viruses. Although the etiological implications of such findings are entirely unclear, such cross-reactive antigens could serve as useful markers.

3.7.1. Mouse Mammary Tumor Virus

Mueller and Grossman[68] described the reaction of 10 of 36 sera from breast cancer patients with an antiserum containing antibodies to mouse mammary tumor virus (MTV) and concluded that the sera contained MTV-associated antigens. Although it appeared in these studies that virus antigens were being detected, the specificity of the reactions was not extensively defined.

3.7.2. Other Viruses

Kryukova *et al.*[69] detected precipitin reactions between human breast tumor and milk samples and an antiserum to HEp-2 virus, which appears to be very similar or identical to the Mason–Pfizer monkey virus (MPMV). Yeh *et al.*[70] also reported the detection of an antigen related to p27 (protein with molecular weight of 27,000 daltons) of MPMV in breast cancer tissues, by inhibition in RIA. However, very large amounts of tissues were needed to partially inhibit.

3.8. Histopathological Markers

3.8.1. Lymphoid Cells

A number of pathologists have noted considerable differences in the morphology among primary breast cancers and axillary lymph nodes. Several have pointed to features, involving infiltrations with lymphoid cells, that appeared to have prognostic implications. Black *et al.*[71] emphasized the lymphoreticuloendothelial responses to breast cancer. Prognostically favorable cases of breast cancer had diffuse and perivenous lymphoid cell infiltrates in the primary tumors and sinus histiocytosis in the axillary lymph nodes. Perivenous infiltrates and sinus histiocytosis appeared to be particularly good prognostic features. In a confirmatory study, Hunter *et al.*[72] found that 16 of 17 patients with dominant sinus histiocytosis in the regional lymph nodes remained free of recurrent disease for more than 5 years. In contrast, 5 of 6 patients with germinal center hyperplasia in the regional lymph nodes died within 5 years. Patients with both histiological features had an intermediate survival rate; 17 of 25 remained tumor-free. Patients with no lymphoid proliferative response in the lymph nodes did poorly, with 5 of 6 dying early. Friedell *et al.*[73] noted sinus histiocytosis in more than one half of the specimens from Japanese women with breast cancer, and rarely found this feature in English cases. They raised the question of the relationship of this finding to the more benign course of breast cancer in Japan. Tsakraklides *et al.*[74] also studied axillary lymph nodes and found that lymphocytic predominance was associated with the highest survival rate, lymphocytic depletion and unstimulated lymph nodes had the poorest survival, and patients with germinal center predominance had intermediate survival rates. Deodhar *et al.*[75] studied the interaction of tumor cells with axillary lymph node cells *in vitro,* by observing clumping of lymphocytes around tumor cells, cytotoxicity, and blast transformation. Such reactions were mainly seen in N_0 patients, and in reactive patients the lymph node cells were more active than their peripheral blood lymphocytes. In contrast to the studies cited above, Flores *et al.*[76] found no significant correlation between prognosis and lymphoid infiltrates in the tumor or sinus histiocytosis in the lymph nodes.

Recently, some attempts have been made to analyze the types of lymphoid cells involved in the axillary lymph node infiltrates. Tsakraklides *et al.*[77] found that axillary lymph nodes of breast cancer patients contained an average of 64% T cells, as measured by rosette formation with sheep erythrocytes (SRBC), with a range of 32–80%; an average of 36% B cells, as measured by the presence of surface immunoglobulins (SIg's), with a range of 14–46%; and an average of 28% of cells with receptors for Ig's, with a range of 10–45%. In lymph nodes with lymphocytic predominance, there was an increased proportion of T cells, high proliferative responses to phytohemagglutinin (PHA), and a low proportion of B cells. In nodes with germinal center predominance, the opposite pattern was seen, with low numbers of T cells and high numbers of B cells. The low lymphoproliferative responses to PHA of such cells contrasted with the high responsiveness noted by Fisher *et al.*[78] with cells from nodes with prominent germinal centers. A large proportion of the cells in malignant pleural effusions in breast cancer patients has also been shown to be T cells. Djeu *et al.*[79] found that most of the T cells in the effusions had high-affinity receptors for SRBC, in contrast to a considerably lower proportion of such cells in the peripheral blood. No prognostic value for such findings was determined. Richters and Kaspersky[80] concentrated on the presence of SIg-positive B cells in primary tumors and regional nodes. In N+ patients, both tumor-free and involved nodes had a significantly higher percentage of B cells than that seen in N_0 patients. Within the primary tumors of 10 patients, half had no detectable B cells and the others had 10–20%. Although more studies of this aspect are needed, it appears that infiltration of lymph nodes with T cells is a prognostically favorable sign, and a predominance of cells with SIg is a poor prognostic sign. Since it has recently become clear that other cell types, in addition to B cells, can bear IgG on their surface via receptors for the Fc portion of IgG and since Fc-positive cells can accumulate in tumors, it will be important in further experiments to distinguish more definitively between B cells and Fc-positive cells.

3.8.2. *Immunoglobulins*

In addition to studies on lymphocyte SIg's, some investigators have studied the deposition or production of Ig's in the tissues. Irie *et al.*[81] described the presence of Ig and complement on a metastasis to the kidney and their absence on normal kidney cells. Richman[82] found IgA and IgM on both benign and malignant breast epithelial cells, but IgG was detected only on the cancer cells. Roberts *et al.*[83] found more IgM to be produced in tumor tissues than in benign breast tissues, and the levels of both IgM and IgG correlated with the degree of round-cell infiltration. As yet, no prognostic relevance for these findings has been reported.

Waldman *et al.*[84] reported preliminary indications of elevations of the secretory component (SC) of IgA in the sera of some patients with epithelial malignancies. Harris *et al.*[85] followed up on these findings in a study on breast cancer, and found that SC was well represented on primary tumors and on metastases. It was suggested that the detection of SC in lymph nodes or other sites could aid in the identification of small metastatic foci.

3.8.3. *Antigens and Other Markers*

Horne *et al.*[29] performed a retrospective study of breast cancer patients, examining biopsies by the immunoperoxidase technique for the presence of HCG, HPL, and pregnancy-specific β_1-glycoprotein. None of 12 benign tissues had any of these markers, but at least one was present in over 80%, and 60% of the cancers contained all three. Patients with tumor negative for these markers had significantly longer survival. The pregnancy β_1-glycoprotein appeared to be the best prognostic indicator.

Savlov *et al.*[86] studied breast cancer tissues for a variety of enzymes and the relationship between their levels and response to chemotherapy. They found that tissues from patients responding to therapy had higher levels of pyruvate kinase, glucose phosphate isomerase, LDH, and isocitrate dehydrogenase, and using a statistical model, could predict the clinical responses in 25 of 32 patients.

3.9. Alterations in Immune Function in Breast Cancer

As part of the immune surveillance theory, one might expect to find decreased immunological functions prior to development of clinical cancer or in the presence of early, localized disease (discussed in Herberman[87]). On the other hand, if the presence of tumor produced decreased immune reactivity, this might be useful for prognosis and monitoring. There have been many studies of alterations of immune function in breast cancer patients, and these studies will be summarized below, with particular emphasis on the possible diagnostic or prognostic value of the findings.

3.9.1. *Serum Immunoglobulin Levels*

Meyer *et al.*[88] measured serum IgA levels in 42 N+ patients. Some patients had elevated levels and others had low values. Those with low IgA levels before mastectomy continued to have low values thereafter. Among patients receiving postoperative radiotherapy, recurrence correlated significantly with IgA levels below 200 mg/100 ml and lymphocyte counts below 1500/mm^3. In contrast, patients receiving irradiation and having high IgA levels remained tumor-free. Since patients not receiving radiotherapy did not display a correlation between their IgA levels and recurrence, it was suggested that irradiation may be detrimental to patients with a low IgA level and may be helpful to patients with high IgA.

3.9.2. *Peripheral Blood Lymphocytes*

Several investigators have described depressed lymphocyte counts in some patients with breast cancer, either because of the disease itself or because of therapy. Papatestas *et al.*[89] found 40–50% of their patients, prior to therapy, to have lymphocyte counts below 2000/mm^3. For stage II and III patients, this finding was associated with a poor prognosis; such patients had a significantly lower 5-year survival than did those with pretherapy counts above 2000. In contrast, Glas *et al.*[90] reported that initial lymphocyte counts were not predictive of subsequent metastases, and they also noted no effect of surgery on the counts. McCredie *et al.*[91] also found no correlation between preoperative lymphocyte counts and recurrence, but found that lymphocyte counts rose after mastectomy, with a significant mean increase of 87% by 21 days. In attempts to characterize the subpopulations of lymphocytes responsible for the alteration in lymphocyte counts, a number of laboratories have measured the proportions of T cells forming rosettes with SRBC. Although Stjernswärd *et al.*[92] failed to detect any depression in rosette-forming cells (RFC), some subsequent studies have found depressed levels in some patients. Using a 29°C assay[93] that measures a subpopulation of T cells with a high affinity for SRBC, our laboratory has found depressed levels in a considerable proportion of breast cancer patients.[94,95] Whitehead *et al.*[96] also found a significantly decreased proportion of RFC in patients with stages I, II and IV, as compared with normal donors or with patients with benign breast diseases. They noted, however, that stage III patients, despite their poor prognosis, did not have significant alterations from normal. In studies on the possible mechanism for the depressed levels of RFC, this group found a reversal by pretreatment of the lymphocytes with papain.[96] More recently, they described the inhibition of rosette formation by sera of breast cancer patients, with 56% of sera from stage III patients and higher proportions from other stages having this effect.[97] Surprisingly, they found no differences in this inhibitory activity between sera obtained preoperatively or at periods with no evident disease or recurrence. Despite the lack of clinical correlation, characterization of the responsible inhibitory factor might lead to a useful diagnostic test. Petrini *et al.*[98] studied the proportions of SIg-bearing cells in untreated breast cancer patients and found a significant increase above normal. These differences did not seem to be related to the presence or absence of metastases. As mentioned earlier, it is not certain whether these data reflect the proportions of B cells or possibly cells with Fc receptors.

Since postsurgical radiotherapy is a commonly used form of adjunctive therapy, several investigators have studied the effects on lymphocyte counts and on subpopulations of lymphocytes. After radiotherapy, lymphocyte counts have remained depressed for longer than 12 months.[92,99] This decrease has been reported to be more profound in patients with metastases.[90] There has been disagreement about the relative sensitivity of T cells and non-T cells to the effects of radiotherapy. One group reported a relative decrease in RFC,[92] whereas others[98,100] found more depression in non-T lymphocytes. These differences have been attributed to the failure of Stjernswärd *et al.*[92] to remove monocytes before testing, and this population increased after radiotherapy.[98,100] At 3 years after radiotherapy, T cells were still depressed, but non-T lymphocytes had returned to normal levels.[98]

Weese *et al.*[95] found that adjuvant chemotherapy, particularly combined therapy with cyclophosphamide, methotrexate, and 5-fluorouracil, caused a substantial decrease in the proportion of lymphocytes forming rosettes with SRBC in the 29°C assay for the subpopulation with high-affinity receptors. These depressive effects were at least partially reversed in patients also receiving *Corynebacterium parvum* subcutaneously.

3.9.3. *Delayed Cutaneous Hypersensitivity*

Skin tests for delayed hypersensitivity reactions have been used widely for evaluation of immune reactivity of cancer patients. Roberts and Jones-Williams[101] found that about one fourth of patients undergoing mastectomy had negative reactions to a low concentration of streptokinase-streptodornase, and the size of skin reactions decreased significantly with advanced disease. There was no correlation, however, between skin reactivity to this antigen and subsequent recurrence or with round-cell infiltration of the tumor or sinus histiocytosis. Cunningham *et al.*[102] studied patients with untreated recurrent breast cancer, to determine their ability to be sensitized to a new antigen, dinitrochlorobenzene. Patients with strong skin reactions survived significantly longer and responded better to therapy. There was also a higher incidence of positive reactions in patients with well-differentiated tumors containing a dense infiltrate of lymphocytes.

3.9.4. *Lymphocyte Proliferation and Other Cellular Reactions in Vitro*

In vitro assays of cell-mediated immune reactivity have also been used to look for decreased reactivity in cancer patients. Decreased proliferation of lymphocytes in response to mitogens has been extensively studied, but clear depression has been largely restricted to patients with advanced or inoperable cancer (see, for example, Whittaker and Clark[103]). It should be noted, however, that significant depression has been seen even with patients with stage I or II disease,[103] and using the relative proliferation index to better quantitate the responses relative to normal donors, Dean *et al.*[104] found depression in about one third of the breast cancer patients. Glas *et al.*[90] studied proliferative responses to PHA and to purified protein derivative (PPD) at various times before and after therapy. They found that depressed responses to either stimulant before therapy were not predictive of recurrence. At the time of recurrence, responses to PPD were lower than those seen initially or those of patients who were still tumor-free. In contrast, responses to PHA did not change from pretherapy values. In this study, radiotherapy led to a decreased response to PHA and PPD, but there was a return to normal by 12 months. McCredie *et al.*,[99] however, described an increased response to PHA after irradiation. In our laboratory, we have observed persistent depressed proliferative responses in patients who received radiotherapy.

Matthews and Whitehead[105] recently described inhibition of antibody-dependent cell-mediated cytotoxicity by sera from more than a third of breast cancer patients. This inhibition apparently was not due to immune complexes, but the inhibitory factor was not characterized. No information was provided on the clinical status of the patients, nor were controls with benign sera performed. Preliminary data on decreased monocyte chemotactic responses in untreated breast cancer patients were also reported,[106] and this was reversed within a week after surgery.

3.10. Immune Responses to Breast-Tumor-Associated and Other Antigens

Many tumor-associated antigens or other antigens can elicit an immune response in the tumor-bearing patient. Antigens can often be recognized by the host when present in very small amounts, and recognition and response could occur prior to release of the antigens into the circulation. It might therefore be expected that immunological reactions would be detected while tumors were still small and localized, and that assays for such reactions might be more sensitive markers for breast cancer than the circulating tumor markers. Detection of immune reactions in breast cancer patients may also serve as means for identifying and isolating new breast-cancer-associated antigens, which could then be used for raising heterologous antibodies and setting up RIAs.

3.10.1. Humoral Antibody Responses

Detection of antibodies to tumor-associated antigens is potentially a very sensitive and logistically simple procedure for diagnosis and monitoring of breast cancer. Unfortunately, despite an appreciable amount of work in this area, there have been relatively few advances, particularly in regard to utilization of these antibodies for diagnosis or monitoring. Some interesting information has accumulated, however, and will be summarized below.

Breast-Tumor-Associated Antigens. Loisillier *et al.*[107] tested the sera of breast cancer patients against tanned erythrocytes coated with breast tumor extracts. Of the patients, 80% were found to have antibody titers in excess of 1 : 160. In contrast, only 11% of patients with benign breast diseases or no patients with other types of cancer had high titers. The antigen detected did not appear to be breast-cancer-specific, since lyophilized milk or extracts of noncancer tissue could absorb the antibodies; however, 8- to 10-fold more of these materials were required as compared with breast cancer extracts.

Taylor and Odili[108] described an autoantibody in the serum of 1 of 11 breast cancer patients tested by complement fixation with a nuclear fraction of autologous and allogeneic breast cancer, but not with normal breast extracts or with extracts of other types of cancer. Two antigens seemed to be involved, one of which was DNAse-sensitive and appeared to be breast-tumor specific. Gentile and Flickinger[109] also described autoantibodies, detecting high-titered reactions of all of 15 breast cancer patients tested in a tanned erythrocyte agglutination test against extracts of autologous tumor. Only rarely did sera from patients with benign breast diseases react, and then only in low titers. Here too the detected antigen was thought to be antigen-specific.

Some studies have involved the testing by immunofluorescence of sera against breast tumor cells or tissue cultured cells derived from breast cancer. Priori *et al.*[110] detected reactions in about two thirds of the breast cancer sera tested against cultured cells derived from breast cancer and also from fibrocystic disease patients. All the normal sera were negative, and both of two sera from patients with fibrocystic disease were negative against the cancer-derived cells. The antigen detected in this study appeared to be common to both the tumor and benign cells. Edynak *et al.*[111] found reactivity against a cytoplasmic antigen in two primary cell cultures with 91% of sera from breast cancer patients, when obtained 7–10 days after mastectomy. In comparison, 20% of normal donor sera, matched for age and parity, reacted; however, no studies with sera from patients with benign breast diseases were reported.

Two groups of investigators have reported precipitating antibodies in the sera of breast cancer patients. Roberts *et al.*[83] found reactivity in the sera of 4 of 16 breast cancer patients against tumor material, but not against extracts of 13 benign or normal tissues. Humphrey *et al.*[112,113] performed more extensive studies and found reactions by either immunodiffusion or complement fixation with 46% of breast cancer patients. Only 1.5% of normal sera reacted, but 34% of sera from patients with fibrocystic disease and 25% from patients with fibroadenomas reacted. Furthermore, the antigen or antigens detected were not tumor-associated, being detected also in extracts of normal breast. Despite these negative features, these workers emphasized the possible prognostic usefulness of these tests. Of 13 N+ patients without detectable antibodies, 11 were dead within 12 months, whereas 15 of 18 N+ patients with antibodies remained free of disease for up to 24 months.

Despite the promise in the data of several of the studies cited above, no subsequent information on the clinical usefulness of these antibodies has been provided.

Springer *et al.*[58] found that anti-T antibodies (see Section 3.5.2), present in virtually all adult human sera, was depressed in 21% of sera from 189 breast cancer patients, compared with 5% of patients with benign breast diseases and 3.6% of persons without cancer. After surgery, two thirds of the cancer patients had an increase in antibody titer, whereas only 1 of 32 patients with benign diseases showed an increase. Although this study involves the quantitative decrease in antibody, rather than the presence of a breast-cancer-associated antibody, this could still be a useful marker for diagnosis or prognosis or both.

The Makari skin test, which has been described as a diagnostic test for a variety of cancers (see the review in Weese[114]), is based on immediate erythematous reactions to intradermal inoculation of autologous serum mixed with tumor extracts. Its use has included the evaluation of patients with breast disease. A high proportion of breast cancer patients reacted, but 45% of patients with benign breast diseases also gave positive results.[115,116] On the basis of some promising reports, an improved version of this test has recently been made available for large-scale testing in the United Kingdom. However, the value of the improved test in breast cancer is not yet clear.

3.10.2. Antibodies to Virus-Associated Antigens

Several investigators have described antibodies in the sera of breast cancer patients that reacted with mouse mammary tumors and with mouse mammary tumor virus (MTV). Charney and Moore[117] performed *in vivo* tests for neutralization of MTV by human sera, and found that 2 of 5 sera from breast cancer patients had weak neutralizing activity. Many sera from breast cancer patients have been found to give positive immunofluorescence with MTV+ mouse mammary tumor cells.[118,119] Most of the reactions could be attributed to heterophile antibodies, but a small number of reactions were thought to be directed against MTV. Newgard *et al.*[120] showed that most human sera could precipitate MTV in a radioimmune precipitation assay, but these reactions could be inhibited by dog milk and lactating mouse mammary glands. They therefore concluded that the observed reactions were due to nonvirus materials. The interpretation to be given the inhibition data is not clear, however, since very large amounts of dog milk were needed to inhibit and this might have been nonspecific.[121]

3.10.3. Cell-Mediated Immune Reactions

a. Skin Window Technique. Black and Leis[122] used a skin window technique to study cellular responses of breast cancer patients. Cryostat sections of autologous breast tissues were placed on an abraded area of skin, and the basophils and other cells in the exudate after 28–30 hr were examined microscopically. Positive responses were found in 40% of tests with cancer tissues and in less than 10% with benign breast tissues. The frequency of reactivity to autologous cancer sections was correlated with stage of disease, with the highest incidence of reactivity in patients with *in situ* carcinoma or precancerous mastopathy and the least in N+ patients. Higher reactivity was also associated with the presence of sinus histiocytosis in the axillary lymph nodes. This procedure was suggested to be useful as an aid in prognosis, but no information has been provided as to whether reactivity would provide information that would add significantly to the prognostic classification of patients by the usual staging procedures.

b. Delayed Cutaneous Hypersensitivity Reactions. Hughes and Lytton[123] first showed in 1964 that 3 of 22 patients with breast cancer had delayed hypersensitivity reactions to crude extracts of autologous tumors and not to extracts of grossly normal breast tissues. Stewart and Orizaga[124] performed similar studies with autologous extracts and found reactions in 12 of 56 breast cancer patients. The reactive patients tended to have anaplastic tumors, positive axillary lymph nodes, and a shorter survival. Reactivity in their tests was therefore a sign of poor prognosis, despite the association of reactivity with infiltration of the tumors with mononuclear cells. Alford *et al.*[125] confirmed the findings of autologous reactivity to autologous crude membrane extracts and showed that allogeneic cancer extracts also elicited skin reactions. Soluble skin reactive antigens were then prepared from the crude extracts by sonication and then separated by Sephadex G-200 chromatography. Sephadex fractions of extracts of normal breast tissue from cancer patients and normal breast tissue from patients with fibrocystic disease, as well as those from breast cancer, gave reactivity in some patients with carcinoma of the breast. These data suggested that some patients had immune reactivity against organ-associated antigens. On further separation by gradient polyacrylamide gel electrophoresis,[126] two adjacent gel regions gave positive reactions. In preliminary tests for skin reactivity, one fraction (region 2b) appeared to contain a breast-cancer-associated antigen and the other (region 2a) a breast-tissue-associated antigen. In these initial studies, region 2a gave positive reactions in patients with localized breast cancer and not in patients with disseminated cancer, whereas the other region gave positive results in most breast cancer patients. In more recent tests with larger numbers of patients and with further splitting of gel regions, region 2a from both a large primary tumor and from MCF-7, a cell line derived from breast cancer, again gave a high incidence of positive reactions with good specificity in patients with localized breast cancer.[127] The region 2b1-3 from MCF-7 also showed good specificity and reacted only in patients with localized disease. The ability to separate breast-tumor-associated skin reactive antigens from an established cell line is very encouraging, since this should be a source for large, standardized batches of antigen for extensive clinical testing. Efforts are also under way to prepare specific antibodies against the isolated skin reactive antigens, which may be used to set up RIAs for these antigens.

c. Leukocyte Migration Inhibition. The leukocyte migration inhibition assay, which is considered a close *in vitro* correlate of delayed cutaneous hypersensitivity, has been widely used to study cell-mediated immunity of breast cancer patients. Anderson *et al.*[128] first showed that the migration of leukocytes of some patients with breast cancer was inhibited on exposure to autologous tumor extracts and not to normal tissues. Decreased reactivity was seen in patients who were free of evident disease after mastectomy, but this was difficult to evaluate because the patients also received radiotherapy. Segall *et al.*[129] confirmed the reactivity of the majority of breast cancer patients to autologous tumor extracts. They looked for cross-reactivity to other breast cancers and obtained only one positive result. Most subsequent studies, however, have found good reactivity of breast cancer patients to allogeneic tumor extracts.[130–132] McCoy *et al.*[133] also found reactivity of the majority of breast cancer patients against extracts of MCF-7, a cell line derived from breast cancer. The occurrence of reactivity of breast cancer patients against common antigens in allogeneic as well as autologous extracts derived from tumors suggests the potential usefulness of this procedure for initial diagnosis. However, although only a small proportion of normal women have shown reactivity in these studies, patients with benign breast diseases have displayed a higher frequency of reactivity.[130–134] Patients whose breast biopsies showed some form of breast disease had reactivity similar to that of breast cancer patients, whereas women whose biopsies showed no evidence of disease failed to react.[134] These results, and the reactivity of some cancer patients against extracts of benign breast lesions, indicate that patients with neoplastic and benign breast diseases may become sensitized to breast tissue antigens. Similar to the findings described above for delayed cutaneous hypersensitivity reactions, Kadish *et al.*[132] reported preliminary evidence for the separation of breast-cancer-associated antigens from breast tissue antigens. A high-molecular-weight fraction elicited reactions in 50% of breast cancer patients and in only 5% of controls. In contrast, a low-molecular-weight fraction elicited reactivity in patients with benign and malignant breast diseases. This finding suggests that two different antigens could be used for clinical testing, one that might be useful to screen for patients with breast diseases and the other that might help to discriminate breast cancer patients from patients with benign breast diseases. It was of interest that the reactivity of women with benign diseases was strongly influenced by surgery, with a drop in reactivity within 1 month after surgery; in contrast, the reactivity of breast cancer patients remained high during this period.[134] Another potential application of this assay might be to assess the prognosis of breast cancer patients. Black *et al.*[135] have examined this issue most extensively, using cryostat sections of tumor tissues as source of antigen. The frequency of reactivity in their study was related to stage, with 90% of patients with *in situ* cancer reacting and only about one third of stage II patients reacting against autologous tumor sections. The relationship between reactivity and disease status at the time of testing has not been well defined in most studies, but recent analysis of data in our laboratory (McCoy and Cannon, unpublished observations) indicates that a higher proportion of patients with recurrent or metastatic disease maintain reactivity than do patients who remain free of detectable disease.

Black *et al.*[136] made the intriguing observation that many patients with breast cancer reacted in leukocyte migration inhibition assays with mouse milk from a high MTV+ strain and usually not with milk from a low MTV strain. A higher incidence of reactivity was noted in patients with other favorable prognostic indicators,[137] and reactivity to the virus-containing milk correlated rather well with reactivity of patients to autologous or allogeneic breast cancer materials.[138] Zachrau *et al.*[139] eluted a protein from the breast tissues that appeared to have a molecular weight similar to that of the gp52 of MTV. This protein itself, however, was not shown to elicit migration inhibition or to be antigenically related to gp52. McCoy *et al.*[140] confirmed that many breast cancer patients can react in leukocyte migration inhibition with MTV+ mouse milk, and also with purified MTV and gp52. In contrast, very few normal donors or patients with other types of cancer have reacted to these materials. The significance of these reactions remains to be determined, since some breast cancer patients have also reacted in migration inhibition to gp69/71 of murine leukemia virus, which does not cross-react seriologically with gp52. The cell-mediated immune response might be recognizing a common determinant on these two viral glycoproteins, or these results might simply reflect a hyperreactivity of breast cancer patients to glycoproteins of various types. Regardless of the nature of the antigenic specificities recognized, however, measurement of reactivity to these viral reagents may be quite useful for monitoring of breast cancer patients.

d. Lymphocyte Proliferative Responses. A number of investigators have studied the proliferative response of lymphocytes from breast cancer patients to tumor cells or to tumor extracts.[141–145] It is difficult to estimate the clinical utility of most of the data, since the numbers of patients in most studies were small, no clinical correlations were made, and allogeneic extracts were usually employed. The use of allogeneic materials has been found to present the potential for response to normal alloantigens,[146] and it is difficult to distinguish this from reactivity to tumor-associated antigens. Such problems in interpretation can be eliminated by the use of autologous tumor cells or extracts. However, this would obviously restrict the usefulness of this test to the study of known cancer patients, after surgical removal of tumor. Dean *et al.*[145] studied 34 patients with breast cancer and observed significant proliferative responses in 12 patients to either intact autologous tumor cells or crude extracts of autologous tumors. The reactivity appeared to be directed against tumor-associated antigens, since normal breast tissue of reactive patients did not stimulate. In a preliminary analysis of the clinical status of reactive patients, no correlation was found with stage of disease or therapy. Patients are now being tested serially, to examine more carefully the usefulness of this test for prognosis or monitoring.

Cunningham-Rundles *et al.*[147] tested lymphocytes from women with breast cancer and from controls for proliferative responses to an extract of MTV+ mouse milk. Many women had positive reactions, with patients with malignant or benign breast diseases having quantitatively higher responses than normal women. Although these data appear to be somewhat similar to those obtained in migration inhibition, one needs to be particularly concerned in this assay that reactions might have been to mouse xenoantigens, which are better represented in MTV+ milk than in MTV − milk, rather than to viral antigens. McCoy *et al.*[140] observed that normal women and men reacted as well as breast cancer patients against MTV.

e. Cytotoxicity. Cell-mediated cytotoxicity assays also have potential applications for diagnosis and monitoring of breast cancer. Several groups reported that the lymphocytes of patients with breast cancer were cytotoxic against tissue culture cells derived from breast cancer.[148–150] If only patients developing breast tumors had cell-mediated reactivity against breast cancer-associated antigens, a sensitive detection method might be available. In several recent studies,[151–154] however, major problems have been noted (discussed in Herberman and Oldham[155]). A considerable number of normal subjects, patients with other types of cancer, and patients with benign breast diseases[156] have been found to react significantly against cultured cells derived from breast tumors, and reactivity of breast cancer patients has been seen against cells derived from dissimilar types of cancer. One possible solution to these problems is to test simultaneously against cultures derived from breast cancer and against cell lines susceptible to the cytotoxicity of normal subjects, and look for relative increases in levels of reactivity against the breast cancer lines. Cannon *et al.*[157] recently reported that breast cancer patients did have relatively higher levels of cytotoxicity against the MCF-7 cell line. Another approach is to physically separate the types of effector cells responsible for each type of cytotoxic activity, and much effort is being expended in this direction. Until some practical solution is found, however, cytotoxicity assays would not appear to be suitable for practical clinical applications. Humoral blocking factors, which can inhibit cell-mediated cytotoxicity, have also been described,[158] and their presence has been associated with progressively growing tumors. The correlations have not been complete (see, for example, Jeejeebhoy[153]), however, and practical applications of the tests for humoral factors will probably have to await further standardization of the basic cell-mediated assays.

f. Leukocyte Adherence Inhibition (LAI). Recently, the LAI technique has attracted considerable attention as a rapid and simple test to detect specific reactivity of breast cancer patients against tumor-associated antigens. Powell *et al.*[159] reported that 95% of 110 breast cancer patients reacted specifically to extracts of breast cancer, as compared with only 1 of 15 patients with benign breast diseases. Although normal blood donors were generally unreactive, 11 of 29 normal subjects who had been extensively exposed to tumor patients or tumor extracts gave positive results. Grosser and Thomson[160] observed reactions in 40 of 47 breast cancer patients, with decreased reactivity or lack of reactivity associated with disseminated disease. In contrast, fewer than 10% of normal donors and none of 7 patients with benign breast diseases reacted. More recently, this same group[161] reported on more extensive evaluation of the correlation of reactivity with clinical status. Reactivity was related to stage of disease, with 88% of patients with stages I and II reacting as compared with 31% of stage IV patients. After mastectomy, half the patients were unreactive during the first 6 months, and subsequent reactivity was related to recurrent disease. Fujisawa *et al.*[162] confirmed the specificity of the procedure, but did not evaluate patients with benign breast diseases. In contrast to the results of Flores *et al.*,[161] they found no differences in reactivity among patients with varying clinical stages or in relation to time after mastectomy. Overall, the LAI test appears quite promising as a diagnostic test. However, the procedure may be subject to unconscious bias, and most of the reported studies have not been performed with a strict double-blind design, with coded specimens.

g. Tests of Macrophage Electrophoretic Mobility (MEM) and Structuredness of Cytoplasmic Matrix (SCM). Caspary and Field[163] originally described the MEM test, in which most patients with a variety of cancers, including breast cancer, reacted against basic proteins derived either from brain or from cancer tissues. Since then, several groups have confirmed these results,[164,165] and because of the very high proportion of cancer patients reacting, the procedure was suggested as a useful means, in conjunction with clinical tests, to exclude the diagnosis of cancer.[166] A very high degree of sensitivity for detection of early cancer was claimed,[166] with reactivity up to 16 years before clinical evidence of disease. However, in a recent trial with coded specimens provided by investigators at a neighboring institution, the MEM test failed to discriminate well between breast cancer patients, patients with benign breast diseases, and normal donors (Pritchard, personal communication).

Cercek *et al.*[167] developed the SCM test, a rapid procedure involving changes in fluorescence polarization after contact of sensitized lymphocytes with antigen. Using the same antigens as used in the MEM test, good differentiation between cancer patients and controls was seen. By using breast cancer antigens, tissue specificity was conferred, with only patients with breast diseases reacting.[168] However, the diagnostic potential of the test was diminished by the finding that the majority of patients with benign breast diseases also reacted.[169] Although other groups have confirmed the original observations (see Bagshawe[165]), it remains to be determined whether this test can withstand the same type of objective evaluation described above for the MEM test.

3.11. Antigen–Antibody Complexes

Most of the tests described above have depended on the specific detection of circulating antigens or immune reactants. If antibodies are being produced against circulating antigens, then antigen–antibody complexes may be formed. Recently, sensitive tests have been developed to detect such circulating immune complexes, and these procedures have been used to examine sera from cancer patients. Theofilopoulos *et al.*[170] detected elevated levels of complexes in the sera of 8 of 20 (40%) patients with breast cancer. Rossen *et al.*[171] had positive results with about 80% of sera from breast cancer patients. The results were not appreciably affected, however, by major differences in clinical status. Although results with this approach are still rather preliminary, detection of immune complexes may be useful clinically and, by analysis of the antigens in the complexes, may also help to identify new immunogenic breast-cancer-associated antigens.

3.12. Use of Multiple Markers

Because of many of the problems described above, no single biological marker may have sufficient sensitivity and specificity for a particular clinical application. One might, however, envision the use of several tests in combination to provide highly discriminatory data. This approach has already been taken by a few groups interested in diagnosis of breast cancer. Tormey *et al.*[27] tested sera and urine of breast cancer patients for CEA, HCG, polyamines, and nucleosides. By using CEA, HCG, and dimethylguanosine, 97% of patients with metastatic disease and two-thirds of those with N+ disease gave positive results. However, this study did not examine the usefulness of the multiple markers to discriminate between the neoplastic and benign breast diseases. Coombes *et al.*[15] evaluated the combined use of a wide variety of markers. Seven markers were elevated in over 50% of patients with metastatic disease, and all these sera could be identified by the use of two markers, e.g., CEA and C-reactive protein, CEA and acid glycoprotein, ferritin and acid glycoprotein, or CEA and ferritin. In patients with localized disease, however, the combined use of seven markers still did not provide sufficient sensitivity. Only 22% of N_0 patients and 45% of N+ patients had elevations of any of these markers. Although patients with benign breast diseases were tested in this study, no information was provided about the use of multiple markers to discriminate between cancer and benign diseases. Franchimont *et al.*[14] employed another group of five markers: CEA, α-fetoprotein, HCG, βHCG, and casein. α-Fetoprotein was never positive and the HCG assays contributed little, but the use of the other markers, CEA and casein, increased the sensitivity of detection. At least one marker was elevated in 69% of patients with localized disease, and in 88% of patients with metastatic disease. In contrast, only 5.5% of patients with benign diseases had any abnormality.

All of the studies cited above used multiple assays for circulating antigens or biochemical markers. Stein *et al.*[172] used multiple assays of immune competence to assess patients with breast cancer. They measured delayed cutaneous hypersensitivity to PPD and dinitrochlorobenzene, lymphoproliferative responses to PHA, lymphocyte counts, and levels of cells forming rosettes with SRBC. Over two thirds of patients with operable breast cancer had some depression in one or more of these tests, as compared with 30% of control subjects. In this study, these assays did not appear to have prognostic value, and some abnormalities persisted for long periods in patients without evident disease. All these patients, however, received postoperative radiotherapy, which may account for much if not all of the persistent immune depression.

Horne *et al.*[29] examined tumor specimens for the presence of several pregnancy-associated markers, HCG, HPL, and pregnancy-specific β_1-glycoprotein. The presence or absence of HCG did not provide useful prognostic information, but the absence of the other two markers was associated with significantly longer survival.

3.13. Conclusions

Among the many markers that have been examined, there are several promising leads for the application of immunological or biochemical tests to the diagnosis and management of breast cancer. From the data collected thus far, it is possible to select a number of markers for more detailed and extensive clinical studies. The very promising circulating markers might include CEA, ferritin, calcitonin, GCDFP, and urinary hydroxyproline. To these might be added some measures of immune function and reactivity, particularly lymphocyte counts and subpopulations of RFC, lymphocyte proliferative responses, and leukocyte migration inhibition. It seems quite likely that some combination of these markers could provide high levels of specificity and sensitivity. Similarly, some combination of procedures for examination of tumor specimens may provide very helpful prognostic information.

One essential issue for initial diagnosis will be to demonstrate clearly that a test or combination of tests will be able to discriminate well between women with cancer and those with benign breast diseases, and that such procedures add to the current armamentarium of diagnostic tests. Women with fibrocystic disease are currently giving positive reactions in many of the procedures described above. Regardless of the cause, this problem will have to be circumvented for the immunological assays to be useful in screening or as adjuncts to early diagnosis of breast cancer.

Many of the techniques that are currently being used need to be refined considerably. Most studies reported to date have concentrated on demonstrating differences between populations of women with breast cancer and control populations. More of the markers need to be thoroughly evaluated for their practical use in staging and monitoring of patients. Procedures still need to be developed for application to the problems of individual patients. One pressing need in this regard is the establishment of standardized reagents, so that the same antigens, antibodies, or other reagents can be used in the study of many patients and in the serial monitoring of individual patients. The identification of antigenic breast-cancer-derived tissue culture lines appears to be very promising in this regard.

It would also be very helpful if simpler tests could be developed for detection and monitoring of breast cancer. The RIAs are probably the most practical tests for large-scale application, and many of the other procedures, like delayed hypersensitivity skin tests or leukocyte migration inhibition, present considerable logistical and technical problems. One approach would be to use tumor-associated antigens, which have been identified by the current techniques, for the production of heterologous antisera and the eventual development of an RIA, which would require for testing only a small volume of serum.

The data accumulated thus far are sufficiently encouraging to lead to a search for new breast-cancer-associated markers and to more extensive evaluation of currently available markers.

4. References

1. S. Shapiro, Evidence on screening for breast cancer from a randomized trial, *Cancer* **39,** 2772–2782 (1977).
2. J. C. Bailar III, Screening for early breast cancer: Pros and cons, *Cancer* **39,** 2783–2795 (1977).
3. L. V. Ackerman and A. L. Katzenstein, The concept of minimal breast cancer and the pathologist's role in the diagnosis of "early carcinoma," *Cancer* **39,** 2755–2763 (1977).
4. D. E. Anderson, Genetic study of breast cancer: Identification of a high risk group, *Cancer* **34,** 1090–1097 (1974).
5. S. Shapiro, J. Goldberg, and G. Hutchison, Lead time in breast cancer detection and implication for periodicity of screening, *Am. J. Epidemiol.* **100,** 357–366 (1974).
6. C. D. Haagensen, *Diseases of the Breast,* 2nd Ed., pp. 155–176, W. B. Saunders, Philadelphia (1971).
7. P. LoGerfo, J. Krupey, and H. J. Hansen, Demonstration of an antigen common to several varieties of neoplasia: Assay using zirconyl phosphate gel, *N. Engl. J. Med.* **285,** 138–141 (1971).
8. G. Reynoso, T. M. Chu, D. Holyoke, E. Cohen, T. Nemoto, J. J. Wang, J. Chuang, P. Guinan, and G. P. Murphy, Carcinoembryonic antigen in patients with different cancers, *J. Am. Med. Assoc.* **220,** 361–365 (1972).
9. T. M. Chu and T. Nemoto, Evaluation of carcinoembryonic antigen in human mammary carcinoma, *J. Natl. Cancer Inst.* **51,** 1119–1122 (1973).
10. D. J. R. Laurence, U. Stevens, R. Bettelheim, D. Darcy, C. Leese, C. Turberville, P. Alexander, E. W. Johns, and A. M. Neville, Role of plasma carcinoembryonic antigen in diagnosis of gastrointestinal, mammary, and bronchial carcinoma, *Br. Med. J.* **3,** 605–609 (1972).
11. A. M. Steward, D. Nixon, N. Zamcheck, and A. Aisenberg, Carcinoembryonic antigen in breast cancer patients: Serum levels and disease progress, *Cancer* **33,** 1246–1252 (1974).

12. D. Y. Wang, R. D. Bulbrook, J. L. Hayward, J. C. Hendrick, and P. Franchimont, Relationship between plasma carcinoembryonic antigen and prognosis in women with breast cancer, *Eur. J. Cancer* **11,** 615–618 (1975).
13. I. C. Henderson, J. Lokich, R. Mayer, A. Skarin, and N. Zamcheck, Carcinoembryonic antigen (CEA) levels in metastatic breast cancer: Quantitative correlation with pattern of metastases, *Proc. Am. Assoc. Cancer Res.* **17,** 202 (1976).
14. P. Franchimont, P. F. Zangerle, J. C. Hendrick, A. Reuter, and C. Colin, Simultaneous assays of cancer associated antigens in benign and malignant breast diseases, *Cancer* **39,** 2806–2812 (1977).
15. R. C. Coombes, T. J. Powles, J. C. Gazet, H. T. Ford, J. P. Sloane, D. J. R. Laurence, and A. M. Neville, Biochemical markers in human breast cancer, *Lancet* **1,** 132–134 (1977).
16. D. C. Tormey, T. P. Waalkes, J. J. Snyder, and R. M. Simon, Biological markers in breast carcinoma. III. Clinical correlations with carcinoembryonic antigen, *Cancer* **39,** 2397–2404 (1977).
17. R. E. Myers, D. J. Sutherland, J. A. Kellen, D. G. Malkin, and A. Malkin, CEA in breast cancer, *Cancer* (in press) (1978).
18. D. E. Haagensen, Jr., S. J. Kister, J. P. Vandevoorde, J. B. Gates, E. K. Smart, H. J. Hansen, and S. A. Wells, Jr., Evaluation of carcinoembryonic antigen as a plasma monitor for human breast carcinoma, *Cancer* (in press) (1978).
19. M. Bordes, R. Michiels, and F. Martin, Detection by immunofluorescence of carcinoembryonic antigen in colonic carcinoma, other malignant or benign tumours, and noncancerous tissues, *Digestion* **9,** 106–115 (1973).
20. E. Heyderman and A. M. Neville, A shorter immunoperoxidase technique for the demonstration of carcinoembryonic antigen and other cell products, *J. Clin. Pathol.* **30,** 138–140 (1977).
21. G. Pusztaszeri and J.-P. Mach, Carcinoembryonic antigen (CEA) in non-digestive cancerous and normal tissues, *Immunochemistry* **10,** 197–204 (1973).
22. P. Sizaret and F. Martin, Carcinoembryonic antigen in extracts of pulmonary cancers, *J. Natl. Cancer Inst.* **50,** 807–810 (1973).
23. H. Denk, G. Tappeiner, R. Eckerstorfer, and J. H. Holzner, Carcinoembryonic antigen (CEA) in gastrointestinal and extragastrointestinal tumors and its relationship to tumor-cell differentiation, *Int. J. Cancer* **10,** 262–272 (1972).
24. D. M. Goldenberg, F. J. Primus, and F. Deland, Tumor detection and localization with purified antibodies to carcinoembryonic antigen, in: *Immunodiagnosis of Cancer* (R. B. Herberman and K. R. McIntire, eds.), Marcel Dekker, New York (1978).
25. G. D. Braunstein, J. L. Vaitukaitis, P. P. Carbone, and G. T. Ross, Ectopic production of human chorionic gonadotrophin by neoplasms, *Ann. Intern. Med.* **78,** 39–48 (1973).
26. N. A. Sheth, J. N. Suraiya, K. J. Ranadive, and A. R. Sheth, Ectopic production of human chorionic gonadotrophin by human breast tumors, *Br. J. Cancer* **30,** 566–570 (1974).
27. D. C. Tormey, T. P. Waalkes, D. Ahmann, C. W. Gehrke, R. W. Zumwatt, J. Snyder, and H. Hansen, Biological markers in breast carcinoma. I. Incidence of abnormalities of CEA, HCG, three polyamines, and three minor nucleosides, *Cancer* **35,** 1095–1100 (1975).
28. D. C. Tormey, T. P. Waalkes, and R. M. Simon, Biological markers in breast carcinoma. II. Clinical correlations with human chorionic gonadotrophin, *Cancer* **39,** 2391–2396 (1977).
29. C. H. W. Horne, I. N. Reid, and G. D. Milne, Prognostic significance of inappropriate production of pregnancy proteins by breast cancers, *Lancet* **2,** 279–282 (1976).

30. N. A. Sheth, J. N. Suraiya, A. R. Sheth, K. J. Ranadive, and D. J. Jussawalla, Ectopic production of human placental lactogen by human breast tumors, *Cancer* **39,** 1693–1699 (1977).
31. W. H. Stimson, J. M. Anderson, and D. M. Farquharson, Pregnancy-associated α_2-glycoprotein, *Lancet* **1,** 542–543 (1977).
32. M. G. Damber, B. Von Schoultz, and T. Stigbrand, Pregnancy-associated α-2-glycoprotein and cancer, *Lancet* **1,** 1182–1184 (1976).
33. W. H. Fishman and L. L. Stolbach, Placental alkaline phosphatase, in: *Immunodiagnosis of Cancer* (R. B. Herberman and K. R. McIntire, eds.), Marcel Dekker, New York (in press).
34. E. M. Edynak, L. J. Old, M. Vrana, and M. P. Lardis, A fetal antigen associated with human neoplasia, *N. Engl. J. Med.* **286,** 1178–1183 (1972).
35. E. M. Edynak, M. P. Lardis, and M. Vrana, Antigenic changes in human breast neoplasia, *Cancer* **28,** 1457–1461 (1971).
36. J. C. Hendrick and P. Franchimont, Radioimmunoassay of casein in the serum of normal subjects and patients with various malignancies, *Eur. J. Cancer* **10,** 725–730 (1974).
37. M. E. Monaco, D. A. Bronzert, D. C. Tormey, P. Waalkes, and M. E. Lippman, Casein production by human breast cancer, *Cancer Res.* **37,** 749–754 (1977).
38. J. Hurlimann, M. Lichea, and L. Ozzello, *In vitro* synthesis of immunoglobulins and other proteins by dysplastic and neoplastic human mammary tissues, *Cancer Res.* **36,** 1284–1292 (1976).
39. P. Franchimont, J. C. Hendrick, A. Thirion, and P. F. Zangerie, Kappa-casein: An index of normal mammary function and a tumor associated antigen, in: *Immunodiagnosis of Cancer* (R. B. Herberman and K. R. McIntire, eds.), Marcel Dekker, New York (in press).
40. D. E. Haagensen, Jr., G. Mazoujian, W. D. Holder, Jr., S. J. Kister, and S. A. Wells, Jr., Evaluation of a breast cyst fluid protein detectable in the plasma of breast carcinoma patients, *Ann. Surg.* **185,** 279–285 (1977).
41. D. E. Haagensen, S. J. Kister, J. Panick, J. Giannola, H. J. Hansen, and S. A. Wells, Jr., Comparative evaluation of carcinoembryonic antigen and gross cystic disease fluid protein as plasma markers for human breast carcinoma, *Cancer* (in press) (1978).
42. D. L. Kleinberg, Human α-lactalbumin: Measurement in serum and in breast cancer organ cultures by radioimmunoassay, *Science,* **190,** 276–278 (1975).
43. P. H. Rümke, D. Bisser, H. G. Kwa, and A. A. M. Hart, Radioimmunoassay of lactoferrins in blood plasma of breast cancer patients, lactating and normal women: Prevention of false high levels caused by leakage from neutrophile leucocytes *in vitro, Folia Med. Neerl.* **14,** 156–160 (1971).
44. J. F. Chiu, L. S. Hnilica, F. Chytil, J. T. Orrahood, and L. W. Rogers, Tissue specific antibodies against human lung and breast carcinoma dehistonized chromatins, *J. Natl. Cancer Inst.* **59,** 151–154 (1977).
45. R. C. Coombes, G. C. Easty, S. I. Detre, C. J. Hillyard, U. Stevens, S. I. Girgis, L. S. Galante, L. Heywood, I. MacIntyre, and A. M. Neville, Secretion of immunoreactive calcitonin by human breast carcinomas, *Br. Med. J.* **4,** 197–199 (1975).
46. A. M. Neville, R. C. Coombes, C. J. Hillyard, and I. MacIntire, Calcitonin as a tumour marker, in: *Cancer Related Antigens* (P. Franchimont, ed.), pp. 151–162, North-Holland Publishing Co., Amsterdam (1976).
47. C. J. Hillyard, R. C. Coombes, P. B. Greenberg, L. S. Galante, and I. MacIntyre, Calcitonin in breast and lung cancer, *Clin. Endocrinol.* **5,** 1–8 (1976).
48. N. A. Sheth, K. J. Ranadive, J. N. Suraiya, and A. R. Sheth, Circulating levels of prolactin in human breast cancer, *Br. J. Cancer* **32,** 160–167 (1975).

49. E. J. Hawrylewicz, W. H. Blair, and L. W. Giltner, Diagnostic value of serum LDH isoenzyme patterns in women with breast lesions, in: *Proceedings of the Third International Symposium on the Detection and Prevention of Cancer,* Marcel Dekker, New York (in press).
50. W. H. Blair, E. J. Hawrylewicz, and L. W. Giltner, Serum LDH isoenzymes as an indicator for the presence and recurrence of breast carcinoma, in: *Proceedings of the Third International Symposium on the Detection and Prevention of Cancer,* Marcel Dekker, New York (in press).
51. H. B. Bosmann and T. C. Hall, Enzyme activity in invasive tumors of human breast and colon, *Proc. Natl. Acad. Sci. U.S.A.* **71,** 1833–1837 (1974).
52. M. Henderson and D. Kessel, Alterations in plasma sialyltransferase levels in patients with neoplastic disease, *Cancer* **39,** 1129–1134 (1977).
53. D. M. Marcus and N. Zinberg, Isolation of ferritin from human mammary and pancreatic carcinomas by means of antibody immunoadsorbents, *Arch. Biochem. Biophys.* **162,** 493–501 (1974).
54. D. M. Marcus and N. Zinberg, Measurement of serum ferritin by radioimmunoassay: Results in normal individuals and patients with breast cancer, *J. Natl. Cancer Inst.* **55,** 791–795 (1975).
55. I. Davidsohn, The loss of blood group antigens A, B and H from cancer cells, in: *Immunodiagnosis of Cancer* (R. B. Herberman and K. R. McIntire, eds.), Marcel Dekker, New York (in press).
56. R. K. Gupta and D. Schuster, Isoantigens A, B and H in benign and malignant lesions of breast, *Am. J. Pathol.* **72,** 253–260 (1973).
57. G. F. Springer, P. R. Desai, and I. Banatwala, Blood group MN antigens and precursors in normal and malignant human breast glandular tissue, *J. Natl. Cancer Inst.* **54,** 335–339 (1975).
58. G. F. Springer, P. R. Desai, and E. F. Scanlon, Blood group MN precursors as human breast carcinoma-associated antigens and "naturally" occurring human cytotoxins against them, *Cancer* **37,** 169–176 (1976).
59. B. Björklund, Review of the immunochemical and clinical significance of TPA, in: *Protides of the Biological Fluids* (H. Peeters, ed.), pp. 505–512, Pergamon Press, Oxford (1976).
60. B. Björklund, V. Björklund, R. Lundström, and G. Eklund, Tissue polypeptide antigen (TPA) in human cancer defense responses, in: *The Reticuloendothelial System in Health and Disease: Immunologic and Pathologic Aspects* (H. Friedman, M. R. Escobar, and S. M. Reichard, eds.), pp. 357–370, Plenum Press, New York (1976).
61. D. Holyoke and T. M. Chu, Tissue polypeptide antigen (TPA), in: *Immunodiagnosis of Cancer* (R. B. Herberman and K. R. McIntire, eds.), Marcel Dekker, New York (in press).
62. C. V. Menandez-Botet and M. K. Schwartz, A pulmonary evaluation of tissue polypeptide antigen (TPA) in serum and/or urine of patients with cancer or benign diseases, *Proceedings of the Folksam Symposium,* Bonniers, Stockholm (1975).
63. W. Mattsson and S. Börjstrom, TPA as a guide in antineoplastic and hormonal treatment of advanced mammary carcinoma, *Proceedings of the Folksam Symposium,* Bonniers, Stockholm (1975).
64. C. E. Guzzo, W. N. Pachas, R. S. Pinals, and M. J. Krant, Urinary hydroxyproline excretion in patients with cancer, *Cancer* **24,** 382–387 (1969).
65. T. J. Powles, C. L. Leese, and P. K. Bondy, Hydroxyproline excretion in patients with breast cancer and response to treatment, *Br. Med. J.* **2,** 164–166 (1975).
66. V. Riley, Breast cancer patients: Substance in blood causing acceleration of erythrocyte sedimentation rate, *Science* **191,** 86–88 (1976).

67. J. G. Roberts, J. W. Keyser, and M. Baum, Serum α-1-acid glycoprotein as an index of dissemination in breast cancer, *Br. J. Surg.* **62,** 816–819 (1975).
68. M. Mueller and H. Grossman, An antigen in human breast cancer sera related to the murine mammary tumor virus, *Nature (London)* **237,** 116–117 (1972).
69. I. N. Kryukova, K. V. Ilyin, N. N. Mazurenko, L. N. Vasilevskaya, and V. M. Zhdanov, Immunologic study of antigens and molecular hybridization of nucleic acids from milk of breast cancer patients, *Neoplasma* **22,** 235–241 (1975).
70. J. Yeh, M. Ahmed, S. A. Mayyasi, and A. A. Alessi, Detection of an antigen related to Mason–Pfizer virus in malignant human breast tumors, *Science* **190,** 583–584 (1975).
71. M. M. Black, T. H. C. Barclay, and B. F. Hankey, Prognosis in breast cancer utilizing histologic characteristics of the primary tumor, *Cancer* **36,** 2048–2055 (1975).
72. R. L. Hunter, D. J. Ferguson, and L. W. Coppleson, Survival with mammary cancer related to the interaction of germinal center hyperplasia and sinus histiocytosis in axillary and internal mammary lymph nodes, *Cancer* **36,** 528–539 (1975).
73. G. H. Friedell, E. A. Soto, S. Kumaoka, O. Abe, J. L. Hayward, and R. D. Bulbrook, Sinus histiocytosis in British and Japanese patients with breast cancer, *Lancet* **2,** 1228–1229 (1974).
74. V. Tsakraklides, P. Olson, J. H. Kersey, and R. A. Good, Prognostic significance of regional lymph node histology in cancer of the breast, *Cancer* **34,** 1259–1267 (1974).
75. S. D. Deodhar, G. Crile, Jr., and C. B. Esselstyn, Jr., Study of the tumor cell–lymphocyte interaction in patients with breast cancer, *Cancer* **29,** 1321–1325 (1972).
76. L. Flores, M. Arlen, A. Elguezabal, S. F. Livingston, and B. S. Levowitz, Host tumor relationships in medullary carcinoma of the breast, *Surg. Gynecol. Obstet.* **139,** 683–688 (1974).
77. E. Tsakraklides, V. Tsakraklides, H. Ashikari, P. P. Rosen, F. P. Siegal, G. F. Robbins, and R. A. Good, *In vitro* studies of axillary lymph node cells in patients with breast cancer, *J. Natl. Cancer Inst.* **54,** 549–556 (1975).
78. E. R. Fisher, E. Saffer, and B. Fisher, Studies concerning the regional lymph node in cancer. VI. Correlation of lymphocyte transformation of regional node cells and some histopathologic discriminants, *Cancer* **32,** 104–111 (1973).
79. J. F. Djeu, J. L. McCoy, G. B. Cannon, W. J. Reeves, W. H. West, and R. B. Herberman, Lymphocytes forming rosettes with sheep erythrocytes in metastatic pleural effusions, *J. Natl. Cancer Inst.* **56,** 1051–1052 (1976).
80. S. A. Richters and C. L. Kaspersky, Surface immunoglobulin positive lymphocytes in human breast cancer tissue and homolateral axillary lymph nodes, *Cancer* **35,** 129–133 (1975).
81. K. Irie, R. F. Irie, and D. L. Morton, Detection of antibody and complement complexed *in vivo* on membranes of human cancer cells by mixed hemoadsorption techniques, *Cancer Res.* **35,** 1244–1248 (1975).
82. A. V. Richman, Immunofluorescence studies of benign and malignant human mammary tissue, *J. Natl. Cancer Inst.* **57,** 263–267 (1976).
83. M. M. Roberts, E. M. Bass, and W. J. Wallace, Antibody production in breast cancer, *Br. J. Surg.* **59,** 904 (1972).
84. R. H. Waldman, J. P. Mach, M. M. Stella, and D. S. Rowe, Secretory IgA in human serum, *J. Immunol.* **105,** 43–47 (1970).
85. J. P. Harris, M. H. Caleb, and M. A. South, Secretory component in human mammary carcinoma, *Cancer Res.* **35,** 1861–1864 (1975).
86. E. D. Savlov, J. L. Witliff, and R. Hilf, Further studies of biochemical predictive tests in breast cancer, *Cancer* **39,** 539–541 (1977).
87. R. B. Herberman, Cell-mediated immunity to tumor cells, in: *Advances in Cancer Research* (G. Klein and S. Weinhouse, eds.), Vol. 19, pp. 207–263, Academic Press, New York (1974).

88. K. K. Meyer, G. L. Mackler, and W. C. Beck, Increased IgA in women free of recurrence after mastectomy and radiation, *Arch. Surg.* **107,** 159–161 (1973).
89. A. E. Papatestas, G. J. Lesnick, G. Genkins, and A. H. Aufses, The prognostic significance of peripheral lymphocyte counts in patients with breast carcinoma, *Cancer* **37,** 164–168 (1976).
90. U. Glas, J. Wasserman, H. Blomgren, and A. De Schryver, Lymphopenia and metastatic breast cancer patients with and without radiation therapy, *Int. J. Radiat. Oncol. Biol. Phys.* **1,** 189–195 (1976).
91. J. A. McCredie, W. R. Inch, and R. M. Sutherland, Peripheral blood lymphocytes and breast cancer: Effects of operation and radiotherapy, *Arch. Surg.* **107,** 162–165 (1973).
92. J. Stjernswärd, M. Jondal, F. Vanky, H. Wigzell, and R. Sealy, Lymphopenia and change in distribution of human B and T lymphocytes in peripheral blood induced by irradiation for mammary carcinoma, *Lancet* **1,** 1352–1356 (1972).
93. W. H. West, C. W. Sienknecht, A. S. Townes, and R. B. Herberman, Performance of a rosette assay between lymphocytes and sheep erythrocytes at 29°C to study patients with cancer and other diseases, *Clin. Immunol. Immunopathol.* **5,** 60–66 (1976).
94. J. Djeu, S. Payne, C. Alford, W. Heim, T. Pomeroy, M. Cohen, R. Oldham, and R. B. Herberman, Detection of decreased proportion of lymphocytes forming rosettes with sheep erythrocytes at 29°C in the blood of cancer patients: Analysis of factors affecting the assay, *Clin. Immunol. Immunopathol.* **8,** 405–419 (1977).
95. J. L. Weese, R. K. Oldham, D. C. Tormey, A. L. Barlock, A. Morales, M. H. Cohen, T. C. Alford, P. E. Shorb, N. T. Tsangaris, W. H. West, C. B. Cannon, J. H. Dean, J. Djeu, J. L. McCoy, and R. B. Herberman, Immunologic monitoring in carcinoma of the breast, *Surg. Gynecol. Obstet.* **145,** 209–218 (1977).
96. R. H. Whitehead, J. Thatcher, C. Teasdale, G. P. Roberts, and L. E. Hughes, T and B lymphocytes in breast cancer: Stage relationship and abrogation of T-lymphocyte depression by enzyme treatment *in vitro, Lancet* **1,** 330–333 (1976).
97. R. H. Whitehead, G. P. Roberts, J. Thatcher, C. Teasdale, and L. E. Hughes, Masking of receptors for sheep erythrocytes on human T-lymphocytes by sera from breast cancer patients, *J. Natl. Cancer Inst.* **58,** 1573–1576 (1977).
98. B. Petrini, J. Wasserman, H. Blomgren, and E. Baral, Blood lymphocyte subpopulations in breast cancer patients following radiotherapy, *Clin. Exp. Immunol.* **29,** 36–42 (1977).
99. J. A. McCredie, W. R. Inch, and R. M. Sutherland, Effect of postoperative radiotherapy on peripheral blood lymphocytes in patients with carcinoma of the breast, *Cancer* **29,** 349–356 (1972).
100. H. Blomgren, R. Berg, J. Wasserman, and U. Glas, Effect of radiotherapy on blood lymphocyte population in mammary carcinoma, *Int. J. Radiat. Oncol. Biol. Phys.* **1,** 177–188 (1976).
101. M. M. Roberts and W. Jones-Williams, The delayed hypersensitivity reaction in breast cancer, *Br. J. Surg.* **61,** 549–552 (1974).
102. T. J. Cunningham, D. Daut, P. E. Wolfgang, M. Mellyn, S. Maciolek, R. W. Sponzo, and J. Horton, A correlation of DNCB-induced delayed cutaneous hypersensitivity reactions and the course of disease in patients with recurrent breast cancer, *Cancer* **37,** 1696–1700 (1976).
103. M. G. Whittaker and C. G. Clark, Depressed lymphocyte function in carcinoma of the breast, *Br. J. Surg.* **58,** 717–720 (1971).
104. J. H. Dean, R. Connor, R. B. Herberman, J. L. McCoy, J. Silva, and R. K. Oldham, The relative proliferation index as a more sensitive parameter for evaluating lymphoproliferative responses of cancer patients to mitogens and alloantigens, *Int. J. Cancer* **20,** 359–370 (1977).

105. N. Matthews and R. H. Whitehead, Inhibition of K cell function by human breast cancer sera, *Br. J. Cancer* **34,** 635–640 (1976).
106. R. Snyderman, M. C. Pike, L. Meadows, G. Hemstreet, and S. Wells, Depression of monocyte chemotaxis by neoplasms, *Clin. Res.* **23,** 297A (1975).
107. F. Loisillier, D. Buffe, K. B. Tan, P. Burtin, and P. Grabar, Etude immunologique des epitheliomas mammaires humains, *Ann. Inst. Pasteur* **109,** 1–21 (1965).
108. G. Taylor and J. L. Odili, Tumour specific T-like antigen of human breast carcinoma, *Br. J. Cancer* **24,** 447–453 (1970).
109. J. M. Gentile and J. T. Flickinger, Isolation of a tumor-specific antigen from adenocarcinoma of the breast, *Surg. Gynecol. Obstet.* **135,** 69–73 (1972).
110. E. S. Priori, G. Seman, L. Dmochowski, H. Gallagher, and D. E. Anderson, Immunofluorescence studies on sera of patients with breast carcinoma, *Cancer* **28,** 1462–1471 (1971).
111. E. M. Edynak, Y. Hirshaut, M. Bernhard, and G. Trempe, Fluorescent antibody studies of human breast cancer, *J. Natl. Cancer Inst.* **48,** 1137–1143 (1972).
112. O. R. Boehm, B. J. Boehm, and L. J. Humphrey, The natural history of the antibody response to breast antigens, *Clin. Exp. Immunol.* **16,** 31–40 (1974).
113. L. J. Humphrey, N. C. Estes, P. A. Morse, Jr., W. R. Jewell, R. A. Boudet, and M. J. K. Hudson, Serum antibody in patients with mammary disease, *Cancer* **34,** 1516–1520 (1974).
114. J. L. Weese, The Makari skin test as an immunodiagnostic technique, in: *Immunodiagnosis of Cancer* (R. B. Herberman and K. R. McIntire, eds.), Marcel Dekker, New York (in press).
115. K. Honda, K. Hoshishima, K. Kato, T. Okazaki, Y. Higuchi, and M. Ikeda, Studies with the Makari skin test in Japan, *Trans. N. Y. Acad. Sci.* **35,** 369–380 (1971).
116. D. E. H. Tee, Clinical evaluation of the Makari tumor skin test, *Br. J. Cancer* **28** (Suppl. I), 187–197 (1973).
117. J. Charney and D. H. Moore, Neutralization of murine mammary tumour virus by sera of women with breast cancer, *Nature (London)* **229,** 627–628 (1971).
118. E. S. Priori, D. E. Anderson, W. C. Williams, and L. Dmochowski, Immunological studies on human breast carcinoma and mouse mammary tumor, *J. Natl. Cancer Inst.* **48,** 1131–1135 (1972).
119. J. M. Bowen, L. Dmochowski, M. F. Miller, E. S. Priori, G. Seman, M. L. Dodson, and K. Maruyama, Implications of humoral antibody in mice and humans to breast tumor and mouse mammary tumor virus-associated antigens, *Cancer Res.* **36,** 759–764 (1976).
120. K. W. Newgard, R. D. Cardiff, and P. B. Blair, Human antibodies binding to the mouse mammary tumor virus: A nonspecific reaction?, *Cancer Res.* **36,** 765–768 (1976).
121. D. Stutman and R. B. Herberman, Immunological control of breast cancer: Discussion, *Cancer Res.* **36,** 781–782 (1976).
122. M. M. Black and H. P. Leis, Jr., Cellular responses to autologous breast cancer tissue: Sequential observations, *Cancer* **32,** 384–389 (1973).
123. L. E. Hughes and B. Lytton, Antigenic properties of human tumours: Delayed cutaneous hypersensitivity reactions, *Br. Med. J.* **1,** 209–212 (1964).
124. T. H. M. Steward and M. Orizaga, The presence of delayed hypersensitivity reactions in patients toward cellular extracts of their malignant tumors. 3. The frequency, duration, and cross reactivity of this phenomenon in patients with breast cancer, and its correlation with survival, *Cancer* **28,** 1472–1478 (1971).
125. C. Alford, A. C. Hollinshead, and R. B. Herberman, Delayed cutaneous hypersensitivity reactions to extracts of malignant and normal human breast cells, *Ann. Surg.* **178,** 20–24 (1973).

126. A. C. Hollinshead, W. T. Jaffurs, L. K. Alpert, J. E. Harris, and R. B. Herberman, Isolation and identification of soluble skin reactive membrane antigens of malignant and normal breast cells, *Cancer Res.* **34,** 2961–2968 (1974).
127. J. Weese, R. B. Herberman, R. B. Hollinshead, G. Cannon, M. Keels, and R. K. Oldham, Specificity of delayed cutaneous hypersensitivity reactions to extracts of human tumor cells, *J. Natl. Cancer Inst.* **60,** 255–263 (1978).
128. V. Andersen, O. Bjerrum, G. Bendixen, T. Schiodt, and I. Dissing, Effect of autologous mammary tumor extracts on human leukocyte migration *in vitro, Int. J. Cancer* **5,** 357–363 (1970).
129. A. Segall, O. Weiler, J. Genin, J. Lacour, and F. Lacour, *In vitro* study of cellular immunity against autochthonous human cancer, *Int. J. Cancer* **9,** 417–425 (1972).
130. A. J. Cochran, R. M. Grant, W. G. Spilg, R. M. Mackie, C. E. Ross, D. E. Hoyle, and J. M. Russell, Sensitization of tumor-associated antigens in human breast carcinoma, *Int. J. Cancer* **14,** 19–25 (1974).
131. J. L. McCoy, L. F. Jerome, J. H. Dean, G. B. Cannon, T. C. Alford, T. Doering, and R. B. Herberman, Inhibition of leukocyte migration by tumor-associated antigens in soluble extracts of human breast carcinoma, *J. Natl. Cancer Inst.* **53,** 11–17 (1974).
132. A. S. Kadish, D. M. Marcus, and B. R. Bloom, Inhibition of leukocyte migration by human breast-cancer-associated antigens, *Int. J. Cancer* **18,** 581–586 (1976).
133. J. L. McCoy, L. F. Jerome, C. Anderson, G. B. Cannon, T. C. Alford, R. J. Connor, R. K. Oldham, and R. B. Herberman, Leukocyte migration inhibition by soluble extracts of MCF-7 tissue culture cell line derived from breast carcinoma, *J. Natl. Cancer Inst.* **57,** 1045–1049 (1976).
134. G. B. Cannon, J. L. McCoy, L. J. Jerome, R. Reddick, C. Alford, V. Tenley, and R. B. Herberman, Immunological relationship between breast carcinoma and benign breast disease as detected by the leukocyte migration inhibition assay, *J. Natl. Cancer Inst.* (in press) (1978).
135. M. M. Black, H. P. Leis, B. Shore, and R. E. Zachrau, Cellular hypersensitivity to breast cancer: Assessment by a leukocyte migration procedure, *Cancer* **33,** 952–958 (1974).
136. M. M. Black, D. H. Moore, B. Shore, R. E. Zachrau, and H. P. Leis, Jr., Effect of murine milk samples and human breast tissues on human leukocyte migration indices, *Cancer Res.* **34,** 1054–1060 (1974).
137. M. M. Black, R. E. Zachrau, B. Shore, D. H. Moore, and H. P. Leis, Jr., Prognostically favorable immunogens of human breast cancer tissue: Antigenic similarity to murine mammary tumor virus, *Cancer* **35,** 121–128 (1975).
138. M. M. Black, R. E. Zachrau, B. Shore, and H. P. Leis, Jr., Biological considerations of tumor-specific and virus-associated antigens of human breast cancer, *Cancer Res.* **36,** 769–774 (1976).
139. R. E. Zachrau, M. M. Black, A. S. Dion, B. Shore, M. Isac, A. M. Andrade, and C. J. Williams, Prognostically significant protein components of human breast cancer tissues, *Cancer Res.* **36,** 3143–3146 (1976).
140. J. L. McCoy, J. H. Dean, G. B. Cannon, T. C. Alford, W. P. Parks, R. V. Gilden, S. T. Oroszlan, and R. B. Herberman, Leukocyte migration inhibition and lymphocyte blastogenesis responses in human breast carcinoma patients to mouse mammary tumor virus, virion gp52 and gp70 antigens, *J. Natl. Cancer Inst.* (in press) (1978).
141. P. Fischer, E. Golub, H. Holzner, and E. Kunze-Muhl, Comparative effects of tumor extracts on lymphocyte transformation in peripheral blood cultures of healthy persons and patients with breast cancer, *Z. Krebsforsch.* **72,** 155–161 (1969).
142. H. Savel, Effect of autologous tumor extracts on cultured human peripheral blood lymphocytes, *Cancer* **24,** 56–63 (1969).

143. C. C. S. Hsu and S. R. Cooperband, *In vitro* responses of lymphocytes from cancer-bearing patients to autochthonous tumor tissues, *Proc. Soc. Exp. Biol. Med.* **136,** 446–448 (1971).
144. G. M. Mavligit, U. Ambus, J. U. Gutterman, and E. M. Hersh, Antigen solubilized from human solid tumors: Lymphocyte stimulation and cutaneous delayed hypersensitivity, *Nature (London) New Biol.* **243,** 188–190 (1973).
145. J. H. Dean, J. L. McCoy, G. B. Cannon, C. M. Leonard, E. Perlin, A. Kreutner, R. K. Oldham, and R. B. Herberman, Cell-mediated immune responses of breast cancer patients to autologous tumor-associated antigens, *J. Natl. Cancer Inst.* **58,** 549–555 (1977).
146. J. H. Dean, J. S. Silva, J. L. McCoy, C. M. Leonard, M. Middleton, G. B. Cannon, and R. B. Herberman, Lymphocyte blastogenesis induced by 3M KCl extracts of allogeneic breast carcinoma and lymphoid cells, *J. Natl. Cancer Inst.* **54,** 1295–1298 (1975).
147. S. Cunningham-Rundles, W. F. Feller, C. Cunningham-Rundles, B. DuPont, H. Wanebo, R. O'Reilly, and R. A. Good, Lymphocyte transformation *in vitro* to RIII mouse milk antigen among women with breast disease, *Cell Immunol.* **25,** 322–327 (1976).
148. I. Hellström, K. E. Hellström, H. O. Sjögren, and G. A. Warner, Demonstration of cell-mediated immunity to human neoplasms of various histological types, *Int. J. Cancer* **7,** 1–16 (1971).
149. G. Fossati, S. Canevari, G. Della Porta, G. P. Balzarini, and U. Veronesi, Cellular immunity to human breast carcinoma, *Int. J. Cancer* **10,** 391–396 (1972).
150. G. Della Porta, S. Canevari, and G. Fossati, Immune responses to tumour and embryo cells in patients with mammary carcinoma, *Br. J. Cancer* **28,** (Suppl. I), 103–107 (1973).
151. R. K. Oldham, J. Y. Djeu, G. B. Cannon, D. Siwarski, and R. B. Herberman, Cellular microcytotoxicity in human tumor systems: Analysis of results, *J. Natl. Cancer Inst.* **55,** 1305–1318 (1975).
152. G. Heppner, E. Henry, L. Stolbach, F. Cummings, E. McDonough, and P. Calabresi, Problems in the clinical use of the microcytotoxicity assay for measuring cell-mediated immunity to tumor cells, *Cancer Res.* **35,** 1931–1937 (1975).
153. H. F. Jeejeebhoy, Immunological studies of women with primary breast carcinoma, *Int. J. Cancer* **15,** 867–878 (1975).
154. S. Canevari, G. Fossati, and G. Della Porta, Cellular immune reaction to human malignant melanoma and breast carcinoma cells, *J. Natl. Cancer Inst.* **56,** 705–709 (1976).
155. R. B. Herberman and R. K. Oldham, Problems associated with study of cell-mediated immunity to human tumors by microcytotoxicity assays, *J. Natl. Cancer Inst.* **55,** 749–753 (1975).
156. F. Avis, I. Mosonov, and G. Haughton, Antigenic cross-reactivity between benign and malignant neoplasms of the human breast, *J. Natl. Cancer Inst.* **52,** 1041–1049 (1974).
157. G. B. Cannon, G. D. Bonnard, J. Djeu, W. H. West, and R. B. Herberman, Relationship of human natural lymphocyte-mediated cytotoxicity to cytotoxicity of breast cancer-derived target cells, *Int. J. Cancer* **19,** 487–497 (1977).
158. I. Hellström, H. O. Sjögren, G. Warner, and K. E. Hellström, Blocking of cell-mediated tumor immunity by sera from patients with growing neoplasms, *Int. J. Cancer* **7,** 226–237 (1971).
159. A. E. Powell, A. M. Sloss, R. N. Smith, J. L. Makley, and C. A. Hubay, Specific responsiveness of leukocytes to soluble extracts of human tumors, *Int. J. Cancer* **16,** 905–913 (1975).

160. N. Grosser and D. M. P. Thomson, Cell-mediated antitumor immunity in breast cancer patients evaluated by antigen-induced leukocyte adherence inhibition in test tubes, *Cancer Res.* **35,** 2571–2579 (1975).
161. M. Flores, J. R. Marti, N. Grosser, J. D. MacFarlane, and D. M. P. Thomson, An overview: Antitumor immunity in breast cancer measured by tube leukocyte adherence inhibition assay, *Cancer* **39,** 494–505 (1977).
162. T. Fujisawa, S. R. Waldman, and R. H. Yonemoto, Leukocyte adherence inhibition by soluble tumor antigens in breast cancer patients, *Cancer* **39,** 506–513 (1977).
163. E. A. Caspary and E. J. Field, Specific lymphocyte sensitization in cancer: Is there a common antigen in human malignant neoplasia?, *Br. Med. J.* **2,** 613–617 (1971).
164. J. A. V. Pritchard, J. L. Moore, W. H. Sutherland, and C. A. F. Joslin, Evaluation and development of the macrophage electrophoretic mobility (MEM) test for malignant disease, *Br. J. Cancer* **27,** 1–9 (1973).
165. K. D. Bagshawe, Workshop on macrophage electrophoretic mobility (MEM) and structuredness of cytoplasmic matrix (SCM) tests, *Br. J. Cancer* **35,** 701–704 (1977).
166. J. A. V. Pritchard, W. H. Sutherland, and T. J. Deeley, Cancer detection, *Lancet* **1,** 637 (1976).
167. L. Cercek, B. Cercek, and C. I. V. Franklin, Biophysical differentiation between lymphocytes from healthy donors, patients with malignant diseases and other disorders, *Br. J. Cancer* **29,** 345–352 (1974).
168. L. Cercek, and B. Cercek, Apparent tumour specificity with the SCM test, *Br. J. Cancer* **31,** 252–253 (1975).
169. L. Cercek and B. Cercek, Changes in the SCM response ratio (RR_{SCM}) after surgical removal of malignant tissue, *Br. J. Cancer* **31,** 250–251 (1975).
170. A. Theofilopoulos, C. B. Wilson, and F. J. Dixon, The Raji cell radioimmune assay for detecting immune complexes in human sera, *J. Clin. Invest.* **57,** 169–182 (1976).
171. R. D. Rossen, M. A. Reisberg, E. M. Hersh, and J. U. Gutterman, The $C1_q$ binding test for soluble immune complexes: Clinical correlations obtained in patients with cancer, *J. Natl. Cancer Inst.* **58,** 1205–1211 (1977).
172. J. A. Stein, A. Adler, S. Ben Efram, and M. Maor, Immunocompetence, immunosuppression, and human breast cancer. I. An analysis of their relationship by known parameters of cell-mediated immunity in well-defined clinical stages of disease, *Cancer* **38,** 1171–1187 (1976).

7

The Cell Kinetics of Mammary Cancers

LEWIS M. SCHIFFER

1. Introduction

The cell kinetics of mammary cancers are not dissimilar from those of other cancers, except perhaps as they are influenced by the hormonal milieu. This chapter will review the available literature of unperturbed breast cancer cell kinetics in both transplantable and spontaneous animal models, as well as the available material on human breast cancers. The data reported here originated from *in vivo* and short-term *in vitro* studies performed directly with the tumor tissue. The reason for this restriction is that tissue cultured breast tumor cells, while they perhaps retain some of the biochemical characteristics of the original specimen, do not necessarily have the same cell kinetic patterns as the primary or metastatic tumors from which they were derived.

The following section on terminology and definitions should be sufficient to orient all but the most inquisitive reader. The latter will find references to each of the experimental studies, and will rapidly become aware of the heterogeneity of techniques and methods of analyses.

1.1. Terminology and Definitions

The cell cycle can be thought of as an orderly progression of a cell from a specific point in the reproductive process through various phases until one of the daughter cells once again reaches that same specific

---

LEWIS M. SCHIFFER • Cancer Biology Section, Cancer Research Unit, Clinical Radiation Therapy Research Center, Allegheny General Hospital, Pittsburgh, Pennsylvania 15212.

point. The median time required to traverse that path is called the cell cycle time (T_C). If the specific point at which one starts is the point in mitosis where telophase ends and two daughter cells are formed, then these daughter cells enter a phase in the cell cycle called G_1. The median time required for a cell to complete G_1 is T_{G_1}. This phase is characterized by increasing production of protein, RNA, and other constituents that are necessary for induction and completion of the next phase of the cell cycle: DNA synthesis, or the S phase. The median time necessary for a cell to double its DNA content is referred to as the DNA synthesis time (T_S). It is during the S phase that cells can incorporate exogenous nucleosides into the newly synthesized DNA. If the specific nucleoside precursor of DNA synthesis (thymidine) is radioactively labeled with tritium (^3H), its incorporation into DNA can be studied by autoradiography or liquid scintillation counting. The former technique proves to be more useful in cell kinetic analyses.

The fraction of cells labeled with tritiated thymidine ([^3H]-TdR) is a measure of the fraction of cells in the S phase at the time of the pulse label, and is defined as the labeling index (^3H]-TdR LI). Naturally, if these cells go on to divide, die, or change in any way, the [^3H]-TdR LI, or thymidine index, will change also. Following completion of DNA synthesis, the cells enter a premitotic G_2 phase. This phase is usually short, but during this time (T_{G_2}), proteins, microtubules and other components necessary for initiation and completion of the division process are synthesized.

When the cell enters prophase, it becomes recognizable as a mitotic figure and remains so during mitosis. The cell rapidly progresses through the remaining mitotic stages to telophase, with the median time referred to as T_M. When telophase ends and the new daughter cells are formed, the cycle, as defined earlier, comes to an end. Each new daughter cell may then traverse the cycle in the same fashion.

Cells such as granulocytes, muscle cells, and villus cells differentiate and go out of the active cell cycle. This usually occurs in the G_1 phase. Loss due to cell death may occur during active cell proliferation or during phases of inactivity. One phase of mitotic inactivity is frequently called the G_0 phase. G_0 originally referred to bone marrow stem cells, considered out-of-cycle, but capable of reentering the cycle when given the proper stimulus. A similar concept, proposed by Mendelsohn,[(1)] is that of the tumor growth fraction (GF). The GF is that fraction of the total tumor cell population that is actively progressing through the cell cycle (G_1, S, G_2, or M), and is thus one minus the fraction of G_0 cells.

It is possible to measure the GF of animal tumors by means of pulse [^3H]-TdR administration and percent labeled mitoses (PLM) analyses, or by other techniques utilizing continuous or intermittent [^{3}H]-TdR administration. These techniques, which require *in vivo* [^{3}]-TdR and multiple samplings, are not generally applicable in a clinical setting, and very little human GF data have been obtained.

An *in vitro* method has recently been described for estimation of the GF of experimental animal tumors, termed the primer-available DNA polymerase (PDP) assay.[2] It is also thought, although not yet proved, to estimate GF in human tumors.[3]

The cell cycle time (T_C) can be calculated from the PLM analysis or from the *in vitro* measurements of [^{3}H]-TdR LI, T_S, and PDP or GF. The potential doubling time (T_P) can also be calculated, and is defined as the time required for the cell population to double in size, in the absence of cell loss.

For all these calculations, one must also consider that the tumor cell populations are not in a steady state. Thus, the mathematics of these calculations involve certain exponential factors to account for growth.

The rate of tumor growth, while frequently difficult to estimate in human tumors, can be measured by calipers in animal tumors and usually follows Gompertzian growth. This type of growth is characterized by an exponential increase interacting with an exponential slowing such that there is a rapid, early increase in the size followed by a slower increase or even a semiplateau phase. The tumor volume doubling time (T_D) for a tumor of any size can be calculated from these equations or measured by tangents at points on this curve. The major cell kinetic parameters that govern the T_D are the cell cycle time, the GF and cell loss (ϕ). The latter can be calculated from the T_P and T_D values, and represent the proportion of cells that are lost as a ratio of those produced.

2. Cell Kinetics of Mouse Mammary Tumors

2.1. Transplantable Tumors

2.1.1. *Tumors of the C3H Strain*

These tumors are derived from a mouse strain known to carry the mouse mammary tumor virus (MTV). The particular lines from which the cell kinetics are reported may or may not have been derived by virus induction, but probably should be so considered.

It is quite clear from the works of Wexler *et al.*[4] and McCredie *et al.*[5] that the rates of growth of early-transplant-generation tumors differ from those of late-transplant generations and spontaneous tumors. This is also reflected in the cell kinetic measurements that will be reported. In addition, and perhaps more important, the individual variations of the cell kinetics of the original donor tumor, while unknown in all these instances, could have very considerable effects on the subsequent cell kinetics of the new transplantable tumor lines.

Denekamp[6] reported cell kinetic results from first-generation transplants derived from a spontaneous tumor. In tumors with a T_D of 108 hr and [^{3}H]-TdR LI of 0.16, she found a T_{G_1} of 15.0 ± 25 hr; T_S, 6.0 ± 1.5 hr; T_{G2}, 2.0 ± 1.0 hr; T_C, 15.2 ± 8.9 hr; GF, 0.37; and ϕ, 0.70. The GF from these data, reported by Steel,[7] was recalculated to 0.46. Other first-generation studies by Fowler *et al.*[8] disclosed longer T_D's 8–9 days and higher ϕ factors of 0.74–0.79. Simpson-Herren and Lloyd[9] described a T_D of 5.5 days and a relatively low [^{3}H]-TdR LI, averaging 0.02.

Tannock[10] studied the cell kinetics of fourth-generation transplants of C3H/He tumors into C3BF$_1$ (C3H × C57BL) mice. He found a mean [^{3}H]-TdR LI of 0.351 ± 3.9 in these hybrid mice. Near areas of necrosis, the [^{3}H]-TdR LI was 0.103 ± 0.034, and regions near blood vessels had a [^{3}H]-TdR LI of 0.500 ± 0.025. Intermediate areas had a [^{3}H]-TdR LI of 0.296 ± 0.048. The PLM curves were not dissimilar, and the cell kinetic parameters were T_{G_1}, 3.0 ± 2.0; T_S, 7.2 ± 1.5 hr; T_{G_2}, 3.0 ± 1.0 hr; and T_C, 12.8 hr. The GF was 0.5.

There are a number of tumor lines of longer transplant generations that have been studied in great detail. Mendelsohn and Dethlefsen[11] reported on three lines derived from the C3H mouse. The S102F had a T_{G_1} of 6.5 ± 4.6 hr; T_S, 7.7 ± 3.4 hr; T_{G_2}, 2.5 ± 1.4 hr; T_M, 0.6 ± 0.3 hr; and T_C, 17.2 ± 5.9 hr. The GF was 0.55. The fast line had a T_{G_1} of 6.7 ± 3.0 hr; T_S, 11.0 ± 5.5 hr; T_{G_2}, 1.3 ± 0.8 hr; T_M, 1.1 ± 0.6 hr; with a T_C of 20.1 ± 6.3 hr and a GF of 0.30. The slow line had a longer T_{G_1} of 17.3 ± 12.3 hr; T_S, 12.5 ± 7.2 hr; T_{G_2}, 2.1 ± 1.5 hr; and T_M, 1.6 ± 1.1 hr. The T_C was 33.5 ± 14.3 hr, and the GF was 0.23.

Studies by Szczepanski and Trott[12] on a C3H line, called 284, of extended transplant generation showed a [^{3}H]-TdR LI of 0.125; mitotic index (MI) of 0.0037; T_{G_1}, 4.7 hr; T_S, 6.8 hr; T_{G_2}, 1.9 hr; T_C, 13.6 hr; GF, 0.25; and ϕ of 0.5 in tumors 0.8–1.0 cm in size.

The C3HBA tumor derived from a spontaneous mammary adenocarcinoma in a C3H/An mouse has been carried for over 30 years by the Jackson Laboratories in C3H/He mice. Nelson *et al.*[13] reported that in small tumors of 56 mg size and T_D's of 61.4 hr, the [^{3}H]-TdR LI was 0.274 ± 0.044 and T_{G_1} was 10.4 ± 13.6 hr; T_S, 8.7 ± 2.9 hr; T_{G_2}, 3.2 ± 2.3 hr; and T_C, 20.3 hr. The GF was 0.70 and ϕ was 0.59.

Still another C3H-derived tumor is the H2712. This tumor was derived from a spontaneous tumor of the C3H/HeHu mouse in 1948 and has been in continued transplantation in the C3H/HeJ mouse ever since that time. Unpublished results[14] on 70-mm^3 tumors with a T_D of 34.7 hr showed a [^{3}H]-TdR LI of 0.33; T_{G_1}, 0.9 hr; T_S, 5.8 hr; T_{G_2}, 2.1 hr; and T_C, 9.0 hr. The GF was between 0.51 and 0.67, and ϕ was 0.57.

The HB mammary tumor in the C3H/Tif/BOM mouse was described[15] as having a [^{3}H]-TdR LI of 0.33 at 21 days post-transplant. The T_{G_1} was 5.9 hr; T_S, 7 ± 0.08 hr; T_{G_2}, 1.4 hr; and T_C, 15.5 hr. The GF was 0.47, and ϕ was 0.84. Oophorectomy produced a lengthening of the T_{G1} to 9.3 hr and reduction of the [^{3}H]-TdR LI to 0.21, but sham surgery had no effect on the cell kinetics.

2.1.2. *Tumors of the DBA Strain*

The DBAH tumor line was derived from a DBA/212 mammary adenocarcinoma by Goldfeder.[16] This epithelial tumor in the 38th passage was found to have a T_{G_1} of 7.5 hr; T_S, 6 hr; T_{G_2}, 2.5 hr; and a T_C of 16 hr. Another tumor derived from the same line was also studied by the PLM method in the 458th passage and found to have a T_{G_1} of 3 hr, a T_S of 12 hr, a T_{G_2} of 1 hr, and a T_C of 16 hr. This latter tumor, called the DBAG, was a spindle-cell tumor histologically, and may not be an adenocarcinoma. Also, it had considerable numbers of tetraploid, octaploid, and aneuploid cells.

The T1699 tumor originally found in, and still passaged in, the DBA/2J mouse was studied by PLM analysis at 14 days post-transplantation when the tumor was 1 cm^3 in size.[2] The [^{3}H]-TdR LI was 0.135, and the T_D was 54.2 hr. The T_{G_1} was 3.1 hr; T_S, 5.8 hr; T_{G_2}, 3.7 hr; and T_C, 13.1 hr. The GF was 0.30. Computer analysis of the data differed only in the T_{G_1}, which was increased to 5.4 hr, and the GF increased to 0.40.

Other DBA/2 tumors are known,[17] having been derived from the DBA/2Ha-DD line, but only growth curves have been described.

2.1.3. *Tumors of the BALB/c Strain*

The KHJJ tumor was derived from a hyperplastic alveolar nodule arising spontaneously in a BALB/cCrgl mouse and passaged in BALB/c mice. Tumors were studied at 100 mm^3 when the T_D was approximately 38 hr.[18] Using Steel's method of analysis, Rockwell and co-workers found that at the tumor periphery, the [^{3}H]-TdR LI was 0.358 ± 0.014; the mean T_{G_1} was 2.3 ± 4.5 hr; T_S, 8.4 ± 0.5 hr; T_{G_2}, 2.0 ± 1.5 hr; and T_C, 13.7 ± 4.3 hr. The GF was 0.52, and ϕ was 0.31. Mendelsohn's method of analysis, however, showed a T_{G_1} of 3.7 ± 1.7 hr; T_S, 9.5 ± 2.8 hr; T_{G_2}, 1.4 ± 0.5 hr; and T_C, 15.2 ± 3.2 hr. At the center of the tumor, by the Steel analysis, the T_{G_1} was 4.6 ± 3.0 hr; T_S, 9.4 ± 2.3 hr; T_{G_2}, 2.2 ± 1.2 hr;

and T_C, 16.1 ± 4.2 hr, with a [^{3}H]-TdR LI of 0.223 ± 0.011 and a GF of 0.38. Mendelsohn's method of analysis of these data showed the T_{G_1} to be 4.7 hr; T_S, 10.1 hr; T_{G_2}, 1.5 hr; and T_C, 17.1 hr.

At the 25th passage of the KHJJ tumor, it was placed into tissue culture, and after 33 passages *in vitro,* a colony was reimplanted into BALB/c mice and was renamed the EMT6 tumor. This tumor was also studied by PLM methods at 100 mm^3 in volume. In regions with no necrotic features, Steel's analysis showed the T_{G_1} to be 7.3 ± 10.9 hr; T_S, 11.9 ± 3.6 hr; T_{G_2}, 1.9 ± 0.7 hr; and T_C, 20.7 ± 7.1 hr. Mendelsohn's analysis showed the T_{G_1} to be 8.0 ± 8.0 hr; T_S, 11.9 ± 4.5 hr; T_{G_2}, 1.7 ± 0.8 hr; and T_C, 22.3 ± 9.1 hr. The [^{3}H]-TdR LI was 0.293 ± 0.012; GF, 0.48; and ϕ, 0.61. In areas on the edge of necrosis, by the Steel analysis, the T_{G_1} was 9.4 ± 10.3 hr; T_S, 11.6 ± 6.4 hr; T_{G_2}, 2.6 ± 1.8 hr; and T_C, 22.6 ± 9.0 hr. By Mendelsohn's analysis, the T_{G_1} was 9.8 hr; T_S, 11.1 hr; T_{G_2}, 2.2 hr; and T_C, 23.9 hr. For this region, the [^{3}H]-TdR LI was 0.239 ± 0.011, and the GF was 0.51.

Watson[19] studied a variant of this tumor, designated EMT6/M/AC, which he considered to be histologically more of an adenocarcinoma than was Rockwell's at the time of her PLM studies. He also studied the tumor at different tumor volumes. At the time when the [^{3}H]-TdR was administered, the smallest tumor was 1.5 mm^3 and had a T_D of 31.5 hr. The largest tumor at the time of the PLM analysis was 175 mm^3, and the T_D was 120 hr. Although the [^{3}H]-TdR LI decreased with increasing size, from 0.320 to 0.241, the compartment time parameters did not change in any great detail. T_{G_1} varied from 5.6 to 7.5 hr; T_S, from 7.3 to 9.9 hr; T_{G_2}, from 1.5 to 1.8 hr; and T_C, from 14.1 to 18.5 hr. GF decreased from 0.682 to 0.516, and ϕ increased from 0.403 to 0.743.

2.1.4. *Tumors of the C57BL Strain*

The Ca775 tumor originated in a primary adenocarcinoma of the C57BL strain. It has been passaged for many years in either that line or the BDF$_1$ (C57BL ♀ × DBA/2 ♂) hybrid. The studies of Laster *et al.*[20] and Simpson-Herren and Lloyd[9] were performed on the latter animals. At 4 days posttransplantation, when the tumor weight averaged 46 mg, the [^{3}H]-TdR LI was 0.27 and the T_D was about 19 hr; the T_{G_1} was 3.5 hr; T_S, 7.0 hr; T_{G_2}, 0.5 hr; and T_C, 12.0 hr. The GF was 0.51. At 8 days posttransplantation, the mean tumor weight was 0.38 g, the [^{3}H]-TdR LI 0.18, and the T_D was 38 hr. At that time, T_{G_1} was 4.5 hr; T_S, 6.0 hr; T_{G_2}, 2.5 hr; and T_C, 14.0 hr. The GF was 0.45. Steel's recalculations of these data[7] increased the T_{G_1} to 7.0 hr and T_C to 15 hr. The 14-day tumors weighed 2.7 g, on average, and had a [^{3}H]-TdR LI of 0.09 and a T_D of over 9.7 days. The T_{G_1} was 7.5 hr; T_S, 6.0 hr; T_{G_2}, 1.5 hr; and T_C, 16 hr. The GF was 0.28.

2.1.5. *Tumors of Other Strains*

Janik *et al.*[21] reported on hormone-dependent tumors induced in GR/FIB mice by progesterone–estrone treatments. The cell kinetics were performed on first-generation transplants from the same primary tumor.

On day 23 after transplantation, while under hormonal influence, the T_D was 67 hr and the [^{3}H]-TdR LI 0.27. At that time, the T_{G_1} was 1.8 hr, and T_C, 15.5 hr. The GF was 0.60. When hormones were eliminated, and the tumor was regressing with a $T_{\frac{1}{2}}$ of 36 hr, the [^{3}H]-TdR LI was 0.075. The T_{G_1} was 9.9 hr; T_S, 9.1 hr; T_{G_2}, 2.6 hr; and T_C, 21.6 hr. The GF had dropped to 0.17.

Tannock[22] studied the cell kinetics of the BICR/SA1 tumor derived from a spontaneous tumor and passaged in the A/St mouse. This tumor had been passaged approximately 100 times when PLM curves were performed. The [^{3}H]-TdR LI in regions near blood vessels was 0.62 ± 0.07; near necrosis, 0.30 ± 0.07; and in intermediate regions, 0.42 ± 0.06. The T_{G1} was 4 hr; T_S, 10.7 hr; and T_{G2}, 1.3 hr, with a T_C of 17 hr. The PLM curves did not differ significantly in the three regions, although the GF did because of the differing [^{3}H]-TdR LI. The T_D was 74 hr.

Further analysis by the method of Steel and Hanes[23] revealed a T_{G_1} of 6.6 ± 6.5 hr; T_S, 9.3 ± 1.9 hr; T_{G_2}, 2.9 ± 1.5 hr; and T_C, 18.8 ± 6.9 hr.[24]

2.2. Spontaneous Tumors

2.2.1. *Tumors of the C3H Strain*

This mouse strain, originally a cross of a Bagg albino ♀ × DBA ♂, has, since its use by Heston, always been found to carry the mouse MTV. Mammary tumors develop in parous animals in a greater incidence than nulliparous, and may in fact reach up to a 90–100% incidence in some series. A recent study[25] indicates that the incidence and latent period of tumors vary according to the amount of stress to which the animals are subjected. Stress, as defined by cage crowding and frequent handling, increased the incidence to 90% in 400-day-old parous mice as opposed to virtually none in mice in a protected environment. In this mouse system, the tumors do not appear to be hormone-dependent.

Numerous investigators have reported data on the growth characteristics of the C3H/He tumor, and some of these data will be documented here. McCredie *et al.*[5] reported the differences in growth rates of these primary tumors, first-generation transplants and long-term syngeneic transplants of the C3HBA in the C3H/He mouse. They found that there was a considerable increase in growth rates as one increased the number of transplant generations. From their data, it appears that a 1-cm^3 tumor, approximately 35 days after appearance, has a T_D of about 10–12 days. Magdon and Winterfeld[26] classified their tumors according to their latent period and established seven growth types. Their results are in general accordance with others reported here. Mendelsohn[27] found that the mean T_D for 115 tumors was 8.5 days, with a wide variation of 2–50. Also, Mendelsohn and Dethlefsen[28] found that selection of fast-growing primary tumors for transplantation did tend to give first-generation tumors with faster growth rates than those from more slowly growing primaries.

In a study by Braunschweiger *et al.*[29] on C3H/He tumors, there was noted a wide variation in individual T_D's, ranging from 3.6 to 73.2 days. The mean was 17.0 days and the median 12.4 days for tumors of mean size 1.58 cm^3. The frequency distribution was found to be log normal in character.

Although a C3H/Bts strain of spontaneous mammary tumors was used, Shewell[30] found a mean T_D of 14.7 days and a median of 13.8 days. All these studies emphasize the wide variation inherent in the growth of spontaneous mouse tumors.

Normal mammary epithelium in the C3H mouse has been studied from a cell kinetic viewpoint. Bresciani[31] found an average [^{3}H]-TdR LI of 0.049 with a range of 0.027–0.076. This was confirmed by Banerjee,[32] who found an average [^{3}H]-TdR LI of 0.048, which decreased to 0.0013 at 5 days after oophorectomy. Banerjee and Walker[33] also found that the T_S was an average 20.7 hr, but this value decreased to 8.2 hr in the hormone-stimulated pregnant animal. Bresciani[34] also reported the value for T_S to be 20.1 hr, with a range of 14.8–27.6 hr, which also decreased to 10.4 hr (9.5–11.9 hr range) on hormone stimulation. In another study, Bresciani[31] found the T_S to be an average 21.7 hr, with a range of 13.0–31.2 hr; a T_{G_1} of approximately 45.7 hr; a T_{G_2} of approximately 3 hr; and a TC of approximately 71 hr.

Hyperplastic alveolar nodules (HANs) were found to have a [^{3}H]-TdR LI of 0.071 by Banerjee[32] and of 0.097, with a range of 0.058–0.182, by Bresciani.[31] In Banerjee's study, there was no change with oophorectomy. He also found a T_S of 13.7 hr, with a range of 10.4–20.0 in these nodules.[33] Bresciani[31] found the HAN T_S to be 15.9 hr, with a range of 12.8–20.1; the T_{G_1} was approximately 26.8 hr; the T_{G_2} was approximately 2 hr; and the T_C was approximately 46 hr.

C3H/He tumor cell kinetic studies really began with Mendelsohn's classic reports[1,35,36] in the early 1960s. Those results, followed by additional animal studies in 1965,[24,27] showed that at a T_D of 204 hr and a [^{3}H]-TdR LI of 0.14, the T_{G_1} was 19.4 ± 8.7 hr; the T_S, 11.7 ± 5.2 hr; the T_{G_2}, 1.9 ± 0.9 hr; the T_C, 34.6 ± 10.2 hr; and the GF, 0.40. Analysis by

Steel[7] of Mendelsohn's 1965 data showed a shorter T_{G_1} of 14 hr, a GF of 0.45, and a ϕ of 0.75.

Braunschweiger *et al.*[29] recently published the data from a large group of individual spontaneous C3H/He tumors. The reason for this study was to validate the usefulness of certain *in vitro* cell kinetic techniques, and therefore only a limited number of cell kinetic parameters were measured. However, these could be performed in a large number of individual animals along with certain clinical correlations. There was no difference in the cell kinetics of type A or type B (Dunn classification) tumors, the [^{3}H]-TdR LI and T_S values being identical. Results were obtained for *in vitro* studies of [^{3}H]-TdR LI and T_S similar to those for the *in vivo* studies of type A tumors.

The GF, calculated by the technique of Mendelsohn,[1] was 0.18 ± 0.013, with a range of 0.117–0.282. This was compared to the *in vitro* analysis by the PDP method of Schiffer *et al.*[2] of 0.167 ± 0.004, with a range of 0.093–0.326. Assuming a value of 2 hr for T_{G_2}, the cell cycle kinetic parameters measured *in vitro* (*in vivo* values in parentheses) are T_{G_1}, 15.2 ± 0.8 S.E.M. (10.6); T_S, 10.9 ± 0.2 S.E.M. hr (10.7); T_C, 28.6 ± 1.6 S.E.M. (32.1); ϕ, 0.572 (0.593); [^{3}H]-TdR LI, 0.074 ± 0.005 S.E.M. (0.069).

The [^{3}H]-TdR LI was shown to be inversely related to the T_D, with the highest [^{3}H]-TdR LI's having the shortest T_D's. There appeared to be little or no relationship of T_D to the PDP index. There was a broad direct relationship of the [^{3}H]-TdR LI to the PDP index, although at the largest values of each, there was wide variation. The T_D was directly related to the potential doubling time (T_P), but again, with wide variations. The T_P was indirectly related to the PDP index such that at high PDP values, the T_P was quite short. The cell loss factor (ϕ) increased with increasing T_D's until it reached a plateau at a T_D of about 15–20 days. Finally, there appeared to be a broad, but definite, direct relationship of T_C to T_D.

2.2.2. *Tumors of the CD8F1 Strain*

This tumor system was developed by Martin and his colleagues by cross-mating BALB/cfC3H ♀ with DBA/8 ♂.[37] The females develop breast cancer at a peak age of 9 months. The mice are all mouse-MTV-positive, and have been used for a variety of surgical, immunological, and chemotherapeutic experiments. Their growth curves are highly variable like the C3H/He, and these mice develop lung metastases spontaneously.

Braunschweiger and Schiffer looked at the cell kinetics of this tumor in collaboration with Martin.[38] For tumors between 0.5 and 1.0 cm^3, the [^{3}H]-TdR LI is 0.093 ± 0.038, the T_S is 10.1 ± 0.5 hr, the PDP index is 0.359 ± 0.162, and the T_C is 36.0 ± 9.6 hr.

3. Cell Kinetics of Rat Mammary Tumors

3.1. Transplantable Tumors

There are a number of transplantable rat mammary tumors, from different strains, that have undergone cytokinetic analysis.

The BICR/A4 mammary adenocarcinoma, which was induced by radiation in an August rat, was studied in the fourth transplant generation when the T_D was 120 hr.[39] Steel[7] found the [^{3}H]-TdR LI to be 0.12, with an intermitotic time of 18 hr; G_1, 6 hr; S, 11 hr; G_2, 2 hr; GF, 0.26; and ϕ, 0.52.

The BICR/A9 fibroadenoma reported by Steel *et al.*[40] originally had a T_D of 32 days with a [^{3}H]-TdR LI of 0.014. The second transplant generation had a T_D of 30 days, a [^{3}H]-TdR LI of 0.014, a T_{G_1} of 26 hr, a T_S of 19 hr, a T_{G2} of 2.6 hr, a T_C of approximately 41 hr, a GF of 0.13, and a ϕ of 0.68.[7] At the fourth transplant generation, the T_D was 4.5 days; the [^{3}H]-TdR LI was 0.158, with a shortened T_{G_1} of 15 hr, a T_C of 30 hr, a GF of 0.38, and a ϕ of 0.38. The tenth-generation transplant had a T_D of 1.7 days, a [^{3}H]-TdR LI of 0.31, a T_{G_1} of 9.3 hr, a T_S of 9.2 hr, a T_{G_2} of 2.7 hr, a T_C of 16 hr, a GF of 0.69, and a ϕ of 0.49. There were ploidy changes associated with the variation in cell kinetics. Recalculation by Mendelsohn[24] did not alter these values significantly.

The BICR/A7 mammary adenocarcinoma was induced by intravenous 7,12-dimethylbenz(a)anthracene (DMBA), and the PLM experiment was performed on the third transplant.[39] The T_D was 432 hr; [^{3}H]-TdR LI, 0.053; T_{G_1}, 25 hr; T_S, 18 hr; T_{G_2}, 3 hr; intermitotic time, 74 hr; and GF, 0.25.[7]

The BICR/A12 mammary adenocarcinoma, which was also induced by an intravenous DMBA injection in the August rat, was studied by Janik and Steel[41] in the 5th and 6th transplant generations. On day 13 posttransplantation, when the T_D was 90 hr, the [^{3}H]-TdR LI was 0.28; T_{G1}, 9.4 hr; T_S, 8.9 hr; T_{G2}, 2.4 hr; T_C, 21 hr; GF, 0.87; and ϕ, 0.72.

The BICR/M1, which originated spontaneously in the Marshall strain of rat, was shown by Steel *et al.*[42] to grow more rapidly with increasing transplant generations. In approximately the 150th generation and after 10 years of transplantation, the [^{3}H]-TdR LI was 0.34 ± 0.4; T_{G_1}, 8.0 ± 4.2 hr; T_S, 8.0 ± 1.5 hr; T_{G_2}, 3.0 ± 1.6 hr; and T_C, 19.0 ± 4.7 hr. There was no cell loss in these small tumors, and the T_D was 22.7 ± 2.5 hr.[7,24]

Simpson-Herren and Griswold reported the cell kinetic parameters of the transplantable mammary adenocarcinoma 1/C, which was originally induced by DMBA in a Fischer rat.[43] The PLM curve was done on 14-day tumors of average weight 0.37 g. These small tumors had a T_D of

1.8 days at this time. The T_C was 17.3 hr; T_{G_1}, 6.3 ± 1.4 hr; T_S, 8.6 ± 4.2 hr; T_{G_2}, 2.4 ± 2.1 hr.[24] The [^{3}H]-TdR LI was 0.15 ± 0.3, with a range of 0.06–0.22. Using the calculation of LI_0/LI_t, the GF value for this tumor would be approximately 0.30. However, the results of labeling tumor animals with multiple doses of [^{3}H]-TdR would lead one to suspect a higher GF in the range of 0.70–0.80.

The 13762 mammary adenocarcinoma was induced with DMBA in Fischer 344 rats by Segaloff and has been carried in these rats for over 10 years. The tumors for a recent study, on 14-day-postimplant animals, were supplied by Dr. Arthur Bogden of the Mason Research Institute and have been subsequently carried at Allegheny General Hospital. The 14-day-postimplant tumors have a [^{3}H]-TdR LI of 0.33; T_{G_1}, 5.2 hr; T_S, 5.6 hr; T_{G_2}, 1.5 hr; and T_C, 12.7 hr.[38] The 25-day and 35-day tumor PLM's are almost identical. These tumors were extremely necrotic, however, and only a shell of viable tissue was usually found at these later times. The PLM studies were performed on this viable shell. The GF of the 14-day tumors was calculated at about 0.75, and this confirms the PDP value of 0.70. These tumors had a T_D of approximately 5 days and were over 2 cm in diameter at the time of study.

3.2. Induced Tumors

Although several chemicals have been shown to produce primary mammary tumors in rats, the only one for which significant cell kinetic data are available is 7,12-dimethylbenz(a)anthracene (DMBA). The technique of Huggins,[44] utilizing the oral administration of DMBA dissolved in sesame oil, has been used to induce these tumors in the Sprague–Dawley rat.

Griswold *et al.*[45] studied the growth rate of these tumors and found that in those animals that had tumors at 2 months post DMBA ingestion, there was a mean of 1.4 tumors per animal with a total mass of about 1 g. This increased to a total average tumor mass of 10.0 g at 4–5 months post DMBA ingestion, with a mean of 3.2 tumors per rat. The tumor incidence at 4 months post DMBA was between 63 and 100%.

Subsequent studies by Simpson-Herren and Griswold,[46] using the same technique of tumor induction, showed that tumors of 1.0-g size had an average T_D of 11 days, which was about 15 days after tumor appearance, and the T_D had increased to about 75 days at 10-g size or about 80 days after appearance. The kinetic analysis of their PLM data revealed, for an average tumor size of 2.7 g, a [^{3}H]-TdR LI of 0.10; T_{G_1}, 11.5 hr; T_S, 8.5 hr; T_{G_2}, 1.0 hr; T_C, 22.0 hr; T_D, 432 hr; T_P, 62 hr; and GF, 0.31. Subsequent analysis of these data by Steel[7] revealed minor changes of T_{G_1}, 14 hr; T_S, 9 hr; and GF, 0.35.

Simpson-Herren and Griswold[43] repeated these studies using tumors of different sizes. They found little difference in the [^{3}H]-TdR LI between small (< 1 g) and large (> 3 g) tumors, with the mean values being 0.060 and 0.069, respectively. There was a large range of [^{3}H]-TdR LI's, however—between 0.001 and 0.15 in each group. There was also little change in the time parameters for various compartments, as calculated by two different computer models. For the Barrett–Steel analysis, the median-sized tumors of 1–3 g and T_D of 240–384 hr; the mean T_C was 18.3 hr; T_{G_1}, 6.6 hr; T_S, 9.3 hr; and T_{G_2}, 2.4 hr. The GF, using the LI_0/LI_t model, ranged from 0.1 to 0.2.

Not all tumors were progressively growing, and static tumors showed much lower [^{3}H]-TdR LI's, with a reduction, in one study, from 0.10 to 0.016. Regressing tumors following oophorectomy showed a progressive decline in [^{3}H]-TdR LI to 0.002 in 2 weeks. This could be reversed, in these hormone-sensitive tumors, by administration of estradiol and progesterone. Although Combs *et al.*[47] did not find any difference in the PLM curves of DMBA-induced tumors in various phases of the estrus cycle, because of the change in the [^{3}H]-TdR LI, the GF was found to reach a maximum in early diestrus.

Hoffman and Post[48] also reported cell kinetic studies on DMBA-induced tumors. Their tumors, however, were late-appearing ones and may have been adenomata. Nevertheless, in tumors over 2 cm in size, they found a T_C of 45 hr; T_S, 10 hr; and [^{3}H]-TdR LI, 0.072.

3.3. Spontaneous Tumors

Braunschweiger and Schiffer[38] studied the cell kinetics of truly spontaneous tumors arising in old Sprague–Dawley/ZM rats. These tumors have been large, generally over 3 cm in diameter, and without any record of T_D. Those mammary tumors that were classified as fibroadenomas and adenomas had a [^{3}H]-TdR LI of 0.044 ± 0.025, a T_S of 10.9 ± 1.4 hr, a PDP index of 0.172 ± 0.068, and a T_C of 42.0 hr. Those tumors classified as mammary adenocarcinomas, most of which were well differentiated, had a [^{3}H]-TdR LI of 0.051 ± 0.013, a T_S of 11.0 ± 0.6 hr, and a PDP index of 0.220 ± 0.033. The calculated T_C was 44.6 hr.

4. Miscellaneous Cell Kinetic Studies

4.1. Canine Mammary Tumors

Owen and Steel,[49] in 1969, reported the results of some cell kinetic studies in three dogs with mammary adenocarcinoma. In one animal, they were able to perform a PLM analysis, and this revealed a T_{G_1} of 61 ± 48 hr, a T_S of 5.3 ± 1.7 hr, and a T_{G_2} of 5.4 ± 1.5 hr, with a median T_C of 50 hr. Another animal had four metastatic lesions with [^{3}H]-TdR LI's ranging between 0.014 and 0.036 with T_D's in the range of 16 days. A third animal had a [^{3}H]-TdR LI of 0.01 with an approximate T_D of 40 days.

4.2. Metastases in Animals

This is very little published data on metastatic mammary tumor models. The group at our laboratory has recently begun to study a tumor model of pulmonary metastases induced by the intravenous injection of minced primary tumors into syngeneic C3H/He mice, or into the host from which the tumor was removed. The results indicate a marked difference in the kinetic parameters of the lung nodules from the cell kinetics of the C3H/He spontaneous tumor.[38] It must be kept in mind, however, that these artificial pulmonary metastases are really first-generation transplants, that they are much smaller than the average primary tumors—measuring from 0.5 to several millimeters in diameter, and that the total tumor burden of the metastatic animal is difficult to calculate. However, several generalizations can be made thus far. The lung tumors show typical Gompertzian growth. The gross results of the first several hundred metastatic nodules showed a [^{3}H]-TdR LI of 0.138 ± 0.046 (compared with 0.07 for the primary tumors), a T_S of 9.0 ± 1.7 hr (compared with 10.7 for the primary tumors), and a PDP index of 0.614 ± 0.196 (compared with 0.169 for primary tumors). Calculation for the T_C was 32.4 hr (compared with 28.4 for the primary tumor). Thus, it appears that in this situation, at least, the GF is markedly increased along with the number of cells in the S phase, but the cell cycle time is approximately the same. The [^{3}H]-TdR LI and PDP index decrease with increasing tumor size between 2 and 6 mm in diameter; however, a more complex relationship is observed at smaller nodule size. There was, however, considerable variation in the cell kinetics for multiple nodules in a single host. Comparison of thse data with those for first-generation tumors implanted subcutaneously[6] reveals that in general, the GF is higher and T_C longer in the pulmonary metastases than values reported for subcutaneous transplants.

Lung metastases were found in two Sprague–Dawley rats with spontaneous mammary tumors.[38] The cell kinetics of the first tumor were [^{3}H]-TdR LI 0.107 (0.06 in the primary); T_S, 8.1 hr (9.7 hr in the primary); PDP, 0.258 (0.259 in the primary); and T_C, 18.8 hr (38.7 hr in the primary). The second tumor had a [^{3}H]-TdR LI of 0.101 (0.022 in the primary); T_S, 11.4 hr (not done in the primary); PDP, 0.292 (0.152 in the primary); and T_C, 30.0 hr (approximately 70 hr in the primary).

As will be described later, there are also cell kinetic differences between primary and metastatic human tumors.

5. Cell Kinetics of Human Mammary Tumors

5.1. Growth Characteristics

5.1.1. *Primary Tumors*

A number of generalizations have been made that relate the rate of growth of primary breast tumors to clinical prognosis.[50] Adenocarcinomas generally have an intermediate growth rate between rapidly growing sarcomas and more slowly growing squamous-cell carcinomas. In general, the T_D is proportional to age, such that older women have T_D's that are longer than younger women. The more rapidly growing tumors appear in women with the shortest survival times.

Gershon-Cohen *et al.*[51] reported, in 18 cases of primary breast tumors, T_D's in the range of 23–209 days. There was no correlation with histology, but in patients without axillary metastases, the average doubling time was 128 days, while in those patients with axillary metastases, the mean T_D was 85 days.

Kusama *et al.*[52] found that the median doubling time of primary breast lesions was between 2.1 and 4.0 months, with a rather wide spread of values.

Charlson and Feinstein,[53] while not giving any significant new data, felt that analysis of tumor growth rate from the patient's detailed history offered an excellent opportunity for staging the disease more accurately. They felt that this information should be incorporated into clinical treatment trials.

Tubiana and Malaise[54] reviewed their data from Charbit *et al.*[55] and found a T_D of 166.3 days (34 patients) in primary breast cancers.

In a recent article, Spratt *et al.*[56] speculated on the definition of acute and chronic breast cancer using the primary tumor data of Gershon-Cohen, but no significant new data were accumulated.

In conclusion, it appears that the reported T_D's for primary breast tumors, in the range of clinical quantitation and without known metastases, averaged between 90 and 166 days.

5.1.2. *Recurrent Tumors*

Philippe and LeGal,[57] noting the difficulties involved in studying the growth rate of primary breast tumors, reported on cutaneous nodules appearing in the mastectomy scar. They found a mean T_D of 40 days, a median of 30 days, and a range of 3–211 days. They also felt that there was a bimodal distribution, one with an average T_D of 25 days and the other with a T_D of 93 days.

Lee and Spratt[58] and Lee[59] reported on growth rates of soft tissue metastases, including nodal, of breast cancer. They found a log normal frequency distribution. In 66 untreated tumors, a mean T_D of 17.1 days was found, with a two-S.D. range of 3.4–86.1 days. There was considerable variation in the growth rates of two or more lesions in the same patient.

A recent study by Pearlman[60] was also conducted on mastectomy scar recurrences and is based on tumors in 82 patients. His data give a median T_D of about 22 days, with extreme ranges of approximately 3–200 days. Furthermore, he divides the growth rate into rapid with a T_D of $<$ 25 days, about 50% of patients; intermediate, 26–75 days, about 35%; and slow, $>$ 76 days, about 15%.

In conclusion, therefore, the mean T_D in mastectomy scar recurrences and skin lesions is around 20–40 days, considerably shorter than the primary tumor.

5.1.3. *Pulmonary Metastases*

A third site at which breast cancer can effectively be measured is the lung, by X-ray techniques. Spratt and Spratt[61] measured 29 pulmonary metastases and found an average T_D of 82 days with a two-S.D. range of 16–426 days. Breur[62] studied 6 patients and found a range of 23–745 days, and Brenner *et al.*[63] found a range of 30–330 days in 4 metastases. Joseph *et al.*,[64] in 9 patients, found a T_D of 11–100 days. Tubiana and Malaise[54] found an average T_D of 82.7 days in the pulmonary metastases of 134 patients. In summary, the minimum doubling time was 11 days and maximum of 745 days, with an average of about 80 days. This value appears to lie between those for the primary tumors and recurrent skin lesions.

5.2. Benign Tumors

Meyer and Bauer[65] recently reported their results on the cell kinetics of 38 benign lesions of the breast. In tumors taken from patients in the first half of the menstrual cycle, the median [^{3}H]-TdR LI for nonneoplastic ducts was 0.0038, with a range of 0.0025–0.016. For fibroadenomas, the median was 0.0042, with a range of 0.0014–0.0085. For the remainder of the menstrual cycle in nonneoplastic ducts, the median [^{3}H]-TdR LI was 0.0150, with a range of 0.0025–0.056, and for the fibroadenomas, 0.0171, with a range of 0.0035–0.047.

Braunschweiger and Schiffer[38] studied 10 patients with fibroadenomas and found a mean [^{3}H]-TdR LI of 0.019, a PDP index of 0.094, a mean T_S of 18 hr, and a calculated T_C of 87.2 hr. Three patients with ductal papillomas had a [^{3}H]-TdR LI of 0.031, PDP index of 0.113, T_S of 18.3 hr, and a calculated T_C of 64.7 hr. Thirty-four patients with fibrocystic disease were also studied. There was a mean [^{3}H]-TdR LI of 0.033, the PDP index was 0.158, and four of these patients had an average T_S of 20.3 hr. The calculated T_C was 92.5 hr.

5.3. Primary Tumors

Johnson and Bond,[66] in 1960, studied one patient with primary breast cancer by the incubated tissue slice method and found a [^{3}H]-TdR LI of 0.0065. Nordenskjöld *et al.*[67] referred to 18 patients, an unknown number of whom had primary cancers of the breast, and found that the [^{3}H]-TdR LI varied from 0 to 0.148, with a median of 0.035. Silvestrini *et al.*[68] reported on 66 primary tumors and found a wide variation of less than 1% to greater than 14%. They demonstrated a relationship to age, the younger patients having higher LI's, but showed no positive relationship to histology. They also found, in their data, that the median T_S value was 7 hr, with a range of 5.5–8.5 hr. The T_P value was about 15.3 days, with premenopausal patients having a median T_P of 9.1 ± 2.1 days and postmenopausal patients 20.5 ± 4.6 days.

Meyer and Bauer[69] reported on the cell kinetics of 39 patients with breast cancer. They found variations in the [^{3}H]-TdR LI from 0.005 to 0.15, with a mean of 0.038. They also found that the smallest tumors had the lowest LI's, and that a high LI was associated with more frequent nodal metastases. In a more recent paper, the same authors increased their number of study patients to 92.[65] The [^{3}H]-TdR LI range was 0.0004–0.186, with a log normal distribution. The mean was 0.037 and the median 0.021. There was no correlation with size of the tumor, but patients with large numbers of positive lymph nodes had higher [^{3}H]-TdR LI's. Patients younger than 50 years of age had higher [^{3}H]-TdR LI's in agreement with Silvestrini's results. Sklarew *et al.*[70] reported on 56 patients with primary tumors. They found a mean [^{3}H]-TdR LI of 0.024 ± 0.021 S.D., with means T_S values of 21.9 ± 4.3 hr. The range for the [^{3}H]-TdR LI was 0.0013–0.0972, and for the T_S, from 11.5 to 33.0 hr. Straus and Moran[71] described one patient with primary breast cancer with [^{3}H]-TdR LI's of 0.107–0.128 from the same lesion.

Tubiana *et al.*[72] reviewed 98 primary breast tumors studied at their institution. The overall [^{3}H]-TdR LI was 0.0099, with tumors 4.5 cm^3 or less having a [^{3}H]-TdR LI of 0.0065 and larger ones an LI of 0.0142. Patients younger than 40 years of age had higher [^{3}H]-TdR LI's, but this difference was not statistically significant.

Ninety-three primary breast cancers were studied *in vitro* by Schiffer and Braunschweiger.[38,73] The mean [^{3}H]-TdR LI was 0.056, with a median of 0.047 and a range of 0.005–0.343. The distribution appeared to be log normal. The mean T_S value was 18.4 hr, with a range of 13.5–23.6 hr and a median of 18.3 hr, with a normal distribution. The PDP index or GF averaged 0.239 with a log normal distribution. The calculated average T_P was 326 hr, and the T_C was 120.6 hr. The corresponding medians were 245 hr and 93 hr. If one compares these results with those found in metastatic lesions, in the same series, the significant difference lies in the lower [^{3}H]-TdR LI with correspondingly shortened T_P and T_C values. The metastatic patients had smaller lesions as measured by the pathologist, but they were estimated at the same size by the clinicians. Also, the tumor burden was considerably less in the primary disease category. The PDP or GF was not different.

Patients less than 49.9 years of age had higher [^{3}H]-TdR LI's and lower PDP indices, resulting in short T_P and T_C values. Those with one to three positive axillary lymph nodes and younger than 49.9 years of age, the same group that appears to respond best to adjuvant chemotherapy, had a high [^{3}H]-TdR LI (0.086) as compared with those with one to three positive nodes and older than 50 (0.044) years of age. Thus, age appears to be a factor in the proliferative potential of these tumors. Those patients with the best clinical prognostic features had lower [^{3}H]-TdR LI's than average. There appeared to be subsets of patients with anomalous results, but the number of patients in each category was too small to enable definitive statements to be made. Those patients with very low [^{3}H]-TdR LI's had large tumors with many positive nodes, but average PDP indices. Conversely, there were a few patients with very small tumors and no positive nodes who had very low [^{3}H]-TdR LI's.

5.4. Metastatic Tumors

Wolberg and Brown[74] reported that the mean [^{3}H]-TdR LI in 13 breast specimens was 0.021, and Coons *et al.*,[75] in one patient, that the [^{3}H]-TdR LI was 0.0074. Titus and Shorter[76] referred to two cases with [^{3}H]-TdR LI's of 0.01 and 0.016, Clarkson *et al.*[77] reported one case of malignant pleural effusion with a [^{3}H]-TdR LI of 0.17, and Fabrikant and Wisseman[78] a case with a [^{3}H]-TdR LI of 0.006. Utilizing a technique of local [^{3}H]-TdR injection, Young and DeVita[79] found [^{3}H]-TdR LI's of 0.21–0.25 in three patients and, additionally, determined that the T_S was 19–24 hr in these cases by the PLM method. Wolberg and Ansfield[80] studied 170 patients, by the tissue slice incubation method, and found a median [^{3}H]-TdR LI of 0.019. Terz *et al.*[81] studied two breast cancer patients by PLM methods. The first had a [^{3}H]-TdR LI of 0.126 $\pm$ 0.017 S.E.M.; T_{G_1}, 19 $\pm$ 2.3 hr; T_S, 13.2 $\pm$ 3.4 hr; T_{G_2}, 6 $\pm$ 3.7 hr; and T_C, 31 hr.

The GF was 0.25. The second patient had a T_S of 9.5 hr with a T_{G_2} of 4.8 hr and a [^{3}H]-TdR LI of 0.16 ± 2.2 S.E.M. Malaise *et al.*[82] reported on 121 cases of adenocarcinoma in the literature, including some from their own institute, 75 of which were breast, and found a geometric mean [^{3}H]-TdR LI of 0.021. Sky-Peck[83] studied 25 patients with breast cancer and found the mean [^{3}H]-TdR LI to be 0.019, with a range of 0.002–0.114.

Murphy *et al.*,[84] using the technique of Livingston *et al.*,[85] studied 19 patients with metastatic breast disease. The median [^{3}H]-TdR LI was 0.13, with a range of 0.0075–0.26. Thirlwell *et al.*,[86] using the same technique, reported that the mean [^{3}H]-TdR LI in 9 patients was 0.105, with a range of 0.040–0.155. Livingston *et al.*[87] reported on 56 patients, some of which may have been included with the comments above. The median [^{3}H]-TdR LI was 0.083, with a range of 0–0.26. Post *et al.*[88] recently reported on 8 patients with breast cancer and found [^{3}H]-TdR LI's of 0.02–0.11, a composite G_2 + M/2 of 4 hr, and T_S of 24 hr.

Schiffer and Braunschweiger[73] studied 18 patients with metastatic breast cancer, most of which were skin or subcutaneous lesions. The mean [^{3}H]-TdR LI, performed by a modified Livingston technique,[85] was 0.086, with a range of 0.037–0.142. The mean T_S value in 7 patients was 19.4 hr, with a range of 15.8–23.7 hr. The PDP index mean was 0.262, with a range of 0.075–0.616, and the calculated T_P was 183 hr, with a range of 107–337 hr. The mean T_C value was 64.5 hr, with a range of 26.3–95.5 hr.

Straus and Moran[71] reported a PLM curve on a breast cancer patient sampling from chest wall lesions. They found a T_S of 18 hr, a T_{G_2} + M of 10 hr, and [^{3}H]-TdR LI's varying from 0.156 to 0.285.

One patient of unusual interest, with malignant cytosarcoma phylloides, was studied by means of cell kinetic techniques.[38] Several months following a simple mastectomy, the patient began to have extremely rapid growth of recurrent chest wall disease. The tumor was growing with a T_D of 4 days at the time that the 15 × 10 × 8 cm mass was biopsied. The [^{3}H]-TdR LI was 0.117; the T_S was 14 hr; the PDP index was 0.457. The T_P was calculated at 90.9 hr, and the T_C at 49.4 hr. The cell loss factor was virtually zero.

6. Conclusions

Certain generalizations about cell kinetics can be made from the preceding data, although they are not necessarily limited to mammary cancers. Transplantable tumors are more highly proliferative (have higher [^{3}H]-TdR LI's and GF's) than their spontaneous counterparts, and this difference is exaggerated with increasing transplant generations. Metastases also appear to be more highly proliferative than their primary tumors. In general, small tumors are more highly proliferative than large ones, and tumors near blood vessels appear to be more proliferative than those farther away. Nonmalignant mammary tissues respond to hormone stimulation, either natural or exogenous, by becoming more highly proliferative, and some hormone-responsive tumors behave in the same fashion.

It should be quite clear, however, that there is no single "breast tumor kinetics." Each system, at each tumor size, and at each transplant generation, has its own characteristic cell kinetic patterns. Even then, however, the variability may be extremely wide. Table I represents examples of cell kinetic variation, within specific tumor systems, that have been studied in our laboratory. The techniques have been virtually the same for all these studies. Additionally, the intratumor variations, when the size of the tumor permitted duplicate studies, have been relatively minor. This is reflected in the results of the transplantable tumors, in which the variation is much less than in the metastatic or primary systems. Not only the cell kinetics but also the volume T_D's are variable, and this is especially so for the primary tumors.

Although there has been very little comment about the perturbed state (changes in cell kinetics with treatment) in this chapter, the reader can readily imagine the variation in response to an S-phase chemothera-

Table I
Examples of Variability of Cell Kinetic Measurements in Mammary Tumors

Mammary tumor system	[³H]-TdR LI		DNA synthesis time (hr)		PDP index	
	Mean	Range	Mean	Range	Mean	Range
Transplantable						
Rat 13762	0.33	0.25–0.35	5.6	5.3– 5.9	0.70	0.63–0.73
Mouse C3H/67-A[a]	0.08	0.05–0.12	11.5	10.7–12.6	0.27	0.20–0.31
Metastatic						
Mouse C3H/He	0.14	0.04–0.33	9.0	5.9–19.9	0.61	0.26–0.94
Human	0.09	0.04–0.14	19.4	15.8–23.7	0.26	0.08–0.62
Primary						
Mouse C3H/He	0.06	0.04–0.10	10.2	9.3–11.7	0.16	0.10–0.19
Mouse $CD8F_1$	0.09	0.03–0.20	10.1	9.7–11.4	0.36	0.16–0.68
Rat Sprague-Dawley/ZM	0.05	0.02–0.14	11.0	8.8–13.2	0.22	0.11–0.44
Human						
Fibroadenoma	0.02	0.01–0.05	18.0	16.5–19.8	0.09	0.06–0.14
Adenocarcinoma	0.06	0.01–0.34	18.4	13.5–23.6	0.24	0.03–0.98

[a] Dethlefsen's slow line.

peutic agent if the [^{3}H]-TdR LI of one tumor varies by a factor of 10 from another. Likewise, the translation of cell kinetic data from an animal model to man is very risky, considering the variations inherent in both. The classic techniques of cell kinetic analyses may have lulled clinicians into thinking that a standard therapy, based on average cell kinetic data, may be possible. In view of the data obtained from individual tumors *in vitro,* as represented in Table I, this possibility of one overall treatment scheme seems to have little chance of success. Rapid *in vitro* techniques, designed so that the results are known in a time frame suitable for clinical utilization, are becoming available, and promise to contribute to the future of individualization of tumor therapy.

ACKNOWLEDGMENT: Part of this work was supported by Grant CA-10438 and Contract NO1-CB-43899 from the National Cancer Institute, National Institutes of Health.

7. References

1. M. L. Mendelsohn, Autoradiographic analysis of cell proliferation in spontaneous breast cancer of C3H mouse. III. The growth fraction, *J. Natl. Cancer Inst.* **28,** 1015–1029 (1962).
2. L. M. Schiffer, A. M. Markoe, and J. S. R. Nelson, Estimation of tumor growth fraction in murine tumors by the primer-available DNA-dependent DNA polymerase assay, *Cancer Res.* **36,** 2415–2418 (1976).
3. P. G. Braunschweiger and L. M. Schiffer, PDP includes in human tumors: Evidence for proliferative correlations, *Proc. Thirteenth Annu. Meet. Am. Soc. Clin. Oncol.* **18,** 276 (1977).
4. H. Wexler, S. K. Orme, and A. S. Ketcham, Biological behavior through successive transplant generations of chemically induced and spontaneous sources in mice, *J. Natl. Cancer Inst.* **40,** 513–523 (1968).
5. J. A. McCredie, W. R. Inch, and R. M. Sutherland, Difference in growth and morphology between the spontaneous C3H mammary carcinoma in the mouse and its syngeneic transplants, *Cancer* **27,** 635–642 (1971).
6. J. Denekamp, The cellular proliferation kinetics of animal tumors, *Cancer Res.* **30,** 393–400 (1970).
7. G. G. Steel, The cell cycle in tumours: An examination of data gained by the technique of labelled mitoses, *Cell Tissue Kinet.* **5,** 87–100 (1972).
8. J. F. Fowler, P. W. Sheldon, A. C. Begg, S. A. Hill, and A. M. Smith, Biological properties and response to x-rays of first-generation transplants of spontaneous mammary carcinomas in C3H mice, *Int. J. Radiat. Biol.* **27,** 463–480 (1975).
9. L. Simpson-Herren and H. H. Lloyd, Kinetic parameters and growth curves for experimental tumor systems, *Cancer Chemother. Rep.* **54,** 143–174 (1970).
10. I. F. Tannock, Population kinetics of carcinoma cells, capillary endothelial cells and fibroblasts in a transplanted mouse mammary tumor, *Cancer Res.* **30,** 2470–2476 (1970).
11. M. L. Mendelsohn and L. A. Dethlefsen, Cell kinetics of breast cancer: The turnover of nonproliferating cells, in: *Recent Results in Cancer Research* (M. L. Griem *et al.*, eds.), Vol. 42, pp. 73–86, Springer-Verlag, Berlin—Heidelberg—New York (1973).

12. L. V. Szczepanski and K. R. Trott, Post-irradiation proliferation kinetics of a serially transplanted murine adenocarcinoma, *Br. J. Radiol.* **48,** 200–208 (1975).
13. J. S. R. Nelson, R. E. Carpenter, and D. Durboraw, Mechanisms underlying reduced growth rate in C3HBA mammary adenocarcinomas recurring after single doses of x-rays or fast neutrons, *Cancer Res.* **36,** 524–532 (1976).
14. L. M. Schiffer, unpublished results.
15. W. Feaux de Lacroix, F. J. Hensen, P. J. Klein, E. Nola, and K. J. Lennartz, Effect of different sex of tumor bearing mice and of ovariectomy on the proliferation kinetics of a solid transplantable mammary carcinoma (C3H mouse), *Z. Krebsforsch.* **87,** 181–192 (1976).
16. A. Goldfeder, Biological properties and radiosensitivity of tumours: Determination of the cell-cycle and time of synthesis of deoxyribonucleic acid using tritiated thymidine and autoradiography, *Nature (London)* **207,** 612–614 (1965).
17. M. Hosokawa, F. Forsini, and E. Mihich, Fast- and slow-growing transplantable tumors derived from spontaneous mammary tumors of the DBA/2Ha-DD mouse, *Cancer Res.* **35,** 2657–2662 (1975).
18. S. C. Rockwell, R. F. Kallman, and L. F. Fajardo, Characteristics of a serially transplanted mouse mammary tumor and its tissue culture-adapted derivative, *J. Natl. Cancer Inst.* **49,** 735–749 (1972).
19. J. V. Watson, The cell proliferation kinetics of the EMT6/M/AC mouse tumour at four volumes during unperturbed growth *in vivo, Cell Tissue Kinet.* **9,** 147–156 (1976).
20. W. R. Laster, Jr., J. G. Mayo, L. Simpson-Herren, D. P. Griswold, Jr., H. H. Lloyd, F. M. Schabel, Jr., and H. E. Skipper, Success and failure in the treatment of solid tumors. II. Kinetic parameters and "cell cure" of moderately advanced carcinoma 755, *Cancer Chemother. Rep.* **53**(1), 169–188 (1969).
21. P. Janik, P. Briand, and N. R. Hartmann, The effect of estrone–progesterone treatment on cell proliferation kinetics of hormone-dependent GR mouse mammary tumors, *Cancer Res.* **35,** 3698–3704 (1975).
22. I. F. Tannock, The relation between cell proliferation and the vascular system in a transplantable mouse mammary tumor, *Br. J. Cancer* **22,** 258–273 (1968).
23. G. G. Steel and S. Hanes, The technique of labelled mitoses: Analyses by automatic curve-fitting, *Cell Tissue Kinet.* **4,** 93–105 (1971).
24. M. L. Mendelsohn, Computer-analyzed transit times for vertebrate cell cycle and phases, in: *Cell Biology I* (P. L. Altman and D. D. Katz, eds.), pp. 11–13, Federation of American Societies for Experimental Biology, Bethesda, Maryland (1976).
25. V. Riley, Mouse mammary tumors: Alteration of incidence of apparent function of stress, *Science* **189,** 465–467 (1975).
26. E. Magdon and G. Winterfeld, Untersuchungen zum Wachstumsverhalten spontaner Maurmakarzinome bei C3H-Inzuchtmäusen, *Arch. Geschwulstforsch.* **45,** 782–794 (1975).
27. M. L. Mendelsohn, The kinetics of tumor cell proliferation, in: *Cellular Radiation Biology,* pp. 498–513, Williams and Wilkins, Baltimore (1965).
28. M. L. Mendelsohn and L. A. Dethlefsen, Effects of selection and passage on volumetric growth rate of mouse mammary tumors, Lawrence Livermore Laboratory Report UCRL-51798 (1975).
29. P. G. Braunschweiger, L. Poulakos, and L. M. Schiffer, Cell kinetics *in-vivo* and *in-vitro* for C3H/He spontaneous mammary tumors, *J. Natl. Cancer Inst.* **59,** 1197–1204 (1977).
30. J. Shewell, The effect of methotrexate on spontaneous mammary adenocarcinomata in female C3H mice, *Br. J. Cancer* **33,** 210–216 (1976).
31. F. Bresciani, A comparison of the cell generative cycle in normal, hyperplastic and neoplastic mammary gland of the C3H mouse, in: *Cellular Radiation Biology,* pp. 547–557, Williams and Wilkins, Baltimore (1965).

32. M. R. Banerjee, Hormonal control of DNA synthesis: Altered responsiveness of hyperplastic alveolar nodules of mouse mammary gland, *J. Natl. Cancer Inst.* **42,** 227–234 (1969).
33. M. R. Banerjee and R. J. Walker, Duration of DNA synthesis in hyperplastic aveolar nodules of C3H/He mouse mammary gland, *J. Natl. Cancer Inst.* **39,** 551–555 (1967).
34. F. Bresciani, Effect of ovarian hormones on duration of DNA synthesis in cells of the C3H mouse mammary gland, *Exp. Cell Res.* **38,** 13–32 (1965).
35. M. L. Mendelsohn, F. C. Dohan, and H. A. Moore, Jr., Autoradiographic analysis of cell proliferation in spontaneous breast cancer of C3H mouse. I. Typical cell cycle and timing of DNA synthesis, *J. Natl. Cancer Inst.* **25,** 477–484 (1960).
36. M. L. Mendelsohn, Autoradiographic analysis of cell proliferation in spontaneous breast cancer of C3H mouse. II. Growth and survival of cells labeled with tritiated thymidine, *J. Natl. Cancer Inst.* **25,** 485–500 (1960).
37. R. L. Stolfi, D. S. Martin, and R. A. Fugmann, Spontaneous murine mammary adenocarcinoma: Model system for evaluation of combined methods of therapy, *Cancer Chemother. Rep.* **55,** 239–251 (1971).
38. P. G. Braunschweiger and L. M. Schiffer, unpublished results.
39. G. G. Steel, The kinetics of cell proliferation in tumors, in: *Time and Dose Relationships in Radiation Biology as Applied to Radiotherapy,* pp. 130–140, Associated Universities, Upton, New York (1969).
40. G. G. Steel, K. Adams, J. Hodgett, and P. Janik, Cell population kinetics of a spontaneous rat tumor during serial transplantation, *Br. J. Cancer* **25,** 802–811 (1971).
41. P. Janik and G. G. Steel, Cell proliferation during immunologic perturbation in three transplanted tumours, *Br. J. Cancer* **26,** 108–114 (1972).
42. G. G. Steel, K. Adams, and J. C. Barrett, Analysis of the cell population kinetics of transplanted tumours of widely-differing growth rate, *Br. J. Cancer* **20,** 784–800 (1966).
43. L. Simpson-Herren and D. P. Griswold, Jr., Studies of the cell population kinetics of induced and transplanted mammary adenocarcinoma in rats, *Cancer Res.* **33,** 2415–2424 (1973).
44. C. Huggins, Methodology of selective induction of cancers in adult rats, in: *Prognostic Factors in Breast Cancer* (A. P. M. Forrest and P. B. Kunkler, eds.), pp. 465–471, Williams and Wilkins, Baltimore (1968).
45. D. P. Griswold, H. E. Skipper, W. R. Laster, W. S. Wilcox, and F. M. Schabel, Jr., Induced mammary carcinoma in the female rat as a drug evaluation system, *Cancer Res.* **26,** 2169–2180 (1966).
46. L. Simpson-Herren and D. P. Griswold, Jr., Studies of the kinetics of growth and regression of 7,12-dimethylbenz(α)anthracene-induced mammary adenocarcinoma in Sprague–Dawley rats, *Cancer Res.* **30,** 813–818 (1970).
47. J. W. Combs, M. F. Mackey, and J. L. Bennington, Modification of the fraction of proliferating cells in DMBA-induced rat mammary tumors by host factors, *Proc. Am. Assoc. Cancer Res.* **12,** 61 (1971).
48. J. Hoffman and J. Post, Replication and 5-iodo-2′-deoxyuridine-^{3}H incorporation by tumor and normal cells, *Cancer Res.* **26,** 1313–1318 (1966).
49. L. N. Owen and G. G. Steel, The growth and cell population kinetics of spontaneous tumors in domestic animals, *Br. J. Cancer* **23,** 493–509 (1969).
50. J. O. Archambeau, M. B. Heller, A. Akanuma, and D. Lubell, Biologic and clinical implications obtained from the analysis of cancer growth curves, *Clin. Obstet. Gynecol.* **13,** 831–856 (1970).
51. J. Gershon-Cohen, S. M. Berger, and H. S. Klickstein, Roentgenography of breast cancer moderating concept of "biologic predeterminism," *Cancer* **16,** 961–964 (1963).
52. S. Kusama, J. S. Spratt, Jr., W. L. Donegan, F. R. Watson, and C. Cunningham, The gross rates of growth of human mammary carcinoma, *Cancer* **30,** 594–599 (1972).

53. M. E. Charlson and A. R. Feinstein, The auxometric dimension: A new method for using rate of growth in prognostic staging of breast cancer, *J. Am. Med. Assoc.* **228,** 180–185 (1974).
54. M. Tubiana and E. P. Malaise, Growth rate and cell kinetics in human tumours: Some prognostic and therapeutic implications, in: *Scientific Foundations of Oncology* (T. Symington and R. L. Carter, eds.), pp. 126–135, William Heinemann, London (1976).
55. A. Charbit, E. P. Malaise, and M. Tubiana, Relation between the pathological nature and the growth rate of human tumors, *Eur. J. Cancer* **7,** 307–315 (1971).
56. J. S. Spratt, Jr., M. L. Kaltenbach, and J. A. Spratt, Cytokinetic definition of acute and chronic breast cancer, *Cancer Res.* **37,** 226–230 (1977).
57. E. Philippe and Y. LeGal, Growth of seventy-eight recurrent mammary cancers, *Cancer* **21,** 461–467 (1968).
58. Y.-T. N. Lee and J. S. Spratt, Jr., Rate of growth of soft tissue metastases of breast cancer, *Cancer* **29,** 344–348 (1972).
59. Y.-T. N. Lee, The lognormal distribution of growth rates of soft tissue metastases of breast cancer, *J. Surg. Oncol.* **4,** 81–88 (1972).
60. A. W. Pearlman, Breast cancer—Influence of growth rate on prognosis and treatment evaluation, *Cancer* **38,** 1826–1833 (1976).
61. J. S. Spratt, Jr. and T. C. Spratt, Rates of growth of pulmonary metastases and host survival, *Ann. Surg.* **159,** 161–171 (1964).
62. K. Breur, Growth rate and radiosensitivity of human tumors. I. Growth rate of human tumors, *Eur. J. Cancer* **2,** 157–171 (1966).
63. M. W. Brenner, L. R., Holsti, and Y. Perttalia, The study by graphical analysis of the growth of human tumours and metastases of the lung, *Br. J. Cancer* **21,** 1–13 (1967).
64. W. L. Joseph, D. L. Morton, and P. C. Adkins, Prognostic significance of tumor doubling time in evaluating operability in pulmonary metastatic disease, *J. Thorac. Cardiovasc. Surg.* **61,** 23–31 (1971).
65. J. S. Meyer and W. C. Bauer, Tritiated thymidine labeling index of benign and malignant breast epithelium, *J. Surg. Oncol.* **8,** 165–181 (1976).
66. H. A. Johnson and V. P. Bond, A method of labeling tissues with tritiated thymidine *in vitro* and its use in comparing rates of cell proliferation in duct epithelium, fibroadenoma and carcinoma of human breast, *Cancer* **14,** 639–643 (1960).
67. B. Nordenskjöld, A. Zetterberg, and T. Löwhagen, Measurement of DNA synthesis by ^{3}H-thymidine incorporation into needle aspirates from human tumors, *Acta Cytol.* **18,** 215–221 (1974).
68. R. Silvestrini, O. Sanfilippo, and G. Tedesco, Kinetics of human mammary carcinomas and their correlation with the cancer and the host characteristics, *Cancer* **34,** 1252–1258 (1974).
69. J. S. Meyer and W. C. Bauer, *In vitro* determination of tritiated thymidine labeling index (LI), *Cancer* **36,** 1374–1380 (1975).
70. R. J. Sklarew, J. Hoffman, and J. Post, A rapid *in-vitro* method for measuring cell proliferation in human breast cancer, *Cancer* **40,** 2299–2302 (1977).
71. M. J. Straus and R. E. Moran, Cell cycle parameters in human solid tumors, *Cancer* **40,** 1453–1461 (1977).
72. M. Tubiana, P. Chauvel, A. Renaud, and E. P. Malaise, Vitesse de Croissance et histoire naturelle du cancer du sein, *Bull. Cancer* **62,** 341–358 (1975).
73. L. M. Schiffer and P. G. Braunschweiger, Cytokinetics of human breast cancer: Primary vs metastatic lesions, *Proc. Twelfth Annu. Meet. Am. Soc. Clin. Oncol.* **17,** 238 (1976).
74. W. H. Wolberg and R. R. Brown, Autoradiographic studies of *in vitro* incorporation of uridine and thymidine by human tumor tissue, *Cancer Res.* **22,** 1113–1119 (1962).

75. H. Coons, A. Norman, and A. Nahum, *In vitro* measurements of human tumor growth, *Cancer* **19,** 1200–1204 (1966).
76. J. Titus and R. Shorter, Labeling of human tumors with tritiated thymidine, *Arch. Pathol.* **79,** 324–328 (1965).
77. B. Clarkson, K. Ota, T. Ohkita, and A. O'Connor, Kinetics of proliferation of cancer cells in neoplastic effusions in man, *Cancer* **18,** 1189–1213 (1965).
78. J. I. Fabrikant and C. L. Wisseman III, *In vitro* incorporation of tritiated thymidine in normal and neoplastic tissues, *Radiology* **90,** 361–363 (1968).
79. R. Young and V. DeVita, Cell cycle characteristics of human solid tumors *in vivo, Cell Tissue Kinet.* **3,** 285–290 (1970).
80. W. H. Wolberg and F. J. Ansfield, The relation of thymidine labeling index in human tumors *in vitro* to the effectiveness of 5-fluorouracil chemotherapy, *Cancer Res.* **31,** 448–450 (1971).
81. J. J. Terz, H. P. Curutchet, and W. Lawrence, Jr., Analysis of the cell kinetics of human solid tumors, *Cancer* **28,** 1100–1110 (1971).
82. E. P. Malaise, N. Chavaudra, and M. Tubiana, The relationship between growth rate, labeling index and histological type of human solid tumors, *Eur. J. Cancer* **9,** 305–312 (1973).
83. H. H. Sky-Peck, Effects of chemotherapy on the incorporation of ^{3}H-thymidine into DNA of human neoplastic tissue, *Natl. Cancer Inst. Monogr.* **34,** 197–203 (1971).
84. W. K. Murphy, R. B. Livingston, V. G. Ruiz, F. G. Gercovich, S. L. George, J. S. Hart, and E. J. Freireich, Serial labeling index determination as a predictor of response in human solid tumors, *Cancer Res.* **35,** 1438–1444 (1975).
85. R. B. Livingston, V. Ambus, S. L. George, E. J. Freireich, and J. S. Hart, *In vitro* determination of thymidine-^{3}H labeling index in human solid tumors, *Cancer Res.* **34,** 1376–1380 (1974).
86. M. P. Thirlwell, R. B. Livingston, W. K. Murphy, and J. S. Hart, A rapid *in vitro* labeling index method for predicting response of human solid tumors to chemotherapy, *Cancer Res.* **36,** 3279–3283 (1976).
87. R. B. Livingston, A. Sulkes, M. P. Thirlwell, W. K. Murphy, and J. S. Hart, Cell kinetic parameters: Correlation with clinical response, in: *Growth Kinetics and Biochemical Regulation of Normal and Malignant Cells,* pp. 767–785, Williams and Wilkins, Baltimore (1977).
88. J. Post, R. J. Sklarew, and J. Hoffman, The proliferative patterns of human breast cancer cells *in vivo, Cancer* **39,** 1500–1507 (1977).

8

Therapy in Experimental Breast Cancer Models

ARTHUR E. BOGDEN

1. Introduction

The growing acceptance that cancer encompasses a disparate group of diseases in which malignant cells tend to retain certain unique characteristics of the tissue or organ of their origin has been paralleled by an increasing interest in experimental animal model systems that not only are representative of human malignancies histologically, but also originate in the organ or tissues, and have the growth and metastasizing characteristics of the particular neoplastic disease for which they serve as a model.

Stressing the need for more animal models for immunotherapeutic trials in syngeneic systems, Carter[1] states, "Such systems would aid in determining the important multiple variables in experimental hosts *before* applying them to man. At the present time we have a mass of trials with a wide range of materials used by schedules, dosage levels, durations, and sequences chosen purely on the empirical instinct of the investigator. The danger in this approach lies in amassing a collection of uninterpretable, noncomparable, and negative data which will impair future developments in the field." The number and types of questions that can be answered by clinical trials, controlled or otherwise, are limited by the resultant logistical and ethical problems. These limitations, however, are the very area in which animal experimentation offers significant advantages.

ARTHUR E. BOGDEN • Mason Research Institute, Worcester, Massachusetts 01608.

Bartlett *et al.*[2] point out that

> animal studies permit new approaches to be tested individually as primary treatments and to be compared with the results of withholding all therapy [in contrast to clinical testing of new treatments initially in patients with advanced disease, and then only in combination with "best available" conventional therapy]. In animals, it is possible to study large numbers of subjects, and thereby attain greater sensitivity in the detection of therapeutic benefits, better control of experimental variables, and greater capacity to evaluate numerous variables. Assessment time in animal models is short, thus there is less "turn-around" time for confirmation and refinement of the study. Overall, these features establish animal models as an appropriate first level for evaluating new therapies. They can provide direction in the selection of agents and procedures for clinical testing and can help to refine the questions that must ultimately be answered by human experimentation.

Clearly, in a field that is as vital and complex as cancer therapy, animal experimentalists must have the help of clinicians in developing animal model tumor systems that will predict the clinical success of combinations of new as well as of existing chemotherapeutic agents and treatment modalities. Current limitations in our knowledge of the mechanisms fundamental to the nature of malignancy as well as to the interaction of drugs with the malignant cell further cautions that the increasing clinical emphasis on the use of combinations of chemotherapeutic agents and therapeutic modalities must have a basis for such selections other than empiricism.

Webster's[3] definition of a model as "a description or analogy used to help visualize something that cannot be directly observed," or as "an example for imitation or emulation," points up the critical consideration in selection of model systems for cancer research: that there is no one animal model for the disparate group of diseases identified as cancer. The age-old argument concerning which animal tumor model is the best model for "human cancer" is a fruitless argument because no single human cancer is a proper model for all human cancer. We are in agreement with Bartlett *et al.*[2] that basically, animal models incorporate features of convenience for the experimentalist, a luxury not available to the clinician. Though the most artificial model has the potential to produce new information about the biology of cancer, if preclinical investigations are to generate guidelines for the realistic design of appropriate clinical studies, it is imperative that the experimentalist not be blinded to the limitations of his or her model systems. By the same token, the clinical oncologist must have a sufficient appreciation of the model to know whether extrapolations can be made.

If, as in Huxley's view of the biology of neoplasia[4] from the standpoint of biological systematics, each tumor cell population has sufficient marks of individuality and heritability of character to be properly considered as a unique species within a new phylum of obligate parasitic eukaryotes derivable from metazoan cells, then selection of the proper animal model for addressing specific clinical problems is especially relevant in breast cancer. This chapter is devoted, therefore, to a discussion of *in vivo* animal tumor models with the objective of indicating the type of experimentally derived and clinically relevant information that can be obtained from each, as well as their inherent limitations as bioassay systems. To minimize redundancy, much information relevant to the use of experimental animal mammary tumor models in breast cancer research will be mentioned only superficially because investigators closely associated with or having done the pioneering work in those areas are authoring chapters on pertinent topics in this volume. Therefore, this presentation is by no means to be considered a critical review of the pertinent literature. Rather, it is the author's humble evaluation of the animal models used in various aspects of breast cancer research. Many of the conclusions are based on personal experience, but some are, admittedly, gleaned from the scientific literature to complete the picture.

The relevance of *in vitro* model systems for the study of hormone-dependent human breast cancer and the type of exquisite fundamental information that can be garnered from these isolated systems was recently summarized by Lippman *et al.*[5] and is further discussed by C. Kent Osborne and Marc E. Lippman in Chapter 4.

In analyzing experimental animal data from the viewpoint of the experimentalist or of the clinician, one should not lose sight of the fact that animal tumor models are, in essence, *in vivo* assay systems used to detect and define biological phenomena. As such, they must have well-defined and reproducible growth patterns and reactivities so that modifications in such parameters, resulting from experimental manipulations, can be correctly interpreted, and, when necessary, duplicated. To be practical as a model, the model itself must be predictable. Ideally, also, its use should be logistically feasible, and the time frame for its use as a test system must be reasonable. The intrinsic value of the data obtainable from such experimental animal systems is critically dependent on how well the experimental variables have been identified and controlled.

2. The Mathematical Model

The seemingly more abstract mathematical approach to predicting tumor response to chemotherapeutic agents is illustrated in the recent reports by Norton and Simon and co-workers[6,7] demonstrating that the entire growth pattern of an individual tumor was predictable on the basis of measurements early in its growth history. Of particular importance are the results obtained with their mathematical model,[7] which bring into question a fundamental concept regarding the relationship between tumor size and sensitivity to therapy, i.e., that small tumors, due to their large fraction of actively dividing cells, are more "sensitive" to cytotoxic therapy than tumors of equivalent histology but of larger size.[8,9] Norton and Simon[7] point out that while L1210 and other experimental tumors may approximate expotential growth, most experimental and clinical tumors do not,[10] but exhibit an exponentially diminished growth with increasing size, so that a limiting asymptotic size, i.e., a plateau size, is approached, a pattern of growth termed "Gompertzian."[11] For Gompertzian growth, the growth rate is slowest for both very small and very large tumors, and is maximum at the "inflection point," i.e., when the tumor is about 37% of its maximum size.

The concept expressed by the equation proposed by Norton and Simon describes a relationship between tumor size and sensitivity to therapy that appears to be more consistent with clinical and laboratory data than the conventional concept. The implications of their equations are that for tumors sufficiently small to be below their inflection point, growth fraction decreases in magnitude with decreasing tumor size. Hence, the sensitivity to therapy in terms of rate of regression for a given level of therapy would decrease with decreasing tumor size, even though the growth fraction is increasing. Though the concept does not necessarily mean that the smaller tumor is less "curable" than a larger one, it does imply that a dose schedule capable of causing a dramatic rate of regression of a tumor of intermediate size may not be sufficient to cure a small tumor when applied over a duration of therapy limited in time by considerations of host tolerance. Thus, making maximum use of their equation, the most efficient therapy for advanced tumors would be to apply moderately intense therapy initially, when the tumor may be above its inflection point, but to intensify therapy after the attainment of complete remission, when the tumor is probably below its inflection point. This conclusion by Norton and Simon is similar to the "late intensification" approach recently proposed for the therapy of acute myelogenous leukemia by Bodey *et al.*[12] Thus, the mathematical model suggests that following complete remission, tumor recurrence may be due not to biochemical resistance (though this possibility is not denied) but to the suboptimal use of the original agents. These intriguing suggestions, indicated as experimental leads by a mathematical model, can be investigated and confirmed with animal tumor models.

3. The Hyperplastic Alveolar Nodule, Model of the Preneoplastic Lesion

Studies on the histogenesis and pathogenesis of mammary tumors have suggested that virus-, hormone-, and carcinogen-induced mammary hyperplasia and dysplasia in mice and rats are preneoplastic with the potential for progressing to form mammary tumors. These lobuloalveolar nodules are more frequent in the mammary glands of strains of mice that have a high frequency of mammary adenocarcinoma than in strains with a low incidence. They increase in frequency with age, and they have been shown by direct experimental means to be precancerous to the common mammary adenocarcinomas of mice.[13,14] That virus-induced hyperplastic alveolar nodules (HANs) in mice develop neoplastic outgrowth on transplantation into the mammary gland free fat pad in the isologous host was first demonstrated by DeOme *et al.*[15] Similar results were later reported by Beuving,[16] who transplanted carcinogen-induced HANs in the rat into the gland free mammary fat pad. Induction by a carcinogen and the tendency to become neoplastic after transplantation support the suggestion that HANs are preneoplastic. The recent work of Sinha and Dao[17] indicates, however, that when mammary cells are exposed to a carcinogen, they are transformed directly to neoplastic cells and require no intermediate steps. A comparison of certain biochemical characteristics of HANs and mammary tumors induced in Sprague–Dawley female rats as well as mammary glands of pregnant rats revealed that HAN cell populations differ from both the normal mammary gland and the mammary tumor cells by their nonresponsiveness to hormonal stimulation for growth and the loss of the capacity to bind estradiol, characteristics of estrogen target tissues. They retained, however, the functional capacity to synthesize casein, a biochemical property that the mammary tumors did not possess.[18]

How closely HANs can serve as an animal model for preneoplastic lesions in the human breast is exemplified by the report by Jensen *et al.*[19] A subgross sampling technique with histological confirmation was used to study the pathology of 119 whole human breasts either containing cancer or contralateral to cancer, or taken from random routine autopsies. Atypical nodules were observed much more frequently in the cancer-associated group than in the group of routine autopsy breasts. The nodules showed varying degrees of anaplasia that formed a continuum between normal epithelium and carcinoma *in situ*, usually of the common ductal type. These investigators suggest that as apparent markers for increased cancer risk, atypical lobules in the human breast may be homologous to the HANs that are abundant in high-mammary-cancer strains of mice. Though their evidence is indirect, it supports the hypothesis that such atypical lesions are common preneoplastic lesions in the human mammary gland.

4. The Canine Model

Mammary tumors are among the most common of canine neoplasms. That mammary tumors occurring spontaneously in the general dog population represent an appropriate clinical model of human breast neoplasia is indicated by the follwoing data:

- The peak reported incidence in man is in the 50- to 58-year age group, and in dogs the highest incidence is observed in animals 8 to 11 years old, a comparable age group.
- Sex distribution is very similar in both man and dogs, being almost entirely a disease of females.
- A significant role of the sex hormones in mammary neoplasia is also evidenced in both man and dog. Animals oophorectomized before puberty have a cancer risk only 0.5% that of intact females, with later castration being less effective in preventing the disease. In humans, early oophorectomy also decreases the risk of breast cancer.
- Nulliparous women have a significantly higher incidence of breast cancer than do their multiparous counterparts. In dogs, a higher mammary tumor incidence has been reported in nulliparous animals as well as in animals having had a few litters compared to those having had a large number of litters.
- Histopathological study of canine mammary lesions has in many respects followed that of human breast disease. Histologically benign lesions that have been found in dog mammary tissue are morphologically very similar to those recognized commonly in the human breast.
- Most malignant human breast lesions are infiltrating carcinomas of the duct system. The dog also has infiltrating malignant epithelial neoplasms of ductal origin that are histologically very similar to those of man. Of all canine mammary tumors, 40% are adenocarcinomas, an incidence equivalent to that in the human. Lobular carcinoma and lobular carcinoma *in situ* occur as 10% of all canine carcinomas, a figure that corresponds to the human incidence. Although other types of neoplastic mammary lesions also occur in dogs, as in man, sarcomas of the breast are uncom-

mon. Thus, malignant epithelial breast neoplasms in the dog represent a close model of human breast cancer in morphology, clinical behavior, and incidence.

Inherently, therefore, the canine model includes all the variables encountered clinically, which require the same definition and control, with the same staging of tumors and careful randomization of subjects, as are applied in human clinical trials. The time frame for carrying out controlled studies for evaluating chemotherapeutic agents and treatment modalities is inordinately long. Bostock[20] found the median survival time of animals with histologically diagnosed carcinomas to be 70 weeks, with only 43% of dogs with carcinoma eventually dying as a result of the tumor. A more accurate prognosis was made possible, however, by subdividing these tumors into their different morphological types. The associated costs for maintaining statistically significant numbers of experimental subjects in clinical-trial-oriented studies is also unacceptably high. Though the validity of the canine model for human breast cancer is unquestioned, its use in breast cancer research is thus primarily cost-limited. The information quoted above was extracted from publications by Mulligan,[21] Mason,[22] Misdorp *et al.*,[23] Strandberg and Goodman,[24] Anderson,[25] and Taylor *et al.*[26]

5. The Hormone-Dependent Autochthonous or Primary Tumor Model

This model of human breast cancer has been most intensively studied in rodents and has been divided into two general categories based on etiology, whether spontaneous (?) or induced. That certain aromatic hydrocarbons induce mammary tumors in some species of rodents[27] has been known for many years. Although tumors develop preferentially in the mammary gland of certain strains of mice and rats, the local application of such carcinogens to the region of the mammary gland is unnecessary for the production of breast tumors.[28–30]

With the demonstration that multiple mammary adenocarcinomas can be easily induced in outbred, female Sprague–Dawley rats by a single administration of the polycyclic hydrocarbon 7,12-dimethylbenz(a)anthracene (DMBA),[31] this tumor system has been most intensively studied and used in investigation on hormone dependency of mammary neoplasia. Hormone dependency is demonstrated by the tendency of these mammary tumors to regress after ovariectomy[31] or hypophysectomy,[32] or after the administration of various steroids or antihormones.[33–36]

The hormone-dependent nature of this experimental tumor system therefore provides a useful model for the evaluation of the endocrine factors that are concerned in the growth and maintenance of the mammary tumor and also for the assessment of various drugs considered to have some potential in the clinical management of human breast cancer. Hormonal manipulation continues to play an important role in the patient with breast cancer, and current ability to quantitate estrogen[37] and progesterone[38,39] receptors in these malignant tissues has permitted a more rational approach to the choice of therapy and has made the outcome of such therapy more predictable.

The experimental design and results obtainable when testing antiestrogens in the DMBA-induced mammary tumor system are illustrated by the study of Nicholson and Golder,[40] in which the effects of tamoxifen and its nitrogen mustard derivative were compared. The availability of nonsteroidal antiestrogens such as nafoxidine and tamoxifen, which exert their action through a competitive binding at the receptor level,[41] offers a useful tool for studying the role of prolactin in stimulating tumor growth when the action of circulating estrogens is blocked. That prolactin is an influential hormone in murine mammary tumorigenesis and is also an important factor in both the development and growth of murine mammary tumors in unequivocal. Welsch and Nagasawa[42] recently reviewed the role of prolactin in murine (rat and mouse) mammary tumorigenesis.

In view of the evidence that prolactin was indispensable for estrogen action in stimulating tumor growth in this experimental model,[43] Manni *et al.*[44] studied DMBA-induced rat mammary tumor growth in an experimental design in which the antiestrogen tamoxifen was used to study the effects of prolactin, and the ergot derivative, lergotrile mesylate, was used to suppress prolactin in order to investigate the effect of estrogens.

Quadri *et al.*[45] demonstrated that the inhibitory effects of pharmacologic doses of androgen on DMBA-induced mammary tumor growth was reversed by the administration of exogenous prolactin. Since the androgen does not reduce serum prolactin levels, it had been proposed that androgens may block the peripheral effects of prolactin at the level of the tumor cell. To determine whether the androgen effect was due to an alteration in the ability of the tumor cell to bind prolactin at the receptor level, Costlow *et al.*[46] used an experimental design that took advantage of the heterogeneity in hormone dependence and responsiveness of DMBA-induced mammary tumors. By observing individual tumor response following treatment with testosterone propionate, tumors could be classified as responders (regression) or nonresponders (continued growth). It was found, for example, that testosterone propionate caused a substantial reduction in the prolactin receptor content in responsive but not in nonresponsive tumors.

The innate variability between DMBA-induced tumors (in the same animal as well as among animals) can be both a boon in terms of permitting a refinement in experimental design and a bane in terms of difficulty in identifying and controlling variables. As one example, DMBA has the capacity to induce two different types of mammary neoplasms, adenocarcinomas and fibroadenomas.[47,48] These may need to be differentiated histologically. Second is the hormone sensitivity, one of the most intriguing features of the DMBA-induced mammary tumor in the rat. This sensitivity, which is demonstrable by alteration of tumor growth and by interference with the tumor-induction process following appropriate steroid therapy and endocrine-organ ablation, with and without specific hormone replacement, is complicated by the apparent subtotal, transient, and sometimes biphasic response in terms of tumor growth or regression following certain therapies or alternative procedures.[48–53] Third, Huggins *et al.*[27] detected hormone-independent cell groups among tumors generally thought to be hormone-dependent. Young *et al.*[53] also recognized this characteristic, as did Daniel and Prichard,[54,55] who suggested that the failure of a tumor to respond to hormonal changes may result either from insufficient hormonal influence or as a response to the presence of neoplastic tissue that is independent of hormonal control. Teller *et al.*[51] subsequently demonstrated that tumor regression could be achieved by massive estrogen treatment in a significant number of rats with tumors that had previously failed to regress following ovariectomy. Finally, differences exist in responsiveness of earlier tumors as compared with tumors arising later. Nicholson and Golder[40] found an 80% response to ovariectomy of tumors occurring 23 weeks after DMBA administration as compared with a 50% response in a second population of tumors existing in older animals. These results are consistent with the data of Griswold and Green,[56] who reported a gradual increase in autonomy in DMBA-induced tumors with increasing time between tumor induction and treatment. There is little question, therefore, that the vagaries of the DMBA-induced mammary tumor system must be thoroughly understood for its effective use as an experimental model.

Similar to the hormone dependency observed by Noble *et al.*[57,58] in Nb rats following prolonged estrogenization, Sluyser and VanNie[59] induced mammary tumors with estrone and progesterone in ovariectomized GR/A mice. Exhibiting the variability and instability observed in the rat, not all the induced tumors were hormone-dependent, and after several transplantations, hormone-dependent tumors became first hormone-responsive and, finally, hormone-independent. Hormone independence was associated with low estrogen-receptor content. Their data were consistent with the assumption that hormone-responsive mammary tumors in GR/A mice are mixed populations of hormone-dependent and hormone-independent cells.

Although the developmental stages of mouse mammary tumorigenesis appear to be markedly influenced by secretory levels of prolactin,[42,60–62] the advanced spontaneous mammary tumors in most strains of mice appear to be prolactin-independent. Treatment of C3H/HeJ female mice bearing advanced spontaneous mammary tumors with a prolactin suppressor did not significantly influence the growth of these tumors,[63] and neither were they significantly influenced by the chronic administration of prolactin[64] or pituitary isografts.[65] It appears, therefore, that spontaneous C3H mouse mammary neoplasms gradually but fairly consistently evolve *in situ*, i.e., without the selective effects of transplantation, from a stage of prolactin responsiveness to a stage of prolactin independence, an event that is much more common in this species than in the rat.[42]

6. Hormone-Dependent/Responsive Mammary Tumors Established in Serial Transplantation

Apropos of the hormone responsiveness of carcinogen-induced mammary tumors is the rare mammary neoplasm that retains its hormone responsiveness after prolonged serial transplantation in syngeneic hosts, and even more rarely, its hormone dependence. There is a difference between hormone *dependence* and hormone *responsiveness* that is sometimes ignored in reference to hormonal effects on tumor growth. Cessation of tumor growth or regression in the absence of a hormone, or following an ablative procedure, exemplifies dependence. In contrast, a modification in growth pattern such as reported by Bogden *et al.*[66] on the responses of the 13762 (growth stimulation) and R-35 (growth retardation) mammary adenocarcinomas to increased endogenous prolactin levels exemplifies hormone responsiveness. The direct effect of prolactin on the R3230AC mammary tumor in producing biochemical changes similar to those resulting from administration of estrogens, reflected by a stimulation of certain metabolic pathways suggestive of lactation induction and a retardation of growth of the tumor, is another example of responsiveness as reported by Hilf *et al.*[67]

A long-transplanted, responsive, mammary carcinoma syngeneic with Fischer 344 strain females is the 13762E subline of the 13762 mammary adenocarcinoma that had been induced with DMBA by Segaloff.[68] The 13762E subline has been maintained in serial transplantation in females treated with estradiol or implanted with a 20-mg pellet of diethylstilbesterol. Hormone responsiveness in this tumor system is defined as tumor growth stimulated by exogenous estrogens. This tumor grows well in the untreated, ovary-intact female, but growth is retarded following ovariectomy.

The best known, and perhaps most thoroughly studied, hormone-dependent mammary tumor that has been successfully carried in serial transplantation without loss of hormone dependency is the MT-W9 of Kim and Furth,[69] which was induced in Wistar–Furth rat strain females by a combined treatment of a subthreshold dose of 3-methylcholanthrene and mammotropic hormones. The mammotropin (prolactin) dependency[70] has been maintained over successive transplant generations by sustained stimulation of exogenous mammotropic hormone such as is provided by coimplantation with the MtTW10, a mammosomatropic pituitary tumor[71,72] that secretes both growth hormone and prolactin.

Progression from hormone dependence to autonomy is a natural sequence of events for mammary tumors established in serial transplantation. Hormone sensitivity can vanish very quickly after one or two successive transplantations; i.e., they will grow well in any syngeneic animal regardless of the functional state of host endocrine organs.[73] The progression from hormone dependence to autonomy in mammary tumors as an *in vivo* manifestation of sequential clonal selection was demonstrated by Kim and Depowski[72] with the MT-W9 tumor. This original, fully mammotropin-dependent tumor gave rise to an estrogen-dependent variant, MT-W9A, that grows only in normal adult female hosts and regresses promptly upon oophorectomy. Subsequently, this tumor produced a subline, the fully autonomous MT-W9B, that grows well in any syngeneic rats regardless of their hormonal status. A third subline derived from the autonomous tumor was designated MT-W9C, the androgen-responsive line, because it grows better in male than in female rats. Thus, this tumor system has served as a model for studies on the natural history of breast cancer with respect to its progression from hormone dependence, to hormone responsiveness, to autonomy.[72]

The ability to maintain the hormonal dependency of the MT-W9 tumor during successive passages in syngeneic rats conditioned with pharmacologic levels of prolactin,[70] as well as the derived sublines, has provided a reproducible test system for studies on prolactin and estrogen binding in hormone-dependent and autonomous rat mammary carcinomas,[71,74] and has permitted an elucidation of the regression process of hormone-dependent tumors following hormone deprivation.[75–78] There is little question that the value of these tumor models to breast cancer research is proportional to the effort that must be devoted to maintaining the stability and reproducibility of the biological characteristics unique to each tumor line.

7. The Autochthonous or Primary Tumor Model (Hormonally Independent)

The designation as "spontaneous" of tumors arising in untreated experimental animals is essentailly a term of convenience commonly used to indicate either neoplasms of unknown etiology, neoplasms in host animals that have been intentionally not manipulated, or those tumors arising in mouse mammary tumor virus (MMTV)-infected, high-tumor-incidence, strains of mice. The term is unsatisfactory, primarily because there is a tendency to project a direct relationship with "spontaneous" tumors of man and conclude that the autochthonous tumor in the rodent, simply because of its apparent spontaneity, is a completely relevant tumor model for cancer in man. Since the term "autochthonous" is defined as "found in the place of formation; not removed to a new site,"[79] it will be used in lieu of spontaneous.

Whether the etiological factors in mammary tumorigenesis, operative in genetically defined rodent populations maintained within isolated and carefully controlled animal facilities, reflect the factors operative in the human female randomly exposed to the stresses and environmental carcinogens of daily living is somewhat of a moot question. At the molecular level, the fundamental mechanisms involved in viral or chemical carcinogenesis may be the same in man as in the rodent. However, the multiplicity of the phenotypic manifestations, resulting from carcinogenic as well as cocarcinogenic effects, that are possible in freely roving, random-bred human populations can hardly be completely duplicated by a single, genetically homogeneous and isolated rodent population. Superimposed on this heterogeneity in the etiology of mammary tumorigenesis, Brennan,[80] in a scholarly dissertation on Huxley's review of the biology of neoplasia,[4] points out that each individual neoplasm is postulated to be a new species, and therefore the individual neoplastic cells constituting a particular neoplasm would also be expected to have functionally significant epigenetic differences from other neoplasms, even those derived from the same organ and species. Thus, in the same animal, no two tumors of one organ would be identical, although they would share, as would also the tumors of isogeneic hosts, a wider range of properties and antigens than would those of unrelated hosts.[81–83] Within this framework, Brennan[80] rightfully concludes that "biological experiments in mammary cancer induction and observations of the natural course of the disease in both the virus-related murine and hydrocarbon-induced rat experimental tumor systems reveal many correspondences within these systems with features observed in clinical breast cancer."

There is also little question that the autochthonous mammary tumor system, whether viral-, hormonal-, or chemical-carcinogen-induced, or resulting from the wrath of God, has many factors in correspondence with human cancer, factors that are adequately reproducible and predictable to be experimentally useful. Despite these favorable aspects, there are nevertheless limitations to the use of autochthonous tumors as a bioassay system that preclude its characterization as the ideal model for human breast cancer.

First, it is etiologically incorrect to label mammary tumors arising in genetically defined, MMTV-infected, high-tumor-incidence murine populations as spontaneous. In view of the multiple factors operative in mammary tumorigenesis, factors controlled by both genetic and hormonal mechanisms,[84] one wonders whether a viral-induced tumor can be representative of all of human breast cancer. Viral etiology is discussed in detail by J. Schlom *et al.* in Chapter 2.

Martin *et al.*[85] present a very complete discussion of the characteristics desired of the ideal animal tumor model for breast cancer, projecting their extensive experience with the spontaneous (?) mammary tumor in the CD8F$_1$ hybrid mouse strain[86–89] as an example of a murine system that meets their criteria as summarized in Table I.

The CD8F$_1$ model tumor system is a good example of the hormone-independent,[85] autochthonous murine tumor. The animals employed in this model are the first-generation hybrid, resulting from the cross-mating of inbred DBA/8 males with inbred BALB/c females. The hybrid females, infected maternally with the MMTV, have a high mammary tumor incidence (about 80%) arising in animals that are 9–10 months of

Table I
Characteristics of Spontaneous (Autochthonous) Murine Mammary Carcinomas Indicative of the Ideal Breast Cancer Model[a]

1. Have a viral etiology.
2. Grow relatively slowly as solid tumors.
3. Metastasize.
4. Are relatively refractive to chemotherapeutic agents.
5. Are immunogenic in their autochthonous hosts.
6. Can be used to evaluate combined-modality therapy in autochthonous hosts.
7. Can be employed for expedient testing as first-generation transplants.

[a] Summarized from Martin *et al.*[85]

age. Typical of the autochthonous systems, there is not a clear-cut 100% tumor incidence, nor is the percentage "takes" of first-generation transplants (90–95%) completely predictable.[85] There is also a question of the immunogenetic relationship between the $CD8F_1$ tumor and $CD8F_1$ recipients that requires a minimum inoculum of 10^5 viable tumor cells to initiate a tumor,[89] although this type of immunogenicity may be a reflection of virus-associated antigen.

The autochthonous animal tumor model does not permit the degree of identification, definition, and control of variables that is obtainable with well-defined, carefully monitored and established, transplantable syngeneic tumor systems. Autochthonous tumor systems are fraught with all the variables and undefinables encountered in clinical breast cancer. Admittedly, the animal model has one distinct advantage over clinical experimentation—there are no ethical problems in randomizing tumor-bearing subjects into appropriate untreated control groups.

Mimicking clinical observations, there is heterogeneity in the growth rates of different autochthonous tumors arising in the same animal as well as in different animals of the same inbred strain. In plotting the growth patterns of 23 individual $CD8F_1$, mammary tumors, Anderson *et al.*[89] found a range in growth rate among individual $CD8F_1$ tumors that required from 3 weeks to as long as 10 weeks to reach a size approximating 4 g. With regard to metastases, it was found that only 62% (13 of 21) of the mice had pulmonary metastases detectable by bioassay by the time their tumors had reached a calculated weight of 4 g.

The use of autochthonous tumors in the first transplant generation as a test system, though admittedly not far removed biologically from the parent autochthonous tumor, is in essence a syngeneic test system without the homogeneity and reproducibility of syngeneic systems. Hosokawa *et al.*[90] established 19 transplantable tumor lines from individual autochthonous mammary tumors of the DBA/2 Ha-DD mouse. Of these lines, 5 were classified as fast-growing, 8 as medium-growing, and 6 as slow-growing. The rate of growth among 5 of these lines remained stable during 11–19 transplant generations. First-transplant-generation test systems are therefore restricted in the scope of testing to the number of animals that can be implanted with tissue obtainable from one tumor donor. Combining more than one autochthonous tumor for first-transplant-generation grafting in syngeneic hosts is a compounding of variables, i.e., preparing a gemisch of tissues with inherent differences in growth rate, immunogenicity, histology, and requirements for vascularization.

Obviously, the autochthonous mammary tumor in the rodent, such as the $CD8F_1$ model, more closely mirrors the innate heterogeneity observed with breast cancer clinically. Its use as an assay system therefore requires the same stringent staging of tumors, with definition, monitoring, and randomization of the individual tumor-bearing animal as the oncologist applies to each patient in clinical trials.

8. Transplantable Syngeneic Mammary Tumor Models

Most commonly, intact animal systems serve as models of human diseases at what might be appropriately termed the clinical level, as contrasted to the cellular or molecular levels. Nonetheless, one should not lose sight of the fact that animal tumor models are assay systems and, as such, must have well-defined and reproducible growth patterns and reactivities so that modifications in such parameters resulting from experimental manipulations can be correctly interpreted. It is an axiom that when attempting to define an unknown, experimental variables must be identified and controlled. In experimental breast cancer, the need for tumors with well-defined growth and metastatic patterns as well as responsiveness to chemotherapy, to ionizing radiation, and to immunotherapy is best met by the syngeneic mammary tumor system that has been established and stabilized in serial transplantation. The heterogeneity that one encounters with autochthonous tumors, animal and human, is also reflected in syngeneic tumor systems, except that the differences are found among established tumor lines rather than among tumors within a line. With careful monitoring, one finds that the characteristics peculiar to a particular tumor line are reproducible in each of the tumors of that line, transplant generation after transplant generation, whether ten animals or a thousand animals are implanted at any particular passage.

Maintaining the stability of the histologically more complex mammary adenocarcinomas while in serial transplantation requires intensive monitoring of a number of factors as well as a "bit of art." The foremost prerequisite is that the tumor will have originated in a member of a highly inbred strain of animals. The purpose of inbred strains, the members of which serve as recipients for syngeneic tumor grafts, is to have available for experimentation a large population of genetically identical animals in which the relationship of each animal to a neoplasm arising in one of its members mimics the autochthonous tumor–host relationship. Thus, malignant tumors should be 100% transplantable in syngeneic hosts, producing progressive and eventually lethal growths.

In large-scale commercial breeding programs, where numerous production breeder lines are maintained, it is essential that such lines be checked periodically for immunogenetic drift from designated parental or reference strains. Such monitoring is carried out under the auspices of the Mammalian Genetics and Animal Production Section of the Division of Cancer Treatment, National Cancer Institute. All rat and mouse strains being bred and maintained under contract with the National Cancer Institute are monitored as a routine quality-control procedure. It behooves the investigator working with syngeneic transplantable systems to know the source and quality of his experimental animals.

Unlike the homogeneous cell populations making up leukemias or fibrosarcomas, mammary adenocarcinomas are generally differentiated, having definite histological structures such as acini, which may or may not be arranged as papillary extensions of tumor growth, and maintain fairly constant epithelial/stromal ratios. Such characteristics must be monitored histologically. The "art" in tumor transplantation is manifested in the ability to maintain these histological characteristics over many transplant generations by "knowing" from gross appearance which tissue is representative of the tumor and should be selected for transplantation. Histology, though necessary, is somewhat of a *post facto* confirmation of the selection made. The most careful technician can unwittingly convert a well-differentiated adenocarcinoma to an undifferentiated carcinoma, or even sarcoma, in a few passages. Where a transplantable tumor is being used for chemotherapy studies, it is also advisable to monitor its responsiveness to a known active compound. That the more complex mammary adenocarcinomas may contain cell populations of different drug sensitivities was demonstrated by Bogden *et al.*[91] with two ascites tumor lines converted from the solid 13762 mammary adenocarcinoma. When the 13762MAT-B and 13762MAT-C sublines were tested against a number of chemotherapeutic agents, there were differences in responsiveness between the two ascites sublines as well as between the ascites sublines and the parent tumor. Cryopreserving a quantity of tissue in the early transplant generations of a syngeneic tumor provides a source for not only replacing lost tumor lines but also renewing long-transplanted lines. The Breast Cancer Task Force of the Division of Cancer Biology and Diagnosis, National Cancer Institute, has established such a tumor bank of animal and human tissues.

It is evident, therefore, that syngeneic tumor systems such as the mammary adenocarcinomas are subject to change during serial transplantations, and characteristics such as histology, growth rates, chemotherapy responsiveness, and metastases need to be continuously monitored.

Transplantable syngeneic mammary tumors (TSMTs) represent a spectrum of growth fractions. Table II compares the growth and metastasizing characteristics of six TSMT systems used for testing in our laboratory. The DMBA 14 adenocarcinoma, for example, has a maximum

Table II
Transplantable Syngeneic Mammary Tumors: Difference in Growth Rate, Incidence of Metastases, and Survival Time Characteristic for Each Tumor System

Tumor designation	Histological type	Rat strain of origin and transplantability	Maximum growth rate (mm/day)[a]	Metastases (% incidence)[b]	Survival time (days)[c]
3M2N	Squamous-cell carcinoma	Fischer 344	2.57	10	45.4 ± 8
13762	Adenocarcinoma	Fischer 344	1.85	93	48.7 ± 8
R3230AC	Adenocarcinoma	Fischer 344	1.35	10	49.8 ± 16
DMBA 1	Adenocarcinoma	Fischer 344	1.39	0	58.9 ± 11
SMT-2A	Poorly differentiated carcinoma	Wistar–Furth	1.06	95	61.1 ± 7
DMBA 14	Adenocarcinoma	Fischer 344	1.06	15	84.7 ± 15

[a] Determined as the average of the longest and shortest diameters.
[b] Detected macroscopically at necropsy; therefore, incidences reflect early rather than late metastases, since only early metastases have sufficient time to develop into macroscopically discernible masses before cachexia and death.
[c] Means ± S.D.

growth rate less than half that of the 3M2N squamous-cell carcinoma. The mean survival time of DMBA 14 is comparable to that of the $CD8F_1$ first-transplant-generation tumor, which when used in advanced tumor experiments has a duration of about 2.5 months, at which time all controls have died from tumors.[85] The metastatic characteristics of these TSMT systems also run the gamut from nonmetastasizing to an incidence of 95%. In addition, it is possible to select the type of metastases desired. Metastases from subcutaneous grafts of the 13762 mammary adenocarcinoma occur via both the blood vascular and lymphatic systems to the lungs and/or other viscera in almost 100% of the animals by day 20 postimplantation. Excision of subcutaneous tumors on days 18–20 prolongs survival, but the surgically treated animals eventually die from metastatic tumor growths, whereas excision prior to day 15 results in a high percentage of cures. On the other hand, the SMT-2A appears to metastasize exclusively by way of the lymphatics involving peripheral, mediastinal, and retroperitoneal lymph nodes and, only occasionally, to the lungs. Surgical excision as early as 3 days after implantation into the right inguinal fat pad does not prevent its metastasis.[92]

Since metastases in these two tumor systems are highly predictable (an incidence > 90%), they lend themselves to realistic studies on combinations of therapeutic modalities directed to the eradication of metastases, the primary problem in the treatment of breast cancer. Bogden *et al.*,[93] using the 13762 mammary tumor system, were able to show that a short, nonimmunosuppressive course of chemotherapy, whether preceding or following surgery, was most effective against metastases when used in combination with a nonspecific immunological adjuvant. Neither prolonged chemotherapy nor immunotherapy alone following surgery was as effective. Kreider *et al.*,[94] in determining the optimal conditions for conducting experiments with the solid and ascites sublines of the 13762 mammary adenocarcinoma, also found that the frequency of axillary lymph node metastases from the subcutaneous site increased as a function of the duration of the time interval between tumor implantation and surgical excision, and that both the solid and ascites tumors were weakly immunogenic. Assessing response of the 13762 tumor to bacillus Calmette-Guérin (BCG) treatment, they found that admixture of tumor cells with BCG suppressed tumor growth, but when given at a remote site, BCG was ineffective. This finding was in accord with the results obtained in the same tumor system by Sparks *et al.*,[95,96] who demonstrated that intralesional injection of BCG before resection of the tumor not only increased median survival, but also resulted in a significant number of long-term cures.

That TSMTs are also amenable for use in studies evaluating combinations of therapeutic modalities involving X-irradiation is exemplified by the following study. The X-irradiation sensitivity of three mammary tumor systems that have shown diverse responsiveness to certain chemotherapeutic agents was examined with the purpose of comparing their responses to a single large irradiation dose, as well as to three fractional doses, in an effort to determine whether these particular mammary tumors exhibit significant and measurable differences in X-irradiation sensitivity, thereby mimicking a problem encountered clinically.

Experimental Design for Comparing X-Irradiation Sensitivity. Animals bearing one of the three TSMT models were randomized into one of three groups: an unirradiated control, a group receiving a single 3000-rad (R) dose, and a group receiving three fractional doses of 1000 R at 2- or 3-day intervals. The experimental design is summarized in Table III.

The following tumor systems were compared in this study: *R3230AC,* a well-differentiated mammary adenocarcinoma of spontaneous origin; *SMT-2A,* a poorly differentiated mammary carcinoma; and *13762,* a differentiated mammary adenocarcinoma.

Taking into consideration the influence that the condition and state of the tumor bed, and the prevailing oxygen tension, have on the radiosensitivity of neoplasms, all tumor systems were irradiated when about the same size, approximately 11.5 mm average diameter. Tumor sizes were determined by vernier caliper and recorded as the average of the longest and shortest diameters in millimeters. Differences among the three tumors in the number of days between tumor implantation and X-irradiation reflect differences in the growth rates of the three tumor systems (Table III).

X-irradiation was provided at a dose rate of 55 R/min by a constant-potential deep-therapy X-ray unit (General Electric Maximar 250-III) with filtration provided by 0.25 mm Cu plus 1.0 mm Al. Tumor-bearing animals, under light chloral hydrate anaesthesia, were placed on a lead sheet with bodies completely lead-shielded so that only the tumors were exposed to irradiation. Tumors were X-irradiated in pairs with exposure monitored in air by a Victoreen model 570 condenser radiometer. For this study, *complete remission* indicates tumor size below measurable limits, i.e., barely palpable; for *partial remission,* tumor-size reduction is reported as a percentage of the size prior to X-irradiation; and *duration of remission* indicates the time (days) required for a tumor to regrow from maximum remission to the size attained prior to X-irradiation.

Table III
Experimental Design for Comparing X-Irradiation Sensitivity of Three Mammary Tumor Models

Group No.	Rat strain	Number of animals	Tumor system	X-irradiation dose	Days postimplant
I	Fischer 344	6	R3230AC	Unirradiated	Control
II		6		3000 R (1×)	14
III		6		1000 R (3×)	15, 18, 20
IV	Wistar–Furth	8	SMT-2A	Unirradiated	Control
V		8		3000 R (1×)	21
VI		8		1000 R (3×)	22, 25, 27
VII	Fischer 344	6	13762	Unirradiated	Control
VIII		6		3000 R (1×)	12
IX		6		1000 R (3×)	11, 13, 15

Results. In Figs. 1–6, the tumor sizes for unirradiated controls have been indicated as means (open circles) plus or minus one standard deviation (solid lines). X-irradiated tumors, on the other hand, have been plotted individually (solid circles) from the day of irradiation.

Figure 1 illustrates the response of R3230AC mammary tumors to a single 3000-R dose of X-irradiation. All tumors but one (85%) responded to irradiation by an average reduction in size of 18.5%. The average duration of remission was 6 days. The one tumor showing no remission showed growth inhibition that persisted for 7 days. It is evident from Fig. 1 that the growth rate of 50% of the tumors fell within the standard deviation of the unirradiated controls. However, a retarded growth rate appeared to persist with the remaining tumors.

When a total dose of 3000 R was administered to R3230AC mammary tumors in three fractional doses of 1000 R, only 1 of 6 tumors (17%) responded with a reduction in tumor size (Fig. 2). All other tumors went into a stationary growth phase that persisted an average of 6.8 days. Although growth of irradiated R3230AC tumors was progressive thereafter, the rates of growth appeared to be slower than the unirradiated controls. It is evident that a single dose of X-irradiation was more effective in this tumor system than three fractional doses.

Figure 3 illustrates the response of the SMT-2A mammary carcinoma to a single 3000-R dose of X-irradiation. The response of this tumor system was dramatic. All tumors (100%) were in complete remission by day 34, i.e., by day 13 postirradiation. Three tumors eventually regrew, having an average remission duration of 15.3 days. Five animals died from massive lung metastases while their irradiated subcutaneous

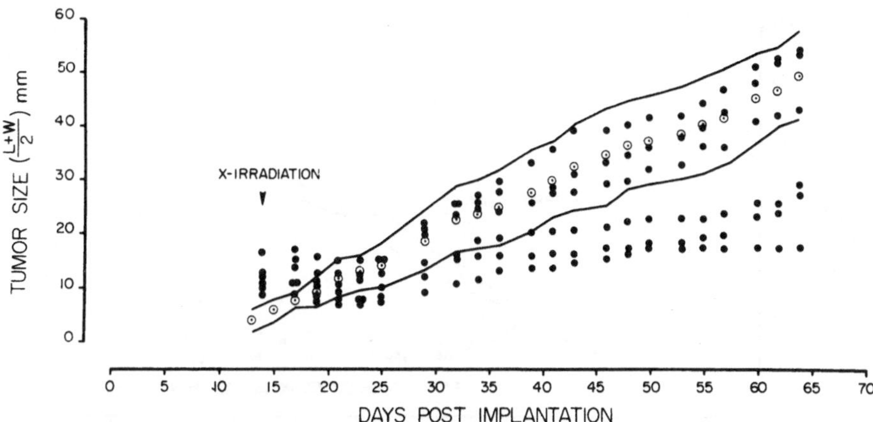

Fig. 1. Response of the R3230AC mammary adenocarcinoma to a single 3000-R dose of X-irradiation.

tumors were in complete remission. This is a highly metastatic tumor system, and it is evident that metastases had occurred prior to what appears to have been curative X-irradiation therapy of the primary subcutaneous tumor.

Figure 4 illustrates the response of the SMT-2A mammary carcinoma to three fractional doses of 1000 R X-irradiation. Though still highly responsive, the three fractional doses were not as effective as a single 3000-R dose. Only 4 tumors (50%) went into complete remission,

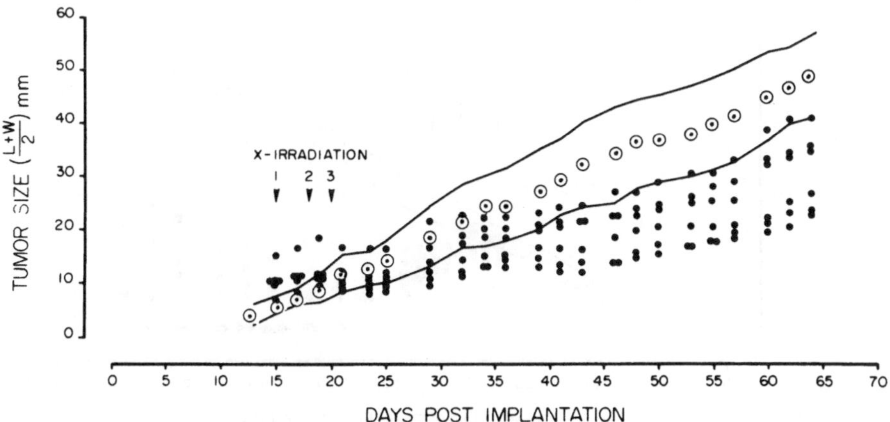

Fig. 2. Response of the R3230AC mammary adenocarcinoma to three fractional doses of 1000 R X-irradiation.

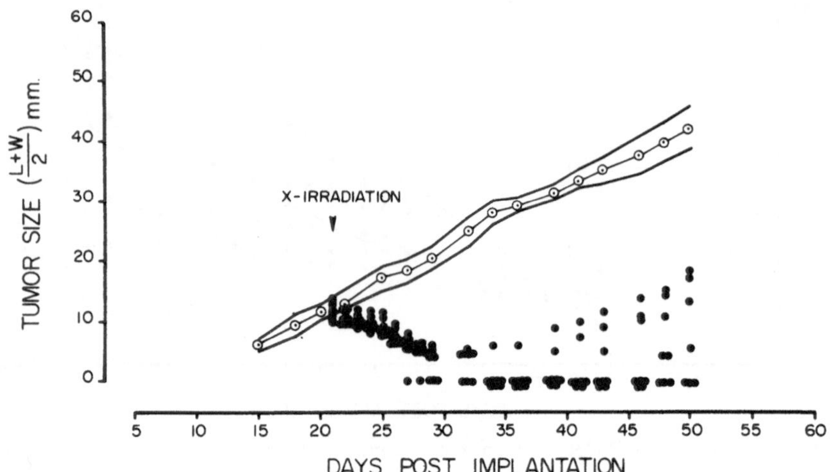

Fig. 3. Response of the SMT-2A mammary carcinoma to a single 3000-R dose of X-irradiation.

with an average remission duration of approximately 13.5 days; 3 tumors had a reduction in size of 50% or greater, with a remission duration of 12.7 days, and 1 tumor had only a slight reduction in size. All animals died with evidence of massive lung metastases, including one animal that had a subcutaneous tumor that was still in complete remission.

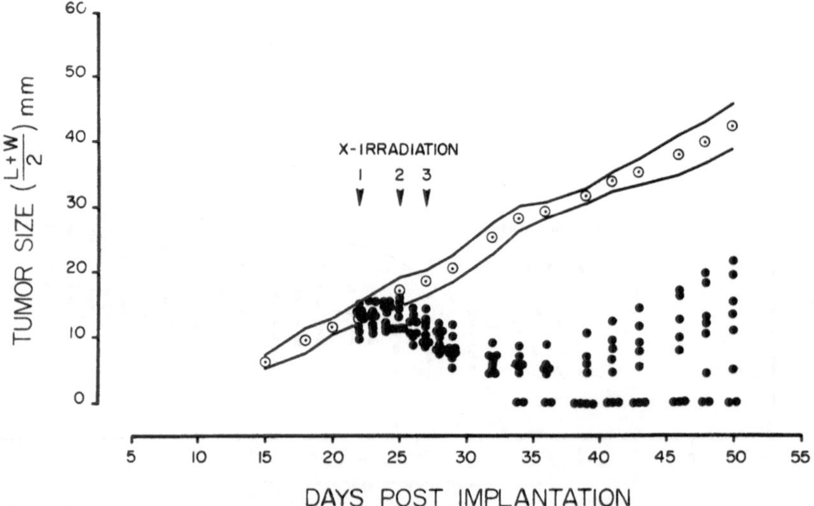

Fig. 4. Response of the SMT-2A mammary carcinoma to three fractional doses of 1000 R X-irradiation.

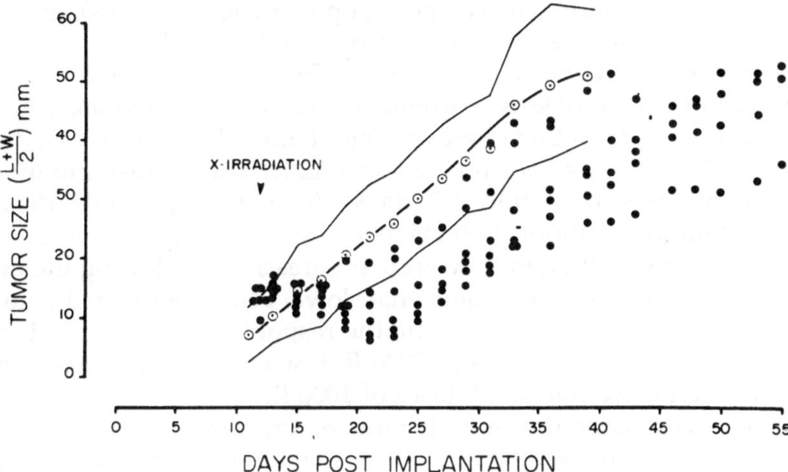

Fig. 5. Response of the 13762 mammary adenocarcinoma to a single 3000-R dose of X-irradiation.

Figure 5 illustrates the response of 13762 mammary tumors to a single 3000-R dose of X-irradiation. Of 6 tumors, 2 (33%) showed little or no measurable response to the single dose of X-irradiation; 1 animal had a 57% reduction in tumor size; two animals, 37% reduction; and 1 animal's tumor was reduced in size by only 12%. The 4 tumors that showed some indication of remission, though they grew progressively after the

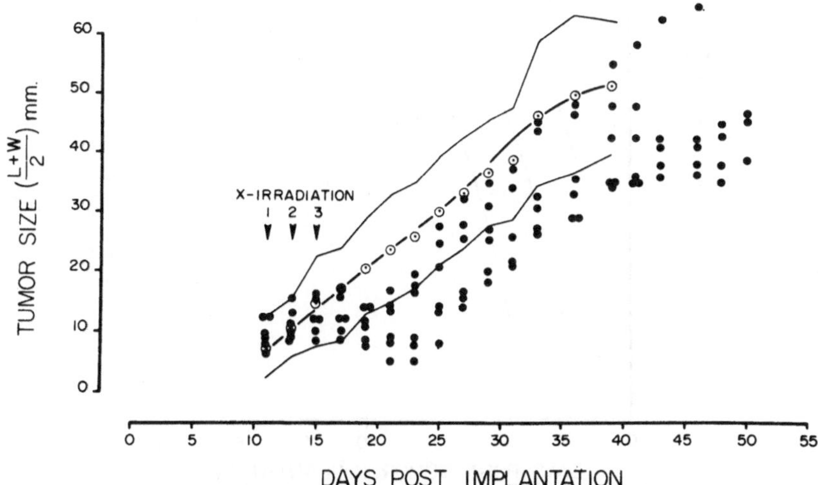

Fig. 6. Response of the 13762 mammary adenocarcinoma to three fractional doses of 1000 R X-irradiation.

initial remission period (an average of approximately 4.8 days), grew at a slower rate as compared with the unirradiated controls.

Figure 6 illustrates the response of 13762 mammary tumors to three fractional doses of 1000 R X-irradiation each. Of 6 tumors, 2 (33%) showed only a retardation in growth rate; 1 animal's tumor was reduced in size by 50%, 1 by 42%, and 2 by only 12%. The average duration of these remissions was only 4.2 days. As with the single dose of X-irradiation, the 4 tumors that evidenced some reduction in tumor size after X-irradiation, though they grew progressively following the initial remission period, grew at a somewhat slower rate as compared with the unirradiated controls. In comparing the responses illustrated in Figs. 5 and 6, it is apparent that a single 3000-R dose of X-irradiation was more effective than three fractional doses of 1000 R.

If one compares the growth rates of the three mammary tumor systems used in this study (Fig. 7), it becomes evident that X-irradiation responsiveness of these mammary tumor models is not directly related to growth rate.

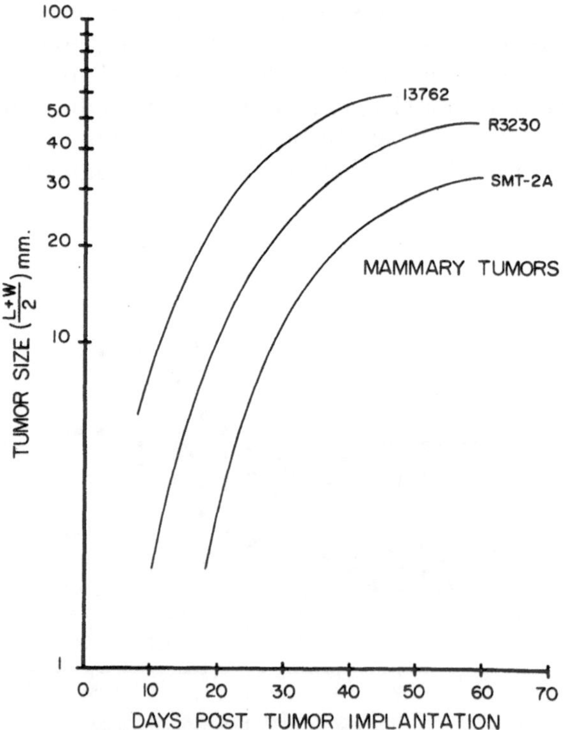

Fig. 7. Comparison of the growth rates (Gompertzian growth curves) of the R3230AC, SMT-2A, and 13762 mammary tumors.

Conclusion. Although all three tumor systems showed some degree of X-irradiation sensitivity, there were measurable differences in responsiveness among the tumor systems, e.g., SMT-2A > 13762 > R3230AC. Differences in irradiation sensitivity did not appear to be growth-rate-related, and a single X-irradiation dose of 3000 R was more effective in producing tumor remissions than three fractional doses of 1000 R each.

The SMT-2A mammary adenocarcinoma responded to X-irradiation with complete remissions. As a metastasizing tumor system, animals died from metastases despite "curative" X-irradiation therapy of the subcutaneous tumor.

Characteristic differences in the growth and metastasizing characteristics as well as the X-irradiation responsiveness of these mammary tumor models permit sophisticated studies on combinations of therapeutic modalities to include X-irradiation.

9. Use of Transplantable Syngeneic Mammary Tumor Systems for Addressing Specific Clinical Problems

Reproducibility of syngeneic tumor systems permits a degree of flexibility in experimental design and control not possible with autochthonous tumors. Though each syngeneic tumor system can serve as a model and be predictive for only a certain class of tumors found within the disease of breast cancer, by careful selection of the tumor system appropriate to the particular clinical problem under study, experiments can be designed to answer specific clinical questions. In the following studies, carried out in our laboratory, the 13762 mammary adenocarcinoma was the tumor model used. It was selected because it best mimics the chemotherapy responsiveness and acquired drug resistance of those human breast cancers that are sensitive to the alkylating agents.[97] It has also shown a degree of selectivity for agents most active in breast cancer, and has been used effectively in integrating immunotherapy with other therapeutic modalities.[93–96] In each of the following three studies, the clinical problem or question will be stated and the experimental design will be described, followed by the results of the experiment and a summary answer provided by the study.

Materials and Methods. In all studies, Fischer 344 strain females, approximately 45–50 days of age, were implanted subcutaneously with a routine 2- to 3-mm^3 piece of the syngeneic 13762 mammary tumor on day 0. Grafts were placed on the right side, midway between the inguinal and axillary areas. On day 10–12, exceptionally large or small tumors were eliminated to have a more homogeneous-sized tumor population, and the remaining animals were then randomized into experimental groups. Tumor-size measurements were determined by caliper 3 times weekly and were plotted as the average of the longest and shortest diameters in millimeters. All compounds were administered per os in a Klucel vehicle.

The following definitions of tumor response are pertinent to all studies: *complete remission* indicates tumor size below measurable limits, i.e., barely palpable; *partial remission,* tumor size reduced by 50% or more of largest size prior to induction of remission; and *duration of remission* indicates the time (days) required for a tumor to regrow from maximum remission to size attained prior to onset of remission; "cure" indicates tumor in complete remission, i.e., animal without evidence of tumor regrowth or free of macroscopically discernible metastases when sacrificed and autopsied on day 218.

Problem 1. Mammary tumors initially responsive to cytoxan (CTX) (NSC 26271) therapy become unresponsive or resistant during chronic or prolonged drug treatment. When is the optimal time, during cytoxan therapy, for a crossover to another alkylating agent such as phenylalanine mustard (PAM) (NSC 8806) to obtain maximum remission induction and duration of remissions?

Experimental Design. The experimental mammary tumor system selected for this study was one that is responsive to both CTX and PAM. It also becomes resistant to both compounds during prolonged or chronic therapy. Tumor grafts were implanted on day 0, and treatment of all experimental groups with CTX was initiated on day 11 when tumors were well established in early log phase of measurable growth. Obtaining tumor-size measurements 3 times weekly permitted establishing tumor growth and regression curves for monitoring the effects of therapy. PAM was substituted for CTX in four experimental groups: (1) on day 18, when tumors were still regressing from CTX therapy; (2) on day 25, when tumors were in maximum remission as a result of CTX therapy; (3) on day 32, early in the regrowth of CTX-resistant tumors; and (4) on day 36, later during the regrowth of CTX-resistant tumors. The designation of experimental groups and the treatments are summarized in Table IV.

Results. The results are most easily visualized from Figs. 8–11. In all figures, solid squares depict the growth curve for the vehicle-treated controls, and open circles with paralleling lines depict the means plus or minus one standard deviation of the group treated only with CTX from day 11 to survival. The solid circles represent individual tumors from the time when PAM treatment was initiated, i.e., when the crossover from CTX to PAM was made. Thus, Fig. 8 represents individual tumor re-

Table IV
Experimental Groups and Treatments for Determining Optimal Time for Drug Crossover

Group No.	Number of animals	Treatment
I	10	Vehicle-treated control, 0.5 ml, 3 Rx/week, day 11–ST
II	10	PAM, 2 mg/kg, 3 Rx/week, day 11–ST
III	22	CTX, 5 mg/kg, 3 Rx/week, day 11–ST
IV	10	CTX, 5 mg/kg, 3 Rx/week, days 11–17
		PAM, 2 mg/kg, 3 Rx/week, day 18–ST
V	10	CTX, 5 mg/kg, 3 Rx/week, days 11–24
		PAM, 2 mg/kg, 3 Rx/week, day 25–ST
VI	10	CTX, 5 mg/kg, 3 Rx/week, days 11–31
		PAM, 2 mg/kg, 3 Rx/week, day 32–ST
VII	10	CTX, 5 mg/kg, 3 Rx/week, days 11–35
		PAM, 2 mg/kg, 3 Rx/week, day 36–ST

sponses resulting from a crossover made during the initial stages of remission induction by CTX.

Figure 9 illustrates individual tumor responses resulting from a CTX-to-PAM crossover made at a time when tumors were in maximum remission as a result of CTX therapy. Figures 10 and 11, on the other hand, show individual tumor responses when the crossover to PAM was made early (Fig. 10) and later (Fig. 11) during the regrowth of

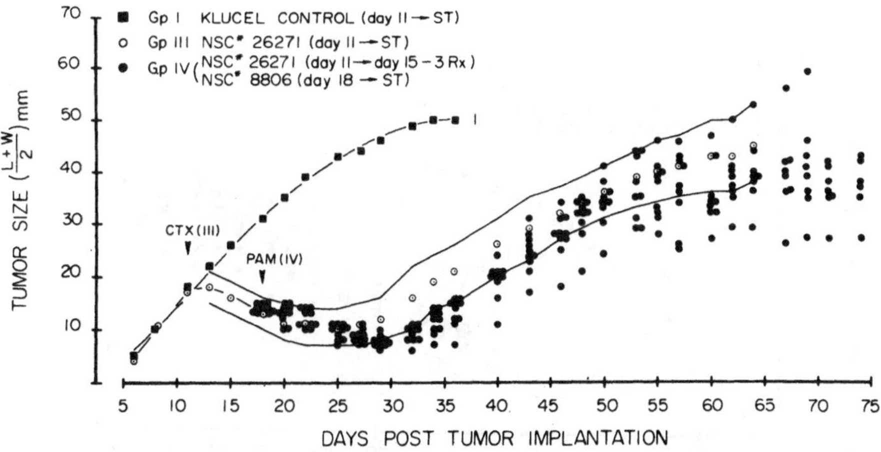

Fig. 8. Individual tumor responses to a crossover from CTX (NSC 26271) to PAM (NSC 8806) initiated during the initial stages of remission induction by CTX.

CTX-resistant tumors. A comparison of the individual tumor responses in the four figures shows that PAM was most effectively substituted for CTX when tumors were in maximum remission resulting from CTX therapy. When the crossover to PAM was delayed until tumors became CTX-resistant, the effectiveness of PAM was lost.

A comparison of remission induction activity and effect on life span is summarized in Table V. Treatment with PAM alone from day 11 to

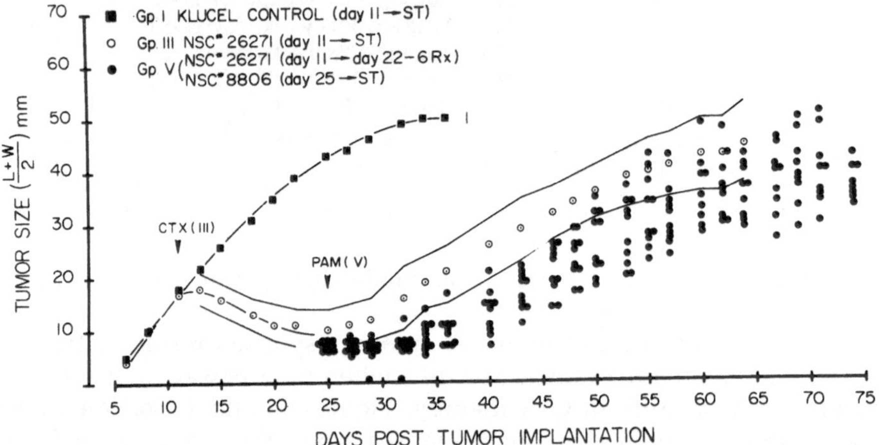

Fig. 9. Individual tumor responses resulting from a CTX (NSC 26271)-to-PAM (NSC 8806) crossover made at a time when tumors were in maximum remission as a result of CTX therapy.

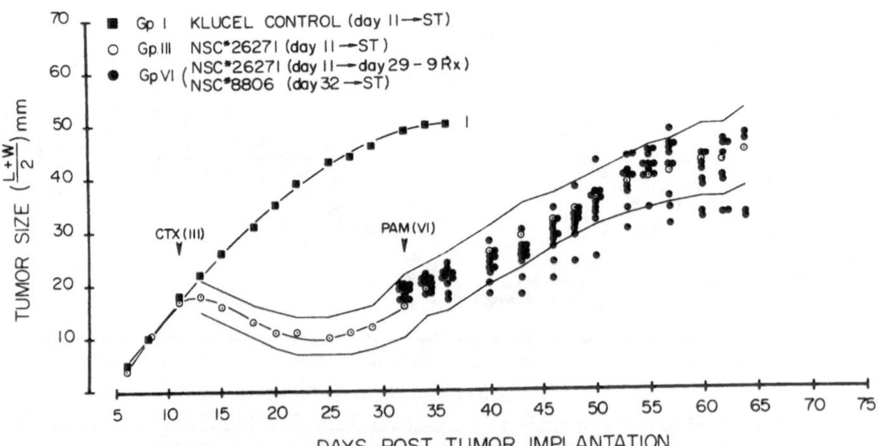

Fig. 10. Individual tumor responses following a crossover from CTX (NSC 26271) to PAM (NSC 8806) *early* during the regrowth of CTX-resistant tumors.

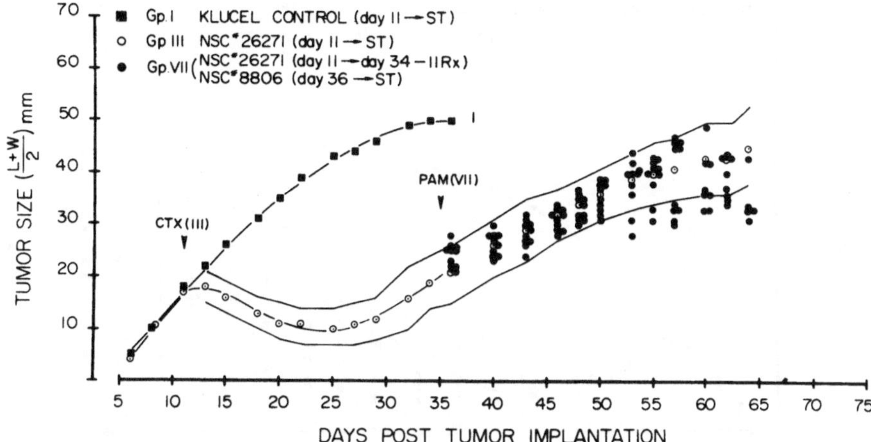

Fig. 11. Individual tumor responses following a crossover from CTX (NSC 26271) to PAM (NSC 8806) *late* during the regrowth of CTX-resistant tumors.

Table V
Comparison of Remission-Induction Activities and Effects on Life Span: Determining the Optimal Time for Crossover

Group No.	Treatment	Remissions (%) Complete	Remissions (%) Partial (50% or more)	Average duration of remissions (days)[a]	Survival time (days)[b]
I	Vehicle control	0	0	0	35.9 ± 6.8
II	PAM, day 11–ST	70	100	17.5 ± 8.5	75.7 ± 12.3
III	CTX, day 11–ST	5	41	9.1 ± 4.8	71.8 ± 9.6
IV	CTX, days 11–17 PAM, day 18–ST	0	100	12.1 ± 3.1	75.3 ± 13.5
V	CTX, days 11–24 PAM, day 25–ST	10	100	14.7 ± 3.4	77.9 ± 8.5
VI	CTX, days 11–31 PAM, day 32–ST	0	20	8.0 ± 1.4	69.4 ± 10.8
VII	CTX, days 11–35 PAM, day 36–ST	0	40	10.5 ± 1.3	71.6 ± 13.8

[a] Duration-of-remissions calculation includes both complete and partial remissions, as means ± S.D.'s.
[b] Survival time calculated as means ± S.D.'s.

survival was most effective, inducing 70% complete remissions. The remaining 30% of animals had tumor remissions of 50% or greater. By contrast, treatment with CTX alone induced only 5% complete and 41% partial remissions. A comparison of these activities is illustrated in Fig. 12. Individual tumor responses for the PAM-treated group are shown

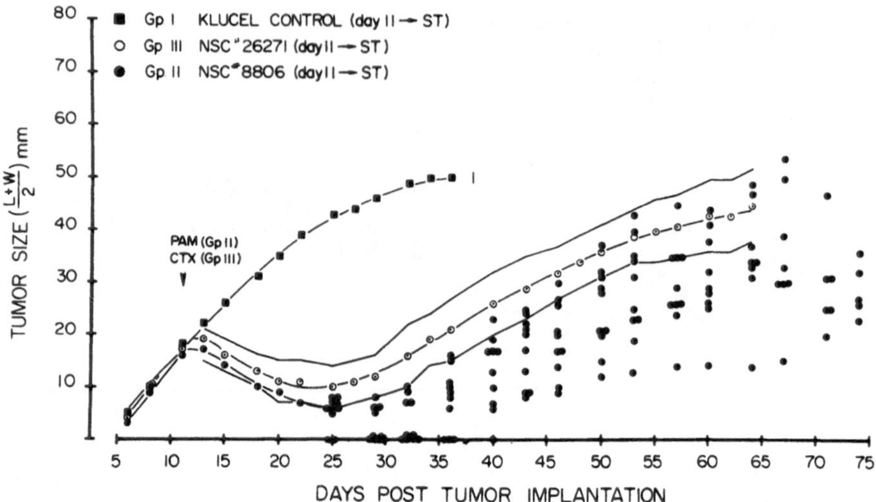

Fig. 12. Relative effectiveness of CTX (NSC 26271) and PAM (NSC 8806) administered as single agents in the 13762 mammary tumor model.

from day 25. Note, however, that mean tumor sizes for the PAM-treated group were already smaller from day 15.

Substituting PAM for CTX when tumors were in maximum remission as a result of CTX therapy induced 10% complete remissions, and prolonged the average duration or remissions as well as the life span over that of the other experimental groups (Table V). In essence, the data summarized in Table V support the conclusion, which can also be drawn from the figures, that there is an optimal time in a chronic treatment regimen to initiate a CTX-to-PAM crossover.

A word of caution is appropriate, however. As demonstrated in this study, the 13762 mammary adenocarcinoma was more responsive to PAM than to CTX; i.e., PAM was more effective when used as a single chemotherapeutic agent in this particular tumor system. Of particular relevance to this consideration, therefore, is the observation that pretreating a PAM-responsive mammary tumor with CTX markedly reduced the efficacy of PAM treatment. The results of this study would indicate that there may be a PAM-susceptible population of tumor cells that, though not responsive to CTX, nevertheless may be conditioned during CTX therapy to become less responsive or resistant to PAM.

Summary Answer. Remissions were maximized and duration of remissions prolonged when PAM was substituted for CTX at a time when tumors were in maximum remission as a result of CTX therapy. If substitution of PAM for CTX was delayed until CTX-resistant tumors began to regrow, there was evidence of cross-resistance that became more pronounced the longer substitution of PAM for CTX was delayed.

Substitution of PAM early during CTX therapy, i.e., while tumors were still undergoing remission induction, though it enhanced remission induction and duration, was not as effective as delaying the substitution of PAM until maximum CTX effects were obtained.

Problem 2. Many mammary tumors are responsive to both cytoxan (CTX) (NSC 26271) and phenylalanine mustard (PAM) (NSC 8806). During prolonged or chronic therapy with either drug alone, tumors become drug-resistant. If the administration of CTX and PAM were alternated in a chronic therapy regimen, would remission induction and remission duration be enhanced over either drug alone, and would the induction of drug resistance be prevented or delayed?

Experimental Design. The experimental mammary tumor selected for this study was one that is responsive to both CTX and PAM. It also becomes resistant to both compounds during prolonged or chronic therapy. Tumor grafts were implanted on day 0, and treatment of all experimental groups was initiated on day 11 when tumors were well established in early log phase of measurable growth. Obtaining tumor-size measurements 3 times weekly permitted establishing tumor growth and regression curves for monitoring the effects of therapy. There were three experimental groups in addition to the vehicle-treated controls: (1) a group treated only with CTX; (2) a group treated only with PAM; and (3) a group in which the administration of CTX and PAM was alternated every 2 or 3 days in a chronic treatment regimen requiring drug administration 3 times weekly. The designation of experimental groups and treatments are summarized in Table VI.

Results. The results are illustrated in Figs. 13 and 14, which reflect the tumor growth and regression curves as well as individual tumor sizes in the various experimental groups. In both figures, the solid squares depict the growth curve for the vehicle-treated controls, and the open circles with paralleling lines depict the means plus or minus one standard deviation of the group treated only with CTX.

Table VI
Experimental Groups for Studying the Effects of Alternating Cytoxan and Phenylalanine Mustard Treatments

Group No.	Number of animals	Treatment
I	10	Vehicle-treated control, 0.5 ml, 3 Rx/week, day 11–ST
II	10	CTX, 3 mg/kg, 3 Rx/week, day 11–ST
III	10	PAM, 1.6 mg/kg, 3 Rx/week, day 11–ST
IV	10	CTX, 3 mg/kg; PAM, 1.6 mg/kg; alternated every 2 or 3 days. Treatment initiated with CTX on day 11 and PAM on day 13, alternating to ST.

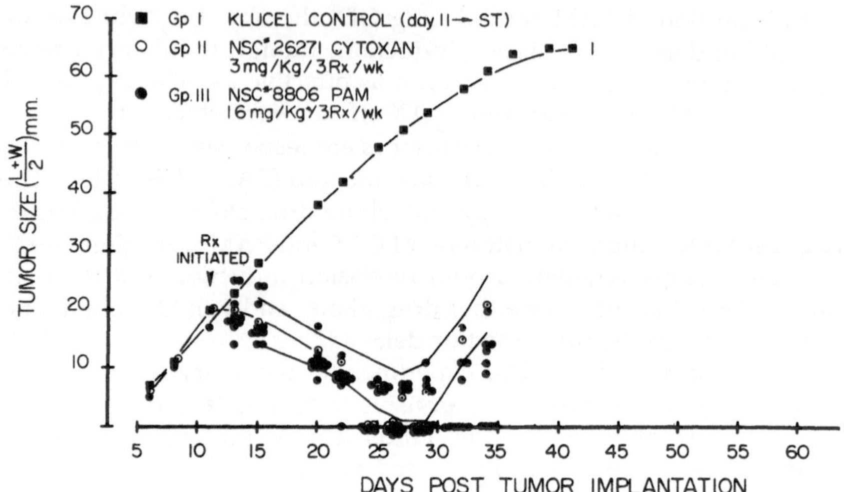

Fig. 13. Relative therapeutic effects of CTX and PAM administered as single agents, i.e., without crossover.

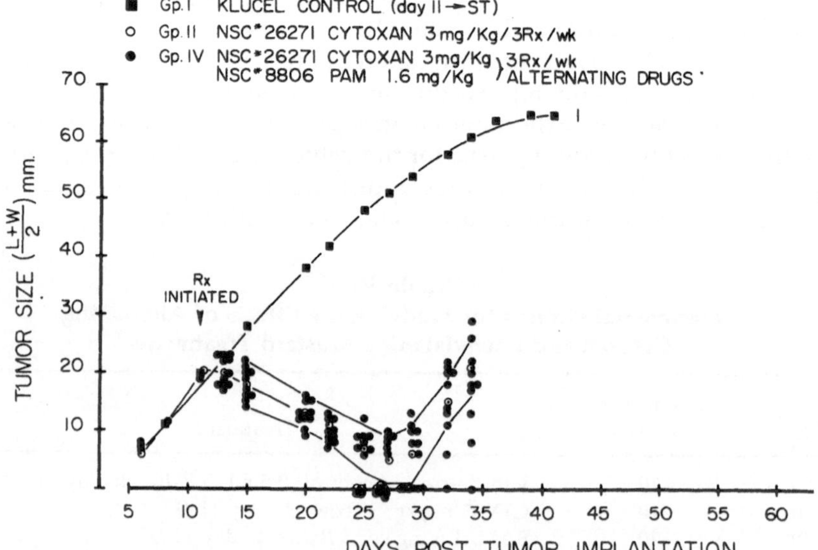

Fig. 14. Therapeutic effect of a treatment regimen in which CTX and PAM were administered in a continuing crossover, i.e., alternated every 3 days.

Figure 13 compares the therapeutic effects of CTX and PAM administered alone, i.e., without crossover. The solid circles represent the individual tumor sizes in the PAM-treated group. The clustering of the open (CTX-treated) and solid (PAM-treated) circles on the zero baseline indicates the number of tumors in each treated group that are in complete remission at any point in time.

Figure 14 illustrates and compares the effect of CTX, administered without crossover (open circles), with that resulting from a treatment regimen in which CTX and PAM were alternated, i.e., crossover every 2 or 3 days (solid circles). The clustering of open and solid circles on the zero baseline represents the numbers of tumors in each experimental group in complete remission.

In comparing Figs. 13 and 14, it is evident that PAM was a more effective chemotherapeutic agent than CTX when each drug was administered without crossover, confirming the results obtained in the first study. A treatment regimen in which CTX and PAM were alternated, i.e., crossed over every 2 or 3 days, was no more effective than CTX alone in terms of remission induction, duration of remissions, or inhibition of the emergence of drug-resistant tumors. A comparison of remission-induction activity and effect on life span, summarized in Table VII, further supports the conclusions drawn from the figures. When administered alone, i.e., without crossover, PAM was most effective, inducing 60% complete remissions, with the average duration of remission being 31 days and an average survival time of 70.9 days. When CTX and PAM were alternated every 2 days, only 40% complete remissions were induced, with an average duration of remission of only 13.6 days and a survival time of only 62.5 days.

Of particular interest is the indication that the inclusion of CTX in an alternating treatment regimen with PAM reduced the therapeutic effectiveness of PAM. This observation supports the results obtained in the studies on Problem 1; i.e., pretreatment with CTX reduced the efficacy of PAM. It would appear, therefore, that in the chemotherapy of PAM-responsive tumors, PAM used as a single agent is more effective than when used as the second agent in a crossover with CTX.

Summary Answer. PAM was more effective as a chemotherapeutic agent than CTX when administered alone, i.e., without crossover. When CTX and PAM were alternated every 2 days, the chemotherapeutic effectiveness of such a continuing crossover was no greater than that of CTX when administered alone, in terms of remission induction, remission duration, life span, or the emergence of drug-resistant tumors.

Problem 3. Cytoxan (CTX) (NSC 26271) is an effective chemotherapeutic agent in breast cancer. Prolonged or chronic treatment with CTX results not only in the induction of remissions but also in the eventual

Table VII
Comparison of Remission-Induction Activities and Effects on Life Span: Alternating Cytoxan and Phenylalanine Mustard Treatments in a Chronic Therapy Regimen

Group No.	Treatment	Remissions (%) Complete	Partial (50% or more)	Average duration of remissions (days)[a]	Survival time (days)[b]
I	Vehicle control	0	0	0	41.0 ± 4.9
II	CTX, day 11–ST	40	100	7.4 ± 1.6	66.1 ± 4.2
III	PAM, day 11–ST	60	100	31.0 ± 24.4	70.9 ± 12.5
IV	CTX and PAM alternated Q2D from day 11 to ST	40	100	13.6 ± 13.0	62.5 ± 8.4

[a] Duration-of-remissions calculation includes both complete and partial remissions as means ± S.D.'s.
[b] Survival time calculated as means ± S.D.'s.

regrowth of drug-resistant tumors. The phenomenon of remission induction followed by the emergence of drug-resistant tumors also holds true for phenylalanine mustard (PAM) (NSC 8806), for dibromodulcitol (DBD) (NSC 104800), and for hexamethylmelamine (HMM) (NSC 13875). Which of the three alkylating agents would be most effective in a crossover with CTX as the initial chemotherapeutic agent?

Experimental Design. The experimental mammary tumor system selected for this study was one that is responsive to all four drugs, CTX, PAM, DBD, and HMM, responsive being defined as measurable tumor regression. The tumor also becomes resistant to all four compounds during prolonged or chronic therapy.

Tumor grafts were implanted on day 0, and treatment of all experimental groups with CTX was initiated on day 12 when tumors were well established in log phase of measurable growth. Obtaining tumor-size measurements 3 times weekly permitted establishing tumor growth and regression curves for monitoring the effects of therapy. CTX therapy was continued until tumors were in maximum remission (day 27), at which time the crossover was made to one of the three alternative drugs, PAM, DBD, or HMM, on day 29. The designation of experimental groups and the treatments are summarized in Table VIII.

Results. The effectiveness of the crossover from CTX is illustrated in Fig. 15 for PAM, in Fig. 16 for DBD, and in Fig. 17 for HMM. In all three figures, solid squares depict the tumor growth curve for the vehicle-treated controls, and open circles with paralleling lines depict

Table VIII
Experimental Groups and Treatments for Crossover Studies with Alkylating Agents

Group No.	Number of animals	Treatment
I	9	Vehicle-treated control, 0.5 ml, 3 Rx/week day 12–ST
VI	9	CTX, 3 mg/kg, 3 Rx/week, day 12–ST
VII	9	CTX, 3 mg/kg, 3 Rx/week, days 12–27 PAM, 2 mg/kg, 3 Rx/week, day 29–ST
VIII	9	CTX, 3 mg/kg, 3 Rx/week, days 12–27 DBD, 60 mg/kg, 3 Rx/week, day 29–ST
IX	9	CTX, 3 mg/kg, 3 Rx/week, days 12–27 HMM, 40 mg/kg, 3 Rx/week, day 29–ST

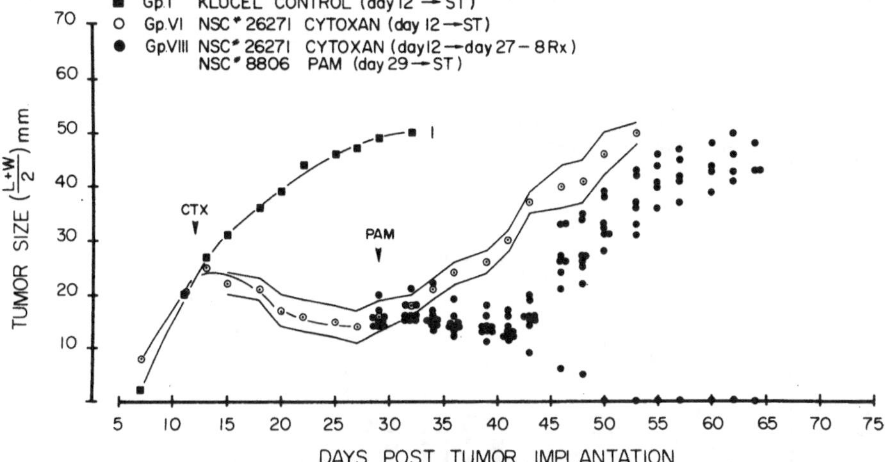

Fig. 15. Comparison of the effectiveness of a crossover from CTX to PAM.

the means plus or minus one standard deviation of the group treated only with CTX from day 12 to survival.

The solid circles in Fig. 15 depict the individual tumor sizes of those tumors in which the crossover from CTX on day 29 was made to PAM. The prolongation of remission duration is evident, as is the induction of one complete remission that persisted to day 215 as a "cure." In Fig. 16, the solid circles depict those tumors in which the crossover from CTX was made to DBD on day 29. The significantly longer duration of remissions as well as the number of complete remissions induced are evident. One complete remission persisted to day 215 as a "cure." The least

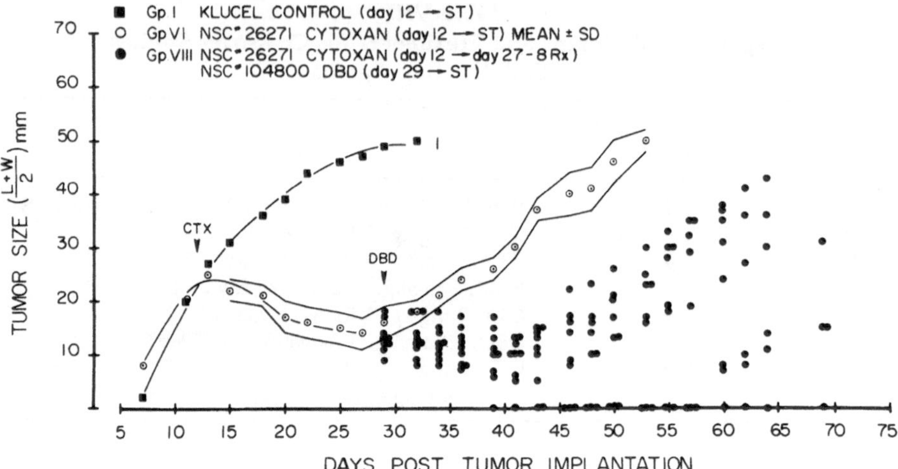

Fig. 16. Comparison of the effectiveness of a crossover from CTX to dibromodulcitol (DBD).

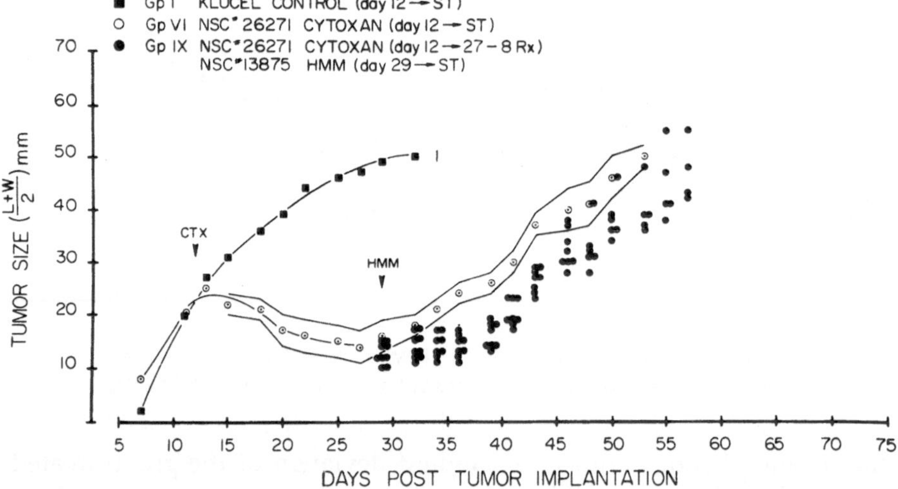

Fig. 17. Comparison of the effectiveness of a crossover from CTX to hexamethylmelamine (HMM).

effective crossover from CTX on day 29 was to HMM (Fig. 17). There was a slight prolongation of remissions, and no complete remissions were induced.

The greater effectiveness of DBD as the drug for crossover from CTX is also shown in Table IX. All tumors were put into 50% or greater

Table IX
Comparison of Remission-Induction Activities and Effects on Life Span in Crossover Studies with Alkylating Agents

Group No.	Treatment	Remissions (%)[a] Complete	Partial (50% or more)	Average duration of remissions (days)[b]	Survival time (days)[c]
I	Vehicle control	0	0	0	33.7 ± 6.6
II	CTX, day 12–ST	0	55	10.8 ± 2.2	50.8 ± 6.4
III	CTX, days 12–27 PAM, day 29–ST	11	44	13.3 ± 9.0	59.1 ± 9.1
IV	CTX, days 12–27 DBD, day 29–ST	33	100	17.1 ± 8.3	71.6 ± 15.0
V	CTX, days 12–27 HMM, day 29–ST	0	66	15.6 ± 0.9	55.1 ± 8.0

[a] Groups III and IV both had 11% "cures."
[b] Duration-of-remissions calculation includes both complete and partial remissions, as means ± S.D.'s.
[c] Survival time calculated as means ± S.D.'s.

remission, 33% of which were put into complete remission resulting in 11% "cures." Extension of survival was also greatest with a Test/Control (T/C) of 212%. The crossover to PAM was next most effective, with 11% complete remissions and 11% "cures." Extension of survival had a T/C of 175%. The least effective crossover was to HMM. Although 66% of the tumors were put into partial remission, there were no complete remissions or "cures," and extension of survival had a T/C of 163%. Treatment with CTX alone, i.e., without crossover, induced 55% partial remissions and a survival extension T/C of 151%.

Summary Answer. The most effective of three alkylating agents for a crossover from CTX, when tumors were in maximum remission, was DBD. PAM was the next most effective agent, and HMM was the least effective.

The clinical questions addressed by the preceding three studies reflect a simple application of the TSMT model in providing specific leads for the cancer chemotherapist. They also reflect the input on clinical problems by members of the Treatment Committee of the Breast Cancer Task Force. If experimental mammary tumor models are to be used effectively in a predictive mode, a close liaison between the clinician and animal experimentalist is mandatory.

In extrapolating such experimental results to the clinic, the physician should be aware of the limitations of the tumor system used. The 13762 mammary adenocarcinoma, for example, can model only for those tumors that show a significant degree of responsiveness to the particular alkylating agents studied, as well as for those tumors that also become drug-resistant following prolonged, chronic therapy.

10. Human Mammary Tumor Xenograft/Nude Athymic Mouse Model

The need for more realistic and clinically extrapolative models for breast cancer research, coincident with demonstration of the athymic condition of the nude mouse by Pantelouris,[98] has focused attention on the human tumor xenograft as a potential candidate for the ideal predictive animal test system. This is a natural assumption, at this point in time, as what could be a better model of human cancer than a human tumor?

The homozygous nude (*nu*/*nu*) mouse is characterized by a dysgenetic thymus and therefore has a defect in the thymus-dependent immunological functions.[99,100] As a consequence, it readily accepts both allo- and xenotransplants. Transplantation of human tumors into the nude mouse has become a generally accepted method for the short- and long-term observation of human tumor behavior and the effects of various treatments.

Establishment in serial transplantation of a number of different human neoplasms, including those arising in the breast, has now been well demonstrated.[101–107] Reinvestigating growth patterns and chromosome constitutions of human malignant tumors after long-term serial transplantation in nude mice, Povlsen *et al.*[108] found that after 27–56 passages over a period of 3½ to 5½ years, two adenocarcinomas of the colon, two malignant melanomas, and one Burkitt's lymphoma retained the cytological and histological appearance of the original inoculated human material, indicating a certain stability in the human tumor/nude athymic mouse model.

However, the nude athymic mouse host is not altogether passive to an implantation of a human tumor xenograft. In contrast to the observations of Povlsen,[108] studies in our laboratory[109] revealed an instability in the histology and chemotherapy responsiveness of two rat mammary adenocarcinomas after 20 transplant generations as xenografts in the nude athymic mouse. Clearly definable acinar structures were markedly reduced in the R3230AC tumor and disappeared in the 13762 tumor. Unresponsiveness to PAM, characteristic of the R3230AC tumor, remained stable, but attempts to reestablish progressive growth in previously syngeneic hosts were unsuccessful. Responsiveness to PAM, a characteristic of the 13762 tumor marked by oncolysis and many complete remissions in syngeneic hosts, was significantly reduced, with no complete remissions after the tenth passage. This type of instability suggests an immunogenetic selective process that may be eliminating the more antigenic cellular clones and possibly the process of hybridization.

Although the nude mouse reportedly lacks the classic cell-mediated rejection mechanisms and has therefore been widely used as an immunologically neutral host in a variety of investigations in tumor immunity and carcinogenesis,[110] this formerly clear-cut distinction has recently been qualified.[111] Human tumors do not always grow when implanted, though the xenograft may not be rejected.[112] While IgG and IgA responses are markedly diminished in the absence of helper T cells, nude athymic mice have a reasonably good IgM antibody response.[110,113] The studies of Ramseier[114] show what appear to be T-cell precursors in spleens of nude mice that can be activated to recognize alloantigens specifically, and Loor[115] demonstrated θ-positive cells in nude mice born from a homozygous *nu/nu* mother.

It should not be surprising, therefore, that all human tumors do not become established in serial transplantation as xenografts, nor do established human tumors produce 100% "takes" (progressive lethal growths) in every transplant generation. Table X illustrates the percentage takes of three human breast tumors that have been established in serial transplantation as xenografts. The lack of 100% predictability of lethal "takes" of subcutaneously implanted xenografts requires that larger numbers of animals be used to ensure statistical validity of assay results. Large numbers of animals per assay, and long-term assays necessitated by the low growth fraction of most human tumor xenograft systems, creates a financial problem of some significance. As of this writing, the cost of a nude athymic mouse is approximately $12 as compared with $0.70 for the normal, haired, immunocompetent animal. In

Table X
"Takes" of Human Breast Tumor Xenografts Established in Serial Transplantation in Nude Athymic Mice

Tumor designation[a]	Origin and histological type	Number of subcutaneous passages in nudes	Takes (%)[b]
Clouser (MX-1)	Breast carcinoma	14	90
Van Natter	Breast carcinoma	7	82
Keilty (MX-2) (MDA-MB-231)	Breast carcinoma	10	83

[a] Data kindly provided by Dr. B. Giovanella.
[b] "Takes": progressive and eventually lethal growth following subcutaneous implantation.

addition, nude athymics require sophisticated isolation facilities for long-term holding.[99,116] Therefore, the initial as well as the maintenance costs of the nude athymic animal are limiting factors in the use of the tumor xenograft for routine preclinical drug testing. There is little question that the evaluation of organ-specific human tumor xenografts as preclinical test systems for drug and therapy evaluation will require time and prove to be expensive, but the potential payoff certainly warrants the effort. If we are to strive for the ideal predictive model for the human disease, then the availability of the human tumor xenograft, albeit in the mouse host, dictates that it be evaluated as a potential candidate. The use of the athymic (nude) mouse in cancer research was recently reviewed at a symposium dealing with this subject.[117]

The work in our laboratory utilizing the nude athymic mouse and human tumor xenografts has been directed to the development of a rapid drug screening method. Our object in developing such a method has been (1) to circumvent the variable of immunocompetency of the xenograft host; (2) to use fewer animals and less time, significantly reducing costs per assay; and (3) at a minimum, to provide a realistic screen for quickly ranking compounds in order of activity for subsequent in-depth testing. The ability to rank compounds in order of activity would, in essence, accelerate those compounds with greatest clinical potential to advanced testing and phase I clinical trials.

The assay that has been developed employs the subrenal capsule implantation site, which provides a rich, vascular bed for nutrient as well as drug delivery, and permits easy visualization of the xenograft *in situ*. A 1-mm^3-size fragment of tumor tissue is selected as the initial graft size because it is small enough to permit permeability of essential nutrients as well as a relatively unrestricted distribution of test drugs throughout the tissue fragment, and yet is large enough to be easily visualized and measured *in situ* under a dissecting microscope. The ability to accurately establish an initial graft size *in situ* permits evaluation of oncolytic effects as well as tumor-growth inhibition.

The assay time frame of 11 days is long enough to permit measurable growth or drug-induced regression of most human tumor xenografts implanted under the renal capsule. Of equal importance, the assay time frame is short enough to evade the complications of an immune response.

To be of value, the parameters of an assay must have a range in units of measurement great enough to permit quantitation. Xenograft measurements *in situ* are made with a dissecting microscope containing an ocular micrometer. Therefore, a 1-mm^3 graft is equivalent to 10 ocular micrometer units (omu), calculated by taking the average of two diameters. The transparency of the renal capsule and clear margination of the tumor through the renal capsule thus permit accurate measurements in terms of ocular micrometer units.

The subrenal capsule assay was designed as a rapid chemotherapy screen employing low growth fraction tumors. This is an important consideration for realistic prediction of clinical activity, since it is the low growth fraction tumors that provide the greatest challenge for chemotherapy. The use of such tumors in an 11-day assay was made possible by simply magnifying the xenograft *in situ* so that 1- or 2-mm changes in size, not measurable by the vernier calipers used for measuring subcutaneously growing tumors, are accurately quantifiable into 10–20 omu. Measurement precision of the initial as well as the *final* graft size *in situ* permits an accurate determination of Δ tumor size, with each tumor-bearing animal essentially serving as its own control. Additional untreated, xenografted control animals indicate the Δ tumor size to be expected from the donor tissue, thereby serving as a tissue quality control.

Examples of assay results obtainable with human mammary tumor xenografts in the "subrenal capsule assay" are summarized in Tables XI and XII. These tumors have been well established in serial transplantation in the nude athymic animal. The number of animals used per compound is noted in each table. Our analysis of assay data that have accumulated, and experience with the assay *per se*, indicate that 3 animals per treated group with 6 in the control group are sufficient for drug evaluation. If necessary, 2 animals per treated group and 3 in the control group will provide comparable data with an acceptable degree of reliability as a screening method for ranking drug activities.

Table XI
Responsiveness of the Clouser Human Breast Carcinoma (MX-1) in a Rapid Chemotherapy Screen[a]

NSC No.	Name	Dose (mg/kg per inj.)[b]	Tumor size (omu)[c] Initial	Final	Δ Tumor size (day 11 − day 0)	% T/C Δ tumor size
19893	5-FU	25	10.0	27.5	17.5	103
13875	HMM (low)	15	11.0	26.5	15.5	91
740	MTX	2	10.0	23.0	13.0	76
125066	BLEO	6.4	10.5	16.5	6.0	35
123127	ADR	1	11.0	13.0	2.0	12
104800	DBD	120	10.5	11.0	0.5	3
13875	HMM (high)	88	11.5	11.5	0	0
8806	L-PAM	3	7.5	0	− 7.5	0
79037	CCNU	8	8.5	0	− 8.5	0
95441	MeCCNU	10	9.5	0	− 9.5	0
26271	CTX	25	13.0	0	− 13.0	0
Saline control			10.0	25.0	15.0 } 17.0	
Saline control			11.5	30.5	19.0 }	

[a] Drugs prepared fresh daily, administered intraperitoneally.
[b] Every day on days 1–10; 1 mouse/compound, 2 mice/control group.
[c] (omu) Ocular micrometer units.

Table XII
Responsiveness of Human Breast Carcinoma SW613 in a Rapid Chemotherapy Screen[a]

NSC No.	Name	Dose (mg/kg per inj.)[b]	Tumor size (omu)[c] Initial	Final	Δ Tumor size (day 12 − day 0)	% T/C Δ tumor size
13875	HMM (low)	15	10.0	24.0	14.0	100
740	MTX	2	11.0	23.0	12.0	86
104800	DBD	120	10.5	19.0	8.5	59
26271	CTX	25	10.0	18.0	8.0	57
123127	ADR	1	11.0	18.0	7.0	50
125066	BLEO	6.4	10.0	16.5	6.5	46
13875	HMM (high)	88	10.0	16.0	6.0	43
8806	L-PAM	3	9.5	15.0	5.5	38
79037	CCNU	8	10.0	12.5	2.5	18
95441	MeCCNU	10	11.0	12.0	1.0	7
19893	5-FU	25	11.0	11.0	0	0
Saline control			10.7	24.7	14.0	

[a] Drugs prepared fresh daily, administered subcutaneously.
[b] Every day on days 1–11; 1 mouse/compound, 2 mice/control group.
[c] (omu) Ocular micrometer units.

In the Clouser breast tumor (Table XI), also designated as MX-1, 5-fluorouracil (5-FU) was completely inactive and CTX was the most active. The difference in antitumor activity between the low and high doses of HMM was also evident. In this tumor system, the low doses of HMM and MTX were inactive, and L-PAM, 1-(2-chloroethyl)-3-cyclohexyl-1-nitrosourea (CCNU), methyl-CCNU, and CTX were actively oncolytic. The ranking of the compounds is clear-cut and is illustrated in Fig. 18. In contrast are the test results obtained with the human breast carcinoma SW613 summarized in Table XII. Although a spectrum of responses was obtained, as well as a clearly defined ranking of compounds and grouping of activity (Fig. 19), it is evident that the SW613 breast carcinoma is not as responsive to agents such as L-PAM and CTX, but is responsive to 5-FU, in contrast to the Clouser breast tumor.

The difference of chemotherapy responsiveness between these two tumor systems is suggestive of the variability encountered clinically. We have found significant differences in the growth rates and responsiveness to a spectrum of chemotherapeutic agents among the human colon adenocarcinomas as well as among the human breast carcinomas.[118] Therefore, one questions the validity of selecting one tumor over another for predicting general clinical response rather than establishing a block of tumor systems reflecting a spectrum of response characteristics. Limiting the human tumor xenograft/nude athymic mouse model to one or two organ-specific systems will be just as restrictive and predict no more realistically than selecting one or two transplantable syngeneic rodent tumor systems.

That the subrenal capsule assay might be effective for predicting the drug sensitivities of *individual* tumors, removed as primary explants at surgery or biopsy, is indicated by the results summarized in Table XIII. The six drugs were tested against primary explants prepared directly from a lung metastasis removed at surgery. DBD was inactive and HMM and L-PAM were the two most active agents, with bleomycin (BLEO) and adriamycin (ADR) showing intermediate activity. Encouragingly, within the 14-day test period, there was measurable growth of the control tissue under the renal capsule, which permitted a ranking of compounds by relative activity (Fig. 20). Studies directed to determining the feasibility of the method for predicting individual tumor response are continuing.

Human tumors established in serial transplantation as xenografts in the athymic (nude) mouse have a growth potential that is similar to that of the "low growth fraction" transplantable, syngeneic tumor systems. That fact, plus the ability to accurately quantify small changes in tumor size occurring over a short time span, coupled with the knowledge that the normal rejection reaction to foreign tissue requires a week to 10 days

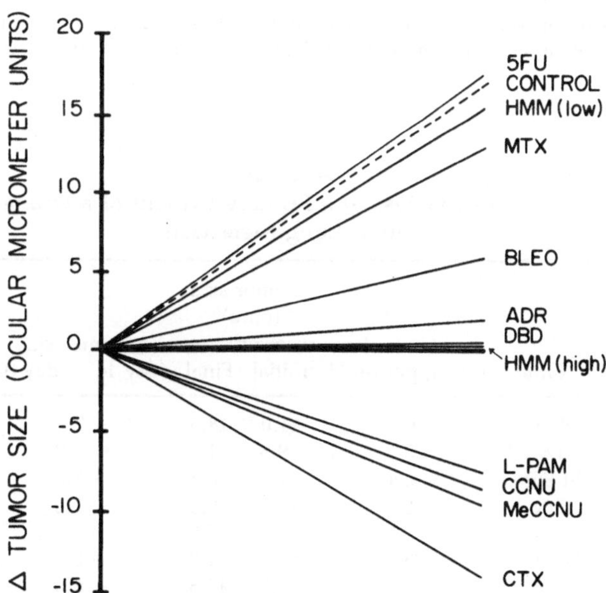

Fig. 18. Ranking of compounds by activity in the Clouser human breast carcinoma model (subrenal capsule assay). Δ Tumor size: tumor size on day 11 − tumor size on day 0.

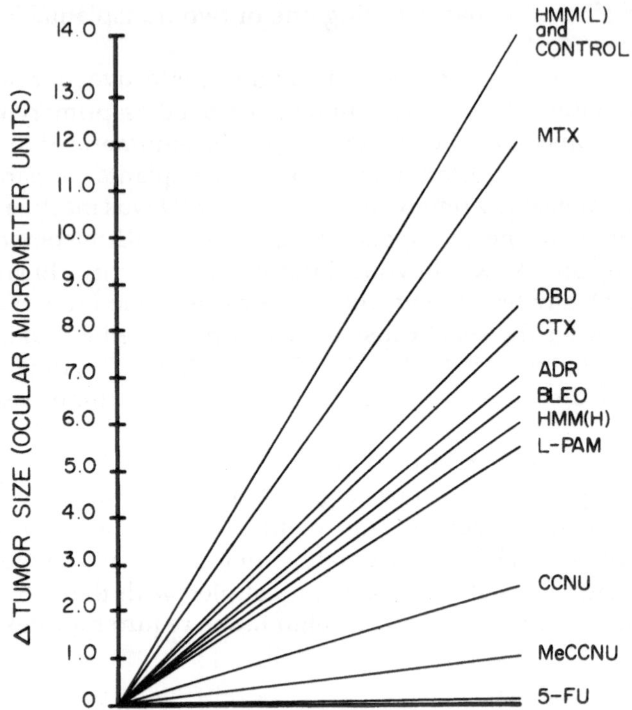

Fig. 19. Ranking of compounds by activity in the SW613 human breast carcinoma model (subrenal capsule assay). Δ Tumor size: tumor size on day 11 − tumor size on day 0.

Table XIII
Chemotherapy Responsiveness of a Primary Explant of a Human Mammary Tumor Lung Metastasis

NSC No.	Name	Dose (mg/kg per inj.)[a]	Tumor size (omu)[b] Initial	Tumor size (omu)[b] Final	Δ Tumor size (day 14 − day 0)	% T/C Δ tumor size
104800	DBD	120	11.0	18.5	+ 7.5	125
26271	CTX	25	9.5	11.5	+ 2.0	33
125066	BLEO	6.4	9.5	8.5	− 1.0	0
123127	ADR	1	11.5	9.5	− 2.0	0
8806	L-PAM	3	12.5	5.0	− 7.5	0
13875	HMM	88	12.0	4.0	− 8.0	0
Saline control			11.5	17.5	+ 6.0	

[a] Every day on days 1–10; 1 mouse/compound, 2 mice/control group.
[b] (omu) Ocular micrometer units.

to develop in normal animals, prompted us to examine the feasibility of using the subrenal capsule technique for implanting human tumor xenografts in immuno-competent (nonnude) mice. BDF_1 mice were implanted with human tumor grafts, and *in situ* initial tumor-size measurements were taken. Groups of 5 mice were sacrificed at intervals after implantation, and final tumor sizes were determined. The growth of eight human tumors that have been established and carried in serial transplantation in the nude athymic mouse were thus studied: three colon tumors, CX-1, CX-2, and Squires; four mammary tumors, MX-1, MX-2, SW613, and Cook; and one lung tumor, LX-1. Five of these tumors, the growth curves of which are plotted as Δ tumor size (Fig. 21, top), exhibited progressive, quantifiable growth for the first 6 days, with regression after 10 days. Two tumors (Fig. 21, bottom), showed progressive growth for 9–10 days after implantation. One tumor, MX-2, which has a transplant interval of greater than 30 days in *nu/nu* mice, grew too slowly to be usable before the rejection developed. Preliminary results of testing chemotherapeutic agents against human tumor xeno-

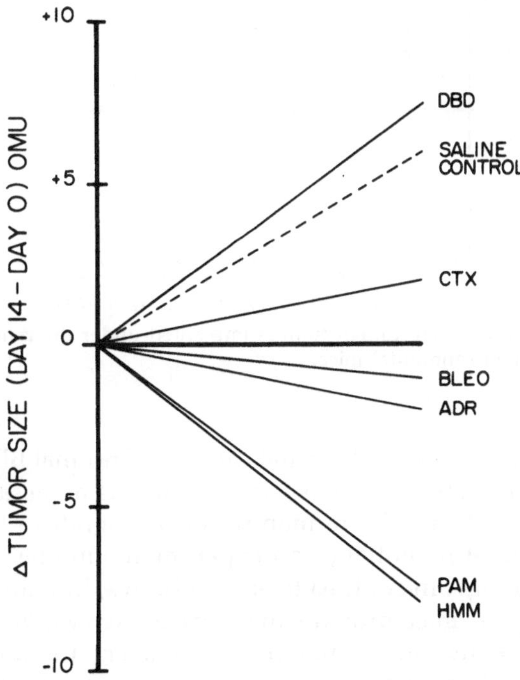

Fig. 20. Predicting individual tumor response. Ranking of compounds by activity in the subrenal capsule assay with primary explants of a mammary tumor lung metastasis obtained at surgery. Δ Tumor size: tumor size on day 14 − tumor size on day 0.

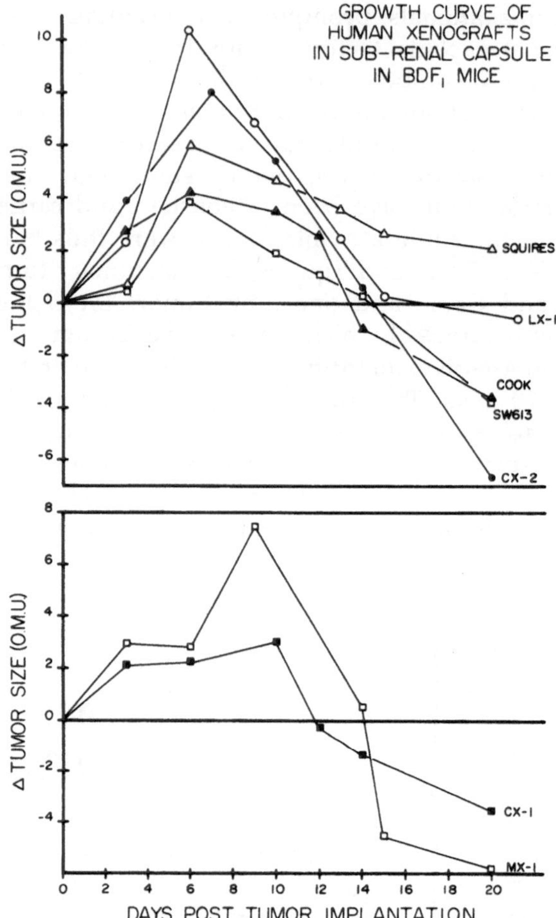

Fig. 21. Growth of seven human tumors implanted under the renal capsule of normal, immunocompetent (nonnude) mice.

grafts, implanted under the renal capsule of normal BDF_1 mice, indicate that the 6-day time frame has practicability as an assay system.

It is evident that the human tumor xenograft in the nude athymic mouse has great potential as an experimental model in both immunology and oncology. In contrast to *in vitro* assays, human tumor xenografts in nude athymic mice provide an *in vivo* assay system that more closely resembles the human situation with respect to morphology, tumor products, kinetics, and drug activation and detoxification mechanisms. It remains to be determined, however, whether organ-specific, human tumor xenografts, established in serial transplantation, prove to be more predictive of clinical activity than similar animal tumors in syngeneic hosts.

11. Conclusions

There is no one animal tumor system that can meet the requirements of serving as *the* experimental model for the complex of diseases identified as breast cancer. Unquestionably, there are animal tumor systems that ideally mimic a particular facet of breast cancer. These facets include the gamut from the preneoplastic lesion to the advanced malignancy, metastasizing and nonmetastasizing; from hormone dependency and responsiveness to the independent tumor—facets that can be studied in experimental designs that provide a tumor–host relationship that is either autochthonous or syngeneic.

Autochthonous mammary tumors, whether as multiples in the same animal, or singly, but in different animals of the same inbred strain, reflect a fundamental variability in endocrine responsiveness, growth characteristics, and immunogenicities similar to that encountered in a clinical population. The autochthonous system is therefore burdened with the same need for stringent identification and control of variables, the same need for statistically valid numbers of experimental subjects, as are required for clinical trials, a significant problem where repetitive duplication of experimental conditions is desired.

Syngeneic mammary tumor systems, on the other hand, offer relatively stable test systems that are established in serial transplantation and that are predictable as to growth and metastasizing characteristics in large populations of experimental animals. The immunogenetic relationship of tumor and host mimics the autochthonous system in terms of histocompatibility. Regardless of the size of the tumor graft recipient population used in a study, however, the syngeneic donor tumor represents, or serves as a model for, only one characteristic of a facet of breast cancer. For example, a syngeneic tumor is generally either hormone-dependent or -independent, metastasizes from subcutaneous implant sites or does not metastasize, is reproducibly responsive, only slightly responsive, or resistant to a number of chemotherapeutic agents. The point emphasized is that each syngeneic system, in contrast to the autochthonous, represents only one variable, albeit in a reproducible and predictive fashion. However, this seeming limitation of the syngeneic system can be overcome. By selecting a tumor panel, one can establish a test system in which the aggregate of individual responses, each of which is clearly defined and reproducible in a specific tumor line, represents a realistic profile or model of the facet of breast cancer under study.

Stability, however, is not an inherent characteristic of serially transplanted syngeneic tumors. Quality-control parameters for monitoring growth and metastasis must be established, especially parameters for monitoring those characteristics peculiar to a tumor as a test system. In this regard, tumor banks as a storage and distribution source for the periodic replacement of tumor lines provide a vital service to the scientific community.

Thus, each model, whether autochthonous or syngeneic, has its variables that must be recognized and adequately controlled to provide significant and predictive experimental results. It has been our modest objective in this chapter to attempt to point out some of the advantages as well as the variables inherent in the mammary tumor models available for breast cancer research. A complete understanding and appreciation of both the advantages and limitations of a desired model on the part of both the animal experimentalist and the clinician who will extrapolate the information gleaned from a model is essential.

In view of the increasing clinical importance of adjuvant chemotherapy and immunotherapy in breast cancer, the use of syngeneic tumor models in experimental designs having the objective of addressing specific clinical problems and questions has been stressed and the feasibility demonstrated.

The appearance of the human tumor xenograft in the athymic (nude) mouse on the horizon of experimental oncology, as a predictive model of cancer in man, has provided the experimentalist with a bioassay system having great potential. Whether organ-specific, transplantable human tumor xenografts prove to be more predictive of clinical activity than similar tumors in syngeneic hosts remains to be determined. That they can be used effectively as test systems for preclinical drug evaluation, and possibly for predicting individual tumor response in direct support of the clinical chemotherapist, has been indicated by examples.

ACKNOWLEDGMENTS: I wish to acknowledge the active collaboration of Dr. D. Jane Taylor of the National Cancer Institute in the studies carried out in our laboratory, and to indicate the importance of her liaison with the clinical groups of the Breast Cancer Task Force and the feedback that has resulted in the development of practicable mammary tumor models. My thanks to a very dedicated and productive professional and technical staff, with special mention to Stacie Speropoulos, who knows exactly how I like records kept and data compiled, to Wendy Grant for her illustrations, and to Carol Ann Zapustas for her patience and perseverance through the many retypings of this manuscript.

12. References

1. S. K. Carter, Immunotherapy in the strategy of cancer treatment (editorial), *Cancer Immunol. Immunother.* **I,** 115–118 (1976).
2. G. L. Bartlett, J. W. Kreider, and D. M. Purnell, Immunotherapy of cancer in animals: Models or muddles? (guest editorial), *J. Natl. Cancer Inst.* **56,** 207–210 (1976).
3. *Webster's New Collegiate Dictionary,* G&C Merriam, Springfield, Massachusetts (1974).
4. J. Huxley, *Biological Aspects of Cancer,* Harcourt, Brace, New York (1958).
5. M. E. Lippman, C. K. Osborne, R. Krazek, and N. Young, *In vitro* model systems for the study of hormone-dependent human breast cancer, *N. Engl. J. Med.* **296**(3), 154–159 (1977).
6. L. Norton, R. Simon, H. D. Brereton, and A. E. Bogden, Predicting the course of Gompertzian growth, *Nature (London)* **264,** 542–545 (1976).
7. L. Norton and R. Simon, Tumor size, sensitivity to therapy, and the design of treatment schedules, *Cancer Treatment Rep.* **61,** 1307–1317 (1977).
8. F. M. Schabel, Jr., Concepts for systemic treatment of micrometastases, *Cancer* **35,** 15–24 (1975).
9. E. Frei, III, Selected considerations regarding chemotherapy as adjuvant in cancer treatment, *Cancer Chemother. Rep.* **50,** 1–5 (1966).
10. A. K. Laird, Dynamics of growth in tumors and normal organisms, *Natl. Cancer Inst. Monogr.* **30,** 15–28 (1969).
11. B. Gompertz, On the nature of the function expressive of the law of human mortality, and on the new mode of determining the value of life contingencies, *Philos. Trans. R. Soc. London* **115,** 513–585 (1825).
12. G. P. Bodey, E. J. Freireich, E. Gehan, K. B. McCredie, Z. Rodriques, J. Gutterman, and A. Burgess, Late intensification therapy for acute leukemia in remission, *J. Am. Med. Assoc.* **235,** 1021–1025 (1976).
13. K. B. DeOme, H. A. Bern, S. Nandi, D. R. Pitelka, and L. J. Faulkin, Jr., *Genetics and Cancer,* pp. 327–348, University of Texas Press, Houston (1959).
14. P. B. Blair and K. B. DeOme, Mammary tumor development in transplanted hyperplastic alveolar nodules of the mouse, *Proc. Soc. Exp. Biol. Med.* **108,** 289–291. (1961).
15. K. B. DeOme, L. J. Faulkin, H. A. Bern, and P. B. Blair, Development of mammary tumors from hyperplastic alveolar nodules transplanted into gland-free mammary fat pads of female C3H mice, *Cancer Res.* **19,** 515–520 (1959).
16. L. J. Beuving, Mammary tumor formation within outgrowth of transplanted hyperplastic alveolar nodules from carcinogen treated rats, *J. Natl. Cancer Inst.* **40,** 1287–1289 (1968).
17. D. Sinha and T. L. Dao, A direct mechanism of mammary carcinogenesis induced by 7,12-dimethylbenz(a)anthracene, *J. Natl. Cancer Inst.* **53,** 841–846 (1974).
18. T. L. Dao, D. Sinha, S. Christakos, and R. Varela, Biochemical characterization of carcinogen-induced mammary hyperplastic alveolar nodule and tumor in the rat, *Cancer Res.* **35,** 1128–1134 (1976).
19. H. M. Jensen, J. R. Rice, and S. R. Wellings, Preneoplastic lesions in the human breast, *Science* **191,** 295–297 (1976).
20. D. E. Bostock, The prognosis following the surgical excision of canine mammary neoplasms, *Eur. J. Cancer* **11,** 389–396 (1975).
21. R. M. Mulligan, Comparative pathology of human and canine cancer, *Ann. N. Y. Acad. Sci.* **108,** 642–690 (1963).
22. M. M. Mason, Canine mammary tumors: A review, prepared for the Endocrinology Evaluation Branch, NCI, under Contract No. PH43-65-6 (May 1967).

23. W. Misdorp, E. Cotchin, J. F. Hampe, A. G. Jabara, and J. Sandersleben, Canine malignant mammary tumors, *Vet. Pathol.* **10,** 241–256 (1973).
24. J. D. Strandberg and D. G. Goodman, Animal model of human disease: Breast cancer, *Am. J. Pathol.* **75,** 225–228 (1974).
25. A. C. Anderson, Parameters of mammary gland tumors in aging beagles, *J. Am. Vet. Assoc.* **147,** 1653–1654 (1965).
26. G. N. Taylor, L. Shabestari, J. Williams, C. W. Mays, W. Angus, and S. McFarland, Mammary neoplasia in a closed beagle colony, *Cancer Res.* **36,** 2740–2743 (1976).
27. C. Huggins, G. Briziarelli, and H. Sutton, Jr., Rapid induction of mammary carcinoma in the rat and the influence of hormones on the tumors, *J. Exp. Med.* **109,** 25–43 (1959).
28. J. Maisin and M. L. Coolen, Au sujet du pouvoir cancerigène du methylcholanthrene, *C. R. Soc. Biol.* **123,** 159–160 (1936).
29. J. V. Proshaska, A. Brunschweig, and H. Wilson, Oral administration of methylcholanthrene to mice, *Arch. Surg.* **38,** 328–333 (1939).
30. J. W. Orr, Mammary carcinoma in mice following the intranasal administration of methylcholanthrene, *J. Pathol. Bacteriol.* **55,** 483–488 (1943).
31. C. Huggins, L. G. Grand, and F. P. Brillantes, Mammary cancer induced by a single feeding of polynuclear hydrocarbons and its suppression, *Nature (London)* **189,** 204–207 (1961).
32. P. M. Daniel and M. M. Pritchard, The response of experimentally induced mammary tumors in rats to hypophysectomy and to pituitary stalk section, *Br. J. Cancer* **17,** 446–453 (1963).
33. M. N. Teller, C. C. Stock, G. Stohr, P. C. Merker, R. J. Kaufman, G. C. Escher, and M. Bowie, Biological characteristics and chemotherapy of 7,12-dimethylbenz(a)anthracene-induced tumors in rats, *Cancer Res.* **26,** 245–252 (1966).
34. D. P. Griswold, H. E. Skipper, W. R. Laster, Jr., W. S. Wilcox, and F. M. Schabel, Jr., Induced mammary carcinoma in the female rat as a drug evaluation system, *Cancer Res.* **26,** 2169–2180 (1966).
35. K. D. Schulz, B. Haselmeier, and F. Holzel, The influence of clomid and its isomers upon dimethylbenzanthracene-induced rat mammary tumors, *Acta Endocrinol. (Copenhagen). Suppl.* **138,** 236 (1969).
36. L. Terenius, Anti-oestrogens and breast cancer, *Eur. J. Cancer* **7,** 57–64 (1971).
37. W. L. McGuire, Current status of estrogen receptors in human breast cancer, *Cancer* **36,** 638–644 (1975).
38. K. B. Horwitz and W. L. McGuire, Specific progesterone receptors in human breast cancer, *Steroids* **25,** 497–505 (1975).
39. M. F. Pichon and E. Milgrom, Characterization and assay of progesterone receptor in human mammary carcinoma, *Cancer Res.* **37,** 464–471 (1977).
40. R. I. Nicholson and M. P. Golder, The effect of synthetic anti-oestrogens on the growth and biochemistry of rat mammary tumors, *Eur. J. Cancer* **11,** 571–579 (1975).
41. E. V. Jensen, H. I. Jacobsen, S. Smith, P. W. Jungblut, and E. R. DeSombre, The use of estrogen antagonists in hormone receptor studies, *Gynecol. Invest.* **3,** 108–123 (1972).
42. C. W. Welsch and H. Nagasawa, Prolactin and murine mammary tumorigenesis: A review, *Cancer Res.* **37,** 951–963 (1977).
43. H. Nasr and O. H. Pearson, Inhibition of prolactin secretion by ergot alkaloids, *Acta Endocrinol.* **80,** 429–443 (1975).
44. A. Manni, J. E. Trujillo, and O. H. Pearson, Predominant role of prolactin in stimulating the growth of 7,12-dimethylbenz(a)anthracene-induced rat mammary tumor, *Cancer Res.* **37,** 1216–1219 (1977).

45. S. K. Quadri, G. S. Kledzik, and J. Meites, Counteraction by prolactin of androgen-induced inhibition of mammary tumor growth in rats, *J. Natl. Cancer Inst.* **52,** 875–878 (1974).
46. M. E. Costlow, R. A. Buschow, and W. L. McGuire, Prolactin receptors and androgen-induced regression of 7,12-dimethylbenz(a)anthracene-induced mammary carcinoma, *Cancer Res.* **36,** 3324–3329 (1976).
47. C. Huggins, R. C. Moon, and S. Morii, Extinction of experimental mammary cancer. I. Estradiol-17β and progesterone, *Proc. Natl. Acad. Sci. U.S.A.* **48,** 379–386 (1962).
48. C. J. Shellabarger and V. A. Soo, Effects of neonatally administered sex steroids on 7,12-dimethylbenz(a)anthracene induced mammary neoplasia in rats, *Cancer Res.* **33,** 1567–1569 (1973).
49. E. Heise and M. Gorlich, Growth and therapy of mammary tumors induced by 7,12-dimethylbenzanthracene in rats, *Br. J. Cancer* **20,** 539–545 (1966).
50. J. Kim, J. Furth, and K. Yannopoulos, Observations on hormonal control of mammary cancer. I. Estrogen and mammatropes, *J. Natl. Cancer Inst.* **31,** 233–259 (1963).
51. M. N. Teller, R. J. Kaufmann, M. Bowie, and C. C. Stock, Influence of estrogens and endocrine ablation on duration of remission produced by ovariectomy or androgen treatment of 7,12-dimethylbenz(a)anthracene-induced rat mammary tumors, *Cancer Res.* **29,** 349–352 (1969).
52. M. N. Teller, C. C. Stock, G. Stohr, P. C. Merker, R. J. Kaufmann, G. C. Escher, and M. Bowie, Biologic characteristics and chemotherapy of 7,12-dimethylbenz(a)anthracene-induced tumors in rats, *Cancer Res.* **26,** 245–252 (1966).
53. S. Young, D. Cowan, and L. Sutherland, The histology of induced mammary tumors in rats, *J. Pathol. Bacteriol.* **85,** 331–340 (1963).
54. P. M. Daniel and M. M. L. Prichard, The response of experimentally induced mammary tumors in rats to ovariectomy, *Br. J. Cancer* **17,** 687–690 (1964).
55. P. M. Daniel and M. M. L. Prichard, Further studies on mammary tumors induced in rats by 7,12-dimethylbenz(a)anthracene (DMBA), *Int. J. Cancer* **2,** 163–177 (1967).
56. D. P. Griswold, Jr., and C. H. Green, Observations on the hormone sensitivity of 7,12-dimethylbenz(a)anthracene-induced mammary tumors in the Sprague–Dawley rat, *Cancer Res.* **30,** 819–826 (1970).
57. R. L. Noble, B. Hochachka, and D. King, Spontaneous and estrogen-produced tumors in Nb rats and their behavior after transplantation, *Cancer Res.* **35,** 766–780 (1975).
58. R. L. Noble and L. Hoover, A classification of transplantable tumors in Nb rats controlled by estrogen from dormancy to autonomy, *Cancer Res.* **35,** 2935–2941 (1975).
59. M. Sluyser and R. VanNie, Estrogen receptor content and hormone responsive growth of mouse mammary tumors, *Cancer Res.* **34,** 3253–3257 (1974).
60. H. A. Bern and S. Nandi, Recent studies of the hormonal influence in mouse mammary tumorigenesis, *Prog. Exp. Tumor Res.* **2,** 90–134 (1961).
61. O. Mühlbock and L. M. Boot, Induction of mammary cancer in mice without the mammary tumor agent by isografts of hypophysis, *Cancer Res.* **19,** 402–412 (1959).
62. N. Haran-Ghera, The role of mammotrophin in mammary tumor induction in mice, *Cancer Res.* **21,** 790–795 (1961).
63. C. W. Welsch and C. Gribler, Prophylaxis of spontaneously developing mammary carcinoma in C3H/HeJ female mice by suppression of prolactin, *Cancer Res.* **33,** 2939–2946 (1973).
64. H. Nagasawa, K. Kuretani, and F. Kanzawa, Effect of prolactin on the growth of spontaneous mammary tumor in mice, *Gann* **57,** 637–640 (1966).
65. H. Nagasawa, R. Yanai, H. Iwahashi, M. Fujimoto, and K. Kuretani, Effect of pituitary isografts on the growth of spontaneous mammary tumor in mice, *Gann* **58,** 337–342 (1967).

66. A. E. Bogden, D. J. Taylor, E. Y. H. Kuo, M. M. Mason, and A. Speropoulos, The effect of perphenazine-induced serum prolactin response on estrogen-primed mammary tumor–host systems, 13762 and R-35 mammary adenocarcinomas, *Cancer Res.* **34,** 3018–3025 (1974).
67. R. Hilf, C. Bell, H. Goldenberg, and I. Michel, Effect of fluphenazine HCl on R3230AC mammary carcinoma and mammary glands of the rat, *Cancer Res.* **31,** 1111–1117 (1971).
68. A. Segaloff, Hormones in breast cancer, *Recent Prog. Horm. Res.* **22,** 351–374 (1966).
69. U. Kim and J. Furth, Relation of mammotropes to mammary tumors. IV. Development of highly hormone dependent mammary tumors, *Proc. Soc. Exp. Biol. Med.* **105,** 490–492 (1960).
70. E. J. Diamond, S. Koprak, S. K. Shen, and V. P. Hollander, The conversion of an ovariectomy-nonresponsive to an ovariectomy-responsive mammary tumor strain, *Cancer Res.* **36,** 77–80 (1976).
71. B. L. Powel, E. J. Diamond, S. Koprak, and V. P. Hollander, Prolactin binding in ovariectomy-responsive and ovariectomy-nonresponsive rat mammary carcinoma, *Cancer Res.* **37,** 1328–1332 (1977).
72. U. Kim and M. J. Depowski, Progression from hormone dependence to autonomy in mammary tumors as an *in vivo* manifestation of sequential clonal selection, *Cancer Res.* **35,** 2068–2077 (1975).
73. U. Kim, J. Furth, and K. H. Clifton, Relation of mammary tumors to mammotropes. III. Hormone responsiveness of transplanted mammary tumors, *Proc. Soc. Exp. Biol. Med.* **103,** 646–650 (1960).
74. M. E. Costlow, R. A. Buschow, N. J. Richert, and W. L. McGuire, Prolactin and estrogen binding in transplantable hormone-dependent and autonomous rat mammary carcinoma, *Cancer Res.* **35,** 970–974 (1975).
75. P. M. Gullino, F. H. Grantham, I. Losonczy, and B. Berghoffer, Mammary tumor regression. I. Physiopathologic characteristics of hormone-dependent tissue, *J. Natl. Cancer Inst.* **49,** 1333–1348 (1972).
76. P. M. Gullino and R. H. Lanzerotti, Mammary tumor regression. II. Autophagy of neoplastic cells, *J. Natl. Cancer Inst.* **49,** 1349–1356 (1972).
77. P. M. Gullino, F. H. Grantham, I. Losonczy, and B. Berghoffer, Mammary tumor regression. III. Uptake and loss of substrates by regressing tumors, *J. Natl. Cancer Inst.* **49,** 1675–1684 (1972).
78. R. H. Lanzerotti and P. M. Gullino, Activities and quantities of lysosomal enzymes during mammary tumor regression, *Cancer Res.* **32,** 2679–2685 (1972).
79. *Dorland's Illustrated Medical Dictionary,* 25th Ed. W. B. Saunders, Philadelphia (1974).
80. M. J. Brennan, Murine and rat mammary tumors as models for the immunological study of human breast cancer, *Cancer Res.* **36,** 728–733 (1976).
81. G. H. Heppner and G. Pierce, *In vitro* demonstration of tumor specific antigens in spontaneous mammary tumors of mice, *Int. J. Cancer* **4,** 212–218 (1969).
82. U. Kim, Metastasizing mammary carcinoma in rats: Induction and study of their immunogenicity, *Science* **167,** 72–74 (1970).
83. S. Nandi and C. M. McGrath, Mammary neoplasia in mice, *Adv. Cancer Res.* **17,** 353–414 (1973).
84. S. Nandi, Interaction among hormonal, viral and genetic factors in mouse mammary tumorigenesis, *Can. Cancer Conf.* **6,** 69–81 (1966).
85. D. S. Martin, R. A. Fugmann, R. L. Stolfi, and P. E. Hayworth, Solid tumor animal model therapeutically predictive for human breast cancer, *Cancer Chemother. Rep.* **5**(2), 89–109 (1975).

86. D. S. Martin, P. E. Hayworth, and R. A. Fugmann, Enhanced cure of spontaneous murine mammary tumor with surgery, combination chemotherapy, and immunotherapy, *Cancer Res.* **30,** 709–716 (1970).
87. R. A. Fugmann, D. S. Martin, P. E. Hayworth, and R. L. Stolfi, Enhanced cures of spontaneous murine mammary carcinomas with surgery and five-compound combination chemotherapy, and their immunotherapeutic interrelationship, *Cancer Res.* **30,** 1931–1936 (1970).
88. R. A. Fugmann, R. L. Stolfi, P. E. Hayworth, and D. S. Martin, Immunologic and chemotherapeutic parameters in a model breast tumor system, *Cancer Chemother. Rep.* **4**(2), 25–32 (1974).
89. J. G. Anderson, R. A. Fugmann, R. L. Stolfi, and D. S. Martin, Metastatic incidence of a spontaneous murine mammary adenocarcinoma, *Cancer Res.* **34,** 1916–1920 (1974).
90. M. Hosokawa, F. Orsini, and E. Mihich, Fast- and slow-growing transplantable tumors derived from spontaneous mammary tumors of the DBA/2 Ha-DD mouse, *Cancer Res.* **35,** 2657–2662 (1975).
91. A. E. Bogden, P. M. Haskell, W. R. Cobb, and D. E. Kelton, Heterogeneity in chemotherapy responsiveness of the solid 13762 rat mammary adenocarcinoma and two derived ascites tumor lines, *Proc. Am. Assoc. Cancer Res.* **17,** 40 (1976).
92. U. Kim, A. Baumler, C. Carruthers, and K. Bielat, Immunological escape mechanism in spontaneously metastasizing mammary tumors, *Proc. Natl. Acad. Sci. U.S.A.* **72,** 1012–1016 (1975).
93. A. E. Bogden, H. J. Esber, D. J. Taylor, and J. H. Gray, Comparative study on the effects of surgery, chemotherapy, and immunotherapy, alone and in combination, on metastases of the 13762 mammary adenocarcinoma, *Cancer Res.* **34,** 1627–1631 (1974).
94. J. W. Kreider, G. L. Bartlett, and D. M. Purnell, Suitability of rat mammary adenocarcinoma 13762 as a model for BCG immunotherapy, *J. Natl. Cancer Inst.* **56,** 797–802 (1976).
95. F. C. Sparks, T. X. O'Connell, and Y.-T. N. Lee, Adjuvant preoperative and postoperative immunochemotherapy for mammary adenocarcinoma in rats, *Surg. Forum* **24,** 118–121 (1973).
96. F. C. Sparks, T. X. O'Connell, Y.-T. N. Lee, and J. H. Breeding, Brief communication: BCG therapy given as an adjuvant to surgery: Prevention of death from metastases from mammary adenocarcinoma in rats, *J. Natl. Cancer Inst.* **53,** 1825–1826 (1974).
97. A. E. Bogden and D. J. Taylor, Predictive mammary tumor test systems for experimental chemotherapy, in: *Breast Cancer: Trends In Research and Treatment* (J. C. Heuson, W. H. Mattheiem, and M. Rozencweig, eds.), pp. 95–110, Raven Press, New York (1976).
98. E. M. Pantelouris, Absence of thymus in a mouse mutant, *Nature (London)* **217,** 370–371 (1968).
99. J. Rygaard and C. O. Povlsen (eds.), *Proceedings of the First International Workshop on Nude Mice,* Gustav Fischer Verlag, Stuttgart, 301 pp. (1974).
100. H. H. Wortis, Immunological studies of nude mice, *Contemp. Top. Immunobiol.* **3,** 243–263 (1974).
101. J. Rygaard and C. O. Povlsen, Heterotransplantation of a human malignant tumor to "nude" mice, *Acta Pathol. Microbiol. Scand.* **77,** 758–760 (1969).
102. C. O. Povlsen and J. Rygaard, Heterotransplantation of human adenocarcinomas of the colon and rectum to the mouse mutant "nude": A study of nine consecutive transplantations, *Acta Pathol. Microbiol. Scand.* **79,** 159–169 (1971).
103. C. O. Povlsen and J. Rygaard, Heterotransplantation of human epidermoid carcinoma to the mouse mutant "nude," *Acta Pathol. Microbiol. Scand.* **80,** 713–717 (1972).

104. B. C. Giovanella, S. O. Yim, A. C. Morgan, J. S. Stehlin, and L. J. Williams, Brief communication: Metastasis of human melanomas transplanted in "nude" mice, *J. Natl. Cancer Inst.* **50,** 1051–1053 (1973).
105. B. C. Giovanella, A. C. Morgan, J. S. Stehlin, L. J. Williams, and D. M. Mumford, Development of invasive tumors in "nude" thymusless mice injected with human cells cultured from Burkitt lymphomas, *Proc. Am. Assoc. Cancer Res.* **14,** 20 (1973).
106. H. C. Outzen and R. P. Custer, Brief communication: Growth of human normal and neoplastic mammary tissue in the cleared mammary fat pad of the nude mouse, *J. Natl. Cancer Inst.* **55,** 1461–1463 (1975).
107. B. C. Giovanella, J. S. Stehlin, and L. J. Williams, Heterotransplantation of human malignant tumor in "nude" thymusless mice. II. Malignant tumors induced by injection of cell cultures derived from human solid tumors, *J. Natl. Cancer Inst.* **52,** 921–930 (1974).
108. C. O. Povlsen, J. Visfeldt, J. Rygaard, and G. Jensen, Growth patterns and chromosome constitutions of human malignant tumors after long-term transplantation in nude mice, *Acta Pathol. Microbiol. Scand. Sect. A,* **83,** 709–716 (1975).
109. A. E. Bogden, D. E. Kelton, W. R. Cobb, T. A. Gulkin, and R. K. Johnson, The effect of serial passage in nude athymic mice on the growth characteristics and chemotherapy responsiveness of rat mammary tumor xenografts, *Cancer Res.* **38,** 59–64 (1978).
110. J. Rygaard, *Thymus and Self Immunology of the Mouse Mutant Nude,* F.A.D.L. Publishers, Copenhagen (1973).
111. A. D. Irving, C. G. D. Brown, G. K. Kanhar, and D. A. Stagg, Comparative growth of bovine lymphosarcoma cells and lymphoid cells injected with *Theileria parva* in athymic (nude) mice, *Nature (London)* **255,** 713–714 (1975).
112. R. T. Prehn, Clinical implications of the data base concerning the tumor-host relationship, in: *Immunobiology of the Tumor–Host Relationship,* (R. T. Smith and M. Landy, eds.), p. 292, Academic Press, New York (1975).
113. W. J. Martin and S. E. Martin, Naturally occurring cytotoxic anti-tumor antibodies in sera of congenitally athymic (nude) mice, *Nature (London)* **249,** 564–565 (1974).
114. H. Ramseier, Specific activation of T lymphocytes from nude mice, *Immunogenetics* **1,** 507–510 (1975).
115. F. Loor, L. B. Hagg, N. S. Mayor, and G. E. Roelants, θ-Positive cells in nude mice born from homozygous *nu/nu* mother, *Nature (London)* **255,** 657–658 (1975).
116. Guide for the care and use of the nude (thymus-deficient) mouse in biomedical research: A report of the committee on care and use of the nude mouse, *ILAR News*, Vol. XIX (2), pp. M3–M20 (1976).
117. D. Houchens and A. Ovejera, *Proceedings of the Symposium on the Use of Athymic (Nude) Mice in Cancer Research* (June, 1977) Gustav Fischer Inc., New York. In press.
118. A. E. Bogden, D. E. Kelton, W. R. Cobb, and H. J. Esber, A rapid screening method for testing chemotherapeutic agents against human tumor xenografts, in: *Proceedings of the Symposium on the Use of Athymic (Nude) Mice in Cancer Research* (June, 1977), (D. Houchens and A. Ovejera, eds.), Gustav Fischer, Inc., New York. In press.

9

Gene Expression in Normal and Neoplastic Breast Tissue

JEFFREY ROSEN

1. Introduction

Most structural gene sequences in mammalian cells represent less than one millionth of the information contained in the genomic DNA. Because of this enormous complexity of genetic information in higher organisms, an understanding of the mechanisms regulating gene expression requires the study of specific genes. Fortunately, considerable progress has been made in the last decade in the development of techniques for the isolation of individual eukaryotic messenger RNAs (mRNAs) and the synthesis of their complementary DNA copies (cDNAs) (for a general review, see Rosen and Monahan[1]). These molecular hybridization probes have been utilized successfully to study gene expression in a number of model systems, notably the control of globin gene expression during erythroid differentiation[2] and steroid hormone induction of ovalbumin mRNA[3] and the other egg-white protein mRNAs[4] in the chick oviduct. These studies have recently been culminated by the determination of the entire nucleic acid sequence of the rabbit β-globin mRNA.[5] However, the precise mechanism controlling the expression of any single eukaryotic gene remains to be established.

Most of the experiments performed in the mammary gland were performed initially within the disciplines of dairy science and endocrine

JEFFREY ROSEN • Department of Cell Biology, Baylor College of Medicine, Houston, Texas 77030.

physiology, and, more recently, breast cancer. In the last few years, however, an increasing number of investigators have realized that the mammary gland also provides an excellent model for studying the regulation of specific genes at the molecular level. It is the intent of this chapter, therefore, to illustrate how the techniques of molecular biology have been applied recently to the study of gene expression in both the normal and the neoplastic mammary gland.

The mammary gland has several characteristics that make it amenable to this type of investigation. It contains two different types of hormonally inducible gene products—one type that is responsive to peptide hormones and a second type that is inducible by a steroid hormone. Thus, the induction of the milk proteins, casein and α-lactalbumin, by the lactogenic hormones, prolactin and placental lactogen, provides a unique model system for studying peptide hormone regulation of gene expression (see Section 6.1). The response to prolactin or placental lactogen is modulated by two steroid hormones, hydrocortisone and progesterone, and can be studied in organ culture using a chemically defined serum-free medium.

Steroid hormones administered *in vivo* have been shown to regulate the accumulation of numerous mRNAs, including ovalbumin mRNA[3] and conalbumin mRNA[4] in the chick oviduct, tryptophan oxygenase mRNA[6] and α-2-U-globulin mRNA[7] in rat liver, and vitellogenin mRNA in *Xenopus* liver.[8] In addition, glucocorticoid induction of mouse mammary tumor viral RNA (MMTV RNA) has been demonstrated *in vitro* in several cloned mammary tumor cell lines.[9,10] The availability of such an *in vitro* system has permitted kinetic studies in which the effect of a steroid hormone on the rates of mRNA transcription and degradation was directly measured.[11,12] These cloned cell lines have the additional advantage of allowing genetic and cell cycle analyses not possible with the mixed cell populations present in other steroid-responsive model systems. The first half of this chapter will therefore be devoted to the isolation and characterization of the individual milk protein mRNAs and MMTV RNA and the synthesis of their respective cDNAs. These molecular probes provide the necessary tools for the study of hormonal regulation of gene expression presented in the latter part of the chapter.

The mammary gland also provides an attractive developmental system, both with respect to the study of the normal process of hormonally induced differentiation and because of viral- and carcinogen-induced neoplastic transformation. The milk proteins, casein and α-lactalbumin, and their mRNAs can be used as specific biochemical markers of differentiated function in the mammary gland. Thus, a careful comparison of the amount of casein and α-lactalbumin mRNAs with the levels of milk protein synthesis and secretion during mammary gland development should yield considerable insight into the mechanism by which hormones regulate differentiated function, i.e., whether they act at the transcriptional, posttranscriptional, translational, or posttranslational level(s). Furthermore, by comparing the expression of the milk protein genes in hormone-dependent tumors with their transcription in normal mammary tissue, it may be possible to understand hormonal control of growth, as well as the regulation of differentiated function. Neoplastic transformation is thought to proceed through an intermediate stage referred to as preneoplasia (see Chapter 3). Thus, a study of MMTV gene expression during the progression from normal tissue to hyperplastic nodules and finally mammary carcinomas may provide additional information about the process of tumorigenesis. Studies in these areas are only in their infancy at the molecular level, and will be briefly described in this chapter.

One additional interesting aspect of the molecular biology of the mammary gland is the multiple interactions of both peptide and steroid hormones in the control of gene expression. There is present among several peptide and steroid hormones a complex interrelationship that influences the ultimate expression of differentiated function. The antagonistic or synergistic effects of one hormone may be mediated by the regulation of the synthesis of a receptor for a second hormone.[13,14] In addition, the increased accumulation of a specific mRNA may reflect, not the direct action of a given hormone at the transcriptional level, but rather an indirect effect on mRNA turnover and degradation (see Section 6.1). These types of interactions will be illustrated, therefore, with particular emphasis on the modulation of prolactin-induced casein synthesis by progesterone and hydrocortisone.

This chapter is not intended to be a comprehensive review of the literature or to discuss all the effects of hormones in the normal and neoplastic mammary gland. Instead, it is designed to highlight recent developments in the control of specific gene expression in the mammary gland. The emphasis is intended to be on studies involving the hormonal control of casein mRNA and casein synthesis in the rat mammary gland, which were performed primarily in the author's laboratory. Several up-to-date and more extensive reviews of the general effects of hormones on the mammary gland have recently been published.[15-17]

2. Isolation of Milk Protein Messenger RNAs

2.1. Detection of mRNA Activity in Heterologous Cell-Free Translation Systems

The initial characterization of a specific mRNA within a population of total cellular or polysomal RNA requires a sensitive cell-free protein-synthesizing system that is capable of translating exogenous mRNA with fidelity. In addition, a careful product analysis is required that rigorously confirms the identity of the newly synthesized protein. This is usually accomplished by specific immunoprecipitation followed by electrophoresis using sodium dodecyl sulfate (SDS)-containing polyacrylamide gels to compare the size of the native and *in vitro* synthesized proteins. There are several precautions and limitations of this type of analysis: First, not all mRNAs are translated with the same efficiency in different cell-free translation assays.[18,19] Second, it is critical to purify and characterize the milk proteins used as antigens and to determine the specificity of the immunoprecipitation assay. Caseins are phosphorylated glycoproteins and have proved to be rather poor antigens.[20,21] Thus, minor contaminants in a putative casein preparation may be more antigenic and lead to the preparation of a nonspecific antibody preparation. This is a particular problem when rennin and Ca^{2+} precipitation or isoelectric precipitation of skim milk are used as the sole methods of preparing purified caseins. Both methods may lead to contamination by noncasein phosphoproteins or whey and serum proteins present in milk.

In our laboratory, purification of the individual rat and mouse caseins was accomplished by chromatography of crude casein obtained by several cycles of isoelectric precipitation at pH 4.5 on DEAE–cellulose in the presence of urea, mercaptoethanol, and EDTA to disrupt protein aggregates.[21] Each casein was then rechromatographed to eliminate cross-contamination and characterized by SDS–polyacrylamide gel electrophoresis (PAGE) before and after iodination. Amino acid analysis and tryptic digests of each casein were also performed. The characterization of the partially purified mixture of rat caseins by SDS-PAGE is illustrated in Fig. 1. The three rat caseins obtained by several cycles of isoelectric precipitation are shown in comparison to bovine α_s-casein and a rat whey protein fraction containing serum proteins and α-lactalbumin. The analysis of the three purified, individual rat caseins following iodination is shown in Fig. 2. Finally, apparent molecular weights of the three caseins of 42,000, 30,000, and 25,000 were determined by electrophoresis in the SDS–phosphate polyacrylamide gel system developed by Weber and Osborn,[22] using appropriate molecular weight standards (Fig. 3). It should be stressed, however, that these are apparent molecular weights, since glycoproteins are known to migrate anomalously during PAGE. For example, in our laboratory, slightly different molecular weights of the rat caseins were observed during PAGE using the SDS–tris–glycine buffer system of Laemmli,[23] illustrating the sensitivity of these phosphorylated glycoproteins to changes in charge even in the presence of SDS. Furthermore, most of the posttranslational

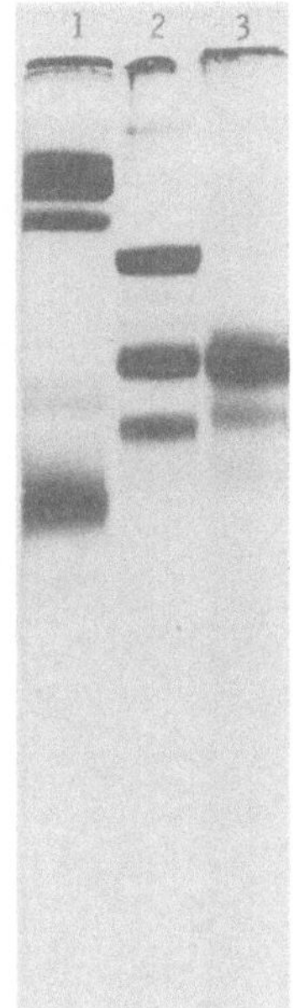

Fig. 1. Analysis of rat casein and whey fractions by SDS-polyacrylamide disc gel electrophoresis. Gel #1, whey fraction; gel #2, acid-precipitated, pH 4.5, rat casein fraction; gel #3, bovine α_S-casein. Electrophoresis was performed as described by Weber and Osborn[22] using 10% polyacrylamide gels.

protein modifications such as glycosylation, phosphorylation, and cleavage of signal peptide sequences (see Section 2.5) do not occur in most cell-free translation systems. Thus, it may be difficult to directly compare the size of the *in vitro* translation products to the fully modified mature proteins. This is the third major problem usually encountered in the analysis of milk protein mRNAs in heterologous cell-free translation systems.

The nonspecific trapping of radioactive protein during immunoprecipitation of cell-free translation products presents another difficulty in the determination of specific milk protein mRNA activity. Immunoprecipitation of newly synthesized milk proteins has been accomplished both by direct, primary antibody–antigen precipitation and by indirect or double antibody precipitation, in which the primary antibody–antigen complex is precipitated by the addition of a second antibody against the initial IgG fraction. Because of the low titer of most anticasein antisera, the latter method usually permits a more quantitative estimate of the level of casein synthesis; i.e., 90–95% of an [^{125}I]casein internal standard is recovered vs. 65–70% in the direct immunoprecipitation method.[24] The trapping level observed using either of these methods may, however, represent 1–3% of the radioactive protein synthesized. Thus, when very low levels of a specific mRNA activity are present, e.g.,

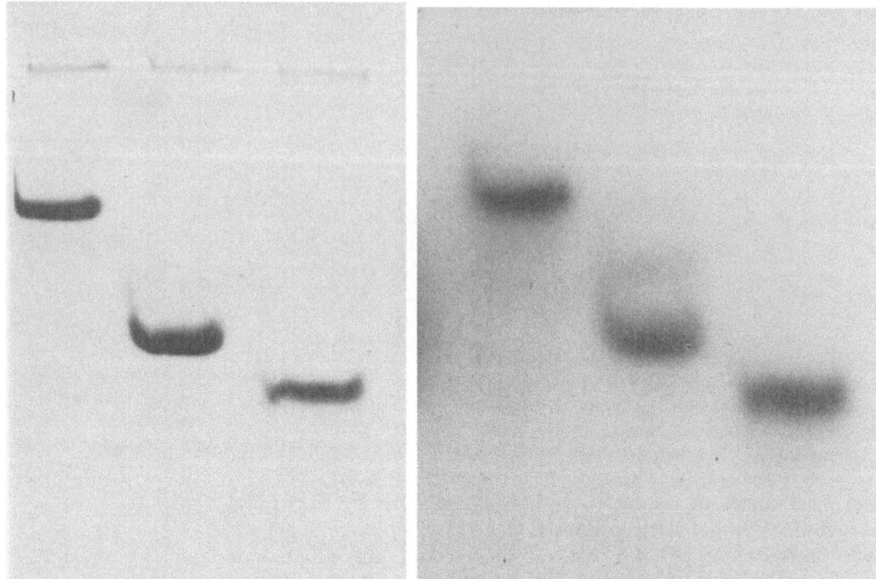

Fig. 2. Analysis of purified rat caseins following iodination by SDS–polyacrylamide slab gel electrophoresis and radioautography. The individual rat caseins were purified as described,[21] iodinated using chloramine-T at 4°C, and analyzed on 10% SDS–polyacrylamide slab gels.[22] *Left panel:* Coomassie blue stained gel; *right panel:* the corresponding radioautogram. The faint stained protein bands present at the top of the gel slots represent bovine serum albumin added as a carrier subsequent to iodination and prior to dialysis of the ^{125}I-labeled caseins.

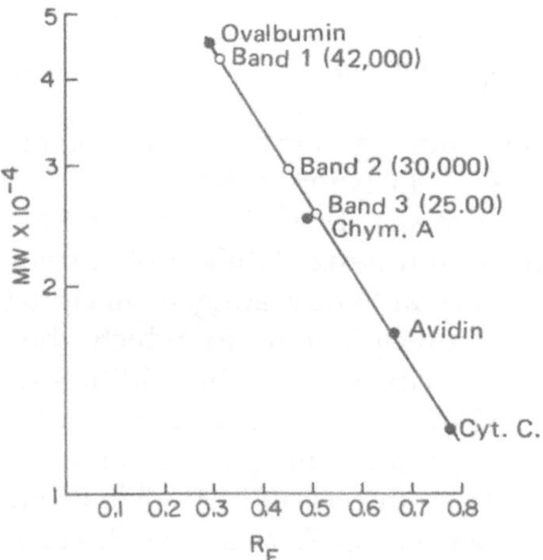

Fig. 3. Molecular-weight determination of rat caseins. Ten percent PAGE containing 1% SDS was performed as described.[22] The molecular weights of the standard proteins were ovalbumin, 45,000; chymotrypsin A, 25,000; avidin subunit, 17,000; cytochrome C, 12,400.

in assays of α-lactalbumin mRNA activity in RNA extracts of mammary tumors or early pregnant tissue, trapping may present a serious problem in quantitating mRNA activity. Control experiments are usually performed by employing a heterologous antibody–antigen complex, preimmune serum, or displacement with an excess of the appropriate, nonradioactive milk protein. In a sensitive and selective assay, greater than 90% displacement should be observed with the appropriate competitor,[24] and the specific mRNA activity should be a linear function of the amount of mRNA assayed. Unfortunately, these limitations of the cell-free translation assay in quantitating mRNA activity have not been recognized in all the studies involving milk protein mRNAs. Many of these problems can be avoided using the techniques of molecular hybridization discussed in Section 3. Cell-free translation systems have been essential, however, for the initial isolation and purification of the individual milk protein mRNAs.

A variety of cell-free translation systems, including those derived from wheat germ,[21,24,25] from Krebs II[25] or Ehrlich ascites cells[26] supplemented with rabbit reticulocyte initiation factors, from rabbit reticulocytes[27] with or without prior treatment with micrococcal nuclease,[28] and by microinjection into *Xenopus* oocytes,[29,30] have been employed to detect various milk protein mRNAs. Each of these translation assays has its own advantages and disadvantages, and may be utilized to answer different questions. For example, the wheat germ translation system is characterized by its ease of preparation and a low level of endogenous protein synthesis, permitting the estimation of both specific and total mRNA activities. However, incomplete chain completion and inefficient translation of some mRNAs, especially those coding for proteins of greater than 50,000 daltons, may complicate product analyses. While the nuclease-treated reticulocyte lysate is more difficult to prepare, it is also more useful for translation of larger mRNAs and the synthesis of discretely sized products, thus facilitating *in vitro* studies of secretory-protein processing. Finally, the microinjection of *Xenopus* oocytes, while it is the most technically difficult assay, also provides the most sensitive method for the detection of nanogram amounts of mRNA.

Casein mRNA activity, and in some cases α-lactalbumin mRNA activity, has now been detected in RNA extracts isolated from sheep,[27] rabbit,[31] guinea pig,[25] mouse,[24,26] and rat[21,24] mammary tissue. The three casein mRNA activities may comprise as much as 50–60% of the total mRNA activity in RNA extracted from 5- to 10-day lactating rat mammary tissue.[24,32] This has greatly facilitated their detection in, and purification from, total cell RNA extracts.[24,32] In addition, milk protein mRNA activity has been detected in RNA isolated from postnuclear cytosols,[25] from membrane-bound polysomes,[31] and from specifically immunoprecipitated polysomes.[33] In our laboratory, however, total cell RNA extracts have been employed for most comparative milk protein mRNA activity determinations, and as the initial source of casein mRNA used for subsequent purification. Direct homogenization of rapidly frozen tissue in a phenol–SDS buffer at pH 8.0 is used to minimize the nuclease activity inherent in most subcellular fractionation procedures. An additional phenol–$CHCl_3$ extraction, followed by treatment with proteinase K and several extractions of the nucleic acid pellet in 3 M sodium acetate, pH 6.0, are usually employed. The resulting total RNA extract is essentially free of any residual nuclease activity, DNA, and protein, and is therefore suitable for subsequent mRNA purification and assay. The specific activity of these total RNA extracts is less than either total polysomal RNA or poly(A)-containing mRNA preparations enriched by affinity chromatography on poly(U)–Sepharose or oligo(dT)–cellulose. Differential losses of mRNA may occur, however, during the isolation of either polysomes or poly(A)-containing mRNA. For example, several laboratories have observed that not all the casein mRNA activity present in RNA extracts isolated from lactating tissue is found in the poly(A)-containing mRNA fraction after affinity chromatography.[21,34] As much as 30–40% of the casein mRNA activity may not be selectively retained, probably due to the presence of very short poly(A) tails on these mRNAs (see Section 2.3). Thus, for comparative purposes, when milk protein mRNA activity is to be determined at different stages of mammary development, or following hormonal administration, these assays are usually performed on total RNA extracts. The sensitivity of the cell-free translation assays is usually not adequate to accurately quantitate milk protein mRNA activity when only a few mRNA molecules per cell are present, or when only a limited number of differentiated alveolar cells is present, such as in virgin mammary tissue or mammary carcinomas. In these cases, molecular hybridization is used to detect milk protein mRNA sequences. However, a comparison of specific mRNA sequences in total cell, nuclear, and polysomal RNA may be helpful in elucidating the potential role of hormones at the posttranscriptional level on RNA processing and mRNA turnover. Once again, these studies usually require specific cDNA hybridization probes for accurate quantitation of mRNA levels, and have only recently been performed in a few experiments (see Section 4.1). Thus, although cell-free translation assays have been of enormous utility for the initial identification and characterization of milk protein mRNAs, most of the studies in our laboratory have been performed using the technique of molecular hybridization. The purification of the individual casein mRNAs and their thorough characterization were necessary prerequisites, however, for the generation of selective cDNA probes.

2.2. Purification of Casein mRNAs

Two different approaches, specific immunoprecipitation of polysomes and precise sizing techniques coupled with poly(A)-RNA affinity chromatography, have been employed successfully for the isolation of purified casein mRNA and the partial resolution of the individual casein mRNAs. Purification in both cases has been facilitated by the fact that the three rat casein mRNAs comprise greater than 50% of the total mRNA activity in RNA isolated from lactating mammary tissue.[24,32] The isolation of total casein mRNA from the bulk of other mRNAs and cellular RNA is therefore reasonably straightforward and could be accomplished by methods previously used for other abundant mRNAs, such as globin,[35] ovalbumin,[36] and fibroin mRNAs.[37] The resolution of the three individual casein mRNAs into distinct chemical entities, however, is extremely difficult and requires more sophisticated techniques.

The technique of specific immunoprecipitation was utilized by Drs. Houdebine and Gaye and their colleagues[33] to purify rabbit casein mRNA from polysomes isolated from lactating mammary tissue. This technique had previously been employed to purify both ovalbumin and albumin mRNAs from hen oviduct and rat liver polysomes, respectively.[38,39] Isolation of large quantities of undegraded polysomes and the preparation of a specific, nuclease-free, high-titer antibody are required in this procedure. Following recognition of the nascent peptide chains by the specific anticasein antibody, recovery of the casein-synthesizing polysomes is then accomplished by the addition of a second antibody prepared against the anticasein IgG fraction. Isolation of total polysomal RNA enriched in casein mRNA is performed by phenol extraction and followed by affinity chromatography to recover the poly(A)-containing casein mRNA, essentially free of ribosomal and transfer RNA.

Although in principle this procedure should be ideal for the isolation of specific mRNAs, the following problems have been encountered: Since caseins are rather poor antigens, low-titer and -affinity antibodies have usually been prepared. This problem may be further complicated when heterologous systems are used; e.g., in the studies of Houdebine and Gaye,[33] purified ewe caseins and their respective antisera were used to isolate rabbit mammary gland casein-synthesizing polysomes. Thus, it was difficult to obtain quantitative precipitation of the casein-synthesizing polysomes. Second, nonspecific trapping of other polysomes in the large precipitates formed may lead to contamination with other mRNAs. In addition, isolation of large amounts of undegraded mammary gland polysomes is difficult in several other species, such as the rat and mouse, and the technique is therefore difficult to use in these cases. Even limited nucleolytic cleavage not evident in polysomal profiles may lead to the isolation of inactive mRNAs. Finally, only small amounts of purified mRNAs, usually only a few micrograms, can be obtained by these procedures. This precludes many of the physical–chemical studies of mRNA purity discussed in Section 2.3. Despite these limitations, sufficient amounts of a mixture of three rabbit casein mRNAs have been prepared using this technique to generate specific cDNA probes[40] (see Section 3.1), and the partial resolution of the individual casein mRNAs has been accomplished.[33]

Our laboratory has successfully utilized the second approach employing precise sizing and affinity chromatography techniques to purify large amounts of rat and mouse casein mRNAs.[21,24] In this procedure, total RNA extracts of lactating mammary tissue are obtained from 8- to 12-day lactating rats or mice. Several hundred micrograms of the purified casein mRNAs can usually be obtained from approximately 200 g of tissue. Purification is accomplished by separating mRNA from the bulk of cellular ribosomal and transfer RNAs by affinity chromatography of the poly(A)-containing mRNA. Individual mRNAs can then be separated based on their unique sizes and conformations if precise sizing techniques, using conditions that prevent mRNA aggregation, are employed. This can be accomplished by the inclusion of denaturing solvents such as formamide, DMSO, and urea during the chromatographic or electrophoretic separation procedures. Alternatively, heating the solution of RNA under conditions of low ionic strength to 70°C followed by rapid cooling will usually prevent mRNA aggregation and increase the resolution of both sizing techniques and affinity chromatography.

The purification of the individual rat casein mRNAs using this approach is illustrated in Fig. 4. The mRNA fractions obtained after each step of purification were analyzed by electrophoresis on 3% agarose gels containing 6 M urea and 0.015 M sodium citrate, pH 3.5. The total RNA extract isolated from lactating tissue was analyzed on gel No. 1. Following a single passage over oligo(dT)–cellulose, analysis of the bound poly(A)–mRNA fraction (gel No. 2) revealed several prominent mRNA bands in the 8–18 S region of the gel. Casein mRNA activity was present as a 15 S mRNA doublet and a 12 S RNA band,[21,24] while α-lactalbumin mRNA electrophoresed as a prominent RNA band at 10 S.[32] A considerable amount of contaminating 18 S and 28 S ribosomal RNA remained in the poly(A)–RNA fraction after only a single passage over oligo(dT)–cellulose. Fractionation of the poly(A)-containing mRNAs and removal of 28 S ribosomal RNA was next accomplished by chromatography on Sepharose 4B. Using this procedure, the casein mRNA fractions could be resolved from most other contaminating mRNAs (gel No. 3). Follow-

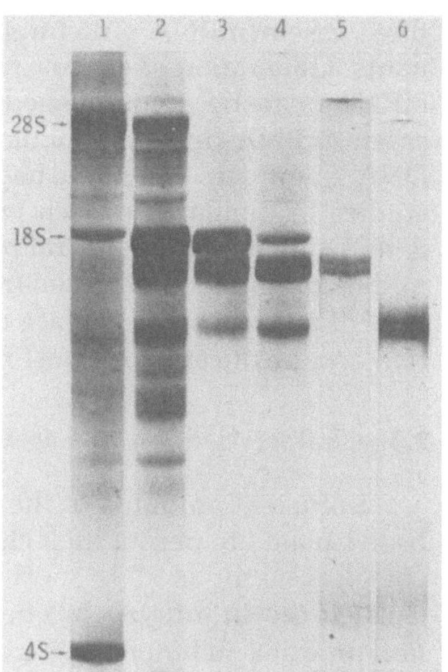

Fig. 4. Purification of 15 S rat casein mRNA determined by agarose–urea gel electrophoresis. RNA obtained during each step of the purification procedure was analyzed by electrophoresis on 3% agarose–urea gels followed by staining with 1% methylene blue in 15% acetic acid as described previously.[36] Gel #1, total RNA extract, 30 μg; #2, dT-cellulose bound poly(A)-enriched RNA, 24 μg; #3, Sepharose 4B peak fraction of casein mRNA activity, 18 μg; #4, rechromatography of Sepharose 4B peak fraction on dT-cellulose, total bound casein mRNA, 12 μg; #5, purified 15 S casein mRNA obtained after preparative agarose gel electrophoresis, 5 μg; #6, purified 12 S casein mRNA obtained after preparative agarose gel electrophoresis, 9 μg.

ing a second passage of the peak fractions of casein mRNA activity obtained by Sepharose 4B chromatography over oligo(dT)-cellulose, only a small amount of contaminating 18 S ribosomal RNA was still present (gel No. 4). Removal of this remaining 18 S ribosomal RNA and partial resolution of the individual casein mRNAs could then be accomplished by preparative agarose–urea gel electrophoresis. Two fractions were obtained, a 15 S mRNA doublet (gel No. 5) and a 12 S casein mRNA fraction that migrated as a single sharp band (gel No. 6). Several hundred micrograms of each of these casein mRNA fractions could be isolated using this method, thereby allowing their detailed physical and chemical characterization.

Although isolation of the casein mRNAs from other contaminating mRNAs and ribosomal RNA is therefore reasonably straightforward, the resolution of the individual casein mRNAs into distinct chemical entities requires more sophisticated techniques. For example, separation of the 15 S mRNA doubtlet into individual mRNAs can be accomplished by slab gel electrophoresis followed by extraction of the separated mRNAs from the individual gel fractions. Similar procedures have been used to separate α- and β-globin mRNAs.[41] However, even the most precise sizing techniques may result in limited sequence cross-contamination due to copurification of similarly sized mRNA fragments. Using recombinant DNA technology, it should be possible to completely resolve the three casein mRNA structural gene sequences free of minor contaminants. Generation of specific hybridization probes for each of the casein mRNAs can be accomplished by cloning a mixture of the three rat-casein-mRNA-derived double-stranded DNA copies. These cloned DNA fragments can then be used in gene mapping and sequencing studies. A similar approach was recently used to isolate the structural gene sequence coding for rat proinsulin mRNA from a partially purified mRNA preparation obtained from islets of Langerhans.[42] No doubt this type of approach will replace those previously mentioned for the isolation and purification of most mRNAs in the future.

2.3. Characterization of Casein mRNAs

Sufficient amounts of the purified rat casein mRNA fractions have been isolated to permit their characterization by biological, physical, and chemical methods.[24] In most other studies, the characterization of the isolated casein mRNAs has been limited to an analysis of their translation products in heterologous cell-free translation systems.[27,33] The use, in many cases, of heterologous translation systems with high levels of endogenous protein synthesis makes a direct assessment of mRNA purity impossible.[27,33] In a limited number of studies, the isolated casein mRNAs have also been characterized by gel electrophoresis, usually employing formamide-containing polyacrylamide gels.[25,33]

The purified rat casein mRNA fractions have been analyzed by translation in both the wheat germ cell-free translation system[21,24] and the mRNA-dependent reticulocyte lysate.[28] Under optimal translation conditions for casein mRNA, at least 90% of the released protein synthesized in response to the 15 S casein mRNA was specifically immunoprecipitable, representing a 178-fold purification compared with the initial lactating RNA extract. Only 60% of the radioactive protein synthesized in response to the 12 S casein mRNA was immunoprecipitable with the total anticasein IgG fraction. However, this may have reflected both the preference of this antibody for the larger rat caseins and its inability to quantitatively precipitate the smaller incomplete fragments that are synthesized *in vitro.* A close correspondence was observed between the total and specifically immunoprecipitated proteins synthesized in the wheat germ system with either the 12 S or the 15 S mRNA fractions as analyzed by fluorography on 5–20% polyacrylamide gradient slab gels. The 15 S mRNA doublet directed the synthesis of the two largest rat caseins, while the 12 S casein mRNA directed the *in vitro* synthesis of the smallest of the three rat caseins. When the 15 S mRNA doublet, which appeared to contain equal amounts of the two largest casein mRNAs as analyzed by agarose–urea gel electrophoresis, was translated in the wheat germ system, only a small amount of the 42,000-molecular-weight casein was synthesized. However, when a similar mRNA preparation was translated in the mRNA-dependent reticulocyte lysate, approximately equal amounts of the two largest rat caseins were synthesized.[28] A similar difference in the proportions of the individual guinea pig caseins synthesized in the wheat germ and ascites cell-free translation systems from the same mRNA preparation was reported.[25] This once again indicates that differences occur in the efficiency of translation of certain mRNAs in cell-free systems. Furthermore, caution should be employed when using mRNA translation as the sole method of assessing mRNA purity.

With these reservations in mind, the purity of several milk protein mRNA fractions have been determined in several heterologous cell-free translation assays. Isolation of either an α_s-casein or a β-casein mRNA fraction was accomplished by specific immunoprecipitation of polysomes followed by poly(U)-Sepharose chromatography as previously described.[33] These fractions were shown to be approximately 75–80% pure with respect to each other, but it was impossible to determine their absolute purities with respect to other noncasein mRNAs, since they were assayed in a reticulocyte lysate translation system. A poly(A)–mRNA fraction isolated from lactating guinea pigs by two cycles of affinity chromatography on oligo(dT)–cellulose has also been shown to contain between 65 and 75% casein mRNA activity and 5–15% α-lactalbumin mRNA activity by translation in either a Krebs ascites or wheat germ cell-free translation system. A similar fraction obtained by poly(U)–Sepharose chromatography of total RNA isolated from lactating rats was reported to contain 40–50% α-lactalbumin mRNA with the remainder as casein mRNA.[32]

In our laboratory, synthesis of cDNA probes and further characterization of the individual casein mRNAs has been performed only with mRNA preparations that are at least 90% pure based on several criteria including cell-free translation.[24] Physical and chemical characterization of the purified rat casein mRNAs has included determination of their poly(A) contents, the length of their poly(A) tails, their molecular weights by both PAGE and direct length measurement by electron microscopy (EM), their base composition, their secondary structure by analysis of derivative thermal denaturation profiles, and an analysis of their putative "cap" structures by inhibition of cell-free translation with "cap" analogues.[24,54] Some of these properties of the 15 S casein mRNA doublet are summarized in Table I.

The molecular weights of the rat casein mRNAs were determined by both agarose–urea gel electrophoresis (Fig. 4) and formamide-containing PAGE,[24] using several well-characterized mRNA standards of compar-

Table I
Properties of 15 S Casein mRNAs

Property	15 S doublet[a]	Total complexity
Molecular weight		
PAGE	1169 ± 78 NT	2484 ± 166
	1315 ± 88 NT	
EM measurement	$\bar{L}_N$ =1190 ± 120 NT	2380 ± 240
Base composition	42.8% GC	
Poly(A) tail	42 A's, number average (15–150 distribution)	
5′ Cap structure	Undetermined sequence	
T_m 0.1 M KCl	46.6° considerable secondary structure	

[a] Abbreviations: (NT) nucleotides; ($\bar{L}_N$) number average length; (GC) guanosine and cytosine; (A) adenosine.

able physical–chemical properties. Overestimation of mRNA molecular weight is routinely observed when ribosomal RNAs are used as molecular weight markers,[43] as has been the case in most analyses of milk protein mRNAs.[25,32,33] The length of the purified 15 S casein mRNA was also directly measured by EM (Fig. 5). The rather broad distribution of mRNA lengths observed was a reflection of the presence of equal amounts of two mRNAs of very similar size in the 15 S mRNA preparation. Heterogenity of poly(A) length may also account for the partial overlap of these mRNAs and the failure to identify two discrete size classes. The results obtained by both PAGE and EM are in close agreement, however, giving a total complexity of the 15 S mRNA doublet of approximately 2400 nucleotides. Using these molecular weights, it can be estimated that the rat casein mRNAs contain between 15 and 40% additional noncoding information. The function and location of these untranslated sequences in casein mRNA remains to be determined. Similar noncoding sequences in rabbit β-globin mRNA have been reported to contain extensive secondary structure and a short region of homology with 18 S mRNA that may play a role in ribosome binding.[5]

Both the 15 S and 12 S casein mRNA contain poly(A) segments at their three prime termini ranging from 15 to 150 adenosines in length.[24] The number average length of the poly(A) tail is approximately 40 adenosines, which is consistent with the steady-state length of poly(A) tracts observed in other mRNAs with long half-lives.[43] The heterogeneous distribution in poly(A) size observed for casein mRNA supports the

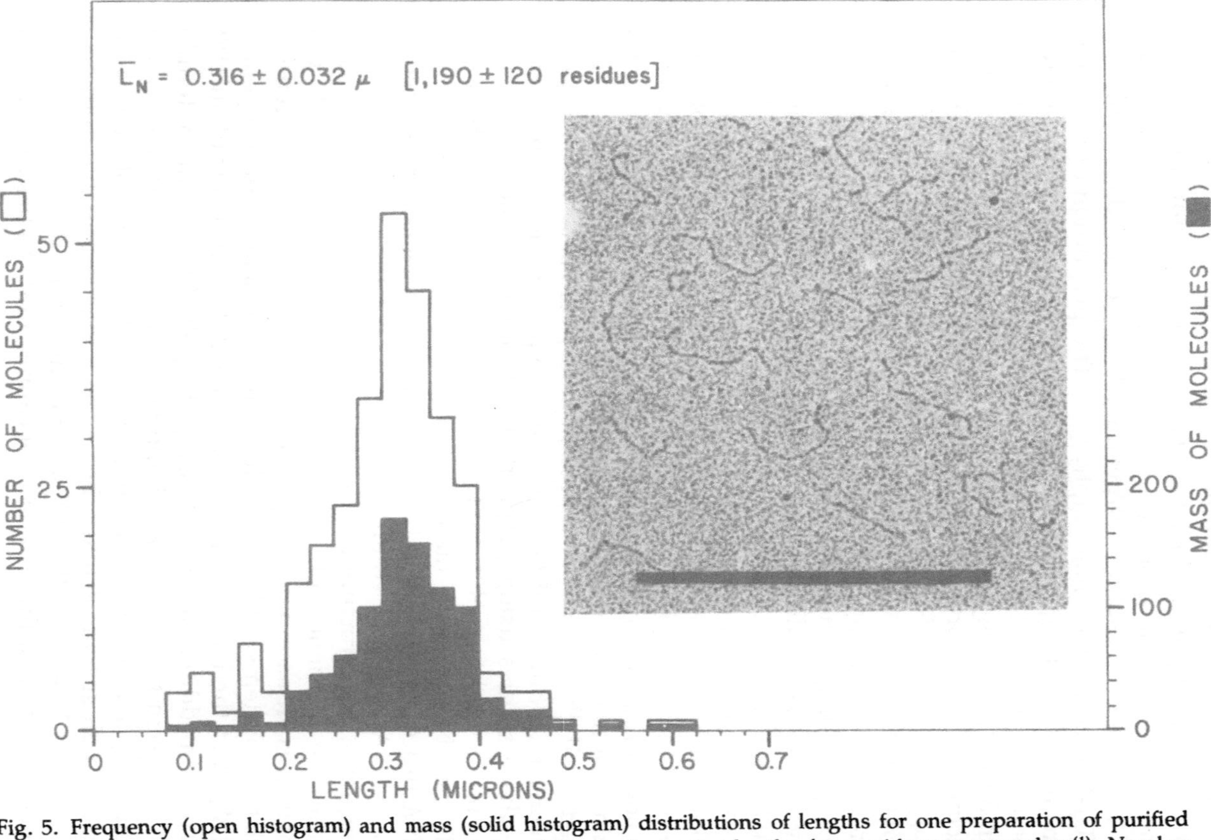

Fig. 5. Frequency (open histogram) and mass (solid histogram) distributions of lengths for one preparation of purified 15 S rat casein mRNA. The RNA was prepared for electron microscopy by the formamide-urea procedure[1]. Number average length was determined for molecules in intervals indicated by bar above histogram. Micrograph inset at right illustrates appearance of casein mRNA with magnification indicated by bar length of 1 μm. The electron microscopy was kindly performed by Dr. Don Robberson at the M. D. Anderson Hospital, Houston, Texas.

hypothesis that a random endonucleolytic attack is involved in the poly(A) shortening observed during the aging of mRNA.[45] The presence of extremely short poly(A) tails (< 20 adenosines) presumably accounts for the lack of binding of a significant amount (20–50%) of the casein mRNA activity in a lactating RNA extract during either (dT)-cellulose or poly(U)-Sepharose chromatography.[21,34] The proportion of polysomal casein mRNA lacking poly(A), i.e., unable to bind to poly(U)-Sepharose, was reported to be 25% of the total casein mRNA even within 3 hr after an eightfold induction of casein mRNA sequences by prolactin and hydrocortisone.[46] However, the conclusion that the processing of casein mRNA does not include a systematic addition of poly(A) to all the mRNAs was not warranted in the absence of more direct information concerning the rates of poly(A) shortening and casein mRNA processing. These results can be obtained only by analyzing newly synthesized pulse-labeled casein mRNA sequences, rather than by studying the steady-state accumulation of mRNA sequences following prolactin administration.

Both mRNA and heterogeneous RNA (HnRNA) of eukaryotic cells contain at their 5′ termini structures of the form $m^7GpppX^mpYp\ldots$ (cap I). In addition, some mRNAs, but no HnRNAs, have been reported to contain a more highly modified terminus, $m^7GpppX^mpY^mpZp\ldots$, designated cap II.[47,49] X^m and Y^m represent methylated bases, which appear to be conserved during mRNA synthesis and processing as suggested by pulse-chase studies using [^{3}H]methylmethionine.[48] Cap II structures are found on the longer-lived mRNA species, suggesting that they may stabilize mRNA in some manner. In addition to its potential role in HnRNA processing, the cap structure appears to facilitate the formation of the mRNA–ribosome complex and increase mRNA stability.[50] The presence of a cap structure in most mRNAs has been shown to greatly increase the efficiency of initiation of protein synthesis.[49,51] Cap analogues, such as m^7Gp, when added to mRNA-dependent cell-free translation systems will also inhibit the translation of exogenous mRNA[52] and the binding of mRNA to the 40 S ribosomal subunit.[53]

Thus, in addition to characterizing the poly(A) tail at the 3′ termini of the rat casein mRNAs, it was important to determine whether a cap structure existed at their 5′ termini. In the absence of labeled casein mRNA of high specific activity, however, it was impossible to directly sequence the 5′ end of the mRNA. Furthermore, direct chemical or enzymatic labeling of the 5′ termini of capped mRNAs has proved to be extremely difficult. An alternative method was therefore used to determine whether a cap structure existed in the purified rat casein mRNAs. The ability of $m^7G^{5'}p$, but not m^7G or $G^{5'}p$, to inhibit completely the translation of casein mRNA in the wheat germ cell-free translation system is shown in Fig. 6. Further studies have demonstrated that addition of the "cap" analogues 10 min after the initiation of protein synthesis resulted in only a 30% inhibition of casein synthesis.(54) This result adds support to the role of the cap structure in the initiation of protein synthesis. Addition of *S*-adenosylhomocysteine, an inhibitor of methylation, to the cell-free system did not affect the initial rate of casein synthesis, suggesting that the putative "cap" structure was methylated prior to translation in the wheat germ system.(54) These data provide an indirect proof that a "cap" structure is present at the 5′ end of the rat casein mRNAs. It had previously been reported that inhibition of translation by "cap" analogues was a function of the type of cell-free system employed rather than the presence of a cap structure at the 5′ end of mRNA.(55) However, more recent studies have confirmed the importance of the 5′ terminal m^7G for mRNA translation regardless of the source of the cell-free extract.(19) Direct proof of the nature of the "cap" structures on the casein mRNAs will, however, require sequence analysis of pulse-labeled mRNA.

In addition to this study of the 3′ and 5′ termini of the purified rat casein mRNAs, a preliminary analysis of their secondary structure has been performed by examining differential melting behavior at 260 and 278 nm, by EM spreading, and by nuclease resistance of [^{125}I]-mRNA. Considerable secondary structure was observed with a T_m of the mRNA

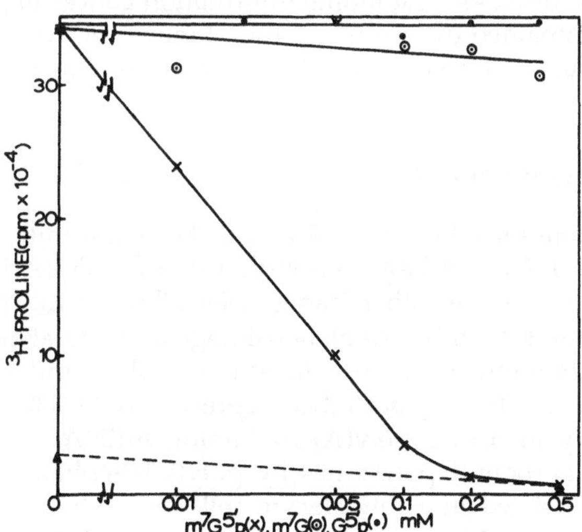

Fig. 6. Inhibition of casein mRNA translation by $m^7G^{5'}p$. Wheat germ translation assays were performed for 2 hr as described previously(24) in the presence of the designated concentrations of $m^7G^{5'}$ p (×), m^7G (⊙), or $G^{5'}$ p (●) and 0.125 μg purified 15 S casein mRNA. (---) Level of endogenous protein synthesis in the absence of added mRNA.

in 0.1 M KCl of 46.6° (Table I). Some evidence of guanosine- and cytosine-rich regions within the mRNA was also found. Casein mRNA exhibited considerably more secondary structure than ovalbumin and globin mRNAs under comparable conditions.[44] In addition, short regions of base-pairing have been observed in casein mRNA by both direct visualization in the electron microscope and sizing of nuclease resistant fragments by PAGE.[54] The relationship of secondary structure to the function of mRNA remains to be determined. Interestingly, considerable secondary structure has been observed in the 5′ and 3′ noncoding regions of rabbit globin mRNA, especially in the region of the AUG initiator codon.[5] Thus, mRNA secondary structure may be an important determinant of the efficiency of mRNA translation, as well as in the regulation of mRNA stability.

The base composition of the rat casein mRNAs has also been determined using microtechniques (Table I). The composition of 43% GC is in agreement with other DNA-like mRNAs.[43] The absence of certain characteristic nucleosides found in ribosomal RNA, such as pseudouridine, suggested that the purified rat casein mRNA was essentially free of rRNA contamination. The rat casein mRNAs have therefore been characterized by biological, physical, and chemical methods, and by all three criteria appeared to be of sufficient purity to allow the generation of selective complementary DNA hybridization probes. It was necessary to utilize several different criteria to determine accurately the purity of these mRNAs. Additional information concerning their purity could also be obtained by examining their kinetics of hybridization with homologous and heterologous cDNA probes, as will be described in Section 3.1.

2.4. α-Lactalbumin mRNA

Casein comprises between 70 and 85% of the milk protein and, accordingly, a relatively large proportion of the mRNA isolated from lactating tissue. On the other hand, α-lactalbumin mRNA would be expected to represent only a small percentage of the total mRNA activity and should therefore be more difficult to isolate and purify. Thus, α-lactalbumin has been reported to represent only 5% of the casein mRNA activity in a ewe poly(A)-containing mRNA fraction isolated from membrane-bound polysomes by poly(U)–Sepharose chromatography and translated in a reticulocyte cell-free system.[27] The *in vitro* product was reported to comigrate with an *in vivo* labeled α-lactalbumin standard with a reported molecular weight of slightly less than 18,000 daltons. In the guinea pig, α-lactalbumin mRNA activity comprised between 5 and 15% of the total mRNA activity obtained by translation of poly(A)-containing mRNA, isolated from a 3- to 6-day lactating, post-nuclear RNA fraction by two cycles of (dT)–cellulose chromatography.[25] Translation in either a wheat germ or Krebs II ascites cell-free system resulted in an *in vitro* protein that was slightly larger than the *in vivo* α-lactalbumin (15,500 vs. 14,500). The identification of α-lactalbumin mRNA activity in the wheat germ translation system using total RNA extracts isolated from 15-day lactating rats was also reported.[56] In these experiments, a direct analysis of the levels of α-lactalbumin and casein mRNA activity was not performed, but the *in vitro* product migrated identically with an [^{125}I]-α-lactalbumin standard of 15,000 apparent molecular weight.

In contrast to the results cited above, a recent study has suggested that α-lactalbumin mRNA activity may represent 20–30% of the total mRNA activity as determined by translating whole cell RNA extracts obtained from 3- to 5-day lactating rats in the wheat germ cell-free translation system.[32] In these studies, two rat α-lactalbumin species of 22,500 and 21,800 daltons were identified, and the *in vitro* product migrated with the 21,800-molecular-weight species. Following poly(U)–Sepharose chromatography, the α-lactalbumin mRNA activity increased to 53% of the total mRNA activity, although only 6% of the total activity was recovered during affinity chromatography. Two subsequent fractionations by sucrose gradient centrifugation of the poly(A)-containing mRNA resulted in a 110-fold overall purification, and 84% of the total mRNA activity was now α-lactalbumin mRNA activity. The purified α-lactalbumin mRNA separated into several bands during composite acrylamide–agarose gel electrophoresis with a peak of activity at 10.5 S. α-Lactalbumin mRNA activity also sedimented at about 8.3 S during sucrose gradient centrifugation. Thus, purified rat α-lactalbumin mRNA was similar in size to a major peak of mRNA activity previously described in our laboratory, which synthesized proteins in the 15,000- to 17,000-molecular-weight range.[21,24] At present, it is not clear why α-lactalbumin mRNA is reported to comprise more than 50% of the total mRNA activity of a poly(A)-containing mRNA fraction in these studies and synthesizes a relatively large protein of 21,800 daltons *in vitro*.[32] In several other studies, α-lactalbumin mRNA represents only 5% of the mRNA activity and synthesizes proteins in the 15,000- to 17,000-molecular-weight range.[17,27,54,56] These differences may reflect the use of different RNA isolation procedures, tissue obtained at different stages of lactation, or different species of animals. Furthermore, in several studies, the comigration of α-lactalbumin synthesized both *in vitro* and *in vivo* was reported.[27,56] This result is not consistent with the synthesis of a precursor containing a signal peptide sequence[25] (see the next section). Further characterization of the purified rat α-lactalbumin mRNA and the cell-free translation products it specifies are required before these studies can be more carefully evaluated.

2.5. Analysis of Cell-Free Translation Products: Synthesis of Preproteins

Both casein and α-lactalbumin are secretory proteins, which are synthesized on polysomes bound to membranes of the endoplasmic reticulum.[57] Recently, it was suggested that a dichotomy may exist between the initiation of casein synthesis and secretion during pregnancy.[58] Thus, prolactin- or placental lactogen-stimulated increases in casein mRNA and casein synthesis may precede the capacity of the tissue to secrete casein. This latter process may be regulated by the increased synthesis of endoplasmic reticulum and the subsequent increase in membrane-bound polysomes. Rough endoplasmic reticulum synthesis may be controlled by the levels of progesterone and hydrocortisone during pregnancy.[59] Thus, the mammary gland has the ability to build up progressively a high capacity for milk protein synthesis prior to lactation and secretion.

Work in a number of laboratories, including that of Blobel and his colleagues, has led to the formation of the signal hypothesis.[60] The essential feature of this hypothesis is the occurrence of a unique sequence of codons located immediately to the right of the initiator codon, which is present only in those mRNAs the translation products of which are to be transferred across a membrane. Translation of the signal codons results in a unique sequence of predominantly hydrophobic amino acid residues on the amino terminus of the nascent chain. This signal sequence triggers attachment of the ribosome to the endoplasmic reticulum, thus allowing the vectorial transport of the protein into the membrane vesicles. Thus, proteins synthesized in certain cell-free translation systems in the absence of the signal peptidase and membranes should contain an additional 20–25 amino acids at their amino termini.

Evidence for the synthesis of a pre-α-lactalbumin was recently obtained by a careful analysis of the sizes of the *in vitro* synthesized proteins specified by a guinea pig milk protein mRNA fraction as compared with native α-lactalbumin secreted in milk.[25] However, no similar evidence of precaseins was obtained by an analysis of the sizes of the cell-free translation products.[24,25] Recent indirect evidence for precursors to the guinea pig caseins was obtained by the studies of Zehavi-Willner and Lane.[30] Injection of a guinea pig total milk protein mRNA fraction into *Xenopus* oocytes resulted in the protection from protease digestion of the newly synthesized caseins within membrane vesicles. No such protection was obtained when the [^{125}I]caseins were injected in a control experiment. This suggested that translation of the mRNA was associated with a processing system, which transfered the proteins across membranes into vesicles. Direct evidence for the existence of precaseins was recently obtained in our laboratory in collaboration with Dennis Shields and Günter Blobel.(28) Thus, when EDTA-stripped dog pancreas membranes were added to an mRNA-dependent reticylocyte lysate cell-free translation system following the translation of the rat casein mRNAs, the newly synthesized rat caseins were shown to be processed and different sized products resulted. This processing was also accompanied by protection from protease digestion, suggesting once again the transfer of the caseins into membrane vesicles. Preliminary amino acid sequences have been determined for each of the signal peptides present in the three rat caseins. Recently, the amino terminal sequences of the precursors of four ovine caseins have also been determined.(148) These sequences are quite similar, but not identical. These studies may provide the information necessary for the elucidation of the mechanisms by which hormones regulate casein secretion.

3. Synthesis and Utilization of Specific Complementary DNA Probes

The synthesis of a high-specific-activity complementary DNA copy of a purified mRNA can be accomplished using either viral RNA-directed DNA polymerase,(61) commonly referred to as reverse transcriptase, or *Escherichia coli* DNA polymerase I.(62) Reverse transcriptases usually purified from avian myeloblastosis virus(63) are capable of using a poly(A)-containing mRNA as a template to synthesize cDNA. This reaction is dependent on the presence of an oligo dT-primer, and under these conditions, RNAs lacking poly(A), such as ribosomal RNA, are poor templates.(61) The conditions necessary for the synthesis of full-length cDNA transcripts using reverse transcriptase were recently established.(64,65) In addition, the self-priming ability of a short hairpin sequence at the 3′ terminus of the single-stranded cDNA greatly facilitates the synthesis of double-stranded DNAs using either *Esch. coli* DNA polymerase I(66) or reverse transcriptase.(42,67) These synthetic eukaryotic genes can then be inserted into bacterial plasmids and amplified in *Esch. coli* using recombinant DNA technology.(5,42) In certain instances where the isolation of very large, intact mRNAs is difficult, e.g., the isolation of MMTV RNA, nonspecific oligodeoxynucleotide primers can be used instead of oligo(dT) to generate representative, albeit not full-length, cDNA hybridization probes of these complex RNAs.(11,68) In these experiments, however, as well as those in which *Esch. coli* DNA polymerase I is utilized to synthesize single-stranded cDNA, both ribosomal and mRNA will be used as templates.[62,68] Therefore, if the mRNA preparation is contaminated with ribosomal RNA, these latter methods are not suitable for the synthesis of selective hybridization probes.

The enormous utility of molecular hybridization probes is now well established. For example, complementary DNA copies of individual mRNAs have been used to establish the number of copies of a specific gene sequence in the cellular genome,[69,70] to quantitate the number of copies of mRNA transcribed *in vivo*[3] or *in vitro* from isolated nuclei or chromatin templates,[71,72] and to estimate the sequence divergence between different mRNAs.[73] In addition, they have been extremely useful tools for the isolation of specific genes[74] and mRNAs[42,75] and for mRNA sequence determination.[76] In the following section, the recent application of this technology to the synthesis of specific hybridization probes for casein and MMTV RNA will be discussed.

3.1. Generation of Specific cDNA Copies of Casein mRNAs

Two methods have been employed to synthesize cDNA probes for the casein mRNAs: Houdebine[40] utilized the Klenow subfragment of *Esch. coli* DNA polymerase I, from which the 5′–3′ exonuclease was removed through proteolysis, to synthesize 200-nucleotide-long cDNA copies of rabbit casein mRNA. This cDNA probe was therefore representative of only approximately 20% of the 3′ end of the template mRNA. When used in hybridization experiments, about 15% of the unhybridized cDNA was resistant to S_1 nuclease digestion, and the hybridizations usually proceeded to approximately 80% completion. The $R_0t_{\frac{1}{2}}$ (RNA concentration in moles/liter × time in seconds at which 50% hybridization occurred) of the hybridization of the cDNA to its purified casein mRNA template was approximately 2×10^{-2} mol sec liter^{-1}.[77] The pooled casein mRNA fraction used for these studies was purified by immunoprecipitation of polysomes followed by poly(U)–Sepharose chromatography (see Section 2.2). On the basis of the hybridization kinetics, it therefore appears to be 120-fold purified over a total lactating RNA extract.[77] The variations in casein mRNA content determined using this cDNA hybridization probe were in good agreement with values previously obtained by cell-free translation, and only a small amount of nonspecific hybridization was observed with heterologous RNAs, such as rabbit reticulocyte polysomal RNA.[77] Houdebine and his colleagues utilized cDNA probes synthesized in this manner to quantitate casein mRNA levels during normal rabbit mammary gland development[78] and following the injection of hormones to pseudopregnant rabbits.[77,79]

Our laboratory has employed a slightly different approach for the generation of selective cDNA hybridization probes for the two purified rat casein mRNA fractions.[80] Using reverse transcriptase, conditions were developed to permit the synthesis of full-length complementary DNA copies of the 15 S and 12 S casein mRNA fractions, of approximately 1300 and 900 nucleotides in length, respectively. Generation of these representative cDNA probes was accomplished by: (1) the use of more highly purified enzyme preparations containing minimal nuclease activity; (2) increasing the incubation temperature to 46 from 37°C and reducing the incubation time to 15–20 min; (3) increasing the deoxynucleotide substrate concentrations from a minimal value of 2 μM to at least 100 μM; and (4) increasing the substrate-to-enzyme ratio. The following modifications of our original procedure are also employed: [^{3}H]-dCTP (100 μM, 20 Ci/mmol) is utilized instead of [H]-dGTP to minimize tritium exchange during long-term hybridizations; the remaining three deoxynucleotide triphosphates are present at 400 μM; sodium pyrophosphate is added at 4 mM final concentration to prevent synthesis of doublestranded DNA and increase the length of the cDNA product[81]; and KCl is present at 50 mM to increase the yield of cDNA.[64]

The cDNA probes synthesized using this protocol selectively hybridized to RNA from lactating tissue, but not to rat liver poly(A)-containing RNA.[80] The resulting hybrids displayed a high T_m, 88.5° in 0.2 M Na^+, characteristic of a well base-paired duplex. In addition, the rate of hybridization of the casein-specific cDNA to various RNA preparations was directly related to the casein mRNA activity of these same preparations determined by a cell-free translation assay. The rate of hybridization of the cDNAs with their template mRNAs was also consistent with the known complexity of these mRNA fractions. Greater than 90% hybridization occurred with cDNAs derived from the 15 S and 12 S mRNA fractions over an R_0t range of one and one half logs with $R_0t_{\frac{1}{2}}$ values of 0.0023 and 0.0032 mol sec liter^{-1}, respectively. Generally, less than 4% of the [^{3}H]-cDNA incubated in the absence of added mammary gland RNA was resistant to S_1 nuclease treatment. Compared with the total RNA isolated from lactating mammary tissue, the kinetics of hybridization indicated that the 15 S and 12 S rat casein mRNAs were purified 166- and 245-fold, respectively. The size of the full-length cDNA probes was determined both by alkaline sucrose gradient centrifugation and by formamide-containing PAGE. Definitive proof of the representative nature of the probe was obtained, however, by nuclease protection experiments in which ^{125}I-labeled 15 S casein mRNA was completely protected from nuclease digestion at a cDNA/mRNA of 2 (Fig. 7). Thus, the sensitivity, specificity, and representativeness of the rat casein

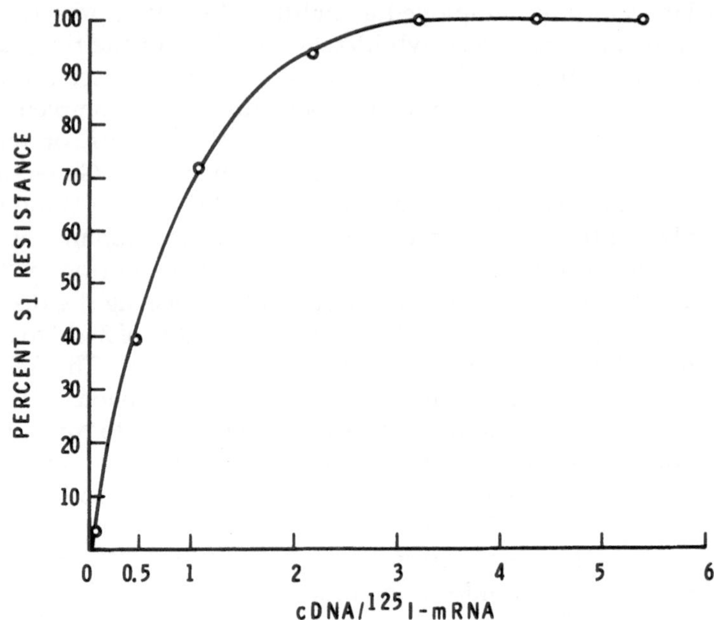

Fig. 7. Protection of [^{125}I]casein mRNA by [^{3}H]casein cDNA. The ability of full-length [^{3}H]-cDNA to protect [^{125}I]casein mRNA from S_1 nuclease digestion was determined following hybridization for 16 hr at 68°C to permit complete reaction at each cDNA/mRNA ratio tested. A control was employed omitting the cDNA.

cDNAs have been established, and they could be used to quantitate accurately the levels of casein mRNA sequences in both cellular RNA populations and in genomic DNA.

Using the self-priming ability of a full-length single-stranded cDNA copy of 15 S casein mRNA, a second strand was synthesized with avian myeloblastosis virus reverse transcriptase.[54] The double-stranded casein gene has been characterized by its resistance to S_1 nuclease (>98%), its chromatographic behavior on hydroxylapatite (85–99% double-stranded), and the ability of the two strands to renature after heat denaturation and quick cooling. The fidelity of base-pairing was demonstrated by its high T_m (91.5° in 0.21 M Na+), and its size was approximately 2600 nucleotides as estimated by formamide-containing PAGE. Preliminary restriction mapping and agarose gel electrophoresis indicated that a number of enzymes were capable of cleaving the double-stranded casein gene. These studies are a necessary prerequisite for the generation of chimeric DNAs and the insertion and amplification of the synthetic casein genes. Using such recombinant DNA technology, it should be possible to resolve the three casein structural gene sequences for use in future gene mapping and DNA sequencing determinations. These techniques will be necessary because cross-hybridization experiments indicated the presence of a limited amount of sequence contamination between the 15 S and 12 S casein mRNAs.[80]

3.2. Synthesis of Representative Probes for MMTV RNA

It is somewhat ironic that while endogenous reverse transcriptase activity was initially characterized and utilized in detergent-disrupted virions for the synthesis of cDNA,[82] it was not until purified eukaryotic mRNAs and exogenous reverse transcriptase were employed that conditions for the synthesis of full-length representative cDNA probes were elucidated.[64,65] Attempts to synthesize viral cDNA copies have been complicated by a poor efficiency of transcription, small DNA products, and unequal representation of the viral RNA sequences in the DNA product.[83] Only recently was the synthesis of long, representative DNA copies of several C-type viral RNAs accomplished.[81,84] In many cases, these procedures may require the isolation of undegraded viral RNA.[81] Application of these techniques to the synthesis of representative copies of MMTV RNA has therefore not been feasible because of the difficulty in isolating undegraded 35 S viral RNA from cell cultures, even using short harvest conditions. Furthermore, the B-type viral reverse transcriptase appears to be particularly labile, and the polymerase activity is especially sensitive to freezing and thawing.[85] The endogenous reaction has usually been employed for the synthesis of MMTV cDNA, resulting in both low yields of cDNA and a probe that was representative of only 60% of the RNA genome even at DNA/RNA ratios as high as 30 : 1.[86] A significant amount of double-stranded cDNA was also synthesized under these conditions, which in some cases may complicate the interpretation of hybridization data.[85]

To generate a probe representative of the whole MMTV genome, it was therefore necessary to develop a technique that did not require the presence of intact viral RNA and that permitted copying of a large, complex RNA (3×10^6 daltons) containing potential regions of extensive secondary structure.[87] This technique was based on the observations of Taylor *et al.*[68] that the efficient transcription of RNA into DNA by avian sarcoma virus polymerase could be accomplished by the use of random oligodeoxynucleotide primers. Fragments of DNase-treated calf thymus DNA (6–8 nucleotides in length) were utilized at a relatively high primer-to-template ratio to initiate the synthesis of short DNA copies along the entire length of the viral genome.[11,85] The resulting cDNA probe was an average of only 450 nucleotides long, but was able to protect 50% of the viral RNA from S_1 nuclease digestion at a cDNA/RNA ratio of 1. At a ratio of 10 : 1, complete protection of the RNA was observed, indicating the presence of a somewhat nonuniform, but representative, DNA copy of the 10,000-base-long viral RNA. This cDNA probe selectively hybridized to its viral RNA template with the expected kinetics of hybridization ($R_0t_{\frac{1}{2}}$ = 1.5 mol sec 1^{-1}), but not to RNA isolated from Moloney leukemia virus.[85] Furthermore, it was shown to be useful for quantitating viral RNA sequences in both producer (dexamethasone-treated-C3H) and nonproducer (BALB/c) cell lines.[85] The T_m of the viral RNA-cDNA hybrid was 88.5° in 0.2 M Na^+, indicating the presence of a well base-paired duplex. These results suggested that both specific and representative cDNA probes could be synthesized even to the large viral RNA using isolated viral RNA, reverse transcriptase, and random oligodeoxynucleotide primers. Thus, another marker of hormone-induced gene expression in the mammary gland, in addition to the milk protein mRNAs, was now available. The ability to quantitate both casein and viral gene expression during normal mammary development and tumorigenesis may help elucidate the mechanisms controlling differential gene expression (see Section 5).

3.3. Uses of cDNA Probes: Gene Dosage and mRNA Homology Studies

In addition to their use in quantitating the levels of specific RNAs in both *in vivo* and *in vitro* synthesized RNA populations, cDNA probes can be employed to determine the gene dosage of a specific sequence in genomic DNA. Selective transcription and mRNA processing rather than gene amplification of unique sequence genes, such as those coding for ovalbumin and globin mRNAs,[70,88] results in the high levels of these mRNAs observed in their respective specialized cell types. However, gene amplification has recently been implicated in the regulation of folate reductase mRNA levels in a methotrexate-resistant cell line.[89] Thus, the casein cDNA probe was utilized to answer the following questions: (1) Is casein mRNA transcribed from unique or repetitive sequences in the rat genome? (2) Does gene amplification or deletion occur during normal mammary differentiation? These questions could be answered by DNA excess hybridization as shown in Fig. 8.

DNA was first isolated from rat liver, lactating rat mammary tissue, and DMBA-induced rat mammary carcinomas. A large mass excess of DNA from each of these tissues was then sheared to a mean size of 350 nucleotides and incubated with high-specific-activity casein $cDNA_{15S}$ of similar size. The kinetics of reassociation of rat DNA and the rate of

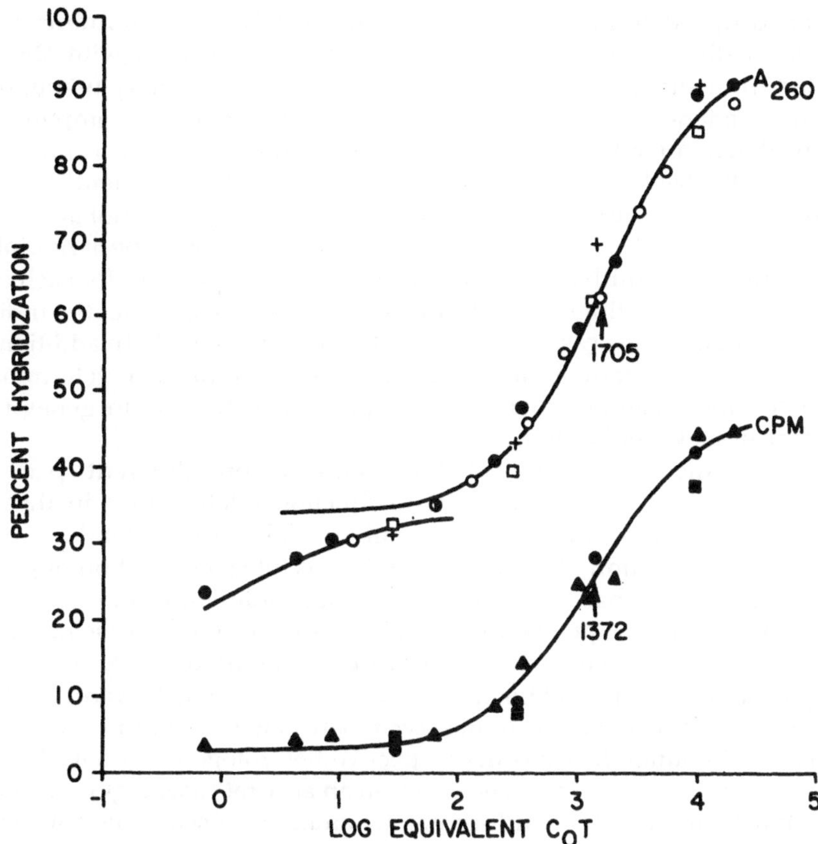

Fig. 8. Determination of the casein gene dosage in rat DNA. DNA excess hybridizations were performed as described using 737 μg rat liver DNA (●,○) and 5.08 × 10^{-5} μg [^{3}H]-cDNA$_{15S}$ (▲, specific activity = 3.9 × 10^6 cpm/μg) in a final volume of either 0.1 or 1.0 ml. The open and solid circles represent the results of two separate experiments. Comparable experiments were also performed with either 476 μg DNA isolated from lactating mammary tissue (+,⊕) or 662 μg DNA isolated from DMBA-induced mammary carcinomas (□,■) using the [^{3}H]-cDNA$_{15S}$ at a complementary sequence ratio similar to that used in the rat liver DNA experiment. The extent of hybridization was analyzed by hydroxylapatite chromatography as described. The A_{260} curve represents the renaturation of the respective nonradioactive DNAs; the CPM curve designates the hybridization of the [^{3}H]-cDNA$_{15S}$. Reprinted with the permission of *Biochemistry*.

[^{3}H]-cDNA hybridization were analyzed on hydroxylapatite (Fig. 8). The observed $C_0t_{\frac{1}{2}}$ of rat unique DNA of 1705 mol sec liter^{-1} was similar to previously published values.[90] The rate of hybridization of the [^{3}H]-cDNA indicated that the 15 S casein mRNA was transcribed from nonrepetitive DNA sequences. The observed $C_0t_{\frac{1}{2}}$ of 1372 mol sec liter^{-1} was similar to that determined for the rat unique DNA. Furthermore, no significant differences were observed in the hybridization with DNA isolated from either normal rat liver, differentiated mammary tissue, or mammary adenocarcinomas. This suggested that major gene amplification or deletion could not account for the alterations in casein mRNA activity observed during normal mammary development[21] or tumorigenesis[91] (see Sections 4.1 and 5). This technique cannot at present distinguish between the presence of one gene copy specifying all three of the casein mRNAs or an individual gene complement for each of the mRNAs. Interestingly, genetic evidence has suggested that the individual α and β bovine casein genes may be closely linked.[20] In addition, more sensitive gene-mapping techniques are required to determine whether small deletions or inserts are present in the casein genes in different tissues (see Section 7).

Comparative studies of mRNA sequences among different species may help elucidate essential structure–function relationships in these RNAs. Such studies have been especially useful in understanding the evolution and function of globin mRNA.[5,92] While conservation of globin gene sequences has been observed during evolution from rabbits to man, considerable sequence divergence was expected for casein. Different apparent molecular weights have been reported for the mouse, rat, guinea pig, sheep, and bovine caseins.[17,20,21,24,27] Furthermore, only a limited cross-reactivity was observed between the specific rat and mouse casein antibodies and the respective heterologous proteins.[24,54] No cross-reactivity was observed between an anti-rat casein IgG fraction and a bovine α_s-casein.[21] Unfortunately, amino acid sequence data are not available to permit a direct comparison of the protein interspecific homologies.

To measure the extent of sequence divergence between the purified rat and mouse 15 S casein mRNAs, both the rates and fidelity of homologous and heterologous hybridization were determined.[80] The rate of hybridization of the rat $cDNA_{15S}$ with mouse 15 S mRNA was approximately one fifth as fast as the homologous rat cDNA–mRNA hybridization, and both reactions went to at least 90% completion. In addition, the T_m of the heterologous hybrid was approximately 11.5° lower than the homologous hybrid, indicating approximately 17% mismatching. Thus, it appears that considerable sequence divergence has occurred between the casein mRNAs of two closely related species. An extension of these studies to other species should yield useful information concerning the evolution of the casein genes and presumably the conservation of specific regions of the mRNAs and proteins that are required for their respective functions.

4. Changes in Casein mRNA Levels and Casein Synthesis during Normal Mammary Gland Development

4.1. Casein mRNA Activity and Sequence Concentration during Normal Mammary Gland Development

Development of the mammary gland during pregnancy is characterized by an increased synthesis of rRNA, polysomes, and tRNA,[93,94] including changes in the ratios of specific isoaccepting species of tRNA.[95] This increase in RNA synthesis is accompanied by the extensive development of endoplasmic reticulum and the appearance of membrane-bound polysomes.[59,96,97] Furthermore, an increased DNA content[98] and the proliferation of alveolar cells occurs,[99] resulting in a highly sophisticated protein-synthetic machinery capable of secreting several grams of casein per day during lactation.[100]

Developmental changes in both mRNA levels and the components of the protein synthetic apparatus have been observed in other hormonally inducible systems, notably during estrogen-induced egg white protein synthesis in the immature chick oviduct[101] and estrogen-induced vitellogenin synthesis in the male *Xenopus* liver.[102] In both of these systems, increases in specific mRNA levels are accompanied by changes in rRNA and tRNA synthesis, endoplasmic reticulum biosynthesis, levels of protein synthesis initiation factors, and membrane-bound polysomes, eventually resulting in the synthesis and secretion of large quantities of a specific protein. These types of coordinated changes are therefore characteristic of many specialized cells containing hormonally inducible gene products. The regulation of casein synthesis and secretion would therefore not be expected to be a "all-or-none" phenomenon dependent solely on the induction of casein mRNA.

Two different methods have been employed in our laboratory to determine the levels of casein mRNA present during normal mammary gland development: cell-free translation and molecular hybridization. In our initial studies, the variations in both casein mRNA activity and total mRNA activity during mammary gland differentiation were determined by assaying total RNA preparations obtained from pregnant, lactating, and regressed (7 days after weaning) mammary tissue in the wheat germ translation system.[21] Glands were pooled from three to five animals in each case. Several concentrations of each total RNA preparation were assayed and the specific activities determined from the initial linear portions of the assay. The casein mRNA specific activity increased 20-fold, while total mRNA specific activity increased only 2-fold between 5 days of pregnancy and 2 days of lactation. Accordingly, casein mRNA activity increased from approximately 7% of the total mRNA activity to 60% during this period. After mammary gland regression following weaning, there was a rapid loss in casein mRNA activity to a barely detectable level, while total mRNA activity was reduced to a value comparable to that found in a 5-day pregnant mammary gland. When the selective increase in casein mRNA activity was corrected for a 2- to 3-fold increase in RNA recovery, reflecting the increased content of RNA present during mammary gland development, a 60-fold overall increase in the total casein mRNA activity resulted between early pregnancy and lactation. During mammary gland development, however, the percentage of fat cells is markedly reduced, and glandular tissue may represent as much as 75% of the total gland during lactation.[99] When the casein mRNA activity was therefore expressed relative to the percentage glandular tissue, the increase in total casein mRNA activity now represented only an 8- to 11-fold change from day 5 of pregnancy until lactation. Thus, the selective induction of casein mRNA activity that occurred during pregnancy and early lactation was accompanied by a proliferation of alveolar cells and resulted in a marked increase in the total concentration of casein mRNA in the lactating rat mammary gland. Moreover, a selective loss in casein mRNA activity was observed during mammary gland involution.

Because of the possibility that different mRNAs may be translated with different efficiencies in cell-free translation systems (see Section 2.1), a more quantitative assay was also used to measure the levels of casein mRNA sequences during normal mammary development. The specific casein $cDNA_{15S}$ probe was hybridized with an excess of RNA isolated from mammary tissue obtained from a 6-month-old virgin animal or at different stages of pregnancy, during lactation, and following regression of the gland after weaning.[80] The extremely sensitive and quantitative cDNA probe was able to detect a limited amount of casein mRNA even in the virgin mammary gland.[103] A series of parallel hybridization curves were generated that displayed progressively faster rates of hybridization. The slowest rate was observed with RNA extracted from virgin mammary tissue, while a maximal rate was found using RNA obtained from 8-day lactating tissue. Using these hybridization data and the equivalent $R_0t_{\frac{1}{2}}$ for the highly purified 15 S casein mRNA, the percentage of casein mRNA in each total RNA extract was determined as follows: 100 × $R_0t_{\frac{1}{2}}$ Pure mRNA Back Hybrid ÷ $R_0t_{\frac{1}{2}}$ Sample RNA. As shown in Fig. 9, casein mRNA sequences represented 0.52% of the total cellular RNA in the 8-day lactating tissue, a 19-fold increase over the amount present at 5 days of pregnancy and an overall 300-fold increase relative to the virgin mammary gland. Since the RNA

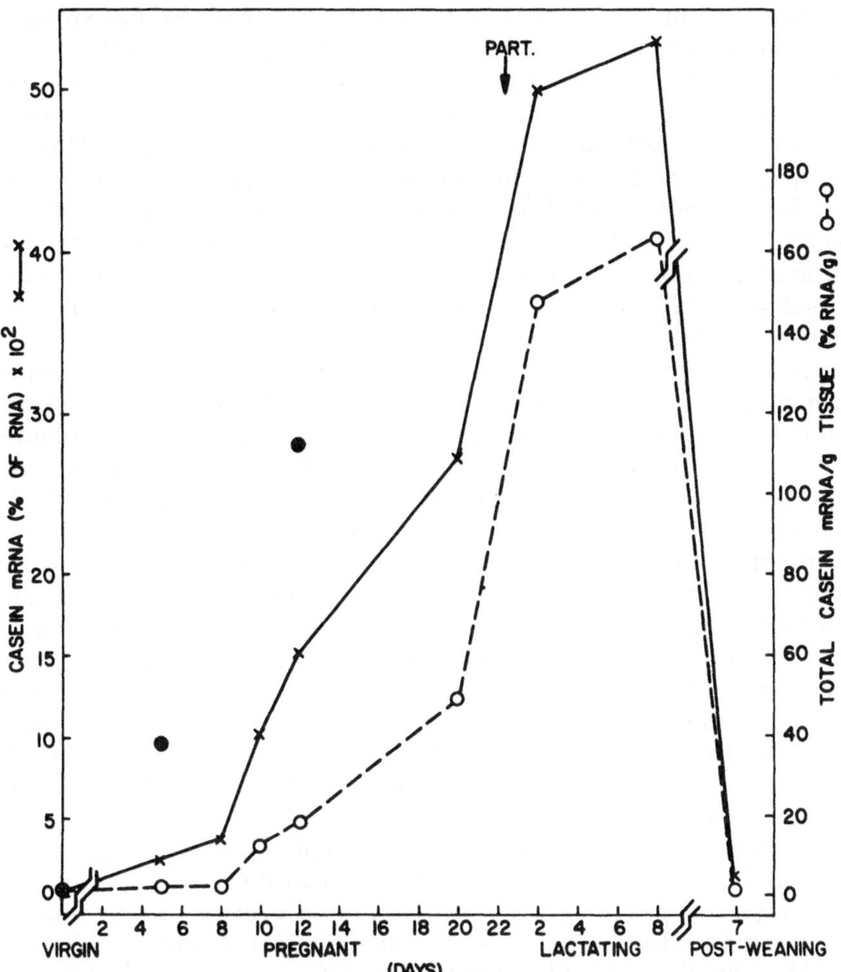

Fig. 9. Casein mRNA levels during rat mammary development. The data are expressed as either the percentage of RNA extracted (×) or the total casein mRNA per gram of tissue (○), which reflects the increasing levels of RNA per gram during mammary development. The closed circles represent the levels of casein mRNA observed 48 hr following ovariectomy of either a 5-day or a 12-day pregnant animal, expressed as the percentage of RNA.

content of the lactating gland is considerably greater than that of the virgin, this resulted in almost a 4000-fold increase in the total number of casein mRNA molecules per gram of tissue during this developmental period.[103] Following regression of the gland, the level of casein mRNA sequences decreased markedly until they comprised only 0.014% of the total RNA.

The amount of 15 S casein mRNA sequences can also be expressed as the number of molecules of casein mRNA per alveolar cell.[80] A 12-fold increase in the number of molecules per alveolar cell was observed between 5 days of pregnancy and 8 days of lactation, reaching a maximal level of 79,000 molecules of the 15 S casein mRNA per cell. Thus, in highly specialized tissues, which are producing large amounts of a given protein, it is not unusual to observe mRNA levels as high as 80,000–100,000 specific mRNA molecules per cell. For example, the levels of casein mRNA observed during lactation were comparable to the number of ovalbumin mRNA molecules determined per tubular gland cell in the fully stimulated hen oviduct.[3]

A surprisingly high level of casein mRNA was, however, found during midpregnancy (10 to 14 days pregnant) by both the translation and hybridization assays. The midpregnant mammary gland contained between one fourth and one third concentration of casein mRNA and approximately 8–16% the total amount of casein mRNA present at 8 days of lactation (Fig. 9). A similar observation was recently reported during pregnancy in the rabbit, where casein mRNA was detected well before the onset of lactation.[78] A low level of casein mRNA was also detected in the virgin rabbit mammary gland in these studies using a 200-nucleotide-long rabbit casein mRNA–cDNA probe. In the rabbit, however, a 900-fold increase in total casein mRNA content was reported between midpregnancy and lactation, whereas in the rat mammary gland, this increase was approximately 10-fold. This 90-fold greater relative difference reflects the decreased amount of casein mRNA observed during mid-pregnancy in the rabbit vs. the rat, rather than increased levels present during lactation. Thus, by comparing the changes in casein mRNA during mammary development in these two species with the known changes in several serum hormone levels, it may be possible to determine which hormones are the principal regulators of casein mRNA accumulation *in vivo* (see Section 4.4).

A coordinate increase in both casein mRNA sequences and translation activity was observed during the development of the rat mammary gland from pregnancy to lactation when the results of both the hybridization and cell-free translation assays were compared.[80] These data suggested that the regulation of casein synthesis most likely did not involve the activation of previously inactive mRNA. However, these results could not rule out the possibility that casein mRNA was sequestered in the nucleus or as inactive mRNP particles in the cytoplasm during pregnancy. A direct analysis of the level of casein mRNA associated with polysomes capable of synthesizing casein *in vitro* was necessary to determine whether "stored mRNA" existed during pregnancy. The results of these studies are described in the following section.

4.2. Relationship of mRNA Levels to Casein Synthesis and Secretion

The exact relationship between the levels of casein mRNA and the rates of casein synthesis *in vivo* is difficult to determine directly. Several problems are inherent in this analysis. First, the inability to incorporate sufficient radioactive amino acids into casein *in vivo* precludes the direct determination of the rate of casein synthesis by specific immunoprecipitation. Second, if the more quantitative double antibody precipitation assay is employed (see Section 2.1), high levels of endogenous nonradioactive caseins may successfully compete for the binding of the radioactive *de novo* synthesized casein. These problems have been circumvented to some extent by using the pseudopregnant rabbit, which has low levels of endogenous casein,[31] by the use of mammary explants in which to perform labeling studies,[31,77–79] and by increasing the titers of the anticasein antibody by supplementation with a purified anticasein IgG fraction.[77–79] The results of such studies[31,77–79] demonstrate that a qualitative agreement usually exists between the levels of casein mRNA, determined either by hybridization or cell-free translation, and the levels of casein synthesis. Thus, increases in casein mRNA observed after hormone treatment or during mammary development are usually accompanied by increased levels of casein synthesis and vice versa. Before the stoichiometry of casein mRNA and casein synthesis can be determined, however, it will be necessary to measure the rates of casein synthesis and turnover, the transit time of casein mRNA on mammary polysomes, and the levels of functional polysomal casein mRNA. These experiments will require an *in vitro* system in which the processes named above can be quantitated following the addition of various peptide and steroid hormones.

Since the previously mentioned problems prevented the direct measurement of casein synthesis in the midpregnant rat, an alternative approach was taken. A procedure was developed for the isolation of functional rat mammary polysomes, which were capable of synthesizing casein *in vitro*.[103] Polysomes were then isolated from either midpregnant or lactating tissue. A comparison of polysome profiles between midpregnancy and lactation indicated a shift from monosomes, disomes, and trisomes to larger polysomes containing 7–11 ribosomes.[103] The concentration of casein mRNA increased only two- to threefold during this period. These results suggested that the polysome size distribution during midpregnancy may not reflect merely the level of casein mRNA, but instead the efficiency of initiation of protein synthesis. The efficiency of initiation of casein synthesis may be influenced by such variables as the specific isoaccepting tRNA population,[104] the availability of endoplasmic reticulum, and an increase in protein synthesis initiation factors.[105] For example, in the rabbit, a shift in the total polysome profile from monomeric forms to polymeric forms was also observed during pregnancy,[78] and this shift was accompanied by a progressive increase in the proportion of polysomes bound to membranes.[106] It was suggested that the actual secretion of casein may be influenced by the proportion of free and bound polysomes.[58] In addition, during prolactin-stimulated lactogenesis in the pseudopregnant rabbit, the capacity of the tissue to synthesize casein mRNA and casein may actually increase more rapidly than the capacity of the tissue to secrete casein.[58] A similar phenomenon is probably occurring during midpregnancy in the rat. However, difficulties in the isolation and quantitation of free and membrane-bound polysomes from pregnant rat mammary tissue preclude the direct assessment of this hypothesis at this time. These studies will require improved homogenization and polysome isolation procedures.

The ability of these isolated rat polysomes to synthesize protein, and specifically casein, was next determined by incubation in a cell-free translation assay containing a salt wash of rabbit reticulocyte polysomes. Polysomal casein synthesis was determined from the linear portion of the total protein synthesis assay by specific immunoprecipitation. Casein synthesis was shown to comprise 21% of the total synthetic capacity of polysomes isolated from 15-day pregnant mammary tissue, and it was greater than 40% of the total protein synthesis in a lactating polysome preparation. Casein mRNA was therefore associated with mammary gland polysomes in the pregnant rat, and these polysomes had the capacity to synthesize casein *in vitro*. Furthermore, the levels of casein mRNA displayed an excellent correlation with the changes in polysomal casein synthesis observed in the midpregnant and lactating rat.[103]

Although mammary polysomes isolated from midpregnant animals had the capability of synthesizing casein *in vitro* when supplemented with heterologous protein synthesis factors and tRNA, this was not proof that casein was being synthesized *in vivo*. Furthermore, if casein synthesis was occurring *in vivo* during pregnancy, the fate of the protein in the absence of secretion was unknown; i.e., was casein being stored in anticipation of subsequent lactation or was the protein turning over prior to lactation and the initiation of secretion? To answer these questions, the levels of casein present in the 105,000*g* supernatant of extracts of pregnant mammary tissue were measured using a radioimmunoassay procedure.[103] The [^{3}H]casein was generated in the wheat germ translation assay using a partially purified rat casein mRNA fraction, rather than using the more commonly employed radioiodinated protein. The assay was capable of detecting casein in amounts as low as 1 ng and was sensitive to 100 ng. Triton-X-100 was added to the homogenization buffer to solubilize any stored casein in the tissue, and a proteolysis inhibitor was utilized to minimize casein degradation. With this method, casein was detected in the soluble protein of a 14-day pregnant mammary gland homogenate. Thus, it appeared that casein synthesis was occurring *in vivo* during pregnancy in the rat, and that at least some of the newly synthesized casein was stored in the pregnant gland prior to lactation.

4.3. Effects of Ovariectomy and Progesterone Administration during Midpregnancy

Ovariectomy of the midpregnant rat and hormone replacement have been used as a model for studying the regulation of both casein mRNA levels and polysomal casein synthesis *in vivo.*[103] Previous studies by Liu and Davis[106] demonstrated that ovariectomy of rats approximately midway through pregnancy induced a lactationlike response, characterized by the appearance of a milklike secretion with the immunological properties of casein. This response could be blocked by the administration of progesterone at the time of ovariectomy.[107] The 12- to 14-day pseudopregnant rabbit has also been used as an alternative model in which to study the hormonal regulation of casein mRNA and casein synthesis.[31,77–79] Administration of prolactin usually twice daily in doses of 12.5 I.U./injection was accompanied by either twice-daily progesterone injections of 5 mg each or hydrocortisone acetate at 2 mg or 7.5 mg per injection.

The overall conclusions obtained using these systems was that the levels of casein mRNA and casein synthesis observed during midpregnancy may reflect both the positive and negative effects of several peptide and steroid hormones. Thus, two peptide hormones, prolactin and placental lactogen, have been shown to induce casein synthesis[108,109] and casein mRNA[31,77,110] both *in vivo* and in mammary gland organ culture. Furthermore, high levels of progesterone during pregnancy may antagonize the lactogenic effects of these hormones.[107,111] Finally, while the continued presence of hydrocortisone is not required for prolactin induction of casein mRNA, hydrocortisone does potentiate the action of prolactin. Results of both *in vivo* and *in vitro* studies demonstrated that hydrocortisone alone does not stimulate casein mRNA synthesis.[79,110]

The role of progesterone as the principal hormone suppressing lactation during pregnancy was given additional support by the experiments performed in our laboratory using the ovariectomized midpregnant rat.[103] Removal of progesterone by ovariectomy resulted in an increased level of casein mRNA, increased polysomal casein synthesis, an increased concentration of intracellular casein, and finally the appearance of a white, milklike secretion. Following ovariectomy, a twofold increase in both casein mRNA activity, determined in a wheat germ translation assay, and casein mRNA sequence concentration, measured using the selective cDNA hybridization probe, was observed (Fig. 9). This effect on mRNA accumulation was first observed within 16 hr following ovariectomy and was maximal between 24 and 48 hr. Progesterone, but not estradiol or hydrocortisone, administered at the time of ovariectomy prevented the increase in casein mRNA. Administration of progesterone after maximal induction of casein mRNA was obtained, either following ovariectomy or during lactation, was unable to significantly reduce the levels of casein mRNA.

Following ovariectomy of a midpregnant rat, a similar shift in polysome profiles from monosomes, disomes, and trisomes to the larger polysomes containing 7–11 ribosomes found during lactation was also observed.[103] Accordingly, a twofold increase in polysomal casein synthesis was detected, reaching a level comparable to that observed in a lactating polysome preparation of greater than 40%. Thus, following ovariectomy, a twofold increase in both mRNA activity and polysomal casein synthesis was observed. Progesterone administration at the time of ovariectomy blocked both these responses and increased the amount of small polysomes present. Ovariectomy also resulted in an increased level of intracellular casein as detected by radioimmunoassay and the appearance of a white, milklike fluid.[103]

Studies using the pseudopregnant rabbit mammary gland also demonstrated that progesterone administered either prior to or simultaneously with prolactin will inhibit the induction of casein mRNA, casein synthesis, and subsequent lactogenesis.[31,77] This suggests that the continuous presence of progesterone is necessary during pregnancy to exert its inhibitory effect on lactogenesis.[112] Once maximal induction is obtained, the marked inhibitory effect of progesterone is not observed. This observation may have important consequences for understanding the mechanism of action of progesterone in inhibiting lactogenesis. Presumably, a direct competition with lactogenic hormones at the transcriptional level to regulate the synthesis of casein mRNA is unlikely.

Progesterone appears to regulate casein synthesis and secretion in a pleiotropic fashion. At least three different potential mechanisms of action have been suggested: (1) Progesterone has been reported to counteract the self-regulated increase in prolactin receptors observed in the mammary gland.[113] This may account for the relatively low levels of prolactin receptors detected in the rabbit mammary gland during pregnancy,[114] and may result in a reduced ability of prolactin or placental lactogen to induce casein mRNA. (2) Progesterone is also an effective competitor for glucocorticoid receptor binding sites in the mammary gland,[115] and therefore may prevent the glucocorticoid-induced development of the rough endoplasmic reticulum necessary for lactogenesis.[59] (3) Finally, a specific, unique progesterone receptor has been demonstrated in mammary carcinomas.[116] However, until the characterization and function of a progesterone receptor in a normal mammary tissue is demonstrated, the possibility that a unique progesterone receptor may mediate direct effects of progesterone distinct from those previously discussed is purely speculative. These potential multiple sites of progesterone action may explain the coordinated responses to ovariectomy observed during midpregnancy, i.e., an increase in casein mRNA levels, an increase in large polysomes and casein synthesis, and finally increased levels of intracellular casein and the initiation of secretion.

4.4. Relationship to Serum Hormonal Variations

The observed alterations in casein mRNA levels (Fig. 9) may be correlated with the serum levels of prolactin and placental lactogen, which undergo marked changes in the rat during pregnancy, lactation, and after weaning.[117,118] Thus, the small amount of casein mRNA activity observed in the early pregnant (0–7 days) mammary gland may result from the increase in serum prolactin (50 ng/ml) that occurs in the rat after coitus, followed by elevated levels (20–30 ng/ml) for the first 3 days of pregnancy.[118] The additional increase in casein mRNA that occurs at approximately day 8 of pregnancy may then be attributable to the vast increase in rat placental lactogen that occurs at this time, reaching levels as high as 1200 ng/ml by day 12 of pregnancy.[119] The 90-fold greater difference in the amount of casein mRNA observed in the rat vs. the rabbit during midpregnancy may reflect these high concentrations of placental lactogen in the rat, compared with very low levels observed in the rabbit; e.g., a maximal level of only 25 ng/ml was found at day 30 of pregnancy.[119] Thus, placental lactogen may be of primary importance in initiating rat mammary gland development during pregnancy.

While serum placental lactogen levels decrease markedly at parturition, prolactin levels are known to increase to approximately 30 ng/ml just prior to parturition. A further increase to levels as high as 300 ng/ml is observed within 9–10 hr postpartum.[120] This may account for the additional rise in casein mRNA reported during early lactation. Following weaning, prolactin levels fall to a basal level of only 10 ng/ml, and accordingly, casein mRNA activity is barely detectable. Thus, prolactin and placental lactogen are performing a dual role in the mammary gland; they initiate alveolar differentiation and proliferation, as well as selectively induce casein synthesis.[109]

These effects of the two peptide hormones are also modulated by several steroid hormones. Thus, in the rat, the plasma level of progesterone increases as early as day 4 of pregnancy to levels of 80 ng/ml and reaches levels of 120 ng/ml during midpregnancy.[118] The inhibitory effect of progesterone observed in our experiments required the administration of several daily injections of the steroid.[103] This was necessary to maintain continuous high serum levels of progesterone in the presence of a high metabolic clearance rate for the steroid,[121] although a comparable inhibitory effect by continuous infusion of low levels of progesterone (12–48 μg/hr) was also reported.[107] Thus, the continuous presence of high serum levels of progesterone during pregnancy may be necessary to inhibit the initiation of lactation. The dramatic fall in serum progesterone that occurs at parturition may then be the signal for the onset of lactation. Finally, plasma corticosteroid levels during midpregnancy in the rat were reported to be comparable to serum progesterone levels approaching levels of 200 ng/ml.[122] Thus, the development of secretory capacity may depend on the relationship between the stimulatory effects of corticosteroids and the inhibitory effect of progesterone during pregnancy.[59]

Because of the complex hormonal interrelationships that occur in the mammary gland, it is extremely difficult to elucidate the exact mechanism by which these hormones control casein gene expression *in vivo*. The preceding discussion was intended to be quite speculative, and many of the relationships discussed above must still be established. Clearly, hormonal effects on casein synthesis may be mediated at several levels within the cell. Furthermore, an increase in casein mRNA observed following the administration of a given hormone *in vivo* should not always be construed to mean that this hormone is acting to stimulate gene transcription. To unravel these complex relationships, we have utilized a well-defined *in vitro* model system (see Section 6.1).

5. Hormonal Regulation of Milk Protein mRNAs in Neoplastic Tissue

The expression of differentiated function in hormone-responsive breast cancer was previously reported in both experimental[59,91,123–125] and human breast cancer.[126–130] Histological examination revealed the presence of secretory activity in a small proportion of 7,12-dimethylbenz(a)anthracene (DMBA)-induced rat mammary tumors[125] and in a transplantable rat mammary adenocarcinoma.[124] In both cases, estradiol treatment caused a lactationlike response in the tumors, and casein was identified in the secretory fluid. Both primary DMBA- and nitrosomethylurea (NMU)-induced rat mammary carcinomas were recently reported to contain α-lactalbumin at levels equal to or less than 10% of the amount found in the 5-day lactating rat mammary gland.[123] Transplantation of a pituitary gland under the kidney capsule of the host, which leads to elevated serum prolactin levels, increased the α-lactalbumin content in the primary DMBA-induced tumors, but unexpectedly reduced the levels observed in the NMU-induced tumors. α-Lactalbumin was also detected in the transplantable R3230AC mammary carcinomas, but levels did not increase with pituitary hormone stimulation.(123)

Studies in our laboratory demonstrated the presence of casein mRNA in approximately 70% of the more than 30 DMBA-induced tumors assayed by molecular hybridization.(91) However, casein mRNA levels were usually only 10% or less of those observed in an 8-day lactating mammary gland. The highest levels of casein mRNA were found in prolactin- and estradiol-treated animals, compared with animals given estradiol treatment alone or in the absence of exogenous hormonal administration. These studies were all performed using the technique of molecular hybridization, because the low levels of casein mRNA present in these tumors could not be accurately measured by cell-free translation assays. In many cases, the levels of activity measured by cell-free translation assay using total tumor RNA extracts were barely above the trapping levels obtained during specific immunoprecipitation. Using the wheat germ cell-free translation assay, prolactin-inducible casein mRNA activity was also demonstrated in the transplantable R3230AC mammary carcinoma.(56) The maximal levels observed were, however, only 1% of the total mRNA activity. Interestingly, in these studies, no effect of prolactin treatment or perpherazine administration was observed on α-lactalbumin mRNA activity, suggesting the possible independent regulation in this tumor line of the milk protein genes specifying casein and α-lactalbumin mRNAs. The significance of these data is uncertain, however, because α-lactalbumin mRNA activity represented less than 0.2% of the total mRNA activity in the R3230AC mammary tumors, and 70–75% of the "specific" α-lactalbumin mRNA activity was not competable with an excess of unlabeled α-lactalbumin. They are in agreement, however, with the previously cited data concerning the lack of pituitary hormone stimulation on tumor α-lactalbumin(123) levels.

At present, it is not possible to determine whether the low levels of casein mRNA observed in experimental breast cancer are due to a limited proportion of the tumor cell population actively synthesizing casein or a decreased response in the entire cell population. However, it is clear from both histological examination and autoradiographic localization of prolactin receptors using [^{125}I]prolactin that heterogeneity of cell types exists within these tumors.(149) Thus, to correlate the hormonal dependency of tumor growth with hormonally induced differentiated function, it will be necessary to identify those cell types that produce casein mRNA and casein. This may be accomplished by either *in situ* hybridization using casein cDNA or immunological localization of casein using fluorescent antibodies.

The absence of casein mRNA in several hormone-independent mammary carcinomas suggested that casein mRNA as detected by cDNA hybridization might be a useful molecular marker for determining hormonal dependence.[91] Whereas the majority of the DMBA-induced mammary adenocarcinomas in Sprague–Dawley rats are dependent on hormones for growth, those that appear in BALB/c mice are characteristically hormone-independent. Using a homologous mouse 15 S casein mRNA–cDNA probe, a number of transplantable mouse mammary tumor lines were also screened for the presence of casein mRNA.[91] In none of these autonomous mouse mammary tumor RNA samples was any significant hybridization detected. These results are illustrated in Table II.

Using probes for both the MMTV and casein mRNAs, we recently quantitated the levels of both the viral and milk protein mRNAs during mouse mammary tumorigenesis.[131] Appreciable levels of casein mRNA were observed in the midpregnant mouse mammary gland. Approximately a 14-fold increase in casein mRNA occurred between midpregnancy and early lactation in the mouse, a slightly greater increase than reported during the same period of rat mammary development. However, very low levels of casein mRNA were detected in the D-2 hyperplastic alveolar nodule tissue and no detectable sequences were found in the autonomous D2 mouse tumors. Thus, the expression of the casein gene appears to be repressed during mouse mammary tumorigenesis. Interestingly, the converse appears to be occurring with respect to MMTV RNA. Thus, very low levels of MMTV RNA were

Table II
Quantitation of MMTV and Casein mRNA during Murine Mammary Tumorigenesis[a]

Strain	Tissue source	MMTV RNA (%)[b]	15 S Casein mRNA (%)[b]
BALB/c	10-day pregnant	N.D.	0.0097
BALB/c	5- to 7-day lactating	N.D.	1.43
BALB/c	D2 nodule	0.119	0.003
BALB/c	D2 tumor	0.058	0.0015

[a] Total nucleic acid extracts were prepared by the SDS–phenol extraction method. This was followed by three extractions with 3 M NaOAc, pH 6.0, which removed DNA and small 4 S and 5 S RNAs. MMTV RNA and casein mRNA sequences were quantitated by cDNA excess hybridization. The results are expressed as the percentage of each RNA preparation composed of MMTV or casein mRNA as determined by a comparison with the rate of hybridization of the pure MMTV and casein mRNAs measured under identical conditions.

[b] Not detectable at levels of less than 0.0005% for MMTV cDNA. These minimal levels of detection were estimated by the background level of hybridization of these probes with mouse liver total RNA.

observed in normal mammary tissue. However, rather high levels of endogenous virus expression were found in the D2 nodules and a significant amount of MMTV RNA sequences were still present in the transplantable D2 tumors. Differential regulation of these two hormone-inducible gene products may therefore be occurring during mouse mammary tumorigenesis. The expression of differentiated function appears to be repressed, while viral gene transcription is stimulated. The relationship between these changes and mammary tumorigenesis remains to be established. These results suggest, however, that casein mRNA may be a useful molecular marker for hormone dependence, at least in experimental breast cancer.

Studies of differentiated function have recently been extended to human breast cancer.[126–130] Detection of casein and in some cases α-lactalbumin by immunofluorescence in areas of structural differentiation of human tumors or by radioimmunoassay of tumor cytosols has been reported. Casein may be not only a useful marker for the presence of hormonal dependency and differentiated function in breast cancer, but also a potential tool for the early detection of mammary tumors. In a preliminary study, radioimmunoassay of casein in the serum of normal and breast cancer patients indicated the presence of casein in patients with both primary and metastatic breast cancer, but not in the serum of nonpregnant women.[133] These studies were recently extended by radioimmunoassay of K-casein and four other putative tumor-specific markers in the serum of breast cancer patients.[132] The K-casein levels were found to be particularly high in the serum of patients with the first clinical stages of breast cancer or metastatic disease. In a recent report from a different laboratory, however, negative results were obtained using a similar radioimmunoassay procedure for total human casein in the serum of breast cancer patients.[130] Thus, the usefulness of these determinations of casein and α-lactalbumin as a diagnostic assay in human breast cancer is currently unresolved. Furthermore, the secretion of large amounts of casein is probably not a characteristic of most human breast cancers.[130] However, studies correlating casein mRNA levels and hormone dependency in human breast cancer have yet to be performed.

6. Model Systems for Studying Hormonal Regulation of Gene Expression

6.1. Prolactin Induction of Casein mRNA in Organ Culture

To elucidate the mechanism by which peptide hormones regulate gene expression, a system is required in which the rapid induction of a specific mRNA can be accurately determined following the addition of the hormone. An ideal model system should permit pulse-labeling studies of the specific mRNA, so that the rates of mRNA synthesis and degradation can be determined. In addition, quantitative measurements will require the availability of a specific cDNA hybridization probe synthesized from a well-characterized purified mRNA. Since mammary gland organ culture is conducted in a serum-free, chemically defined medium, this system allows a detailed examination of the mechanism by which a peptide hormone regulates the synthesis of a specific mRNA, and by which this response is modulated by the interaction with other hormones. Furthermore, since no cloned, prolactin-responsive mammary epithelial cell line is available for study at present, organ culture becomes the system of choice in which to study peptide hormone regulation of a specific mRNA. Few other such systems are currently available, although recent progress in the study of growth-hormone-inducible α-2-U-globulin mRNA in rat liver[134] suggests that this may be another useful experimental model. In this section, the initial characterization of the prolactin induction of casein mRNA and the effect of two steroid hormones, hydrocortisone and progesterone, on this response are described.

Prolactin-induced casein synthesis in virgin explants has been reported to require mammary differentiation prior to the appearance of casein synthesis.[136] Thus, midpregnant mammary gland explants were chosen rather than explants derived from virgin tissue to study the early effects of prolactin in preexisting differentiated alveolar cells. Because of the high levels of casein mRNA previously reported during midpregnancy in the rat,[21,80] explants were initially exposed for 48 hr to a medium containing insulin and hydrocortisone. This was necessary to allow the turnover of the endogenous casein mRNA prior to the addition of prolactin. Following this preincubation with insulin and hydrocortisone alone, prolactin was added, and the levels of casein mRNA determined by cDNA hybridization after an additional 24 or 48 hr.

The response to prolactin was dependent on the day of pregnancy at which the tissue was removed and placed in organ culture. Thus, a 3.5-, 8.5-, and 12-fold induction of casein mRNA, as compared with baseline, was observed when organ explants were obtained from 7-, 10-, and 15-day pregnant rats, respectively.[135] Within 48 hr after the addition of prolactin, a 12- and 25-fold induction of casein mRNA was observed when organ explants were obtained from 10- and 15-day pregnant rats, respectively. These results were consistent with the differentiation of the mammary gland that occurs during pregnancy and the increased percentage of alveolar cells present at day 15 of pregnancy compared with day 7. The observed fold induction was also dependent on the levels of casein mRNA remaining in the insulin–hydrocortisone controls, which had not received prolactin, presumably reflecting the turnover of preexisting casein mRNA during the time in culture.

The kinetics of the prolactin effect on the synthesis of casein mRNA were next examined using the casein cDNA hybridization probe (Fig. 10). A 1.3-fold induction of casein mRNA was observed within 1 hr after the addition of prolactin. Casein mRNA sequences continued to accumulate for 48 hr, reaching a maximal level 13.4-fold greater than the

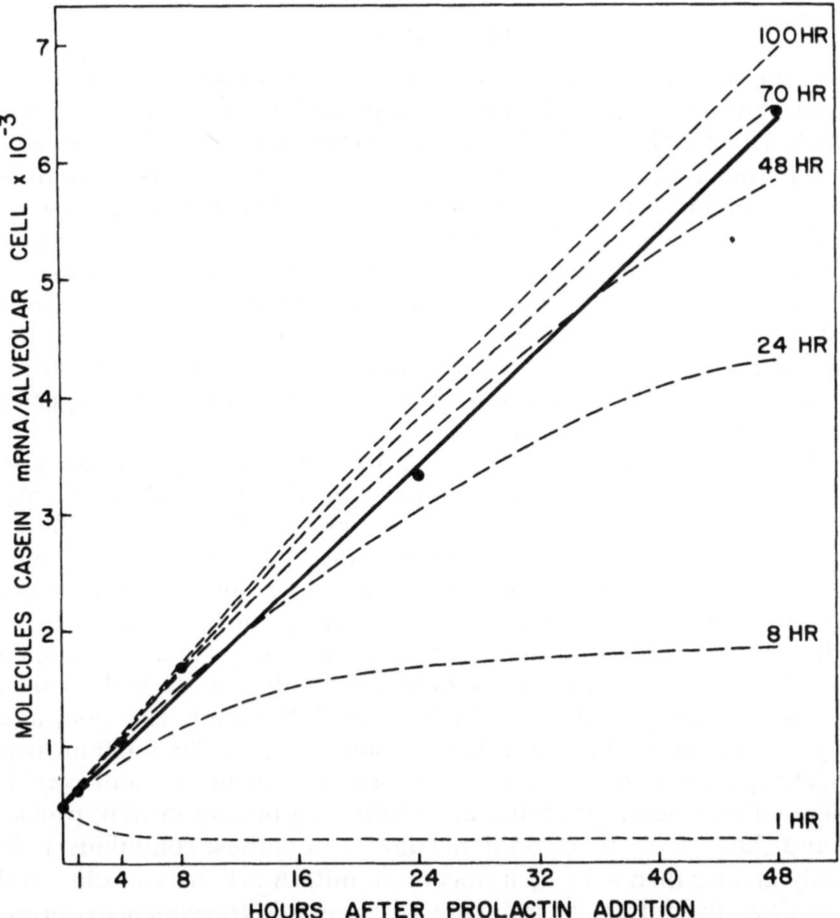

Fig. 10. Analysis of prolactin-induced casein mRNA accumulation. The observed kinetics of accumulation (●——●) were analyzed and compared with theoretical mRNA accumulation curves predicted for different casein mRNA half-lives (---). The theoretical accumulation curves assume a constant rate of synthesis (0.045 molecules/sec per cell) with changing rates of degradation ($t_{\frac{1}{2}}$ = 1–100 hr). The best-fitting theoretical curves indicate a long casein mRNA half-life when compared with the experimental accumulation of the mRNA.

controls. These results suggested that prolactin has a rapid effect on either the transcription or turnover of casein mRNA.

Using the data cited above for the prolactin induction of casein mRNA, it was possible to estimate the initial rate of synthesis and the half-life of casein mRNA.[135] The kinetics of casein mRNA induction were compared with theoretical mRNA accumulation curves predicted for different casein mRNA half-lives (Fig. 10) using the following equation[3,137]:

$$C_t = T/D - (T/D - C_0)\, e^{-Dt}$$

where the final accumulated concentration C_t is a function of both the rate of synthesis T and the rate of degradation ($D = \ln 2/t_{\frac{1}{2}}$) of the mRNA. The initial concentration C_0 of casein mRNA was 478 molecules per cell. The observed accumulation of casein mRNA was best approximated by a half-life of greater than 24 hr, most likely falling between 48 and 70 hr. Additional data are required, however, for a more precise estimate. For example, a slight increase in the epithelial cell number at the later times after prolactin addition, not evident by measuring total DNA content, will influence the number of molecules per cell and the shape of the curve. At present, therefore, these results are best interpreted conservatively, and a long half-life of casein mRNA of greater than 24 hr is a valid approximation.

The present data indicated that prolactin will specifically and rapidly induce casein mRNA accumulation *in vitro*. This is the first demonstration of the rapid *in vitro* effect of a peptide hormone on the accumulation of a specific mRNA. Recent studies by Terry *et al.*[138] also demonstrated the induction of casein mRNA in organ culture as detected in an Ehrlich ascites cell-free protein synthesis assay. In these studies, the entire second thoracic mouse mammary gland of estrogen and progesterone-primed virgin mice were cultured for 6 days in a medium containing insulin, prolactin, growth hormone, estrogen, and progesterone, thus allowing differentiation to occur. The medium was then changed, and the cultures exposed to a lactogenic hormone combination, either insulin, prolactin, and hydrocortisone, or insulin, prolactin, and aldosterone, for 6 additional days. Under these conditions, only slightly greater than a 3-fold induction in mRNA activity was observed after 6 days. In contrast, an induction of casein mRNA sequence concentration as large as 45-fold has been observed within 48 hr in our experiments. This difference may reflect the use of midpregnant tissue in our experiments, the increased sensitivity of the cDNA hybridization assay, or most likely the presence of prolactin in the initial culture medium used by Terry *et al.*[138] Our data also suggested that a minimal hormonal requirement for casein mRNA induction in the midpregnant mammary gland may include only prolactin and insulin.[135] Thus, if the mammary gland explants were incubated for 48 hr in the presence of insulin and hydrocortisone, and the medium was then changed to one containing insulin and prolactin, induction of casein mRNA did occur. These studies were confirmed and extended by Houdebine and his colleagues.[139] They demonstrated the induction of casein mRNA and casein synthesis[140] using a reduced O_2 atmosphere (57 vs. 95%) with prolactin alone, i.e., in the absence of insulin and hydrocortisone. Thus, prolactin appears to be the critical hormone in the regulation of casein mRNA synthesis. However, hydrocortisone does appear to play at least a permissive role in the accumulation of casein mRNA. For example, insulin- and hydrocortisone-treated explants had higher levels of casein mRNA than insulin alone, and insulin-, hydrocortisone-, and prolactin-treated tissue had higher levels of casein mRNA than insulin- and prolactin-treated explants. Thus, hydrocortisone may modulate the rate of prolactin-induced mRNA synthesis, the degradation of casein mRNA, or both. As previously mentioned, glucocorticoids have also been shown to amplify the capacity of prolactin to increase the concentration of casein mRNA available for translation *in vivo,* but were totally ineffective when administered alone.[79] Thus, both *in vitro* and *in vivo,* this steroid hormone has been demonstrated to potentiate the action of prolactin on casein gene expression. It remains to be determined whether the initial requirement for hydrocortisone is to promote rough endoplasmic reticulum biosynthesis[59] and the stabilization of casein mRNA on membrane-bound polysomes.

Since progesterone has been reported to inhibit casein mRNA accumulation, casein synthesis, and secretion *in vivo,*[77,103] the possibility that progesterone might directly prevent prolactin-induced casein mRNA synthesis *in vitro* was tested. Progesterone added simultaneously with prolactin reduced the level of casein mRNA, relative to the prolactin-induced control, in a concentration-dependent manner, i.e., by 30, 50, and 100% at concentrations of 0.5, 1.0, and 5 μg/ml, respectively.[135] Thus, progesterone, in a dose-related fashion, inhibited prolactin induction of casein mRNA in agreement with its suggested role during pregnancy in the rat. The inhibitory effect of progesterone, at the lower doses of 0.5 and 1.0 μg/ml, does not appear to be due to a nonspecific toxic effect of the steroid *in vitro,* since insulin, hydrocortisone, and progesterone (1 μg/ml) controls had amounts of casein mRNA identical to those of insulin and hydrocortisone controls. Alterations in the insulin–hydrocortisone control baseline did appear, however, when progesterone was added at 5 μg/ml.

At present, we cannot distinguish a direct effect of prolactin on mRNA transcription from an indirect action on mRNA turnover. Although the mRNA accumulation data suggest that prolactin must be at least partially acting to increase the rate of casein mRNA synthesis, this should be determined directly by measuring the rates of casein mRNA synthesis in the presence and absence of prolactin (see Section 7). Finally, the ability to rapidly induce the accumulation of a specific mRNA *in vitro* may prove useful in the study of eukaryotic mRNA metabolism in general. For example, this system may allow the investigation of the relationship between the synthesis and processing of any putative casein HnRNA precursors and the appearance of mature polysomal casein mRNA sequences.

6.2. Dexamethasone Induction of MMTV RNA in Cell Culture

Hormonal effects on MMTV RNA accumulation and virus production *in vivo* and *in vitro* have been well documented.[9–12,131,141] Recently, it was demonstrated that dexamethasone, a synthetic glucocorticoid, will stimulate the accumulation of MMTV RNA 10- to 20-fold over the levels observed in untreated cultures.[9,10] This effect was observed in cloned cell lines established from both GR and C3H mouse mammary tumors. The increase in viral RNA accumulation occurred rapidly within 15–30 min after hormone addition.[9,10] In GR cells, following an apparent lag of about 15 min, the viral RNA concentration increased with a half-time of about 2.5 hr, reaching a new steady-state level by 5–6 hr after hormone addition.[10] The induction of viral RNA was sensitive to inhibitors of RNA synthesis, but not of protein or DNA synthesis.[10]

While these results were consistent with a model in which the steroid–hormone receptor complex stimulated viral RNA synthesis, they did not rule out an action of the hormone at the level of RNA processing or turnover. Several recent experiments have suggested that stabilization of specific mRNAs may play an important function in regulating their accumulation in cells. For example, the half-life of ovalbumin mRNA is preferentially and markedly increased in the presence of estrogen.[142] Furthermore, histone mRNA synthesis occurs throughout the entire cell cycle, but accumulates in the cytoplasm only in the presence of DNA synthesis in HeLa cells.[143] Thus, it is important to distinguish hormonal effects on RNA synthesis from those on RNA processing and turnover. Two recent studies using the glucocorticoid–MMTV model system have attempted to analyze this problem directly. The following discussion is presented to illustrate the types of techniques used in these studies, rather than to review the expression of MMTV in the mammary gland. A more comprehensive review of viruses and breast cancer is presented in Chapter 2.

A similar approach was employed by both Ringold *et al.*[12] and Young *et al.*[11] to study dexamethasone induction of MMTV RNA in two cloned mammary tumor cell lines. Cells were pulse-labeled with

[^{3}H]uridine for 15 min at various times after addition of the hormone and the [^{3}H]-RNA hybridized to an excess of unlabeled MMTV cDNA in solution. In the studies of Ringold *et al.*,[12] the hybridization probe employed was "tailed duplex" MMTV, which was composed of double-stranded DNA containing single-stranded tails. Following hybridization to MMTV RNA, the "tailed duplex" hybrids were separated by chromatography on hydroxylapatite in the presence of 8 M urea, 1% SDS, and 0.2 M sodium phosphate. Young *et al.*[11] utilized a cDNA probe containing a poly(dC) tail that was added using the enzyme terminal transferase. Following hybridization in solution, the "tailed" cDNA–RNA hybrids were separated by chromatography on poly(I)–Sephadex. Using these assays and an internal [^{32}P]-MMTV RNA hybridization standard, it could be estimated that after 30- to 60-min exposure to dexamethasone, viral RNA synthesis represented between 1.5 and 3% of the total RNA synthesized in the studies of Young *et al.*[11] and 0.5% of the total RNA synthesis in GR cells.[12] In both cases, the hormone caused a rapid increase in the rate of RNA synthesis, 10-fold within 15 min in GR cells, and 3-fold within 10 min in the 34Ic1101 cell line. These results represent the first demonstration of a rapid effect of a steroid hormone on the synthesis of a specific mRNA.

To rule out an effect of the hormone on the rate of decay of the viral RNA, an actinomycin D chase experiment was performed.[11] A half-life of greater than 8 hr was estimated for the MMTV RNA in both the presence and the absence of the hormone. Interestingly, in the presence of dexamethasone, a slightly increased rate of turnover was observed. These results should be viewed with caution, however, since the half-life and processing of mRNA are known to be affected in the absence of continued RNA synthesis. Thus, in the presence of actinomycin D, the processing of nuclear RNA appears to be interrupted, and mRNA appearance in the cytoplasm is prevented.[144] Recent development of a pulse-chase technique, in which glucosamine and high concentrations of unlabeled uridine are employed to halt [^{3}H]uridine incorporation, almost instantaneously, should permit a reevaluation of these data.[144] These experiments do suggest, however, that the rapid effect of steroid hormones is primarily at the level of gene transcription.

Ringold *et al.*[12] also compared the effects of dexamethasone on MMTV RNA synthesis in several cell lines. In both the GR cell line and an Ml-19 rat hepatoma cell infected with exogenous MMTV, a rapid effect of the steroid hormone on MMTV RNA synthesis was observed. However, no effect of dexamethasone on the rate of viral RNA synthesis was observed in S49 lymphoma cells, which are killed by physiologic concentrations of glucocortocoids. Thus, the MMTV genes appear to be constitutively expressed in S49 cells, and the hormone has no effect on the accumulation of viral RNA. While dexamethasone resulted in a 2.5-hr half-time of viral RNA accumulation in GR cells, 10–12 hr were required to reach half-maximal levels in the hepatoma cell lines. Thus, in both cell lines, the steroid hormone rapidly increases the rate of viral RNA synthesis, but the rates of viral RNA accumulation are markedly different in these cells. These data illustrate the importance of analyzing the effects of hormones on the rates of both synthesis and turnover of mRNA. The accumulation of specific mRNA sequences in a given cell is clearly the result of both processes. For example, globin mRNA may represent only 0.2% of the initial nuclear RNA synthesis, but more than 95% of the cytoplasmic poly(A)-containing mRNA in a cultured, erythroleukemic Friend cell line.[145] Finally, these cloned mammary tumor cell lines provide an excellent model system for studying steroid hormone regulation of gene expression.

7. Conclusions and Future Approaches

It should be apparent from the preceding discussion that enormous progress has been made in the past few years in studying gene expression in both normal and neoplastic mammary tissue. The purification of specific mRNAs and the synthesis of molecular hybridization probes have greatly facilitated the study of the mechanisms of both peptide and steroid hormone action in the mammary gland. Considerable information is still required, however, before the precise details of hormone action in the mammary gland are elucidated. Thus, although the initial binding of prolactin to a membrane receptor has been shown to be an obligatory requirement for the subsequent induction of casein synthesis,[146] the exact mechanism by which any peptide hormone regulates specific gene expression is essentially unknown.

The availability of two well-defined model systems for the study of both steroid and peptide hormone action in the mammary gland is the first prerequisite for future studies. The ability to rapidly induce specific mRNAs and to perform pulse-chase experiments in mammalian cells will permit an elucidation of the primary site of steroid and peptide hormone action, i.e., whether they act at the transcriptional, post-transcriptional, or translational level(s). These studies should also help better define the relationship between the milk protein mRNAs and the regulation of milk protein synthesis and secretion. Finally, with the advent of recombinant DNA technology, it should be possible to isolate pure probes for each of the milk protein mRNAs. These specific DNA probes can then be used to characterize the nature of the casein genes and their primary transcription products. Ultimately, the isolation of a larger piece of genomic DNA containing the casein structural gene sequences should be possible. Because of the recent discovery of RNA splicing, however, it should not be assumed that presumptive regulatory regions or promoters will always be adjacent to these structural gene sequences.[147] An understanding of the primary gene structure and the availability of larger genomic DNAs containing specific genes should provide the basis for future studies of the precise mechanism of hormone action in the mammary gland. An understanding of the mechanism by which hormones regulate gene expression in the normal mammary gland may then help elucidate how these regulatory mechanisms have deviated in hormone-dependent mammary cancer. Clearly, the techniques of molecular biology can now be effectively applied to studying gene regulation in the mammary gland. No doubt considerable progress will have been made in this area even before the appearance of this chapter.

ACKNOWLEDGMENTS: I would like to thank my colleagues Drs. Susan Socher, Robert Matusik, William Guyette, and Robert Pauley for their many helpful suggestions and experimental contributions. The excellent technical assistance of Ms. Carol Waugh and Mr. D. O'Neal was also appreciated. I would also like to acknowledge the use of the library at the Marine Biological Laboratory in Woods Hole, Massachusetts, and Dr. Sheldon Segal for providing space in his laboratory, while this manuscript was being written. This work was supported in part by NIH-CA-16303 and HEW NO-1-CP-43385. I am the recipient of NIH career development award KO-4-CA-00154.

8. References

1. J. M. Rosen and J. Monahan, Messenger RNA isolation, characterization and hybridization analysis, in: *Laboratory Methods Manual for Hormone Action and Molecular Endocrinology* (W. T. Schrader and B. W. O'Malley, eds.), pp. 4-1–4-52, Department of Cell Biology, Baylor College of Medicine, Houston (1977).
2. J. Ross, Y. Ikawa, and P. Leder, Globin messenger-RNA induction during erythroid differentiation of cultured leukemia cells, *Proc. Natl. Acad. Sci. U.S.A.* **69**, 3620–3623 (1972).
3. S. E. Harris, J. M. Rosen, A. R. Means, and B. W. O'Malley, Use of a specific probe for ovalbumin messenger RNA to quantitate estrogen-induced gene transcripts, *Biochemistry* **14**, 2072–2080 (1975).
4. R. D. Palmiter, P. B. Moore, E. R. Mulvihill, and S. Emtage, A significant lag in the induction of ovalbumin messenger RNA by steroid hormones: A receptor translocation hypothesis, *Cell* **8**, 557–572 (1976).
5. A. Efstratiadis, F. C. Kafatos, and T. Maniatis, The primary structure of rabbit β-globin mRNA as determined from cloned DNA, *Cell* **10**, 571–585 (1977).

6. G. Schutz, M. Beato, and P. Feigelson, Messenger RNA for hepatic tryptophan oxygenase: Its partial purification, its translation in a heterologous cell-free system, and its control by glucocorticoid hormones, *Proc. Natl. Acad. Sci. U.S.A.* **70,** 1218–1221 (1973).
7. A. E. Sippel, P. Feigelson, and A. K. Roy, Hormonal regulation of hepatic messenger RNA levels for α2U globulin, *Biochemistry* **14,** 825–829 (1975).
8. G. U. Ryffel, W. Wahli, and R. Weber, Quantitation of vitellogenin messenger RNA in the liver of male *Xenopus* toads during primary and secondary stimulation by estrogen, *Cell* **11,** 213–221 (1977).
9. W. P. Parks, J. C. Ranson, H. A. Young, and E. M. Scolnick, Mammary tumor virus induction by glucocorticoids: Characterization of specific transcriptional regulation, *J. Biol. Chem.* **250,** 3330–3336 (1975).
10. G. M. Ringold, K. R. Yamamoto, G. M. Tomkins, J. M. Bishop, and H. E. Varmus, Dexamethasone-mediated induction of mouse mammary tumor virus RNA: A system for studying glucocorticoid action, *Cell* **6,** 299–305 (1975).
11. H. A. Young, T. Y. Shih, E. M. Scolnick, and W. P. Parks, Steroid induction of mouse mammary tumor virus: Effect upon synthesis and degradation of viral RNA, *J. Virol.* **21,** 139–146 (1977).
12. G. M. Ringold, K. R. Yamamoto, J. M. Bishop, and H. E. Varmus, Glucocorticoid-stimulated accumulation of mouse mammary tumor virus RNA: Increased rate of synthesis of viral RNA, *Proc. Natl. Acad. Sci. U.S.A.* **74,** 2879–2883 (1977).
13. F. Vignon and H. Rochefort, Regulation of estrogen receptors in ovarian-dependent rat mammary tumors. I. Effects of castration and prolactin, *Endocrinology* **98,** 722–729 (1976).
14. B. S. Leung and G. H. Sasaki, Prolactin and progesterone effect on specific estradiol binding in uterine and mammary tissues *in vitro, Biochem. Biophys. Res. Commun.* **55,** 1180–1187 (1973).
15. M. R. Banerjee, Responses of mammary cells to hormones, *Int. Rev. Cytol.* **47,** 1–97 (1976).
16. R. Hilf, J. T. Harmon, R. J. Matusik, and M. B. Ringler, Hormonal control of mammary cancer, in: *Control Mechanisms in Cancer* (W. E. Criss, T. Ono, and J. R. Sabine, eds.), pp. 1–24, Raven Press, New York (1976).
17. R. K. Craig and P. N. Campbell, Molecular aspects of milk protein biosynthesis, in: *Lactation,* Vol. 4 (B. Larsen, ed.), Academic Press, New York (in press).
18. T. P. H. Tse and J. M. Taylor, Translation of albumin messenger RNA in a cell-free protein-synthesizing system derived from wheat germ, *J. Biol. Chem.* **252,** 1272–1278 (1977).
19. L. A. Weber, E. D. Hickey, D. L. Nuss, and C. Baglioni, 5′-Terminal 7-methylguanosine and mRNA function: Influence of potassium concentration on translation *in vitro, Proc. Natl. Acad. Sci. U.S.A.* **74,** 3254–3258 (1977).
20. G. Taborsky, Phosphoproteins, in: *Advances in Protein Chemistry* (C. B. Anfinsen, J. T. Edsall, and F. M. Richards, eds.), Vol. 28, pp. 91–125, Academic Press, New York (1974).
21. J. M. Rosen, S. L. C. Woo, and J. P. Comstock, Regulation of casein messenger RNA during the development of the rat mammary gland, *Biochemistry* **14,** 2895–2902 (1975).
22. K. Weber and M. Osborn, The reliability of molecular weight determinations by dodecyl sulfate polyacrylamide gel electrophoresis, *J. Biol. Chem.* **244,** 4406–4412 (1969).
23. U. Laemmli, Cleavage of structural proteins during the assembly of the head of bacteriophage T_4, *Nature (London)* **227,** 680–685 (1970).

24. J. M. Rosen, Isolation and characterization of purified rat casein messenger ribonucleic acids, *Biochemistry* **15**, 5263–5271 (1976).
25. R. K. Craig, P. A. Brown, O. S. Harrison, D. McIlreavy, and P. N. Campbell, Isolation and characterization of messenger ribonucleic acids from lactating mammary gland and identification of caseins and pre-α-lactalbumin as translation products in heterologous cell-free systems, *Biochem. J.* **160**, 57–74 (1976).
26. P. M. Terry, R. Ganguly, E. M. Ball, and M. R. Banerjee, Murine mammary gland RNA directed synthesis of casein in a heterologous cell-free protein synthesis system, *Cell Differ.* **4**, 113–122 (1975).
27. P. Gaye and L. M. Houdebine, Isolation and characterization of casein mRNAs from lactating ewe mammary glands, *Nucleic Acids Res.* **2**, 707–722 (1975).
28. D. S. Shields, G. Blobel, and J. M. Rosen, unpublished observations.
29. P. N. Campbell, D. McIlreavy, and D. Tarin, The detection of the messenger ribonucleic acid for the α-lactalbumin of guinea-pig milk, *Biochem. J.* **134**, 345–347 (1973).
30. T. Zehavi-Willner and C. Lane, Subcellular compartmentation of albumin and globin made in oocytes under the direction of injected messenger RNA, *Cell* **11**, 683–693 (1977).
31. L.-M. Houdebine and P. Gaye, Regulation of casein synthesis in the rabbit mammary gland: Titration of mRNA activity for casein under prolactin and progesterone treatments, *Mol. Cell. Endocrinol.* **3**, 37–55 (1975).
32. P. K. Chakrabartty and P. K. Qasba, Partial purification of rat α-lactalbumin mRNA, *Nucleic Acids Res.* **4**, 2065–2074 (1977).
33. L.-M. Houdebine and P. Gaye, Purification of mRNAs for ewe αs-casein and β-casein by immunoprecipitation of polysomes, *Eur. J. Biochem.* **63**, 9–14 (1976).
34. L.-M. Houdebine, P. Gaye, and A. Favre, Lack of poly(A) sequence in half of the messenger RNA coding for ewe αs-casein, *Nucleic Acids Res.* **1**, 413–425 (1974).
35. J. Gielen, H. Aviv, and P. Leder, Characteristics of rabbit globin mRNA purification by oligo(dT) cellulose chromatography, *Arch. Biochem. Biophys.* **163**, 146–154 (1974).
36. J. M. Rosen, S. L. C. Woo, J. W. Holder, A. R. Means, and B. W. O'Malley, Preparation and preliminary characterization of purified ovalbumin messenger RNA from the hen oviduct, *Biochemistry* **14**, 69–78 (1975).
37. Y. Suzuki and D. D. Brown, Isolation and identification of the messenger RNA for silk fibroin from *Bombyx mori*, *J. Mol. Biol.* **63**, 409–429 (1972).
38. R. Palacios, D. Sullivan, N. M. Summers, M. L. Kiely, and R. T. Schimke, Purification of ovalbumin messenger ribonucleic acid by specific immunoadsorption of ovalbumin-synthesizing polysomes and millipore partition of ribonucleic acid, *J. Biol. Chem.* **248**, 540–548 (1973).
39. J. M. Taylor and T. P. H. Tse, Isolation of rat liver albumin messenger RNA, *J. Biol. Chem.* **251**, 7461–7467 (1976).
40. L.-M. Houdebine, Synthesis of DNA complementary to the mRNAs for milk proteins by *E. coli* DNA polymerase I, *Nucleic Acids Res.* **3**, 615–630 (1976).
41. B. G. Forget, D. Housman, E. J., Benz, Jr., and R. P. McCaffrey, Synthesis of DNA complementary to separated human alpha and beta globin messenger RNAs, *Proc. Natl. Acad. Sci. U.S.A.* **72**, 984–988 (1975).
42. A. Ullrich, J. Shine, J. Chirgwin, R. Pictet, E. Tischer, W. J. Rutter, and H. M. Goodman, Rat insulin genes: Construction of plasmids containing the coding sequences, *Science* **196**, 1313–1318 (1977).
43. S. L. C. Woo, J. M. Rosen, C. D. Liarakos, Y. C. Choi, H. Busch, A. R. Means, B. W. O'Malley, and D. L. Robberson, Physical and chemical characterization of purified ovalbumin messenger RNA, *J. Biol. Chem.* **250**, 7027–7039 (1975).

44. N. T. Van, J. W. Holder, S. L. C. Woo, A. R. Means, and B. W. O'Malley, Secondary structure of ovalbumin messenger RNA, *Biochemistry* **15**, 2054–2061 (1976).
45. D. Sheiness, L. Puckett, and J. E. Darnell, Possible relationship of poly(A)-shortening to mRNA turnover, *Proc. Natl. Acad. Sci. U.S.A.* **72**, 1077–1081 (1975).
46. L.-M. Houdebine, Absence of poly(A) in a large part of newly synthesized casein mRNA, *FEBS Lett.* **66**, 110–113 (1976).
47. M. Salditt-Georgieff, W. Jelinek, J. E. Darnell, Y. Furuichi, M. Morgan, and A. Shatkin, Methyl labeling of HeLa cell hnRNA: A comparison with mRNA, *Cell* **7**, 227–237 (1976).
48. R. P. Perry and D. E. Kelley, Kinetics of formation of 5′ terminal caps in mRNA, *Cell* **8**, 433–442 (1976).
49. A. J. Shatkin, Capping of eucaryotic mRNAs, *Cell* **9**, 645–653 (1976).
50. Y. Furuichi, A. LaFiandra, and A. J. Shatkin, 5′ Terminal structure and mRNA stability, *Nature (London)* **266**, 235–239 (1977).
51. G. W. Both, A. K. Banerjee, and A. J. Shatkin, Methylation-dependent translation of viral messenger RNAs *in vitro*, *Proc. Natl. Acad. Sci. U.S.A.* **72**, 1189–1193 (1975).
52. E. D. Hickey, L. A. Weber, and C. Baglioni, Inhibition of initiation of protein synthesis by 7-methylguanosine-5′-monophosphate, *Proc. Natl. Acad. Sci. U.S.A.* **73**, 19–23 (1976).
53. D. A. Shafritz, J. A. Weinstein, B. Safer, W. C. Merrick, L. A. Weber, E. D. Hickey, and C. Baglioni, Evidence for role of $m^7G^{5'}$-phosphate group in recognition of eukaryotic mRNA by initiation factor IF-M_3, *Nature (London)* **261**, 291–294 (1976).
54. J. M. Rosen, unpublished observations.
55. H. F. Lodish and J. K. Rose, Relative importance of 7-methylguanosine in ribosome binding and translation of vesicular stomatitis virus mRNA in wheat germ and reticulocyte cell-free systems, *J. Biol. Chem.* **252**, 1181–1188 (1977).
56. N. J. Nardacci and W. L. McGuire, Casein and α-lactalbumin mRNA in experimental breast cancer, *Cancer Res.* **37**, 1186–1190 (1977).
57. P. Gaye, N. Viennot, and R. Denamur, *In vitro* synthesis of α-lactalbumin and β-lactoglobulin by microsomes and bound polyribosomes from the mammary gland of lactating sheep, *Biochim. Biophys. Acta* **262**, 371–380 (1972).
58. L.-M. Houdebine, Distribution of casein mRNA between free and membrane-bound polysomes during the induction of lactogenesis in the rabbit, *Mol. Cell. Endocrinol.* **7**, 125–135 (1977).
59. R. M. Wynn, J. A. Harris, and R. T. Chatterton, Interaction of progesterone and adrenocorticoids in ultrastructural development of the mammary gland of the rat, *Am. J. Obstet. Gynecol.* **126**, 920–930 (1976).
60. G. Blobel and B. Dobberstein, Transfer of proteins across membranes, *J. Cell Biol.* **67**, 835–851 (1975).
61. I. M. Verma, G. F. Temple, H. Fan, and D. Baltimore, *In vitro* synthesis of DNA complementary to rabbit reticulocyte 10S RNA, *Nature (London) New Biol.* **235**, 163–166 (1972).
62. L. A. Loeb, K. D. Tartof, and E. C. Travaglini, Copying natural RNAs with *E. coli* DNA polymerase I, *Nature (London) New Biol.* **242**, 66–69 (1973).
63. D. L. Kacian, K. F. Watson, A. Burny, and S. Spiegelman, Purification of the DNA polymerase of avian myeloblastosis virus, *Biochim. Biophys. Acta* **246**, 365–383 (1971).
64. J. J. Monahan, S. E. Harris, S. L. C. Woo, D. L. Robberson, and B. W. O'Malley, The synthesis and properties of the complete complementary DNA transcript of ovalbumin mRNA, *Biochemistry* **15**, 223–233 (1976).

65. A. Efstratiadis, T. Maniatis, F. C. Kafatos, A. Jeffrey, and J. N. Vournakis, Full length and discrete partial reverse transcripts of globin and chorion mRNAs, *Cell* **4,** 367–378 (1975).
66. A. Efstratiadis, F. C. Kafatos, A. Maxam, and T. Maniatis, Enzymatic *in vitro* synthesis of globin genes, *Cell* **7,** 279–288 (1976).
67. J. J. Monahan, L. A. McReynolds, and B. W. O'Malley, The ovalbumin gene *in vitro* enzymatic synthesis and characterization, *J. Biol. Chem.* **251,** 7355–7362 (1976).
68. J. M. Taylor, R. Illmansee, and J. Summers, Efficient transcription of RNA into DNA by avian sarcoma virus polymerase, *Biochim. Biophys. Acta* **442,** 324–330 (1976).
69. S. Packman, H. Aviv, J. Ross, and P. Leder, A comparison of globin genes in duck reticulocytes and liver cells, *Biochem. Biophys. Res. Commun.* **49,** 813–819 (1972).
70. P. R. Harrison, G. D. Birnie, A. Hell, S. Humphries, B. D. Young, and J. Paul, Kinetic studies of gene frequency. I. Use of a DNA copy of reticulocyte 9S RNA to estimate globin gene dosage in mouse tissues, *J. Mol. Biol.* **84,** 539–554 (1974).
71. R. S. Gilmour and J. Paul, Tissue-specific transcription of the globin gene in isolated chromatin, *Proc. Natl. Acad. Sci. U.S.A.* **70,** 3440–3442 (1973).
72. M. M. Smith and R. C. C. Huang, Transcription *in vitro* of immunoglobulin kappa light chain genes in isolated mouse myeloma nuclei and chromatin, *Proc. Natl. Acad. Sci. U.S.A.* **73,** 775–779 (1976).
73. P. Leder, J. Ross, J. Gielen, S. Packman, Y. Ikawa, H. Aviv, and D. Swan, Regulated expression of mammalian genes: Globin and immunoglobulin as model systems, *Cold Spring Harbor Symp. Quant. Biol.* **38,** 753–761 (1973).
74. S. L. C. Woo, J. J. Monahan, and B. W. O'Malley, The ovalbumin gene purification of the anticoding strand, *J. Biol. Chem.* **252,** 5789–5797 (1977).
75. P. Venetianer and P. Leder, Enzymatic synthesis of solid phase-bound DNA sequences corresponding to specific mammalian genes, *Proc. Natl. Acad. Sci. U.S.A.* **71,** 3892–3895 (1974).
76. C. A. Marotta, B. G. Forget, S. M. Weissman, I. M. Verma, R. P. McCaffrey, and D. Baltimore, Nucleotide sequence of human globin messenger RNA, *Proc. Natl. Acad. Sci. U.S.A.* **71,** 2300–2304 (1974).
77. L.-M. Houdebine, Effects of prolactin and progesterone on expression of casein genes: Titration of casein mRNA by hybridization with complementary DNA, *Eur. J. Biochem.* **68,** 219–225 (1976).
78. R. C. Shuster, L.-M. Houdebine, and P. Gaye, Studies on the synthesis of casein messenger RNA during pregnancy in the rabbit, *Eur. J. Biochem.* **71,** 193–199 (1976).
79. E. Devinoy and L.-M. Houdebine, Effects of glucocorticoids on casein gene expression in the rabbit, *Eur. J. Biochem.* **75,** 411–416 (1977).
80. J. M. Rosen and S. W. Barker, Quantitation of casein messenger ribonucleic acid sequences using a specific complementary DNA hybridization probe, *Biochemistry* **15,** 5272–5280 (1976).
81. J. C. Myers, S. Spiegelman, and D. L. Kacian, Synthesis of full-length DNA copies of avian myeloblastosis virus RNA in high yields, *Proc. Natl. Acad. Sci. U.S.A.* **74,** 2840–2843 (1977).
82. D. Baltimore, RNA-dependent DNA polymerase in virions of RNA tumor viruses, *Nature (London)* **226,** 1209–1211 (1970).
83. H. Temin and D. Baltimore, RNA-directed DNA synthesis and RNA tumor viruses, in: *Advances in Virus Research* (K. M. Smith and M. A. Lauffer, eds.), Vol. 17, pp. 129–186, Academic Press, New York (1972).
84. E. Rothenberg and D. Baltimore, Synthesis of long, representative DNA copies of the murine RNA tumor virus genome, *J. Virol.* **17,** 168–174 (1976).

85. J. P. Dudley, J. S. Butel, S. H. Socher, and J. M. Rosen, MMTV expression in several cloned BALB/c mammary tumor cell lines (in preparation).
86. H. E. Varmus, N. Quintrell, E. Medeiros, J. M. Bishop, R. C. Nowinski, and N. H. Sarker, Transcription of mouse mammary tumor virus genes in tissues from high and low incidence mouse strains, *J. Mol. Biol.* **79,** 663–679 (1973).
87. H.-J. Kung, S. Hu, W. Bender, J. M. Bailey, N. Davidson, M. O. Nicolson, and R. M. McAllister, RD-114, baboon, and wooley monkey viral RNAs compared in size and structure, *Cell* **7,** 609–620 (1976).
88. J. M. Rosen, S. E. Harris, G. C. Rosenfeld, C. D. Liarakos, and B. W. O'Malley, Effect of estrogen on gene expression in the chick oviduct. III. Hybridization studies with [^{3}H]messenger RNA and [^{3}H]complementary DNA under conditions of DNA excess, *Cell Differ.* **3,** 103–116 (1974).
89. F. W. Alt, R. E. Kellems, J. R. Bertino, and R. T. Schimke, Selective multiplication of dihydrofolate reductase genes in methotrexate-resistant variants of cultured murine cells, *J. Biol. Chem.* **253,** 1357–1370 (1978).
90. D. S. Holmes and J. Bonner, Sequence composition of rat nuclear deoxyribonucleic acid and high molecular weight nuclear ribonucleic acid, *Biochemistry* **13,** 841–848 (1974).
91. J. M. Rosen and S. H. Socher, Detection of casein messenger RNA in hormone-dependent mammary cancer by molecular hybridization, *Nature (London)* **269,** 83–86 (1977).
92. W. Salser and J. S. Isaacson, Mutation rates in globin genes: The genetic load and Haldane's dilemma, *Prog. Nucleic Acids Mol. Biol.* **19,** 205–220 (1976).
93. P. Gaye, L.-M. Houdebine, G. Petrissant, and R. Denamur, Protein synthesis in mammary gland, *Acta Endocrinol., VI,* Karolinska Symposium in Research Methods in Reproductive Endocrinology, pp. 426–448 (1973).
94. D. N. Banerjee and M. R. Banerjee, Rapidly-labelled RNA in the mouse mammary gland before and during lactation, *J. Endocrinol.* **56,** 145–152 (1973).
95. A. Elska, G. Matsuka, U. Matiash, I. Nasarenko, and N. Jemenova, tRNA and aminoacyl-tRNA synthetases during differentiation and various functional states of the mammary gland, *Biochim. Biophys. Acta* **247,** 430–440 (1971).
96. T. Oka and Y. J. Topper, Hormone-dependent accumulation of rough endoplasmic reticulum in mouse mammary cells *in vitro, J. Biol. Chem.* **246,** 7701–7709 (1971).
97. R. W. Turkington and M. Riddle, Hormone-dependent formation of polysomes in mammary cells *in vitro, J. Biol. Chem.* **245,** 5145–5152 (1970).
98. R. R. Anderson and C. W. Turner, Mammary gland growth during pseudopregnancy and pregnancy in the rat, *Proc. Soc. Exp. Biol. Med.* **128,** 210–214 (1968).
99. R. E. Munford, Changes in the mammary glands of rats and mice during pregnancy, lactation and involution. I. Histological structure., *J. Endocrinol.* **28,** 1–15 (1963).
100. R. Jenness, Biosynthesis and composition of milk, *J. Invest. Dermatol.* **63,** 109–118 (1974).
101. J. M. Rosen and B. W. O'Malley, Hormonal regulation of specific gene expression in the chick oviduct, *Biochem. Action Horm.* **3,** 271–315 (1975).
102. J. R. Tata, The expression of the vitellogenin gene, *Cell* **9,** 1–14 (1976).
103. J. M. Rosen, D. L. O'Neal, J. E. McHugh, and J. P. Comstock, Progesterone-mediated inhibition of casein mRNA and polysomal casein synthesis in the rat mammary gland during pregnancy, *Biochemistry* **17,** 290–297 (1978).
104. M.-A. Le Meur, P. Gerlinger, and J.-P. Ebel, Messenger RNA translation in the presence of homologous and heterologous tRNA, *Eur. J. Biochem.* **67,** 519–529 (1976).
105. J. P. Comstock, G. C. Rosenfeld, B. W. O'Malley, and A. R. Means, Estrogen-induced changes in translation, and specific messenger RNA levels during oviduct differentiation, *Proc. Natl. Acad. Sci. U.S.A.* **69,** 2377–2380 (1972).

106. T. M. Y. Liu and J. W. Davis, Induction of lactation by ovariectomy of pregnant rats, *Endocrinology* **80,** 1043–1050 (1967).
107. J. W. Davis, J. Wikman-Coffelt, and C. L. Eddington, The effect of progesterone on biosynthetic pathways in mammary tissue, *Endocrinology* **91,** 1011–1019 (1972).
108. D. H. Lockwood, R. W. Turkington, and Y. J. Topper, Hormone-dependent development of milk protein synthesis in mammary gland *in vitro, Biochim. Biophys. Acta* **130,** 493–501 (1966).
109. R. W. Turkington, Induction of milk protein synthesis by placental lactogen and prolactin *in vitro, Endocrinology* **82,** 575–583 (1968).
110. R. J. Matusik and J. M. Rosen, Hormonal regulation of casein mRNA, *Proc. 59th Meeting of the Endocrine Society,* p. 122 (1977).
111. L. Assairi, C. Delouis, P. Gaye, L.-M. Houdebine, M. Ollivier-Bousquet, and R. Denamur, Inhibition by progesterone of the lactogenic effect of prolactin in the pseudopregnant rabbit, *Biochem. J.* **144,** 245–252 (1974).
112. R. T. Chatterton, Jr., W. J. King, D. A. Ward, and J. L. Chien, Differential responses of prelactating and lactating mammary gland to similar tissue concentrations of progesterone, *Endocrinology* **96,** 861–868 (1975).
113. J. Djiane and P. Durand, Prolactin-progesterone antagonism in self regulation of prolactin receptors in the mammary gland, *Natüre (London)* **266,** 641–643 (1977).
114. J. Djiane, P. Durand, and P. A. Kelly, Evolution of prolactin receptors in rabbit mammary gland during pregnancy and lactation, *Endocrinology* **100,** 1348–1356 (1977).
115. G. Shyamala, Specific cytoplasmic glucocorticoid hormone receptors in lactating mammary glands, *Biochemistry* **12,** 3085–3090 (1973).
116. K. B. Horwitz, W. L. McGuire, O. H. Pearson, and A. Segaloff, Predicting response to endocrine therapy in human breast cancer: A hypothesis, *Science* **189,** 726–727 (1975).
117. R. P. C. Shiu, P. A. Kelly, and H. G. Friesen, Radioreceptor assay for prolactin and other lactogenic hormones, *Science* **180,** 968–971 (1973).
118. W. K. Morishige, G. J. Pepe, and I. Rothchild, Serum luteinizing hormone, prolactin and progesterone levels during pregnancy in the rat, *Endocrinology* **92,** 1527–1530 (1973).
119. P. A. Kelly, T. Tsushima, R. P. C. Shiu, and H. G. Friesen, Lactogenic and growth hormone-like activities in pregnancy determined by radioreceptor assays, *Endocrinology* **99,** 765–774 (1976).
120. Y. Amenomori, C. L. Chen, and J. Meites, Serum prolactin levels in rats during different reproductive states, *Endocrinology* **86,** 506–510 (1970).
121. G. J. Pepe and I. Rothchild, Metabolic clearance rate of progesterone: Comparison between ovariectomized, pregnant, pseudopregnant and deciduoma-bearing pseudopregnant rats, *Endocrinology* **93,** 1200–1205 (1973).
122. A. A. Simpson, M. H. W. Simpson, Y. N. Sinha, and G. H. Schmidt, Changes in concentrations of prolactin and adrenal corticosteroids in rat plasma during pregnancy and lactation, *J. Endocrinol.* **58,** 675–676 (1973).
123, P. K. Qasba and P. M. Guillino, α-Lactalbumin content of rat mammary carcinomas and the effect of pituitary stimulation, *Cancer Res.* **37,** 3792–3795 (1977).
124. R. Hilf, Milk-like fluid in a mammary adenocarcinoma: Biochemical characterization, *Science* **155,** 826–827 (1967).
125. F. L. Archer, Fine structure of spontaneous and estrogen-induced secretion in breast tumors in the rat induced by 7,12-dimethylbenz(α)anthracene, *J. Natl. Cancer Inst.* **42,** 347–362 (1969).
126. S. Young, L. S. C. Pang, and I. Goldsmith, Differentiation in breast cancer, *J. Clin. Pathol.* **27,** 94–102 (1974).

127. G. Bussolati, A. Pich, and V. Alfani, Immunofluorescence detection of casein in human mammary dysplastic and neoplastic tissues, *Virchows Arch. Anat. Histol.* **365,** 15–21 (1975).
128. H. N. Rose and C. M. McGrath, α-Lactalbumin production in human mammary carcinomas, *Science* **190,** 673–676 (1975).
129. A. Pich, G. Bussolati, and F. DiCarlo, Production of casein and the presence of estrogen receptors in human breast cancer, *J. Natl. Cancer Inst.* **58,** 1483–1484 (1977).
130. M. E. Monaco, D. A. Bronzert, D. C. Tormey, P. Waalkes, and M. E. Lippman, Casein production by human breast cancer, *Cancer Res.* **37,** 749–754 (1977).
131. R. J. Pauley, J. M. Rosen, and S. H. Socher, MMTV and casein expression during normal and neoplastic mammary tissue development, *J. Cell Biol.* **75,** 350a (1977).
132. P. Franchimont, P. F. Zangerle, J. C. Hendrick, A. Reuter, and C. Colin, Simultaneous assays of cancer associated antigens in benign and malignant breast diseases, *Cancer* **39,** 2806–2812 (1977).
133. J. C. Hendrick and P. Franchimont, Radio-immunoassay of casein in the serum of normal subjects and of patients with various malignancies, *Eur. J. Cancer* **10,** 725–730 (1974).
134. A. K. Roy and D. J. Dowbenko, Role of growth hormone in the multihormonal regulation of messenger RNA for α2U-globulin in the liver of hypophysectomized rats, *Biochemistry* **16,** 3918–3921 (1977).
135. R. J. Matusik and J. M. Rosen, Prolactin induction of casein mRNA in organ culture: A model system for studying peptide hormone regulation of gene expression *J. Biol. Chem.* (in press).
136. I. S. Owens, B. K. Vonderhaar, and Y. J. Topper, Concerning the necessary coupling of development to proliferation of mouse mammary epithelial cells, *J. Biol. Chem.* **248,** 472–477 (1973).
137. F. C. Kafatos, mRNA stability and cellular differentiation, in: *Gene Transcription in Reproductive Tissue* (E. Diczfalusy, ed.), Vol. 5, pp. 319–345, Karolinska Institute, Stockholm (1972).
138. P. M. Terry, M. R. Banerjee, and R. M. Lui, Hormone-inducible casein messenger RNA in a serum-free organ culture of whole mammary gland, *Proc. Natl. Acad. Sci. U.S.A.* **74,** 2441–2445 (1977).
139. E. Devinoy, L.-M. Houdebine, and C. Delouis, Role of prolactin and glucocorticoids in the expression of casein genes in rabbit mammary gland organ culture, quantification of casein mRNA, *Biochim. Biophys. Acta* **517,** 360–366 (1978).
140. C. Delouis and M.-L. Combaud, Lack of mitotic effects of insulin during synthesis of casein induced by prolactin in pseudopregnant rabbit mammary gland organ cultures, *J. Endocrinol.* **72,** 393–394 (1977).
141. S. Nandi and C. M. McGrath, Mammary neoplasia in mice, *Adv. Cancer Res.* **17,** 353–413 (1973).
142. R. F. Cox, Estrogen withdrawal in chick oviduct: Selective loss of high abundance classes of polyadenylated messenger RNA, *Biochemistry* **16,** 3433–3442 (1977).
143. M. Melli, G. Spinelli, and E. Arnold, Synthesis of histone messenger RNA of HeLa cells during the cell cycle, *Cell* **12,** 167–174 (1977).
144. R. Levis and S. Penman, The metabolism of poly(A)$^+$ and poly(A)$^-$ hnRNA in cultured *Drosophila* cells studied with a rapid uridine pulse-chase, *Cell* **11,** 105–113 (1977).
145. H. Aviv, Z. Voloch, R. Bastos, and S. Levy, Biosynthesis and stability of globin mRNA in cultured erythroleukemic Friend cells, *Cell* **8,** 495–503 (1977).
146. R. P. C. Shiu and H. G. Friesen, Blockage of prolactin action by an antiserum to its receptors, *Science* **192,** 259–261 (1976).

147. S. M. Berget, C. Moore, and P. A. Sharp, Spliced segments at the 5′ terminus of adenovirus 2 late mRNA, *Proc. Natl. Acad. Sci. U.S.A.* **74,** 3171–3175 (1977).
148. P. Gaye, J.-P. Gautron, J.-C. Mercier, and G. Hazé, Amino terminal sequences of the precursors of ovine caseins, *Biochem. Biophys. Res. Commun.* **79,** 903–911 (1977).
149. M. E. Costlow and W. E. McGuire, Autoradiographic localization of prolactin receptors in 7,12-dimethyl-benz(*a*)anthracene-induced rat mammary carcinoma, *J. Natl. Cancer Inst.* **58,** 1173–1175 (1977).

Index